# Ecosystem Function in Heterogeneous Landscapes

Gary M. Lovett
Clive G. Jones

Monica G. Turner
Kathleen C. Weathers

Editors

# Ecosystem Function in Heterogeneous Landscapes

With 96 Illustrations

 Springer

*Editors*

Gary M. Lovett
Institute of Ecosystem Studies
P.O. Box AB
65 Sharon Turnpike
Millbrook, NY 12545-0129
USA
lovettg@ecostudies.org

Monica G. Turner
Department of Zoology
University of Wisconsin
430 Lincoln Drive
Madison, WI 53706-1381
USA
turnermg@wisc.edu

Clive G. Jones
Institute of Ecosystem Studies
P.O. Box AB
65 Sharon Turnpike
Millbrook, NY 12545-0129
USA
jonesc@ecostudies.org

Kathleen C. Weathers
Institute of Ecosystem Studies
P.O. Box AB
65 Sharon Turnpike
Millbrook, NY 12545-0129
USA
weathersk@ecostudies.org

Library of Congress Control Number:
2005925186 (hard cover); 2005923444 (soft cover)

ISBN-10: 0-387-24089-6 (hard cover)
ISBN-10: 0-387-24090-X (soft cover)
ISBN-13: 978-0387-24089-3 (hard cover)
ISBN-13: 978-0387-24090-9 (soft cover)
e-ISBN: 0-387-24091-8

Printed on acid-free paper.

Printed in the United States of America. (Techbooks/EB)

9 8 7 6 5 4 3 2 1

springer.com

# Foreword

Among the most difficult problems in the life sciences is the challenge to understand the details of how ecosystems/watersheds/landscapes function. Yet, the welfare of all life, not just the human species, depends upon the successful functioning of diverse and complicated ecosystems, each with various dimensions and compositions. Central to this "working" is the dominance, and to a major extent control, of ecosystems by organisms, which means that these systems are constantly changing as the component organisms change and evolve. Such changes increase the challenge to understand the functioning of ecosystems and landscapes. Moreover, understanding the interactions among the myriad components of these systems is mind-boggling as there are scores of biotic (probably many thousands of species when the microbial components are fully enumerated through genomics) and countless abiotic (ions, molecules, and compounds) entities, all simultaneously interacting and responding to diverse external factors to produce functional or dysfunctional environments for life.

This book focuses on the problems of connectedness and ecosystem functioning. It is difficult enough to understand how an ecosystem functions when it is considered in isolation, but all ecosystems are open and connected to everything else. Clearly, the inputs to any ecosystem are the outputs from others and vice versa, and as such the fluxes represent major, if not critical, points for managing or changing the overall functioning of an ecosystem or landscape. A major challenge is to find appropriate conceptual frameworks to address these complicated problems. Understanding spatial heterogeneity is now recognized as one of the most significant aspects of this challenge. However, because ecologists have ignored spatial heterogeneity for so long, there is a pressing need to integrate it into their studies, theories, and models. With new frameworks and tools, ecology is now poised to make important strides forward in the focused study of heterogeneity from an ecosystem and landscape perspective. Ecology has accepted the challenge of understanding these complicated systems overall, and is making good progress toward doing so. Such knowledge is vital to guide conservation initiatives, sustainable management, mitigation of environmental impacts, and future breakthroughs in understanding.

With funding from The Andrew W. Mellon Foundation, the Institute of Ecosystem Studies (IES) launched a study of "*Ecosystem Function in Mosaic Landscapes: Boundaries, Fluxes, and Transformations*" in 1999. We proposed that our research would advance the understanding of how heterogeneity influences ecosystem function by:

"1) rigorously assess[ing] the degree of ecosystem heterogeneity at different scales . . .;
2) determin[ing] how ecosystem heterogeneity affects long-term change in the mosaics of which they are a part;
3) focus[ing] on the role of boundaries between and within ecosystems in governing ecosystem function; and
4) discover[ing] how fluxes across mosaics affect the organismal, material, and energetic transformations [within and among] ecosystems."

The 2003 Cary Conference, "Ecosystem Function in Heterogeneous Landscapes," addressed many of these challenges and the results are brought together in this book. Cary Conferences, started at IES in 1985, have identified and addressed such major "cutting edge" questions and challenges in an effort to provide leadership in the field. This Conference was no exception.

With the leadership of Drs. Lovett, Jones, Turner, and Weathers, the authors of this volume have brought their diverse talents and experiences to bear on the topic of how interactions among ecosystems affect not only their own functioning, but the function of the larger landscape or region in which they are embedded, and have done so in new and enlightened ways. By evaluating the linkages at different scales, the authors of this volume have progressed toward building the "suspension bridge" between ecosystem and landscape ecology, a major goal of the editors of this volume.

There is an important need for revised models, conceptual as well as mechanistic, that will allow ecologists to bring the many aspects of heterogeneity together under one framework. As ecologists continue to develop these new frameworks for understanding how ecological systems function, the ideas put forward in this book hopefully will catalyze new studies that will lead to a more synthetic and unified understanding of heterogeneity, and in the process, a greater understanding of how ecosystems and landscapes "work."

Gene E. Likens
President and Director
Institute of Ecosystem Studies
July 2005

# Acknowledgments

This book is an outcome of the Tenth Cary Conference held at the Institute of Ecosystem Studies (IES) in Millbrook, NY, April 29-May 1, 2003. Many people helped to make the conference a success, and we sincerely appreciate their efforts. In particular, we are grateful to all the conference participants for contributing the ideas and enthusiasm that made the conference exciting and intellectually challenging. The conference Steering Committee–Lenore Fahrig, Timothy Kratz, and Gene Likens–provided important guidance in the development of the conference program. Our IES Advisory Committee, consisting of Peter Groffman, Michael Pace, Steward Pickett and David Strayer, generously lent their insight and experience from past Cary Conferences to the planning of this one. The entire staff of IES worked together to make the conference run smoothly and to provide a relaxed and stimulating atmosphere for the participants. Eight graduate students—Brian Allen, Darren Bade, Olga Barbosa, Jennifer Fraterrigo, Noel Gurwick, Jay Lennon, Michael Papaik, and Katie Predick—provided logistical support throughout the conference and conveyed their enthusiastic and upbeat attitude to all the participants. Most importantly, our Conference Coordinator, Claudia Rosen, provided us with her organizational talent, unflappable personality, style and good humor. It is because of her efforts that we were able to focus on the science and trust that the myriad problems of conference organization were solved behind the scenes; we thank her sincerely for that.

This book is, in many ways, a separate effort, and numerous individuals generously provided assistance. We thank the authors of the chapters for gamely taking on the broad subject areas assigned to them, giving excellent presentations at the conference, tolerating our nagging, and producing thoughtful and stimulating papers. We appreciate the effort and insight provided by the reviewers of the chapter manuscripts, who provided excellent advice on a demanding schedule. We are especially grateful to the organizations that provided financial support for both the conference and the book, including the National Science Foundation (through grant DEB0243867),

The USDA Forest Service, the Environmental Protection Agency, the A.W. Mellon Foundation, and the Institute of Ecosystem Studies.

Gary M. Lovett
Clive G. Jones
Monica G. Turner
Kathleen C. Weathers
*Editors*

# Contents

# Contributors

*Estelle V. Balian*
School of Aquatic and Fishery Sciences, Box 355020, University of Washington, Seattle, WA 98195, USA. Current Address: 14, rue des laitières, 94300 Vincennes, France.

*Larry E. Band*
University of North Carolina, Chapel Hill, NC 27599, USA

*J. Scott Bechtold*
School of Aquatic and Fishery Sciences, Box 355020, University of Washington, Seattle, WA 98195, USA

*James H. Brown*
Department of Biology, University of New Mexico, Albuquerque, NM 87131, USA

*Mary L. Cadenasso*
Hixon Center for Urban Ecology, School of Forestry and Environmental Studies, Yale University, 205 Prospect Street, New Haven, CT 06511, USA

*F. Stuart Chapin III*
Institute of Arctic Biology, University of Alaska, Fairbanks, AK 99775, USA

*Rodney Denning*
Annis Water Resources Institute, Grand Valley State University, 740 W. Shoreline Drive, Muskegon, MI 49441, USA

*Deanne C. Drake*
School of Aquatic and Fishery Sciences, Box 355020, University of Washington, Seattle, WA 98195, USA. Current Address: The Ecosystem Center, Marine Biological Laboratory, Woods Hole, MA 02543, USA

*Lenore Fahrig*
Landscape Ecology Lab, Department of Biology, Carleton University, 1125 Colonel By Drive, Ottawa, Ontario, Canada K1S 5B6

*Stuart G. Fisher*
School of Life Sciences, Arizona State University, Tempe, AZ 85287, USA

*Janet Franklin*
Department of Biology, San Diego University, 5500 Campanile Drive, San Diego, CA 92182-4614, USA

*Jerry F. Franklin*
College of Forest Resources, University of Washington, Seattle, WA 98195, USA

*C. Susan Grimmond*
Indiana University, Bloomington, IN 47405, USA

*J. Morgan Grove*
USDA Forest Service, Northeastern Research Station, 705 Spear Street, South Burlington, VT 05403, USA

*Jennifer W. Harden*
US Geological Survey, 345 Middlefield Road, MS 962, Menlo Park, CA 94025 USA

*Clive G. Jones*
Institute of Ecosystem Studies, Box AB, Millbrook, NY 12545, USA

*Timothy K. Kratz*
Trout Lake Station, Center for Limnology, University of Wisconsin-Madison, 10810 County Highway N, Boulder Junction, WI 54568, USA

*Joshua J. Latterell*
School of Aquatic and Fishery Sciences, Box 355020, University of Washington, Seattle, WA 98195, USA

*Gary M. Lovett*
Institute of Ecosystem Studies, Box AB, Millbrook, NY 12545, USA

*John A. Ludwig*
CSIRO Sustainable Ecosystems, PO Box 780, Atherton, 4883, Queensland Australia.

*Sally MacIntyre*
Department of Ecology, Evolution and Marine Biology and Marine Science
Institute, University of California, Santa Barbara, CA 93106-6150, USA

*Michelle C. Mack*
Department of Botany, 220 Bartram Hall, University of Florida, Gainesville,
FL 32611, USA

*Amala Mahadevan*
Department of Earth Sciences, Boston University, 685 Commonwealth
Avenue, Room 127, Boston, MA 02215, USA

*Kristen L. Manies*
US Geological Survey, 345 Middlefield Road, MS 962, Menlo Park, CA
94025 USA

*Marcel Meinders*
Laboratory of Soil Science and Geology Wageningen University, PO Box
37, 6700 AA Wageningen, The Netherlands

*Judy L. Meyer*
Institute of Ecology and River Basin Science and Policy Center, University
of Georgia, Athens, GA 30602-2602, USA

*Robert J. Naiman*
School of Aquatic and Fishery Sciences, Box 355020, University of
Washington, Seattle, WA 98195, USA

*William K. Nuttle*
11 Craig Street, Ottawa, Ontario, Canada K1S 4B6

*Thomas C. O'Keefe*
School of Aquatic and Fishery Sciences, Box 355020, University of
Washington, Seattle, WA 98195, USA

*John Pastor*
Department of Biology and Natural Resources Research Institute,
University of Minnesota, Duluth, MN 55812, USA

*Steward T.A. Pickett*
Institute of Ecosystem Studies, Box AB, Millbrook, NY 12545, USA

*Hugh P. Possingham*
The Ecology Centre, University of Queensland, Brisbane, QLD 4072, Australia

*Tracey J. Regan*
The Ecology Centre, University of Queensland, Brisbane, QLD 4072, Australia

*William A. Reiners*
Department of Botany, University of Wyoming, Laramie, WY 82701, USA

*William H. Romme*
Department of Forest, Rangeland, and Watershed Stewardship, Colorado State University, Fort Collins, CO 80523, USA

*Gaius R. Shaver*
The Ecosystems Center, Marine Biological Laboratory, Woods Hole, MA 02543, USA

*David L. Smith*
Epidemiology and Preventative Medicine, University of Maryland School of Medicine, Baltimore, MD 21201, USA, and Fogarty International Center, National Institutes of Health, Bethesda, MD 20892

*Alan D. Steinman*
Annis Water Resources Institute, Grand Valley State University, 740 W. Shoreline Drive, Muskegon, MI 49441, USA

*David L. Strayer*
Institute of Ecosystem Studies, Box AB, Millbrook, NY 12545, USA

*Christina Tague*
Department of Geography, San Diego State University, San Diego, CA 92181-4493, USA

*David J. Tongway*
CSIRO Sustainable Ecosystems, GPO 284 Canberra, 2601, Australian Capital Territory, Australia

*Merritt R. Turetsky*
Department of Plant Biology, Department of Fisheries and Wildlife, Michigan State University, East Lansing, MI 48824, USA

*Monica G. Turner*
Department of Zoology, University of Wisconsin, Madison, WI 53706, USA

*Nico van Breemen*
Laboratory of Soil Science and Geology, Wageningen University, PO Box 37, 6700 AA Wageningen, The Netherlands

*Kathleen C. Weathers*
Institute of Ecosystem Studies, Box AB, Millbrook, NY 12545, USA

*Katherine E. Webster*
Department of Biological Sciences, University of Maine, Orono, ME 04469-5751, USA

*Jill R. Welter*
School of Life Sciences, Arizona State University, Tempe, AZ 85287, USA

*Ethan P. White*
Department of Biology, University of New Mexico, Albuquerque, NM 87131, USA

*Kerrie Wilson*
The Ecology Centre, University of Queensland, Brisbane, QLD 4072, Australia

# Participants in the 2003 Cary Conference
## *With their affiliations at the time of the conference*

*Mr. Brian F. Allan*
Rutgers University

*Mr. Edward A. Ames*
Mary Flagler Cary Charitable Trust

*Dr. Juan J. Armesto*
Universidad de Chile, Santiago, Chile, and Institute of Ecosystem Studies

*Dr. Amy T. Austin*
University of Buenos Aires and Instituto de Investigaciones Fisiológicas y Ecológicas Vinculadas a la Agricultura (IFEVA), Argentina

*Mr. Darren L. Bade*
University of Wisconsin–Madison

*Dr. Larry E. Band*
University of North Carolina–Chapel Hill

*Ms. Olga Barbosa*
Pontificia Universidad Católica de Chile, Chile

*Dr. Susan S. Bell*
University of South Florida

*Dr. Tracy L. Benning*
University of San Francisco

*Dr. Alan R. Berkowitz*
Institute of Ecosystem Studies

*Dr. Harry C. Biggs*
South African National Parks

*Dr. Elizabeth W. Boyer*
State University of New York, Syracuse

*Dr. James H. Brown*
University of New Mexico

*Dr. Mary L. Cadenasso*
Institute of Ecosystem Studies

*Dr. Charles D. Canham*
Institute of Ecosystem Studies

*Dr. Nina F. Caraco*
Institute of Ecosystem Studies

*Dr. Jonathan J. Cole*
Institute of Ecosystem Studies

*Dr. Jana E. Compton*
U.S. Environmental Protection Agency

*Dr. Graeme S. Cumming*
University of Florida

*Dr. Peter J. Dillon*
Trent University, Canada

*Dr. Valerie T. Eviner*
Institute of Ecosystem Studies

*Dr. Holly A. Ewing*
Institute of Ecosystem Studies

*Dr. Lenore Fahrig*
Carleton University, Canada

*Dr. Stuart E.G. Findlay*
Institute of Ecosystem Studies

*Dr. Mary K. Firestone*
University of California–Berkeley

*Dr. Stuart G. Fisher*
Arizona State University

*Dr. Marie-Josée Fortin*
University of Toronto, Canada

*Dr. Douglas A. Frank*
Syracuse University

*Dr. Janet Franklin*
San Diego State University

*Dr. Jerry F. Frankin*
University of Washington

*Ms. Jennifer Fraterrigo*
University of Wisconsin–Madison

*Dr. Christine L. Goodale*
The Woods Hole Research Center

*Dr. Peter M. Groffman*
Institute of Ecosystem Studies

*Mr. Noel Gurwick*
Cornell University

*Dr. Mark E. Harmon*
Oregon State University

*Dr. Sarah E. Hobbie*
University of Minnesota–Saint Paul

*Dr. Jeff E. Houlahan*
University of Ottawa and
University of New Mexico

*Dr. Carol A. Johnston*
University of Minnesota–Duluth

*Dr. Clive G. Jones*
Institute of Ecosystem Studies

*Dr. K. Bruce Jones*
U.S. Environmental Protection Agency–Las Vegas

*Dr. John A. Kelmelis*
U.S. Geological Survey

*Dr. Timothy K. Kratz*
University of Wisconsin–Madison

*Mr. Jay T. Lennon*
Dartmouth College

*Dr. Luc Lens*
University of Ghent, Belgium

*Dr. Gene E. Likens*
Institute of Ecosystem Studies

*Dr. Gary M. Lovett*
Institute of Ecosystem Studies

*Dr. Winsor H. Lowe*
Institute of Ecosystem Studies

*Dr. Amala Mahadevan*
University of Cambridge, United Kingdom, and
University of New Hampshire

*Dr. Judy L. Meyer*
University of Georgia

*Dr. Robert J. Naiman*
University of Washington

*Dr. Richard S. Ostfeld*
Institute of Ecosystem Studies

*Dr. Michael L. Pace*
Institute of Ecosystem Studies

*Mr. Michael J. Papaik*
University of Massachusetts

*Dr. John J. Pastor*
University of Minnesota–Duluth

*Dr. Steward T.A. Pickett*
Institute of Ecosystem Studies

*Dr. Gilles Pinay*
Centre National de la Recherche Scientifique, France

*Dr. Hugh P. Possingham*
University of Queensland, Australia

*Dr. Mary E. Power*
University of California–Berkeley

*Ms. Katie Predick*
University of Wisconsin–Madison

*Dr. Edward B. Rastetter*
The Ecosystems Center, Marine Biological Laboratory

*Dr. William A. Reiners*
University of Wyoming

*Mr. William Robertson IV*
The Andrew W. Mellon Foundation

*Dr. William H. Romme*
Colorado State University

*Dr. Osvaldo E. Sala*
University of Buenos Aires

*Dr. Steven W. Seagle*
University of Maryland

*Dr. Moshe Shachak*
Ben Gurion University, Israel

*Dr. Gaius R. Shaver*
The Ecosystems Center, Marine Biological Laboratory

*Dr. David L. Smith*
University of Maryland School of Medicine

*Dr. Jonathan H. Smith*
U.S. Environmental Protection Agency

*Dr. Alan D. Steinman*
Grand Valley State University

*Dr. David L. Strayer*
Institute of Ecosystem Studies

*Dr. Fred J. Swanson*
USDA Forest Service

*Dr. Christina Tague*
San Diego State University

*Dr. Jennifer L. Tank*
University of Notre Dame

*Dr. David J. Tongway*
Commonwealth Scientific & Industrial
Research Organisation (CSIRO), Australia

*Dr. Merritt R. Turetsky*
U.S. Geological Survey

*Dr. Monica G. Turner*
University of Wisconsin–Madison

*Dr. Maria Uriarte*
Institute of Ecosystem Studies

*Dr. Nico van Breemen*
Wageningen University

*Mr. Joseph S. Warner*
Institute of Ecosystem Studies

*Dr. Kathleen C. Weathers*
Institute of Ecosystem Studies

*Dr. Donald E. Weller*
Smithsonian Environmental Research Center (SERC)

# 1
# Ecosystem Function in Heterogeneous Landscapes

GARY M. LOVETT, CLIVE G. JONES, MONICA G. TURNER, and KATHLEEN C. WEATHERS

## Introduction

The ecosystem concept has been a powerful tool in ecology, as it allows the use of the quantitative and rigorous laws of conservation of mass and energy in the analysis of entire ecological systems. These laws require delimiting an ecosystem by specifying its boundaries; however, we know that these boundaries are porous and that all ecosystems are open systems that exchange matter, energy, information, and organisms with their surroundings. This openness means that ecosystems defined as spatially separate are in fact interconnected parts of a larger landscape. Once we begin to ask about the source of the inputs or the fate of the outputs, we need to consider the ecosystem in its landscape context.

The role of landscape context in ecosystem functioning has historically received rather short shrift, and we believe the subject is ripe for synthesis and conceptual progress. Consequently, the goal of this book is to focus the attention of the ecosystem science research community on how interactions among ecosystems affect the functioning of individual ecosystems and the larger landscape in which they reside. This subject is becoming increasingly important as ecosystem scientists are being asked to provide information on environmental problems at local, regional, and global scales—a task that cannot be accomplished by examining ecosystems in isolation. Fundamentally, the problem of scaling up from individual ecosystems to larger spatial scales depends on how we conceptualize heterogeneity in a landscape composed of multiple, potentially interacting ecosystems.

This book is an outgrowth of the Tenth Cary Conference, held April 29–May 1, 2003, in Millbrook, New York. As with all Cary Conferences, this conference focused on a difficult conceptual and practical problem in ecosystem science and brought together leading thinkers and practitioners to offer different perspectives and try to advance understanding of the issue. This book brings the same approach to print. It reflects the challenges and problems identified by the participants in the conference as well as different perspectives on solutions to those problems, both conceptual and practical.

1

Although ecosystem ecology has focused on ecosystem function, particularly the flows of mass and energy, the spatial structure of landscapes has largely been the province of landscape ecology. Historically, landscape ecologists have tended to focus on the quantification of landscape structure, often to understand its influence on animal movement, population persistence, or disturbance dynamics. It is only recently that landscape ecologists have begun to consider other ecosystem processes such as mass and energy transfer. Thus, in some ways, this book is a bridge between ecosystem and landscape ecology, encompassing both the landscape ecologists' knowledge of spatial structure and the ecosystem ecologists' knowledge of system function. In this book, we take a broad view of the term *landscape*, with no particular spatial scale implied, and we include heterogeneous aquatic as well as terrestrial systems.

We embarked on this project knowing full well that the existence of spatial heterogeneity would not be a startling revelation to ecologists. Heterogeneity is everywhere, and most ecosystem ecologists deal with it on a daily basis in designing their experiments and analyzing their data. Sometimes, ecologists use heterogeneity as a tool, such as when we contrast riffles and pools in a stream or forests on different soil types. Other times, we see spatial heterogeneity as noise obscuring the pattern we wish to observe. Accounting for spatial heterogeneity in ecosystem processes costs us dearly in time, money, and statistical agony. The goal of this book is to move beyond the quantification and description of heterogeneity to understand when it matters to ecosystem function and when it does not. When can we ignore it, when should we deal with it, and, if we need to deal with it, what are the best conceptual tools for doing so?

## Concepts and Definitions

A few key concepts recur throughout the book and require some introduction. First, many of the chapters refer to a scheme for organizing different approaches to spatial heterogeneity proposed by Shugart (1998). Shugart discussed modeling approaches for terrestrial ecosystems, which he classified as "homogeneous," meaning no spatial heterogeneity is represented; "mosaic," meaning that spatial heterogeneity is present in that different spatial units in the model have different characteristics, but there is no interchange between the units; and "interactive," meaning that spatial units are distinct and exchange mass, energy, organisms, or information with one another (Figure 1.1). We found this a useful way to categorize general conceptual approaches to heterogeneity, and this terminology appears repeatedly in the book, beginning with Chapter 2 by Turner and Chapin. Our goal was to understand the circumstances under which each of these approaches is appropriate.

A second concept that occurs throughout the book is that of compositional versus configurational heterogeneity. Compositional heterogeneity refers to the number, type, and abundance of spatial units in the landscape, whereas configurational heterogeneity refers to the spatial arrangement of those units.

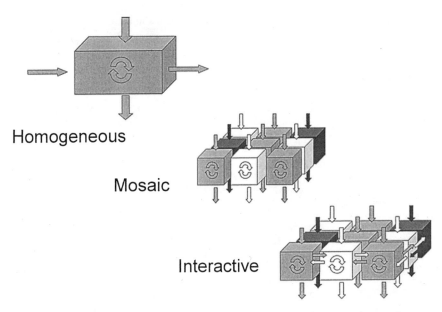

FIGURE 1.1. Schematic representation of three conceptual approaches to heterogeneity. Classification follows Shugart (1998).

A third concept concerns the representation of heterogeneity in data and models. In some cases, heterogeneity is expressed in discrete units, usually called patches. In other cases, heterogeneity is expressed as continuous variation across the landscape; if this variation is monotonic, it is called a gradient. There is also a middle ground between these two end-points, for instance "neighborhood" models in which the properties of a given patch are influenced by its surroundings and the influence often declines with distance from the focal patch, and "networks," which are hierarchically arranged, interconnected series of patches (see White and Brown, Chapter 3).

Finally, there are a number of terms used in the book that may cause confusion because they have different meanings to different people. In an effort to minimize semantic confusion, we have defined several important terms in Table 1.1. These definitions are not meant to be restrictive; rather, they represent what we consider the most common usage of these terms. We asked the authors to make it clear in their papers if they used any of these terms differently.

## Organization of the Book

The book has five sections. Section I ("Challenges and Conceptual Approaches") contains four chapters that describe the problem of dealing with spatial heterogeneity in ecosystem science and offer conceptual

TABLE 1.1.  Definitions of Some Commonly Used Terms in the Book

**Configuration**: A specific spatial arrangement of elements or entities (biotic or abiotic); often used synonymously with spatial structure or patch structure.

**Connectivity**: The spatial continuity of an entity or function.

**Ecosystem**: A spatially explicit unit of the earth that includes all of the organisms, along with all components of the abiotic environment, within its boundaries.

**Ecosystem Function**: Attribute related to the performance of an ecosystem that is the consequence of one or of multiple ecosystem processes. Examples include nutrient retention, biomass production, and maintenance of species diversity.

**Ecosystem Process**: Transfer of energy, material, or organisms among pools in an ecosystem. Examples include primary production, decomposition, heterotrophic respiration, flux and cycling of elements, and evapotranspiration.

**Gradient**: Change in a property across a defined spatial extent.

**Heterogeneity**: The quality or state of encompassing variation in a property of interest, as with mixed habitats or environmental gradients occurring on a landscape; opposite of homogeneity, in which variation in the property is negligible.

**Landscape**: An area that is spatially heterogeneous in at least one factor of interest.

**Patch**: A surface area that differs from its surroundings in structure or function.

**Scale**: Spatial or temporal dimension of an object or process, characterized by both grain and extent.

frameworks to help address the problem. Section II ("Perspectives from Different Disciplines") has four chapters that explore various conceptual and modeling approaches used in other spatial disciplines, specifically population biology, hydrology, epidemiology, and oceanography. Section III ("Illustrations of Heterogeneity and Ecosystem Function") contains seven chapters that treat the role of spatial heterogeneity in a diverse assortment of landscapes, such as arid systems, lakes, and boreal forests, with specific attention to the fundamental issues of what causes spatial heterogeneity, and when it does—and does not—matter for the functioning of the ecosystem or landscape. Section IV ("Application of Frameworks and Concepts") consists of three chapters that treat the need for knowledge about spatial heterogeneity in practical resource management issues pertaining to fire, water, and the design of biological reserves. In the final section, (Section V, "Synthesis"), five chapters (including a final chapter by the editors) tie together the various threads of the book, providing synthetic views of the problem and describing progress in developing overarching conceptual frameworks.

## Reference

Shugart, H.H. 1998. Terrestrial ecosystems in changing environments. Cambridge, UK: Cambridge University Press.

# Section I

## Challenges and Conceptual Approaches

# Editors' Introduction to Section I: Challenges and Conceptual Approaches

The first step toward building a complete understanding of landscape heterogeneity and ecosystem function is to develop a conceptual framework and identify the challenges that need to be overcome. This is no simple task. There are many interactions between spatial heterogeneity and ecosystem processes that occur on multiple temporal and spatial scales; how to structure our thinking in a way that promises new insights is not readily apparent. This first section of the book offers four different perspectives that address this daunting topic, perhaps suggesting some of the structural elements needed for a solid framework.

Monica Turner and Terry Chapin (Chapter 2) briefly describe the background of research on spatial heterogeneity and ecosystem function in both ecosystem and landscape ecology. They introduce the concepts of point processes and lateral transfers to describe situations in which horizontal movement between units in a landscape is or is not important, respectively. They discuss ways of conceptualizing heterogeneity (homogeneous, mosaic, and interactive models) and offer insights to when spatial heterogeneity may be important in ecosystem studies. This chapter presents the basis of a conceptual framework that allows ecologists to sort out when heterogeneity may be important to consider.

Ethan White and Jim Brown (Chapter 3) consider the template upon which ecosystems function and begin by posing the question, "How and why is the landscape heterogeneous?" They argue that it is necessary to have a quantitative understanding of heterogeneity before its functional importance can be understood, and they present three general categories (gradients, patches, and networks) of environmental heterogeneity. They further suggest that these different types of spatial heterogeneity reflect different causal mechanisms, and they illustrate these with selected examples. This chapter offers a conceptual and mathematical framework for characterizing patterns of heterogeneity and understanding the processes underlying those patterns.

In Chapter 4, John Pastor focuses on three processes that generate pattern in the landscape: physical disturbance, directional transport of energy and materials, and diffusive instability. He discusses both the conceptual basis and the mathematical modeling of these phenomena, using many chapters from this book as case studies.

Bill Reiners (Chapter 5) offers a very general and comprehensive conceptual framework for understanding the  transport of mass, energy, organisms, and information on the landscape. He discusses how these transport phenomena are influenced by spatial heterogeneity and how in turn heterogeneity alters the transport. This conceptual framework should be particularly helpful for developing models of fluxes between ecosystems on a landscape, as it describes the fundamental concepts behind transport phenomena.

# 2
# Causes and Consequences of Spatial Heterogeneity in Ecosystem Function

MONICA G. TURNER and F. STUART CHAPIN III

## Abstract

Understanding the causes and consequences of spatial heterogeneity in ecosystem function represents a frontier in both ecosystem and landscape ecology. Ecology lacks a theory of ecosystem function that is spatially explicit, and there are few empirical studies from which to infer general conclusions. We present an organizing framework that clarifies consideration of ecosystem processes in heterogeneous landscapes; consider when spatial heterogeneity is important; discuss methods for incorporating spatial heterogeneity in ecosystem function; and identify challenges and opportunities for progress. Two general classes of ecosystem processes are distinguished. *Point processes* represent rates measured at a particular location; lateral transfers are assumed to be small relative to the measured response and are ignored. Spatial heterogeneity is important for point processes when (1) the average rate must be determined over an area that is spatially heterogeneous or (2) understanding or predicting the spatial pattern of process rates is an objective, for example, to identify areas of high or low rates, or to quantify the spatial pattern or scale of variability in rates. *Lateral transfers* are flows of materials, energy, or information from one location to another represented in a two-dimensional space. Spatial heterogeneity may be important for understanding lateral transfers when (1) the pattern of heterogeneity influences net lateral transfer and potentially the behavior of the whole system, (2) the spatial heterogeneity itself produces lateral transfers, or (3) the lateral transfers produce or alter patterns of spatial heterogeneity. We discuss homogeneous, mosaic, and interacting element approaches for dealing with space and identify both challenges and opportunities. Embracing spatial heterogeneity in ecosystem ecology will enhance understanding of pools, fluxes, and regulating factors in ecosystems; produce a more complete understanding of landscape function; and improve the ability to scale up or down.

## Introduction

Understanding the causes and consequences of spatial heterogeneity in ecosystem function represents a frontier in both ecosystem and landscape ecology (Turner et al. 2001; Chapin et al. 2002), and it is recognized as important in a variety of other disciplines; for example, biological oceanography (Platt and Sathyendranath 1999), limnology (Soranno et al. 1999), soil ecology (Burke et al. 1999), conservation (Pastor et al. 1999), and global change studies (Shugart 1998; Canadell et al. 2000). Ecosystems do not exist in isolation, and interactions among patches on the landscape influence the functioning of individual ecosystems and of the overall landscape. Efforts to estimate the cumulative effect of ecosystem processes at regional and global scales have contributed to the increased recognition of the importance of landscape processes in ecosystem dynamics (Chapin et al. 2002). Transfers among patches, representing losses from donor ecosystems and subsidies to recipient ecosystems, are important to the long-term sustainability of ecosystems (Polis and Hurd 1996; Naiman 1996; Carpenter et al. 1999; Chapin et al. 2002).

Ecology lacks a theory of ecosystem function that is spatially explicit, and there are few empirical studies from which to infer general conclusions. Ecosystem ecology focuses on the flow of energy and matter through organisms and their environment. As such, it addresses pools, fluxes, and regulating factors. Spatially, ecosystem ecology encompasses bounded systems like watersheds, spatially complex landscapes, and even the biosphere; temporally, it crosses scales ranging from seconds to millennia (Carpenter and Turner 1998). From its initial descriptions of the structure and function of a diverse variety of ecosystems, ecosystem ecology moved toward increasingly sophisticated analyses of function; for example, food web analyses, biogeochemistry, regulation of productivity, and so forth (Golley 1993; Pace and Groffman 1998; Chapin et al. 2002). Typically, ecosystem studies are conducted within a single ecosystem, such as a lake or a forest stand, and homogeneous sites are generally chosen to minimize the complications associated with spatial heterogeneity. From ecosystem studies, ecology has gained an excellent understanding of the mechanisms underlying many processes and of temporal dynamics in function. However, understanding patterns, causes, and consequences of spatial heterogeneity in ecosystem function remains a frontier.

Landscape ecology explicitly addresses the importance of spatial configuration for ecological processes (Turner et al. 2001), and, in North America, landscape studies were strongly promoted by ecosystem ecologists (Risser et al. 1984). Landscape ecology often, but not always, focuses on spatial extents that are much larger than those traditionally studied in ecosystem ecology. Early research in landscape ecology emphasized methods to describe and quantify spatial heterogeneity, spatially explicit models to relate pattern and process, and understanding of scale effects. Indeed, there are numerous metrics for quantifying spatial heterogeneity (e.g., Baskent and Jordan 1995; McGarigal and Marks 1995; Gustafson 1998; Gergel and Turner 2002), although the functional interpretation of pattern metrics has proved challenging (Turner et al. 2001). From landscape studies, ecology has gained new insights into how disturbances create and respond to landscape pattern and of population dynamics on heterogeneous landscapes. However, with a few exceptions, the consideration of ecosystem function has poorly been represented. This is surprising, given the initial strong links from ecosystem to landscape ecology (e.g., Risser et al. 1984; Turner 1989). In this paper, we (1) present an organizing framework that clarifies consideration of ecosystem processes in heterogeneous landscapes; (2) consider when spatial heterogeneity is important; (3) discuss methods for incorporating spatial heterogeneity in ecosystem function; and (4) identify challenges and opportunities for progress.

# When Does Space Matter? A Conceptual Framework

Ecosystem processes are heterogeneous. The basic causes of this have been well-known for a long time (Jenny 1941). Heterogeneity is derived from the abiotic template, including factors such as climate, topography, and substrate. In addition, ecosystem processes vary with the biotic assemblage, disturbance events (including long-term legacies), and the activities of humans (Chapin et al. 1996; Amundson and Jenny 1997). However, despite this recognition, most ecosystem ecologists have focused on knowing the mean rates, in spite of the "noise" that results from spatial heterogeneity.

## Organizing Ecosystem Processes

We suggest distinguishing between two general classes of ecosystem process when considering ecosystem function in heterogeneous landscapes. *Point processes* represent rates measured at a particular location (Figure 2.1a). Lateral transfers are assumed to be small relative to the measured response and are ignored. Examples of point processes include site-specific measurements of net primary production, net ecosystem production, denitrification, or nitrogen mineralization. *Lateral transfers* are flows of materials, energy, or information from one location to another represented in a two-dimensional space (Figure 2.1b). Examples of lateral transfers include the flow of nitrogen or phosphorus from land to water or the movements of nutrients across a landscape by herbivores.

Spatial heterogeneity can be considered in both the drivers and the ecosystem response variables (Figure 2.2). For the drivers, one can consider the spatial heterogeneity of the template—which often is multivariate—and of spatial processes, such as disturbance, that alter the template (Foster et al. 1998). For the process, one can consider the spatial pattern of occurrence

(a)

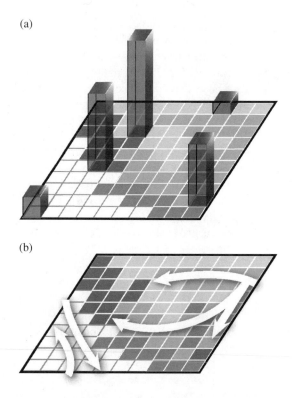

(b)

FIGURE 2.1. Schematic illustration of two general classes of ecosystem processes: (a) point processes and (b) lateral transfers.

(e.g., where denitrification does or does not occur or where there is nutrient movement; Figure 2.2a) or of the magnitude of the rates (Figure 2.2b). For lateral transfers, one can consider the actual pathways of flow (Figure 2.2b). For both point processes and lateral transfers, an aggregate measure of the function of the heterogeneous system (e.g., total P input to a lake) can be considered. When seeking general relationships, it is important to be explicit about both the type of ecosystem process being considered and the variable or response for which spatial heterogeneity is being considered.

## When Is Spatial Heterogeneity Important?

Understanding the relationship between spatial heterogeneity and ecosystem processes is important in at least the following five situations.

(1) For point processes, spatial heterogeneity matters when it is necessary to know the average rate of a process over an area that is spatially heterogeneous. This is of particular importance when there is a nonlinear relationship between the process and a driver that is spatially variable. Although

(a)

(b)

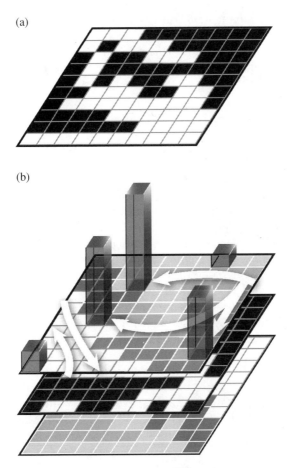

FIGURE 2.2. Spatial heterogeneity can be considered in (a) the occurrence of a process, (b) the magnitude of the rate or flux and the template, which is usually multivariate.

this is largely a sampling issue—knowing how to stratify measurements spatially based on the important driver(s)—it is not trivial.

Estimating methane production from a Siberian landscape that is a mosaic of land and lakes provides an example (Zimov et al. 1997). Lakes dominate the flux of methane within the landscape, but there is substantial heterogeneity of $CH_4$ flux within lakes. Bubbles of methane that form in ice over winter give visual evidence of hot spots of methane release from sediments. Here, the ebullition flux is several orders of magnitude larger than the diffusive flux, which is the main pathway of $CH_4$ flux between areas of bubbling. Therefore, to estimate the $CH_4$ flux from the lake, one must be aware of these different pathways and the spatial distribution of areas of ebullition. These hot spots dominate the fluxes of methane within the lake, and lakes, in turn, dominate fluxes from landscapes. Estimates of the average rate of methane flux from this landscape would be inaccurate if the spatial heterogeneity was ignored. This general class of problems is of great practical importance; ecosystem ecologists remain challenged by developing regional and global budgets for carbon and nutrient fluxes in heterogeneous regions.

(2) Spatial heterogeneity matters when one wants to understand or predict the spatial pattern of process rates. In so doing, one may want to identify locations that are qualitatively different in their processing rates from other areas, or use the spatial pattern or spatial scale of variation as a response variable of direct interest.

Understanding and predicting the spatial pattern of aboveground net primary production (ANPP) following the 1988 fires in Yellowstone National Park, Wyoming, provides an example. Postfire lodgepole pine densities varied from 0 to >500,000 stems ha$^{-1}$ in response to spatial variation fire severity and in pre-fire serotiny within the stand, rather than from variation in soils, topography, or climate (Turner et al. 2004). In turn, ANPP varied from 1 to 15 Mg ha$^{-1}$ yr$^{-1}$ 10 years after the fires and was explained primarily by lodgepole pine sapling density. Compared to "classic" curves of NPP through time (e.g., depicted by Ryan et al. 1997 for spruce in Russia), these patterns indicate that the spatial variation observed in a single age class can equal or exceed the range of mean ANPP through successional time.

The spatial pattern or scale of variation in a process rate may be more informative than the mean, but few studies have explored this. Approaches derived from spatial statistics can be particularly useful in evaluating the scale of spatial variation. For example, the importance of land-use legacies for contemporary forest ecosystems has received increasing attention (e.g., Pearson et al. 1998; Foster et al. 1999; Currie and Nadelhoffer 2002; Dupouey et al. 2002; Mitchell et al. 2002; Turner et al. 2003). Fraterrigo et al. (2005) used a cyclic sampling design derived from spatial statistics (Clinger and Van Ness 1976) to determine whether prior land use influenced the spatial variability of soil chemical properties. Cyclic sampling designs use a repeated pattern of sampled plots that minimizes the number of samples but provides sample pairs separated by any distance (Burrows et al. 2002). Thus, this design is efficient for analyses such as semivariograms, correlograms, and spatial regression. Fraterrigo et al. (2005) hypothesized that soil properties would vary over fine scales in old-growth forest and over coarse scales in areas of past agriculture, which would have homogenized local variation. Results showed that prior land use did homogenize the variability in forest soils, and that the scales of variation for several response variables depended on past land use as hypothesized.

(3) If the occurrence or rate of a lateral transfer responds directly to spatial heterogeneity, then the spatial pattern (composition and configuration) becomes one of the independent variables in the analysis. Many examples can be found in studies of the flux of nutrients from upland to aquatic

ecosystems (e.g., Richards et al. 1996; Johnson et al. 1997; Jones et al. 2001). For example, the amount and arrangement of crop fields and riparian forests influences the delivery of nitrogen and phosphorus to streams (Peterjohn and Correll 1984; Reed and Carpenter 2002). Both the amount and spatial arrangement of land cover types must be considered to predict nutrient delivery. On boreal shield ridges in northwestern Ontario, the spatial arrangement of *Pinus mariana-Pinus banksiana* forest islands relative to patches of lichen, moss, and grass influenced N retention in a 2-yr $NO_3$ addition study (Lamontagne and Schiff 1999). These patches have characteristically different N cycles, with the forest patches being N limited and the lichen patches N saturated; the location of patches in the landscape was important for N export from the catchment.

(4) Spatial heterogeneity may also generate lateral transfers. For example, clearing of natural vegetation for agriculture in western Australia created a new landscape pattern that altered climate. A large block of newly cleared agricultural land was separated from the original heath vegetation by a rabbit fence, producing a new patch type that had a higher albedo and therefore absorbed less solar radiation than the adjacent heath (Chambers 1998). The greater sensible heat flux of the darker native heath vegetation caused the surface air to warm, become more buoyant, and rise. The rising air over the heath was replaced by moist air advected from the adjacent croplands, which in turn was replaced by dry subsiding air from aloft. Thus, the changes in spatial heterogeneity produced a small-scale circulation cell, analogous to a land-sea breeze, that increased precipitation by 10% over the heathlands and reduced it by 30% over the croplands, fundamentally changing this landscape. At a finer spatial scale, the juxtaposition of substrates with different C:N ratios, such as carbon-rich straw adjacent to nitrogen-rich mineral soil, may result in nutrient transfers (Mary et al. 1996). Fungi transport nitrogen to the log so they can produce enzymes to decompose the log. In these examples, spatial configuration is actually producing flows, which otherwise would not have occurred. Thus, understanding spatial heterogeneity is fundamental to understanding these lateral transfers and point processes.

(5) Finally, lateral transfers may produce, amplify, or moderate heterogeneity in patterns. The Alaska coastal current is an example of lateral transfers creating patterns. Ocean waters flow counterclockwise parallel to the coast while fresh water, derived from orographic precipitation as moist marine air strikes the coastal mountains, flows from the land to the ocean. This produces two relatively distinct and stable water masses: a low-density (warm, low salinity), low-nutrient fresh water mass that is adjacent to and above a dense eutrophic ocean water mass (Royer 1981). The front between these two water masses generates conditions that maximize productivity of phytoplankton, zooplankton, and fish. At this boundary, the oligotrophic ocean water provides nutrients, and the sharp density gradient minimizes vertical mixing of phytoplankton out of the photic zone. This boundary is

readily visible from the air from the high chlorophyll content and the concentration of foraging sea birds at the frontal zone. Spatial heterogeneity is a direct consequence of lateral flows.

The lateral transfers of nutrients by animals can also produce spatial patterns in nutrient pools, cycling rates, and productivity. Anadromous fish transport large quantities of marine-derived nutrients to streams and lakes. Otters, bears, and other piscivores move these nutrients to riparian forests, where they can contribute substantially to productivity (Willson et al. 1998; Naiman et al. 2002). The characteristic $^{15}$N signature of marine-derived nitrogen is often detectable up to a kilometer from the river, suggesting a broad corridor of lateral nutrient transfer adjacent to streams with anadromous fisheries. Grazing ungulates also contribute to lateral nutrient transfers. In Switzerland, for example, the patchy distribution of cattle generated sharp nutrient gradients between forests and fields (Schutz et al. 2000). When cattle grazing ceased in national parks, these nutrient gradients became less pronounced, as native ungulates slowly redistributed these nutrients into the forests. Even random lateral movements that differ between predators and prey can generate spatial heterogeneity in ecosystem processes (Pastor, this volume).

## Approaches for Dealing with Spatial Heterogeneity

Given that spatial heterogeneity is frequently important but poorly quantified, how should we begin to incorporate it into ecosystem studies? Shugart's (1998) classification of ecosystem models is also a useful classification for our discussion; we also acknowledge a similar classification of models in Baker's (1989) review of models of landscape change.

### Homogenous Space

The simplest approach has been to assume homogeneity in rates across space—every point can be represented by the mean value of the rate (Figure 2.3a). Although this book focuses on spatial heterogeneity, the assumption of spatial homogeneity remains a valuable starting point or null model. This assumption is particularly useful for approximating pools or fluxes to order of magnitude; for some spatial extrapolations; and when physically averaging a response variable across variability at finer scales than the scale of interest.

Some processes can be extrapolated to large scales without explicitly considering landscape interactions. The extrapolation of carbon flux, for example, may adequately be represented in the short term from an understanding of its response to climate, vegetation, and stand age (Chapin et al. 2002: 329). The simulation of global net ecosystem production (NEP) by the terrestrial

(a)

(b)

(c)

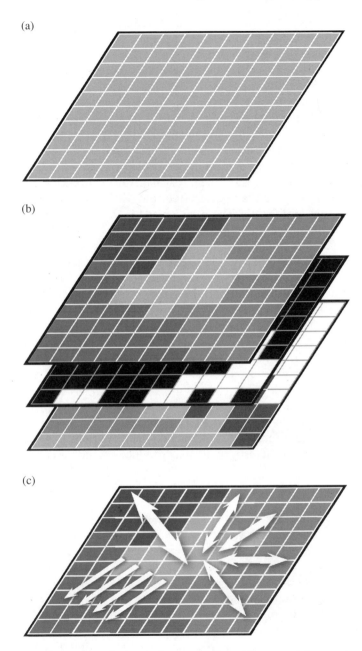

FIGURE 2.3. Three general approaches to dealing with space: (a) assuming spatial homogeneity, (b) the mosaic approach, which is often multivariate, and (c) interacting elements.

ecosystem model (TEM; McGuire et al. 1995) assumes homogeneity of environmental response within biomes to predict global patterns of NEP. This assumption allows the development of global databases even in areas where information is sparse or absent. Comparison of the output of these carbon flux models with seasonal and spatial patterns of atmospheric $CO_2$ identifies areas where assumptions of homogeneity are least justified and where additional information on spatial heterogeneity is most needed.

Eddy flux towers physically average measurements over an area of about 1 km$^2$. The heterogeneity in carbon fluxes resulting from fine-scale variation in soil aeration and other important ecosystem controls within the tower footprint is invisible because of the physical mixing of air. Consequently, the towers provide an accurate integration of the overall flux from the ecosystem (Davidson et al. 2002). These integrated landscape measures may be more useful than fine-scale information if extrapolation to large areas is based on satellite imagery that cannot resolve the fine-scale detail in ecosystem controls. Similarly, ecosystem ecologists frequently measure soil parameters and microbial processes on composite samples that physically average much of the fine-scale heterogeneity present in the ecosystem.

Of course, understanding the situations in which the assumption of spatial heterogeneity is likely to fail is important. Smithwick et al. (2003) used a forest process model to explore the assumption that carbon dynamics can be modeled within homogenous patches (e.g., even-aged forest stands) and then summed to predict broad-scale dynamics. Their results suggested that the additive approach might not capture C dynamics in fragmented landscapes because of edge-induced effects on tree mortality (primarily due to wind) and light limitations (Smithwick et al. 2003). This study nicely illustrates a systematic approach for identifying the conditions under which the assumption of spatial heterogeneity may produce erroneous conclusions.

## Mosaics

Spatial mosaics are the simplest representation of spatial heterogeneity in ecological processes (Figure 2.3b). Mosaics are particularly useful for documenting and predicting spatial heterogeneity in point processes and for spatial extrapolation. It is important to recognize that the mosaic represents not only vegetation or land-cover types; more often, it is a complex multivariate mosaic of underlying controls. The rate of a process at a given location may depend on many factors, such as vegetation type, soil conditions, slope, aspect, elevation, or time since disturbance.

Mosaic effects on ecosystem processes can be represented using a "paint-by-numbers" approach that assumes no interaction among spatial elements. However, this approach is not trivial; it can be very complicated when the relationship is nonlinear, there are multiple drivers of a process, or the distributions of drivers change through time. Practically, regression or classification and regression tree (CART) techniques are often used with empirical data for this approach, with the relationship between a process rate and its drivers represented at each location across a landscape. The most common representation of spatial mosaics is a raster, or grid-cell, approach with resolution (or grain size) appropriate for the process of interest (Turner et al. 2001). Employing this approach requires knowing the spatial distribution of each driver. However, the prediction for each site is based only on the suite of independent variables associated with that location. Ecosystem simulation models can also be used to make predictions across a landscape mosaic. For example, Running et al. (1989) combined simulation models with remotely sensed data to predict photosynthesis, leaf area index, and evapotranspiration rate in grid cells representing the landscape of western Montana.

Many studies in which ecosystem process rates are extrapolated spatially use a mosaic approach. For example, Hansen et al. (2000) predicted rates of ANPP over the western portion of the Greater Yellowstone Ecosystem using a multiple regression model in the mosaic; Turner et al. (2004) used multiple regression within the areas of the 1988 Yellowstone fires to predict spatial variation in ANPP and leaf area index (LAI) within the burn. Similar approaches have been used for nitrogen mineralization rates (Fan et al. 1998), denitrification rates (Groffman et al. 1992), and other responses.

A mosaic approach may employ static or dynamic representations of spatial patterns. In the latter case, model estimates at each time step must account for any changes in spatial pattern that have occurred in at least one driver. These changes in pattern may result from feedbacks between the rate of the ecosystem process being measured or predicted and the occurrence of events that alter the pattern of the drivers—fire is an example of this. The point process rate, however, is still predicted without considering neighbors.

An "advanced paint-by-numbers" approach considers the context of the landscape surrounding a point at which measurements are made. This variant of the paint-by-numbers approach uses the characteristics of the point and the surrounding landscape (i.e., the landscape context) to determine the behavior of a point. In this case, the spatial distribution/pattern of each of the important driving variables must be known. The predicted value at a given site depends not only on the values of the predictor variables at that site, but also on the values of predictor variables in the surrounding area. There is a large literature using this approach to understand the effects of landscape context on the presence and/or abundance of organisms (e.g., Pearson 1993; Mazerolle and Villard 1999). The approach has also been useful in estimating ecosystem processes. For example, the concentration of dissolved organic carbon in lakes and rivers was predicted by the proportion of wetlands in the surrounding landscape (Gergel et al. 1999).

Ecosystem and landscape ecology have made reasonable progress in using the mosaic approach to represent variation in process rates, although the number of studies explicitly sampling for spatial variance remains relatively small. However, this approach is limited in its capacity to address dynamic space, complex feedbacks, or nonlinearities in responses. These components require a more dynamic consideration of interacting elements.

## Interacting Elements

An interacting element approach is required to address lateral transfers. Typically, composition and configuration must both be considered. Ecosystem ecology does not yet have a comprehensive or even a well-developed approach for dealing with lateral transfers (Figure 2.3c). Empirical methods are frequently used to determine whether and when spatial pattern influences lateral transfer rates. Often, the response variable is an indicator of lateral transfer rather than a direct measurement of the transfer rate itself; for example, $NO_3$ concentration in soil water (e.g., lysimeter studies) may be used to track the movement and fate of N as it is transported from one ecosystem type to another. Labeled substances may be used as tracers to track directly the flow paths and rates or areas that differ in the composition and configuration of land cover types may be compared. Simulation models are also employed to predict the consequences of alternative spatial arrangements of cover types on lateral transfers. We consider three approaches of increasing complexity.

### Static Spatial Pattern–Dynamic Lateral Transfers

The simplest approach to exploring the consequences of spatial pattern for lateral flows is to evaluate the consequences of a static landscape pattern on lateral transfers. This approach has been used particularly for studies of land-water interactions. Shaver et al. (1991), for example, tracked nutrient flows in a toposequence in Alaska based on the typical configuration of landscape elements. A comparative empirical approach can be used in which, for example, the spatial arrangements of land cover in a variety of watersheds is related to stream nutrient concentrations (e.g., Hunsaker and Levine 1995; Jones et al. 2001). The flows themselves are not measured directly, and concentration or loading is the index of magnitude of flow. Models are also helpful in this arena; for example, Weller et al. (1998) explored the effects of length, width, and number of gaps in a riparian buffer on nutrient delivery to a stream by using a simulation model. However, common to all of these approaches is the absence of feedback from the lateral transfer to the spatial pattern.

### Dynamic Spatial Pattern–Dynamic Lateral Transfers

Here, spatial patterns are not stationary, and flows are assumed to respond to changes in the landscape template. Landscapes are constantly altered by natural disturbances and anthropogenic activities, and temporal changes in the spatial patterns of drivers can be represented. Horizontal flows respond to changes in these spatial patterns. For example, in the watershed of Lake Mendota, Wisconsin, land cover shifted from agricultural to urban uses between the 1930s and 1990s. The runoff of water from the terrestrial surface to the lake following storm events has become much more "flashy" during this period (Wegener 2001), illustrating how lateral transfers can respond to dynamic patterns over 60 years. Again, the lateral transfers do not alter the spatial pattern, but they respond to its temporal change.

### Dynamic Spatial Pattern–Dynamic Lateral Transfers–Feedbacks Between Pattern and Process

Here, spatial patterns change, altering flows, which, in turn, alter the template itself. This complex set of relationships is perhaps most interesting, but poorly understood; again, both empirical and modeling approaches are informative. On Isle Royale, for example, moose (*Alces alces*) selectively browse on hardwood trees and balsam fir (*Abies balsamea*), which leads to domination of the landscape by conifers such as spruce. In turn, spruce domination alters patterns of productivity and nutrient cycling across the landscape, which then influences moose foraging patterns. These reciprocal interactions between moose and vegetation have been elucidated through a combination of intensive studies of moose movement and foraging patterns, vegetation dynamics, and nutrient cycling, along with models that explore the possible behaviors of the system (e.g., McInnes et al. 1992; Jeffries et al. 1994; Moen et al. 1997, 1998; Pastor et al. 1999). Similar complex relationships between ungulates and vegetation patterns have been observed in African landscapes (e.g., Seagle and McNaughton 1992; Augustine 2003).

In river-floodplain ecosystems, we also see reciprocal interactions between the water and the land. Floodplains and rivers are linked as integrated ecosystems through the exchange of particulate and dissolved matter (Tockner et al. 1999). The spatial patterns of geomorphology and vegetation in a floodplain can influence flooding and flow velocity, at least in years that are not extreme. Geomorphological and biological processes are inherently linked in a functional hierarchy (van Coller et al. 2000). A reciprocal interaction approach has also been used to model fire-vegetation in interior Alaska (Rupp et al. 2000, 2002). The landscape template (vegetation configuration and composition) determine both fire spread and subsequent seed dispersal and regeneration pattern. These processes, in turn, determine the vegetation template on the landscape, which influences fire probability and spread. Inclusion of these dynamic interactions allows an evaluation of potential impacts of external factors on either landscape pattern (e.g., land-use effects on vegetation pattern) or process (e.g., climate effects on fire probability). This dynamic approach is particularly important under circumstances where either pattern or process is undergoing directional change.

## Challenges and Opportunities

If ecologists have recognized for 60 years that ecosystem processes are spatially heterogeneous (Jenny 1941), why is this topic relatively unexplored? We suggest there are several fundamental reasons then discuss some approaches for making sustained progress.

One challenge is that the interface between ecosystem and spatial ecology lacks a well developed theory (White and Brown, this volume). There is relatively little to guide us in our empirical studies, so our developing understanding has largely been empirical. However, even in empirical studies, the form of the relationship between response and driver variables is poorly understood and may well be nonlinear.

The technical sophistication and costs required to sample many ecosystem processes is relatively high. Sophisticated, expensive equipment is needed for many biogeochemical analyses, sample analysis is costly, and field sampling is labor-intensive. Adding the spatial dimension to a study design can substantially increase the number of samples needed. If a study attempts to understand spatial variance in rates over a large area, the logistics of conducting the sampling become quite challenging. As is true for many studies of broad-scale patterns, there are few opportunities to conduct experiments, although there are many opportunities for studying natural events or management actions from an experimental viewpoint. Even so, many people trained in ecosystem process studies lack advanced training in landscape ecology, spatial statistics, and spatially explicit models. Likewise, many people trained in landscape ecology lack the technical training in ecosystem ecology and biogeochemistry to address these questions.

Lack of understanding also results, in part, from inherent challenges related to variance and scale. For example, variance at fine spatial scales is extremely high for most biogeochemical processes, many of which are regulated by microorganisms. Relatively little is known about how microbial communities vary through both time and space. Because process rates may be measured at scales different from those of the controls, noise in the data can be overwhelming. Sampling adequately to obtain a general trend is already challenging without the added goal of understanding spatial variation.

Statistical considerations have also prompted ecosystem ecologists to avoid studies of spatial variation. In an effort to be rigorous, most ecosystem ecologists design observational or experimental studies that test for statistical differences between ecosystem types or treatments. This motivates experimental designs that minimize spatial variation (e.g., one- or two-way ANOVAs). Pastor (1995 and this volume) argues that this statistical preoccupation has done a disservice to ecosystem ecology, particularly modeling, where it is often more important to know the shape of a relationship between control and ecosystem response (e.g., between water availability and NPP) than to ask a simple yes/no question. Astute spatial sampling designs that incorporate heterogeneity in presumed control variables can provide valuable insights into nonlinearities and thresholds in controls over ecosystem processes that will never emerge from simple ANOVA designs.

Despite these challenges, there are ways to make progress, as described below.

## Exploit Heterogeneity to Enhance Understanding of Processes

We urge ecologists to embrace spatial complexity and to treat it as an opportunity! Variance may be an important clue to our understanding of processes. For example, the fine-scale variation in microbial activity from one unit of soil to another could reflect important differences between processes within versus outside of soil aggregates, just as at larger scales we know that urine patches differ functionally from the matrix or that lakes differ from the terrestrial matrix. The extent to which ecosystem ecologists tend to think of heterogeneity as a nuisance rather than a reflection of important process controls is still problematic.

The spatial variability in tree N uptake within a small catchment was evaluated by Barker et al. (2002) by measuring major fluxes in the N cycle in 50 plots (20 m × 20 m). Results showed that overstory N uptake varied spatially in the watershed with stand structure, although the variance among different calculations was even greater. Nonetheless, uptake was correlated with stand structure. These results also underscore the intensive sampling required and some of the methodological challenges associated with estimating spatial structure in complex processes.

## Conduct Studies at Multiple Scales

It is not possible to measure intensively everywhere, so sampling designs must be strategic. For example, intensive measurements at a small number of sites based on hypotheses can provide insights into mechanisms. However, these studies benefit from extensive measurements of simple integrative indices of these mechanisms at a larger number of sites to provide context. Nested sampling designs (Webster and Oliver 2001) are also useful. In addition, "smart" sampling designs derived from spatial statistics can maximize the power of the data. For example, a cyclic sampling design was used by Burrows et al. (2002) to maximize information about the variance of vegetation characteristics surrounding an eddy flux tower at Park Falls, Wisconsin. The data were also used to derive a spatial map of leaf area index (LAI) along with a map of spatial error measures for the study area (Burrows et al. 2002). Such methods afford the ability to quantify the scales of variation along with mean values of factors hypothesized to be important. Even though there is now a well developed statistical methodology to assess process controls at multiple scales, it has seldom been applied in ecosystem studies. The combination of intensive studies with spatially extensive measurements can also be used to see how well the knowledge at fine scales can be applied more broadly.

## Use Empirical and/or Simulation Models for Extrapolation

Modeling can be a powerful tool for exploring the range of conditions under which a given set of process controls leads to plausible outcomes. The simulation results can then be tested against field observations. These extrapolations represent testable hypotheses about our understanding of the system, and they should be used more widely as such (Miller et al. 2004). For example, the extrapolation of a hypothesized relationship using paint-by-numbers can be tested in the field to determine the limits of the validity of this presumed relationship. Models provide context and permit exploration of more combinations of conditions than we can assess in the field. Statistical models can also be used to extrapolate to broad scales and can be tested with remote sensing data and/or extensive field measurements to see whether they are consistent with predictions.

## Be Creative About When and How to Use Discrete versus Continuous Representations of Space

There are a variety of ways in which space may be represented in both drivers and response variables. The two most common representations of spatial heterogeneity include categorical maps and point data (Gustafson 1998). In categorical maps, variables are mapped in space, and both composition and configuration can be quantified. A wide variety of metrics is available to quantify such patterns (e.g., McGarigal and Marks 1995). Although categorical maps are often created from continuous data (e.g., forest cover is often mapped based on the proportion of a cell occupied by trees), this approach ignores spatial variation within the units (Gustafson 1998). Point-data analysis, in contrast, assumes the system property is spatially continuous, and an area is sampled to generate spatially referenced information about the system. Analysis techniques include trend-surface analysis, various techniques that address spatial autocorrelation (e.g., correlograms, semivariograms), and interpolation. Platt and Sathyendranath (1999) correctly note, however, that universal functions for continuous variation of environmental properties generally have not been discovered.

Careful consideration of how and why space should be represented is crucial, and the representation of heterogeneity should match the question and be scaled correctly. Point data are required for interpolation methods (e.g., kriging) or for using scales of variation as a response variable. However, a categorical approach might simplify the analysis of biogeochemical hot spots by eliminating the need to treat all variation in processing rates. For example, one might predict locations where a process like denitrification occurs in a floodplain (or where the rate exceeds some meaningful threshold) rather than predicting the actual rates. We echo Gustafson's (1998) plea for moving beyond the patch-based view of spatial heterogeneity and for recognition of the complementarity between categorical and continuous representations of space.

## Collaborate and Explore Other Bodies of Theory

Intra- and interdisciplinary collaboration often produces new insights, and we encourage ecologists to look beyond their research specialty. What theories developed in other disciplines within or outside of ecology might be helpful? Percolation theory (Stauffer 1985; Stauffer and Aharony 1992), a branch of physics, offered new modeling and analysis techniques that were applied in landscape ecology (Gardner et al. 1987) and led to new insights about crucial thresholds in connectivity (With and King 1997). Within ecology, there is an extensive body of literature on source-sink dynamics for populations—might that theory be relevant for lateral transfers of matter or energy? Gases and particulates emitted from managed or natural ecosystems (sources) can be transported great distances, altering the recipient (sink) ecosystems. Boerner and Kooser (1989) studied redistribution of leaf litter within a 73-ha watershed in Ohio and used donor and sink terminology. Donor sites lost 4.5–5.7 ka ha$^{-1}$ yr$^{-1}$ of N and 0.3–0.5 kg ha$^{-1}$ yr$^{-1}$ of P through redistributed litter; sink areas received subsidies of 2.2–6.1 kg ha$^{-1}$ yr$^{-1}$ N and 0.2–0.4 kg ha$^{-1}$ yr$^{-1}$ of P. Pastor (this volume) also suggests that cross-fertilization between ecosystem ecology and evolutionary studies is likely to produce new understanding about ecosystem function in time and space.

## Looking Ahead

Understanding spatial heterogeneity has been referred to as "the final frontier" in other areas within ecology (e.g., Kareiva 1994). Although new challenges will continually arise, understanding the causes and consequences of ecosystem function in heterogeneous landscapes is a challenge that will be present for some time. Methods to quantify spatial heterogeneity abound; gaining a functional understanding of spatial pattern should be the priority rather than the development of new pattern metrics. If knowledge of spatial heterogeneity and ecosystem function improves, it is appropriate to consider the significance of this enhanced understanding. There are at least three areas in which advances will be significant to our science.

First, understanding of pools, fluxes, and regulating factors in ecosystems will be enhanced—and this defines the purview of ecosystem ecology. By understanding heterogeneity, what causes it, and when it matters, we will have a much better understanding of fundamental ecosystem processes. Broad-scale estimates of biogeochemical processes, which are key for understanding regional to global phenomena, require spatial understanding (e.g., Groffman et al. 1992). Factors such as disturbance frequency and size, species distributions, and exotic species invasions that are inherently spatial may influence not only the magnitude but also the sign of currently observed ecosystem fluxes within the next century (Canadell et al. 2000). Second, we will gain a more complete understanding of landscape function. At present, there is greater knowledge about how certain populations respond to patterns, the role of disturbance dynamics, and even the perceptions and effects of humans. However, this list conspicuously excludes knowledge of ecosystem function in both natural and anthropogenic landscapes. Indeed, understanding spatial heterogeneity and disturbance is one of the key needs for global studies (Schimel et al. 1997). Third, the ability to scale up or down will be improved. Using spatial models and spatial extrapolations as hypotheses should help identify the domains through which certain relationships do and do not scale (Miller et al. 2004). Ultimately, these gains should lead to improved predictions of changes in regional systems that involve multiple feedbacks between pattern and process at multiple scales.

*Acknowledgments.* For feedback and spirited discussions of the ideas presented in this paper, we especially thank Gary Lovett, Clive Jones, Kathy Weathers, and the 2002–2003 Turner lab group (Dean Anderson, Jeff Cardille, Alysa Darcy, James Forester, Jen Fraterrigo, Kris Metzger, Katie Predick, Erica Smithwick, and Anna Sugden-Newbury). We thank Michael G. Turner for creating the graphics. The manuscript benefited from comments by Jennifer Fraterrigo, Katie Predick, Gary Lovett, and two anonymous reviewers. This research was supported by grants to M.G.T. from the Andrew W. Mellon Foundation and the National Science Foundation (NSF) and sabbatical funding from the University of Wisconsin Graduate School, and by grants to F.S.C. from NSF and the U.S. Forest Service for the Bonanza Creek LTER program and from NSF studies in Arctic System Science.

## References

Amundson, R., and Jenny, H. 1997. On a state factor model of ecosystems. BioScience 47: 536–543.

Augustine, D.J. 2003. Long-term, livestock-mediated redistribution of nitrogen and phosphorus in an East African savanna. J. Appl. Ecol. 40: 137–149.

Baker, W.L. 1989. A review of models of landscape change. Landscape Ecol. 2: 111–133.

Barker, M., Van Miegrot, H., Nicholas, N.S., and Creed, I.F. 2002. Variation in overstory nitrogen uptake in a small, high-elevation southern Appalachian spruce-fir watershed. CJFR 32: 1741–1752.

Baskent, E.Z., and Jordan, G.A. 1995. Characterizing spatial structure of forest landscapes. Can. J. Forest Res. 25: 1830–1849.

Boerner, R.E.J., and Kooser, J.G. 1989. Leaf litter redistribution among forest patches within an Alleghany Plateau watershed. Landscape Ecol. 2: 81–92.

Burke, I.C., Lauenroth, W.K., Riggle, R., Brannen, P., Madigan, B., and Beard, S. 1999. Spatial variability of soil properties in the shortgrass steppe: the relative importance of topography, grazing, microsite, and plant species in controlling spatial patterns. Ecosystems 2: 422–438.

Burrows, S.N., Gower, S.T., Clayton, M.K., Mackay, D.S., Ahl, D.E., Norman J.M., and Diak, G. 2002. Application of geostatistics to characterize leaf area index (LAI) from flux tower to landscape scales using a cyclic sampling design. Ecosystems 5: 667–679.

Canadell, J.G., Mooney, H.A., Baldocchi, D.D., Berry, J.A., Ehleringer, J.R., Field, C.B., Gower, S.T., Hollinger, D.Y., Hunt, J.E., Jackson, R.B., Running, S.W., Shaver, G.R., Steffen, W., Trumbore, S.E., Valentini, R., and Bond, B.Y. 2000. Carbon metabolism of the terrestrial biosphere: a multitechnique approach for improved understanding. Ecosystems 3: 115–120.

Carpenter, S.R., and Turner, M.G. 1998. At last: a journal devoted to ecosystem science. Ecosystems 1: 1–5.

Carpenter, S.R., Ludwig, D., and Brock, W.A. 1999. Management of eutrophication for lakes subject to potentially irreversible change. Ecol. Applications 9: 751–771.

Chambers, S. 1998. Short- and long-term effects of clearing native vegetation for agricultural purposes. Ph.D. Thesis, Flinders University of South Australia, Adelaide, Australia.

Chapin III, F.S., Torn, M.S., and Tateno, M. 1996. Principles of ecosystem sustainability. Am. Naturalist 148: 1016–1037.

Chapin III, F.S., Matson, P.A., and Mooney, H.A. 2002. Principles of terrestrial ecosystem ecology. New York: Springer-Verlag.

Clinger, W., and Van Ness, J.W. 1976. On unequally spaced time points in time series. Ann. Stat. 4: 736–745.

Currie, W.S., and Nadelhoffer, K.J. 2002. The imprint of land-use history: patterns of carbon and nitrogen in downed woody debris at the Harvard Forest. Ecosystems 5: 446–460.

Davidson, E.A., Savage, K., Verchot, L.V., and Navarro, R. 2002. Minimizing artifacts and biases in chamber-based measurements of soil respiration. Agric. Forest Meteorol. 113: 21–37.

Dupouey, J.L., Dambrine, E., Laffite, J.D., and Moares, C. 2002. Irreversible impact of past land use on forest soils and biodiversity. Ecology 83: 2978–2984.

Fan, W., Randolph, J.C., and Ehman, J.L. 1998. Regional estimation of nitrogen mineralization in forest ecosystems using Geographic Information Systems. Ecol. Applications 8: 734–747.

Foster, D.R., Knight, D.H., and Franklin, J.F. 1998. Landscape patterns and legacies resulting from large infrequent forest disturbances. Ecosystems 1: 497–510.

Foster, D.R., Fluet, M., and Boose, E.R. 1999. Human or natural disturbance: landscape-scale dynamics of the tropical forests of Puerto Rico. Ecol. Applications 9: 555–572.

Fraterrigo, J., Turner, M.G., Pearson, S.M., and Dixon, P. 2005. Effects of past land use on spatial heterogeneity of soil nutrients in Southern Appalachian forests. Ecological Monographs 75: 215–230.

Gardner, R.H., Milne, B.T., Turner, M.G., and O'Neill, R.V. 1987. Neutral models for the analysis of broad-scale landscape patterns. Landscape Ecol. 1: 19–28.

Gergel, S.E., Turner, M.G., and Kratz, T.K. 1999. Scale-dependent landscape effects on north temperate lakes and rivers. Ecol. Applications 9: 1377–1390.

Gergel, S.E., and Turner, M.G., eds. 2002. Learning landscape ecology: a practical guide to concepts and techniques. New York: Springer-Verlag.

Golley, F.B. 1993. A history of the ecosystem concept in ecology: more than the sum of the parts. New Haven, CT: Yale University Press.

Groffman, P.M., Tiedje, T.M., Mokma, D.L., and Simkins, S. 1992. Regional-scale analysis of denitrification in north temperate forest soils. Landscape Ecol. 7: 45–54.

Gustafson, E.J. 1998. Quantifying landscape spatial pattern: what is the state of the art? Ecosystems 1: 143–156.

Hansen, A.J., Rotella, J.J., Kraska, M.P.V., and Brown, D. 2000. Spatial patterns of primary productivity in the Greater Yellowstone Ecosystem. Landscape Ecol. 15: 505–522.

Hunsaker, C.T., and Levine, D.A. 1995. Hierarchical approaches to the study of water quality in rivers. BioScience 45: 193–203.

Jeffries, R.L., Klein, D.R., and Shaver, G.R. 1994. Vertebrate herbivores and northern plant communities: reciprocal influences and responses. Oikos 71: 193–206.

Jenny, H. 1941. Factors of soil formation. New York: McGraw-Hill.

Johnson, L.B., Richards, C., Host, G., and Arthur, J.W. 1997. Landscape influences on water chemistry in midwestern streams. Freshwater Biol. 37: 209–217.

Jones, K.B., Neale, A.C., Nash, M.S., Van Remortel, R.D., Wickham, J.D., Riitters, K.H., and O'Neill, R.V. 2001. Predicting nutrient and sediment loadings to streams from landscape metrics: a multiple watershed study from the United States Mid-Atlantic Region. Landscape Ecol. 16: 301–312.

Kareiva, P. 1994. Space: the final frontier for ecological theory. Ecology 75: 1.

Lamontagne, S., and Schiff, S.L. 1999. The response of a heterogeneous upland boreal shield catchment to a short term $NO_3^-$ addition. Ecosystems 2: 460–473.

Mary, B., Recous, S., Darwis, D., and Robin, D. 1996. Interactions between decomposition of plant residues and nitrogen cycling in soil. Plant Soil 181: 71–82.

Mazerolle, M.J., and Villard, M.A. 1999. Patch characteristics and landscape context as predictors of species presence and abundance: a review. Ecoscience 6: 117–124.

McGarigal, K., and Marks, B.J. 1995. FRAGSTATS. Spatial analysis program for quantifying landscape structure. USDA Forest Service General Technical Report PNW-GTR-351.

McGuire, A.D., Melillo J.W., Kicklighter, D.W., and Joyce, L.A. 1995. Equilibrium responses of soil carbon to climate change: empirical and process-based estimates. J. Biogeogr. 22: 785–796.

McInnes, P.F., Naiman, R.J., Pastor, J., and Cohen, Y. 1992. Effects of moose browsing on vegetation and litter of the boreal forest, Isle Royale, Michigan, USA. Ecology 75: 478–488.

Miller, J.R., Turner, M.G., Smithwick, E.A.H., Stanley, E.H., and Dent, L.C., 2004. Spatial extrapolation: the science of predicting ecological patterns and processes. Bio Science 54: 310–320.

Mitchell, C.E., Turner, M.G., and Pearson, S.M. 2002. Effects of historical land use and forest patch size on myrmecochores and ant communities. Ecol. Applications 12: 1364–1377.

Moen, R., Cohen, Y., and Pastor, J. 1997. A spatially explicit model of moose foraging and energetics. Ecology 78: 505–521.

Moen, R., Pastor, J., and Cohen, Y. 1998. Linking moose population and plant growth models with a moose energetics model. Ecosystems 1: 52–63.

Naiman, R.J. 1996. Water, society and landscape ecology. Landscape Ecol. 11: 193–196.

Naiman, R.J., Bilby, R.F., Schindler, D.E., and Helfield, J.M. 2002. Pacific salmon, nutrients, and the dynamics of freshwater and riparian ecosystems. Ecosystems 5: 399–417.

Pace, M.L., and Groffman, P.M., eds. 1998. Successes, limitations and frontiers in ecosystem science. New York: Springer.

Pastor, J. 1995. Diversity of biomass and nitrogen distribution among plant species in arctic and alpine tundra ecosystems. In Arctic and alpine biodiversity: patterns, causes and ecosystem consequences, eds. F.S. Chapin III, and C. Korner, pp. 255–269. Berlin: Springer-Verlag.

Pastor, J., Cohen, Y., and Moen, R. 1999. Generation of spatial patterns in boreal forest landscapes. Ecosystems 2: 439–450.

Pearson, S.M. 1993. The spatial extent and relative influence of landscape-level factors on wintering bird populations. Landscape Ecol. 8: 3–18.

Pearson, S.M., Smith, A.B., and Turner, M.G. 1998. Forest fragmentation, land use, and cove-forest herbs in the French Broad River Basin. Castanea 63: 382–395.

Peterjohn, W.T., and Correll. D.L. 1984. Nutrient dynamics in an agricultural watershed: observations on the role of a riparian forest. Ecology. 65: 1466–75.

Platt, T., and Sathyendranath, S. 1999. Spatial structure of pelagic ecosystem processes in the global ocean. Ecosystems 2: 384–394.

Polis, G.A., and Hurd, S.D. 1996. Linking marine and terrestrial food webs: allochthonous input from the ocean supports high secondary productivity on small islands and coastal land communities. Am. Naturalist 147: 396–423.

Reed, T., and Carpenter, S.R. 2002. Comparisons of P-yield, riparian buffer strips, and land cover in six agricultural watersheds. Ecosystems 5: 568–577.

Richards, D., Johnson, L.B., and Host, G. 1996. Landscape-scale influences on stream habitats and biota. Can. J. Fisheries Aquatic Sci. 53(Suppl. 1): 295–311.

Risser, P.G., Karr, J.R., and Forman, R.T.T. 1984. Landscape ecology: directions and approaches. Special Publication Number 2. Champaign, IL: Illinois Natural History Survey.

Royer, T.C. 1981. Baroclinic transport in the Gulf of Alaska. Part II. A fresh water driven coastal current. J. Marine Res. 39: 251–266.

Running, S.W., Nemani, R.R., Peterson, D.L., Band, L.E., Potts, D.F., Pierce, L.L., and Spanner, M.A. 1989. Mapping regional forest evapotranspiration and photosynthesis by coupling satellite data with ecosystem simulation, Ecology 70: 1090–1101.

Rupp, T.S., Chapin, F.S., and Starfield, A.M. 2000. Response of subarctic vegetation to transient climatic change on the Seward Peninsula in northwest Alaska. Global Change Biology 6: 541–555.

Rupp, T.S., Starfield, A.M., Chapin III, F.S., and Duffy, P. 2002. Modeling the impact of black spruce on the fire regime of Alaskan boreal forest. Climatic Change 55: 213–233.

Ryan, M.G., Binkley, D., and Fownes, J.H. 1997. Age-related decline in forest productivity: pattern and process. Adv. Ecol. Res. 27: 213–262.

Seagle, S.W., and McNaughton, S.J. 1992. Spatial variation in forage nutrient concentrations and the distribution of Serengeti grazing ungulates. Landscape Ecol. 7: 229–241.

Schimel, D.S., VEMAP Participants, and Braswell, B. H. 1997. Continental scale variability in ecosystem processes: models, data, and the role of disturbance. Ecol. Monogr. 67: 251–271.

Schutz, M., Krusi, B.O., Edwards, P.J., eds. 2000. Succession research in the Swiss National Park. National Park-Forschung in der Schweiz, No. 89.

Shaver, G.R., Knadelhoffer, K.J., and Giblin, A.E. 1991. Biogeochemical diversity and element transport in a heterogeneous landscape, the north slope of Alaska. In Quantitative methods in landscape ecology, eds. M.G. Turner, and R.H. Gardner, pp. 105–125. New York: Springer-Verlag.

Shugart, H.H. 1998. Terrestrial ecosystems in changing environments. Cambridge, UK: Cambridge University Press.

Smithwick, E.A.H., Harmon, M.E., and Domingo, J.B. 2003. Modeling multiscale effects of light limitations and edge-induced mortality on carbon stores in forest landscapes. Landscape Ecol. 18: 701–721.

Soranno, P.A., Webster, K.E., Riera, J.L., Kratz, T.K., Baron, J.S., Bukaveckas, P.A., Kling, G.W., White, D.S., Caine, N., Lathrop, R.C., and Leavitt, P.R. 1999. Spatial variation among lakes within landscapes: ecological organization along lake chains. Ecosystems 2: 395–410.

Stauffer, D. 1985. Introduction to percolation theory. London: Taylor & Francis.

Stauffer, D., and Aharony, A. 1992. Introduction to percolation theory, 2nd ed. London: Taylor & Francis.

Tockner K., Pennetzdorfer, D., Reiner, N., Schiemer, F., and Ward, J.V. 1999. Hydrological connectivity, and the exchange of organic matter and nutrients in a dynamic river floodplain system (Danube, Austria). Freshwater Biol. 41: 521–535.

Turner, M.G. 1989. Landscape ecology: the effect of pattern on process. Annu. Rev. Ecol. Systematics 20: 171–197.

Turner, M.G., Gardner, R.H., and O'Neill, R.V. 2001. Landscape ecology in theory and practice. New York: Springer-Verlag.

Turner, M.G., Romme, W.H., Tinker, D.B., and Kashian, D.M. 2004. Landscape patterns of sapling density, leaf area, and aboveground net primary production in postfire lodgepole pine forests, Yellowstone National Park (USA). Ecosystems 7: 751–775.

Turner, M.G., Pearson, S.M., Bolstad, P., and Wear. D.N. 2003. Effects of land-cover change on spatial pattern of forest communities in the southern Appalachian Mountains (USA). Landscape Ecol 18: 449–464.

van Coller, A.L., Rogers, K.H., and Heritage, G.L. 2000. Riparian vegetation–environment relationships: complementarity of gradients versus patch hierarchy approaches. J. Veg. Sci. 11: 337–350.

Webster, R., and Oliver, M.A. 2001. Geostatistics for environmental scientists. Chichester, UK: John Wiley & Sons, Ltd.

Wegener, M.W. 2001. Long-term land use/cover change patterns in the Yahara Lakes region and their impact on runoff volume to Lake Mendota. M.S. Thesis, University of Wisconsin, Madison, WI.

Weller, D.E., Jordan, T.E., and Correll, D.L. 1998. Heuristic models for material discharge from landscapes with riparian buffers. Ecol. Applications 8: 1156–1169.

Willson, M.F., Gende, S.M., and Marston, B.H. 1998. Fishes and the forest. BioScience 48: 455–462.

With, K.A., and King, A.W. 1997. The use and misuse of neutral landscape models in ecology. Oikos 97: 219–229.

Zimov, S.A., Voropaev, Y.V., Semiletov, I.P., Davidov, S.P., Prosiannikov, S.F., Chapin III, F.S., Chapin, M.C., Trumbore, S., and Tyler, S. 1997. North Siberian lakes: a methane source fueled by Pleistocene carbon. Science 277: 800–802.

# 3
# The Template: Patterns and Processes of Spatial Variation

ETHAN P. WHITE and JAMES H. BROWN

## Abstract

Ecosystem processes are inherently variable in space and time, in part because they occur on a spatially heterogeneous template or landscape. For many purposes, the patterns of heterogeneity can be characterized as gradients, patchworks, or networks—or some combination of these fundamental patterns. Each class of landscape pattern implies that it has been generated by certain kinds of abiotic or biotic mechanisms, which can be described by particular mathematical formulations. We illustrate these points with a few selected, ecologically relevant examples. Quantitatively characterizing the patterns of variation in the template and understanding their causes, correlates, and consequences are important steps in investigating the influence of spatial heterogeneity on the structure and function of ecological systems at all scales from molecular to global.

## Introduction

Before getting too far into the consideration of the spatial heterogeneity of ecological processes, it is usually necessary to ask: How and why is the landscape heterogeneous? To understand how ecological processes play out on an underlying template of abiotic and biotic environmental variation, it is first necessary to understand that variation. At any given time, this template sets the initial conditions for the subsequent structural development and dynamic interactions of the system. So how is the template structured, why is it organized this way, and how does it change over time? These are big, complicated questions. The answers draw from many disciplines and remain incomplete.

Nevertheless, we will attempt to provide a conceptual framework to characterize some of the fundamental features of spatial environmental heterogeneity. We should make it clear from the outset that we do not consider ourselves to be either ecosystem or landscape ecologists. We hope to offer an outsider's perspective on characterizing and understanding heterogeneity.

What we have done is to collect in one place ideas stretching from physics and the earth sciences to biology and ecology and to suggest that we can use these concepts and mathematical tools to begin to characterize heterogeneity in a more general framework. We define heterogeneity simply as spatial variation in the environment. We suggest that this environmental variation can be characterized as a combination of gradients, patches, and networks. We discuss how these patterns can be characterized mathematically, how they are formed, and some of the consequences for the ecological processes that play out on these templates. Finally, we attempt to illustrate the potential utility of a centralized approach to dealing with heterogeneity by providing several examples from the literature.

## Patterns and Their Causes

We recognize three categories of patterns: gradients, patchworks, and networks. We do this with some trepidation. We are well aware of the pitfalls of dividing the natural world, and the frameworks that we use to study it, into compartments that may be artifactual human constructs. Nevertheless, such a classification seems appropriate in this case for several reasons. First, the processes that usually create these patterns are often distinct and operate at different scales. Second, the qualitative differences in the patterns and their causal processes mean that different mathematical and analytical methods are necessary to characterize them. Third, some degree of simplification is appropriate, even desirable, to study ecological processes on complex landscapes. The search for syntheses and mechanistic explanations based on first principles will require some simplifications, but ones that capture the essence of the phenomena.

## *Gradients*

We define gradients as patterns of continuous variation, typically of a single focal variable. Under this definition, there can be no more independent variables than there are Euclidean dimensions of the system. For two-dimensional space, therefore, there can be only two gradients of orthogonal variation. If more than two gradients occur on the earth's surface, there will be some degree of correlation among them. This can make gradients difficult to disentangle, especially because several gradients can simultaneously influence an observed pattern. In practice, we are often concerned with one-dimensional gradients: for example, with patterns such as temperature varying with latitude or elevation, temperature and pressure varying with water depth, and time of exposure varying with height in the intertidal. As in most of these examples, the pattern of variation itself may be curvilinear, just as long as it is continuous.

Gradients are fairly common. They tend to occur whenever there are strong polar differences in one or more correlated variables with some kind of averaging, homogenizing process operating in between. They are most apparent at large spatial scales where physical factors operate over substantial distances to generate relatively continuous variation in temperature, light, pressure, solute concentrations, and other important features of the biosphere. For example, the latitudinal gradient of temperature is due to the position of the earth in relation to the sun, and to the homogenizing effects of air and water movement. The elevational gradient of temperature is due to adiabatic heat exchange in response to variation in air pressure and again to the homogenizing effects of air movement. The gradient concept is fundamental to ecology and has been well developed for some time (Whittaker 1967). Where the process generating the gradient is known, it should be possible to use first principles to describe the quantitative pattern of variation.

Gradients tend to be best behaved at relatively large scales where the generating process dominates the variability in the observed values. As one "zooms in" to smaller scales within the gradient, additional processes become dominant, and the continuous gradient pattern becomes swamped by the now dominant local processes. Examined in detail on sufficiently small scales, temperature does not vary smoothly and monotonically with either latitude or elevation. An example is a thermal inversion in air temperature with elevation, a fairly common phenomenon. Nevertheless, a gradient described by a simple monotonic function usually captures most of the variation of temperature with respect to latitude, elevation, and water depth, at scales over which the impacts of the major process (solar incandescence, adiabatic cooling, and solar penetration) operate. At smaller scales, other processes dominate, and the previously smooth relationship appears increasingly patchy.

## Patches

Patches are the pattern that most biologists consider when talking about spatial heterogeneity. In principle, patches can be defined as discrete units of area that are more similar to one another in one or more variables than to their neighbors (Kotliar and Wiens 1990). For example, a patch type could be defined by an area of some size either containing or lacking nitrogen-fixing plants. In practice, many patch types must be based on artificial cutoffs (e.g., high nitrogen vs. low nitrogen, lowlands vs. highlands), and resulting arbitrary boundaries. Sometimes, the borders between patch types are effectively steep gradients, more continuous than discrete (Gustafson 1998).

Much of this type of discrete spatial heterogeneity is, at its core, due to the three-dimensional complexity of the earth's surface. If the earth were a simple plane or a perfectly smooth sphere, environmental variation would likely be characterized by simple gradients, with a maximum of two truly independent axes. However, the real landscape is heterogeneous and discontinuous because of geological and biological processes. The geological processes of tectonics and erosion have created a crumpled, dimpled, and layered surface, which interacts with the predictable gradients of solar energy input, air and water pressure, tidal exposure, and other factors to create a complex discontinuous abiotic template.

Biological processes modify this already complex template in several ways. First, as discrete entities with unique combinations of variables, individual organisms serve as patchy environments for other organisms. The most obvious example of this is hosts serving as patchy environments for parasites and symbionts. However, this phenomenon is actually much more general. Gradients and topographical features influence local climate and soil conditions. This patchy local abiotic environment determines the flora that can inhabit the area, and the flora, which is patchy as a result of the climate and soils, combines with the abiotic template to influence the abundances and distributions of animals at the site. Feedbacks between the animals, plants, and the abiotic environment can then occur, causing additional variation. For example, organisms can act as engineers, moving materials or altering flows to create new patches or alter existing ones (e.g., Jones et al. 1994). Examples include plant canopies creating unique microenvironments by altering the flows of energy, water, and nutrients. Burrowing animals can alter soil properties and create unique structures that are used by still other organisms (e.g., Reichman and Seabloom 2002).

Given the enormous variety of patch types, and of the processes that produce them, can we draw any generalizations about their properties? Patchy environments have traditionally proven difficult to describe quantitatively and thus to model. Perhaps the most promising approach is based on the application of fractal geometry (Mandelbrot 1983). Interestingly, it appears that many different kinds of patches have self-similar or fractal-like distributions. This means that, over at least some substantial range of scales, patterns of covariation can be characterized by power laws of the form

$$Y = Y_0 X^b, \tag{3.1}$$

where $Y$ is some variable that can be considered the dependent variable, $Y_0$ is a normalization constant, $X$ is the independent variable, and $b$ is another constant, the scaling exponent. Power laws have the useful property of being linearized by taking the logarithms of both sides of Equation (3.1),

$$\ln(Y) = \ln(Y_0) + b \ln(X), \tag{3.2}$$

such that a plot of $\ln(Y)$ as a function of $\ln(X)$ is a straight line with a slope of $b$ and an intercept of $\ln(Y_0)$. The slope, $b$, can take on a wide range of values that produce a wide variety of curves when plotted on linear axes. These curves can be increasing ($b > 0$), decreasing ($b < 0$), or invariant ($b = 0$), and the increasing curve can be concave up ($b > 1$), concave down ($0 < b < 1$), or linear ($b = 1$; Figure 3.1). The variation described by Equation (3.1) is

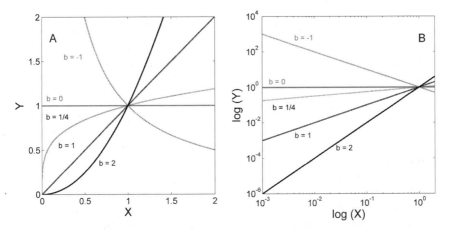

FIGURE 3.1. Example plot of power functions with different exponents, $b$ (A, linear axes; B, logarithmic axes). For all functions, $Y_0 = 1$. Note that except when $b = 1$, relationships are curvilinear when plotted on linear axes, but all are linear when plotted on logarithmic axes.

called self-similar or fractal, because the ratios of variables at any scale have a constant relationship to each other. That is

$$Y_1/Y_2 = (X_1/X_2)^b, \tag{3.3}$$

where $Y_1$, $Y_2$, $X_1$, and $X_2$ represent measurements of $Y$ and $X$ at two different scales, 1 and 2, respectively.

Multiple approaches to characterizing the shape and distribution of patches based on fractal-like behavior of particular features have been proposed (Milne 1991b). These approaches include the standard box counting and mass fractal dimensions (approximations of the Hausdorff dimension), the perimeter-area fractal dimension, and many others. These different fractal dimensions characterize different aspects of the patchy environment (Milne 1991b).

Many patches in nature, although they may be characterized in a variety of different ways, appear to have fundamentally fractal-like properties. This is true of patches and other landforms created by abiotic geological processes. The classic case is that of a coastline, which appears self-similar over a wide range of scales so long as the geological parent material and formative process is essentially the same (Richardson 1961; Mandelbrot 1983). As the length of the ruler used to measure the coastline gets smaller, the total length of the coastline increases (*coast length* $\propto$ *ruler length*$^{-D}$, where $D$ is the fractal dimension). Although the coastline is continuous and therefore not necessarily patchy in a traditional sense, it is "patchy" in a mathematical sense when compared to a straight line (i.e., it is not smooth). More obvious patchiness occurs when a complex geological landscape is partially filled with water, creating either lakes on land or islands in water.

One characterization of the fractal nature of patches that we find particularly intriguing is the scaling of frequency *versus* magnitude. It is well
established that, for earthquakes, the area involved in a seismic event is
approximately inversely proportional to the number of those events
observed (i.e., there are more smaller events). This is called the Guttenberg-
Richter law and it is the basis for measuring the magnitude of earthquakes
on a logarithmic Richter scale. This relationship between frequency and the
area involved is described by a power-function relationship, with a slope of
approximately −1. This general pattern between frequency and magnitude
has been observed in other systems, in particular forest fires (Malamud et al.
1998) and financial markets (Mandelbrot 1997). Although relatively poorly
studied in ecological systems, there is some evidence that ecological patches
may follow a similar power-function distribution. In particular, lakes,
islands, and vegetation patches have frequency-magnitude distributions
with $b \approx -1$ (Figure 3.2; see also Korcak 1938; Hastings et al. 1982; Wetzel

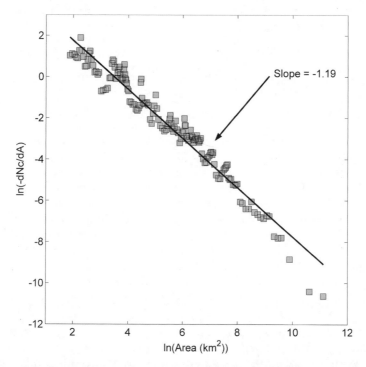

FIGURE 3.2. Plot of the frequency of the Southwest Pacific and Moluccan islands by
island area. Binning method (linear or logarithmic) and bin size complicate estimation of the precise underlying distribution. Thus, we generate the inverse cumulative
distribution function (cdf) for the observed data and then estimate the underlying
probability density function (pdf) by calculating the slope of the cdf using a sliding
window with a 5-point width (Malamud et al. 1998). This approach provides equivalent results to those based entirely on the cdf (Hastings and Sugihara 1993). Data on
island area was taken from Flannery (1995).

1991; Hastings and Sugihara 1993). This slope is approximately the same as that for earthquakes and implies that the total magnitude of all events in any given logarithmic magnitude class is approximately equal. For example, for the islands in Figure 3.2, the total area of small islands (1 to 10 km$^2$ in area) should be approximately equal to the total area of large islands (100 to 1000 km$^2$ in area). Similar slopes have been observed for forest fire frequency (Malamud et al. 1998). Peninsulas (Milne 1991b) and forest patches (N. Baum, unpublished data) also appear to have a power-function relationship between frequency and area, though the reported exponents are closer to $-2$. This suggests that the general form of the power-law relationship holds for different landscape features but that the specific exponent depends on the particular feature being observed. Consequently, differences in exponents may suggest important differences in the processes generating the patterns and in their effects on biological systems. Lakes, islands, vegetation patches, peninsulas, and burns all represent heterogeneously distributed ecological patches that have important consequences for ecological processes at scales of organization from the individual to the ecosystem.

Organisms are the source of additional patchiness. And again, some of the patterns may be fractal-like. For example, most deserts can be characterized as a mosaic of two patch types: vegetation and bare soil. Figure 3.3 shows the pattern of perennial vegetative cover on Brown's long-term study site in the Chihuahuan desert. Analysis of these patches using the box-counting method reveals a fractal-like distribution, similar to that for coastlines, with the area of occupied grid cells increasing as larger cells are used to characterize patches (Figure 3.3B, inset). This relationship is traditionally presented as a negative relationship between the number of cells occupied with vegetation and the size of the cells (Figure 3.3B). In addition to broad taxonomic groups like plants, individual species exhibit similar patterns of presence and absence (e.g., Virkkala 1993; Kunin 1998; Lennon et al. 2002; Olff and Ritchie 2002; Green et al. 2003).

Many other power laws are related to plant and animal body size. They are the subject of the large literature on biological allometry (Peters 1983; Calder 1984; Schmidt-Nielsen 1984; Brown and West 2000; Brown et al. 2002). Within functional groups, such as trees in a forest or animals in a habitat, total population density or number of individuals per unit area, $N$, often appears to scale with body mass, $M$, as

$$N = N_0 M^{-3/4}, \tag{3.4}$$

a power-law scaling relation that appears to reflect the scaling of whole-organism metabolic rate and hence per-individual resource requirements (e.g., Damuth 1981; Enquist et al. 1998; Li 2002). In pelagic lake and marine ecosystems, there are somewhat different scaling relations that hold across an enormous range of organisms, from unicellular phytoplankton and prokaryotes to the largest fish and whales. Total density scales as $M^{-1}$, so that total biomass is invariant or scales as $M^0$ (e.g., Sheldon et al. 1972; Cyr et al. 1997; Kerr and Dickie 2001). It is interesting to note that the scaling of

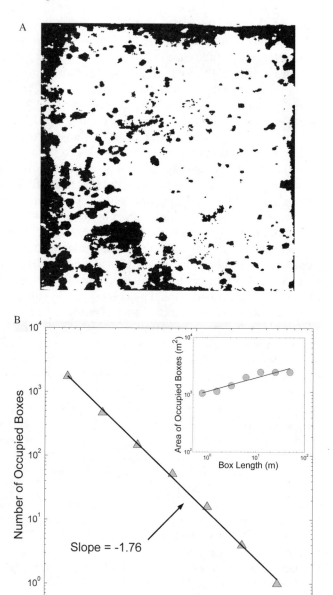

FIGURE 3.3. Fractal-like pattern of vegetation patchiness at Brown's long-term research site near Portal, AZ. (A) Map of the vegetation cover (black) on a 50 × 50 m plot. (B) Fractal dimension plot, using the box-counting method, of the number of grid cells on the map occupied by vegetation as a function of the length of the edge of a grid cell. Insert shows the same data plotted in a different way, with the total area of occupied grid cells replacing the number of occupied grid cells.

population density could be considered to be a form of frequency-magnitude scaling. These scaling relations mean that organisms are distributed on the landscape with predictable relationships among density, size, and other correlated variables, such as nearest neighbor distance, stem diameter, canopy height and radius, and water, mineral, and energy flux for plants; and nearest neighbor distance, home range size, movement distance, food requirement, and excretion rate for animals. So, to the extent that organisms constitute patchy environments or resources for other organisms, these scaling relations can be used to predict important characteristics of patch structure and dynamics. In addition, these patterns (e.g., home range size $\propto M^1$) suggest that organisms of different size interact with the environment at different scales (e.g., Morse et al. 1985). This should have important consequences for the scales at which heterogeneity impacts organisms.

We have listed but a few of the possible patchy distributions in ecological systems. It is clearly important to begin to catalogue and understand how other attributes of the geological and biological templates scale and to integrate these patterns into ecological research. For example, what are the relations among perimeter, area, and elevation for islands or comparable dimensions of perimeter, area, volume, and depth for lakes? What is the nature of the distribution of distinctive soil patches, such as serpentine or gypsum, and, if they can be described as fractal-like, how do the normalization constants and scaling exponents change across different geological settings? Some of the answers to these questions are probably available in the geological literature, but they have not generally been picked up and used by ecosystem and landscape ecologists.

Another important question is how these varied fractal-like patterns are related to one another. We stated earlier that there are different fractal dimensions that characterize different features of patchy environments. These different fractal dimensions each appear to describe multiple phenomena. It may be that the components of this diverse assemblage of self-similar relationships are connected to one another in much the same way as has recently been shown for hydrologic networks and biological allometries (see "Networks," below; some relationships between dimensions are understood, e.g., Hastings and Sugihara 1993). If this is true, then the confusing labyrinth of fractal landscape metrics might condense to a small number of important underlying variables.

## Networks

Our final pattern is the network, which we define as a system of connected, hierarchically branching elements of structure and function. Networks represent combinations of both relatively continuous and discrete variation. Along the direction of flow, when measured at coarse scales, the variation appears relatively continuous. For example, the variation in stream properties from headwaters to mouth are the basis of the river continuum concept (Vannote et al. 1980). On the other hand, when viewed at a smaller scale, the variation is more discrete. So, for example, the properties of a stream change abruptly when two similar-sized branches join.

The application of networks to heterogeneity is twofold. First, the properties of a network determine the values of important parameters within that network. For example, the width, depth, and nutrient loading of a stream network depend on the order of the stream (Rodriguez-Iturbe and Rinaldo 1997). This creates predictable heterogeneity for processes and organisms operating within the network. Second, networks often flow over non-network templates (e.g., streams over land), and in doing so they create a particular distribution of the materials that they are fluxing across the landscape (water, nutrients, sediments, etc.). It is believed that many natural networks are produced by some process of self-organization, and many of them seem to be fractal-like, at least over some range of scales (Rodriguez-Iturbe and Rinaldo 1997). Given this self-similarity we can begin to describe patterns in the network quantitatively.

Some of these natural networks are abiotic. The classic examples are streams and related networks such as river deltas, desert alluvial fans, and tidal drains. These branched hierarchies are formed by the physical forces generated by flowing water, and the resulting continuous reconfiguration of the channel due to erosion of substrates and deposition of sediments during both extreme flood events and more usual flows. Geologists and hydrologists have studied stream networks and their self-organizational formation. The famous Horton-Strahler system of characterizing the order of branches was developed for streams (Horton 1945; Strahler 1957). This system describes the hierarchy of the network and can be illustrated most simply by thinking about pruning the source (outermost) branches of the stream network sequentially. First prune the source branches. By definition, these are the first-order branches. Using the pruned network, prune the terminal branches again. These branches become second order and so on until only the trunk remains (Melton 1959).

Ordered in this way, networks exhibit fractal-like properties. Examples include Horton's ratios (Horton 1945)

$$n_{w+1}/n_w = R_n$$
$$l_{w+1}/l_w = R_l \qquad (3.5)$$
$$a_{w+1}/a_w = R_a,$$

where $n_w$ is the number of streams of order $w$, $l_w$ is the average length of those streams, $a_w$ is the average area of those streams, and $R_n$, $R_l$, and $R_a$ are constant ratios between those values at order $w + 1$ and order $w$ (invariant ratios across hierarchical levels are characteristic of self-similar patterns). Another example is Hack's law (Hack 1957),

$$L \propto A^h, \qquad (3.6)$$

which characterizes the relation between the length of the main channel in a drainage basin, $L$, and the area of that basin upstream, $A$, in terms of a scaling exponent, $h$. For a visual description and a complete list of stream network scaling relationships, see Dodds and Rothman (1999).

The numerous patterns in streams have recently been shown to be related to one another, thus simplifying the description of multiple empirical scaling relations to two simple quantitative descriptors: the fractal dimension of individual streams ($D$ similar to that of the coastline example) and the ratio of the logarithms of $R_l$ and $R_n$ (Dodds and Rothman 1999). These patterns may be explained mechanistically based on the stream networks minimizing their global energy expenditure (Rodriguez-Iturbe and Rinaldo 1997; Rinaldo et al. 1998), providing a more process-oriented explanation for these observed patterns. For an in-depth treatment of river network scaling, see Rodriguez-Iturbe and Rinaldo (1997) and references above.

Organisms also form hierarchically branching networks. The most obvious are the fractal-like architectures of both the roots and shoots of most land plants (Morse et al. 1985; Tatsumi et al. 1989; Fitter and Strickland 1992; Neilson et al. 1997). Structural and functional properties of some of these networks are described by scaling laws, which have been used to characterize their self-similar organization and the relationships between structural and functional variables. Most of the work to date has focused on plant architecture and vascular systems (e.g., McMahon and Kronauer 1976; Niklas 1994; Neilson et al. 1997; West et al. 1997, 1999; Horn 2000). These networks form fractal-like habitat for terrestrial and subterranean organisms that use plants (e.g., Morse et al. 1985).

## Why a Quantitative Framework?

So far, we have suggested that environmental variability can be divided into three major categories, and that each of them can, at least in some cases, be described using a relatively simple quantitative framework. One great advantage of a having such a quantitative framework for studying heterogeneity is that these characterizations can be incorporated into models for ecological processes (e.g., Ludwig et al. 2000). Consequently, it should often be possible not only to predict whether heterogeneity is important for the question being studied, but also to understand precisely how the organization of spatial variation affects ecological processes. This provides the potential to move beyond purely correlative studies to understand the operation of different processes at different spatial scales (Milne 1991a). It should be useful in determining which habitat variables, and their associated patterns of heterogeneity, are important for a particular process. Such a framework may eventually answer a question that we have been asked to address: At what scale does heterogeneity become unimportant (i.e., when can it be ignored)? The answer will surely be that this scale depends on the question of interest, the type of heterogeneity considered, and the inherent scale of the units and processes. Gradients, patches, and networks at the scale of micrometers are important for microbes but probably unimportant for elephants and whales (e.g., Addicott et al. 1987; With and Crist 1996). The way to define this scale for a particular process may be through a combination of quantitative modeling and empirical analysis. Determining this scale is simply a special case of using these descriptors of heterogeneity to make quantitative predictions about their effects on ecological systems. There are several good examples of these quantifications being used to model and understand ecological processes.

## Examples

An example of the use of a quantified gradient for studying patterns of species coexistence is provided by Yamamura (1976), who used a theoretical gradient to explore patterns of the spatial distribution of plant communities. He showed that by introducing simple continuous gradients into basic population dynamics models (through the influence of spatial position on the growth rate and competition parameters), he could generate patterns of species distributions reflecting different combinations of competitive exclusions and coexistence. Studies of diversity maintenance based on spatial and/or temporal variability in environmental conditions have benefited from taking a similar quantitative approach (e.g., Chesson 2000). Patterns of compositional change along a gradient can be explained by combining the relatively continuous change in one or more key environmental variables with the impacts of those changes on important population variables for the species involved. An example of this is provided by Arris and Eagleson (1994), who used the response of tree species productivity to changes in the length of the growing season, photosynthetic capacity, potential evapotranspiration, and soil moisture availability along a latitudinal gradient to predict the location of the ecotone transition between boreal and deciduous forests in the eastern United States. By quantifying the gradients, they were able to show that through their influence on the rates of production, the gradients should lead to a transition in the dominant forest type at approximately the latitude observed. This suggests that the broad-scale heterogeneity in the environment (i.e., the gradients) produced the ecotone pattern through influences of abiotic environmental variables on net primary production.

An excellent example of the use of the fractal-like nature of patches to describe model ecological systems is provided by Ritchie and Olff (1999). They suggest that due to the fractal-like clustering of resources (Milne 1992, 1997), herbivores of different sizes will see the patchiness of landscapes differently and thus respond differently to the patchy pattern of resource. Small, dense patches can be used by small species, whereas large, less concentrated patches are more appropriate for large species. Because resources are patchily distributed and the different patches are used differently by different body sizes, these relationships can be used to predict body size distributions of coexisting herbivorous mammals and to help understand how variation in body size facilitates the maintenance of biodiversity (Hutchinson 1959; Hutchinson and MacArthur 1959). The authors use the observed fractal-like nature of resource distributions to make specific quantitative predictions about the frequency distribution of body sizes and the number of species that can be supported by a habitat. This example illustrates how a quantification of heterogeneity can provide explicit predictions about its impacts on ecological systems.

We are less familiar with the use of networks for characterizing ecological heterogeneity. They have proved useful in understanding and quantifying the effects of resource distribution networks on metabolic rates of animals and plants, and these effects cascade through ecological systems, having effects at scales from individuals to entire ecosystems (Enquist et al. 2003; Brown et al. 2004). One area where networks will likely prove important for characterizing heterogeneity is in aquatic and riparian ecology. The increasingly well quantified and mechanistically understood scaling relations for stream networks have many obvious ecological implications.

One effort in aquatic ecology is to understand how stream properties, and hence ecological patterns and processes, vary from source streams to the main channel. Geologists and hydrologists have developed a solid understanding of abiotic variation as a function of stream order through a quantitative approach that uses scaling laws to characterize the hierarchical self-similarity of river networks (see "Networks," above). This approach does not explain all of the important patterns, but it does provide robust, quantitative characterizations of a suite of important variables (e.g., flow rate, stream length and width, etc.), thereby providing a first-order model of abiotic heterogeneity from headwaters to main channels. The next challenge for stream ecologists is to begin to understand how these abiotic patterns influence biotic processes. The river continuum concept (Vannote et al. 1980) and the flood pulse concept (Junk et al. 1989) attempt implicitly to understand how the regular abiotic scaling properties of streams affect the ecology of riverine and riparian ecosystems. These concepts would seemingly benefit from the explicit incorporation of the quantitative framework describing the changes in the abiotic template as a function of stream order.

## Conclusions

The emphasis of this book and of the Cary Conference that spawned it is on the extent to which, and the mechanisms by which, spatial heterogeneity affects ecosystem function. We define ecosystem function as the fluxes and transformations of energy, materials, and information (and of organisms containing those currencies) that occur within and between ecosystems and other ecological subsystems. These flows and transformations are inherently heterogeneous. They occur in specific places on the landscape, and they are driven by abiotic and biotic processes that are heterogeneously distributed. These fluxes and transformations are also inherently heterogeneous at all spatial scales. Some processes, such as biotic weathering of rock surfaces and microbial uptake of organic compounds, occur at the molecular level of organization and at the scale of nanometers to micrometers. Other processes, such as the circulation of the atmosphere and oceans, occur at regional to global levels of organization and on the scale of $10^n$ kilometers. The structure and dynamics of these flows are governed largely by the geometric, physical, and biological characteristics of the spatial template.

An essential task for understanding how habitat heterogeneity affects ecosystem processes is to characterize the patterns of heterogeneity and to understand the processes underlying those patterns. A useful framework is to recognize that patterns of environmental variability across landscapes can generally be separated into three major categories: gradients, patches, and networks. Each of these categories can often be described using a relatively simple quantitative framework. By incorporating these quantifications into the study of biological systems, it should be possible to predict not only if heterogeneity will have an effect on ecosystem function, but also precisely what the nature and magnitude of the effect should be. Progress toward increased understanding, precision, and predictability will also benefit from incorporating advances from other disciplines, including physics, chemistry, biology, and the earth sciences, on the laws, principles, and factors that generate the gradients, patches, and networks and that govern the flows and transformations of energy, materials, information, and organisms within and between these heterogeneous landscape elements.

*Acknowledgments.* We would like to thank S.K.M. Ernest, A.H. Hurlbert, M.G. Turner, and two anonymous reviewers for helpful comments on previous versions of this manuscript. We would like to thank B.T. Milne for exposing us to some of the ideas and papers related to gradients and patches, M. Harner for valuable discussion of river networks, H. Olff for conducting the box-counting for Figure 3.3, N. Baum for sharing her preliminary results, and W. Jetz for helpful discussion of temperature gradients. J.P. White, J.T. White, and J.C. White provided a pleasant atmosphere in which to complete this chapter. Financial support was provided by an NSF Graduate Research Fellowship and by an NSF Biocomplexity grant (DEB-0083422).

## References

Addicott, J.F., Aho, J.M., Antolin, M.F., Padilla, M.F., Richardson, J.S., and Soluk, D.A. 1987. Ecological neighborhoods: scaling environmental patterns. Oikos 49: 340–346.

Arris, L.L., and Eagleson, P.S. 1994. A water-use model for locating the boreal deciduous forest ecotone in eastern North-America. Water Resources Res. 30: 1–9.

Brown, J.H., Gillooly, J.F., Allen, A.P., Savage, V.M., and West, G.B. 2004. Toward a metabolic theory of ecology. Ecology 85: 1771–1784.

Brown, J.H., Gupta, V.K., Li, B.L., Milne, B.T., Restrepo, C., and West, G.B. 2002. The fractal nature of nature: power laws, ecological complexity and biodiversity. Philos. Trans. R. Soc. London Ser. B Biol. Sci. 357: 619–626.

Brown, J.H., and West, G.B., eds. 2000. Scaling in biology. Oxford: Oxford University Press.

Calder, W.A. 1984. Size, function, and life history. Mineola, NY: Dover.

Chesson, P. 2000. Mechanisms of maintenance of species diversity. Annu. Rev. Ecol. Systematics 31: 343–366.

Cyr, H., Peters, R.H., and Downing, J.A. 1997. Population density and community size structure: comparison of aquatic and terrestrial systems. Oikos 80: 139–149.

Damuth, J. 1981. Population-density and body size in mammals. Nature 290: 699–700.

Dodds, P., and Rothman, D. 1999. Unified view of scaling laws for river networks. Phys. Rev. E 59: 4865–4877.

Enquist, B.J., Brown, J.H., and West, G.B. 1998. Allometric scaling of plant energetics and population density. Nature 395: 163–165.

Enquist, B., Economo, E., Huxman, T., Allen, A., Ignace, D., and Gillooly, J. 2003. Scaling metabolism from organisms to ecosystems. Nature 423: 639–642.

Fitter, A.H., and Strickland, T.R. 1992. Fractal characterization of root system architecture. Funct. Ecol. 6: 632–635.

Flannery, T. 1995. Mammals of the South-West Pacific and Moluccan Islands. Ithaca, NY: Cornell University Press.

Green, J., Harte, J., and Ostling, A. 2003. Species richness, endemism and abundance patterns: tests of two fractal models in a serpentine grassland. Ecol. Lett. 6: 919–928.

Gustafson, E. 1998. Quantifying landscape spatial pattern: what is the state of the art? Ecosystems 1: 143–156.

Hack, J.T. 1957. Studies of longitudinal stream profiles in Virginia and Maryland. U.S. Geological Survey professional paper 294-B: 45–97.

Hastings, H.M., Pekelney, R., Monticciolo, R., vun Kannon, D., and del Monte, D. 1982. Time scales, persistence and patchiness. BioSystems 15: 281–289.

Hastings, H.M., and Sugihara, G. 1993. Fractals: a user's guide for the natural sciences. Oxford: Oxford University Press.

Horn, H.S. 2000. Twigs, trees, and the dynamics of carbon in the landscape. In Scaling in biology, eds. J.H. Brown and G.B. West. pp. 199–200. Oxford: Oxford University Press.

Horton, R.E. 1945. Erosional development of streams and their drainage basins: Hydrophysical approach to quantitative morphology. Geological Soc. Am. Bull. 56: 275–370.

Hutchinson, G.E. 1959. Homage to Santa Rosalia; or, why are there so many animals? Am. Naturalist 93: 145–159.

Hutchinson, G.E., and MacArthur, R.H. 1959. A theoretical ecological model of size distributions among species of animals. Am. Naturalist 93: 117–125.

Jones, C.G., Lawton, J.H., and Shachak, M. 1994. Organisms as ecosystem engineers. Oikos 69: 373–386.

Junk, W.J., Bayley, P.B., and Sparks, R.E. 1989. The flood pulse concept in river-floodplain systems. Canadian Special Publication of Fisheries and Aquatic Sciences 106: 110–127.

Kerr, S.R., and Dickie, L.M. 2001. Biomass spectrum. New York: Columbia University Press.

Korcak, J. 1938. Deux types fondamentaux de distribution statistique. Bull. Inst. Int. Stat. 3: 295–299.

Kotliar, N.B., and Wiens, J.A. 1990. Multiple scales of patchiness and patch structure: a hierarchical framework for the study of heterogeneity. Oikos 59: 253–260.

Kunin, W.E. 1998. Extrapolating species abundance across spatial scales. Science 281: 1513–1515.

Lennon, J.J., Kunin, W.E., and Hartley, S. 2002. Fractal species distributions do not produce power-law species-area relationships. Oikos 97: 378–386.

Li, W.K.W. 2002. Macroecological patterns of phytoplankton in the northwestern North Atlantic Ocean. Nature 419: 154–157.

Ludwig, J.A., Wiens, J.A., and Tongway, D.J. 2000. A scaling rule for landscape patches and how it applies to conserving soil resources in savannas. Ecosystems 3: 84–97.

Malamud, B., Morein, G., and Turcotte, D. 1998. Forest fires: an example of self-organized critical behavior. Science 281: 1840–1842.

Mandelbrot, B.B. 1983. The fractal geometry of nature. New York: W.H. Freeman and Company.

Mandelbrot, B.B. 1997. Fractals and scaling in finance: discontinuity, concentration, risk. New York: Springer-Verlag.

McMahon, T.A., and Kronauer. R.E. 1976. Tree structures: deducing the principle of mechanical design. J. Theor. Biol. 59: 443–466.

Melton, M.A. 1959. A derivation of Strahler's channel-ordering system. J. Geol. 67: 345–346.

Milne, B.T. 1991a. Heterogeneity as a multiscale characteristic of landscapes. In Ecological heterogeneity, eds. J. Kolasa and S.T.A. Pickett, pp. 69–84. New York: Springer-Verlag.

Milne, B.T. 1991b. Lessons from applying fractal models to landscape patterns. In Quantitative methods in landscape ecology, eds. M.G. Turner and R.H. Gardner, pp. 199–235. New York: Springer.

Milne, B.T. 1992. Spatial aggregation and neutral models in fractal landscapes. Am. Naturalist 139: 32–57.

Milne, B.T. 1997. Applications of fractal geometry in wildlife biology. In Wildlife and landscape ecology: effects of pattern and scale, ed. J.A. Bissonette, pp. 32–69. New York: Springer.

Morse, D.R., Lawton, J.H., Dodson, M.M., and Williamson, M.H. 1985. Fractal dimension of begetation and the distribution of arthropod body lengths. Nature 314: 731–733.

Neilson, K.L., Lynch, J.P., and Weiss, H.N. 1997. Fractal geometry of bean root systems: Correlations between spatial and fractal dimension. Am. J. Botany 84: 26–33.

Niklas, K.J. 1994. Plant allometry: the scaling of form and process. Chicago: The University of Chicago Press.

Olff, H., and Ritchie, M.E. 2002. Fragmented nature: consequences for biodiversity. Landscape Urban Planning 58: 83–92.

Peters, R.H. 1983. The ecological implications of body size. New York: Cambridge University Press.

Reichman, O., and Seabloom, E. 2002. The role of pocket gophers as subterranean ecosystem engineers. Trends Ecol. Evol. 17: 44–49.

Richardson, L.F. 1961. The problem of contiguity: an appendix of statistics of deadly quarrels. General Systems Yearbook 6: 139–187.

Rinaldo, A., RodriguezIturbe, I., and Rigon, R. 1998. Channel networks. Annu. Rev. Earth Planetary Sci. 26: 289–327.

Ritchie, M.E., and Olff, H. 1999. Spatial scaling laws yield a synthetic theory of biodiversity. Nature 400: 557–560.

Rodriguez-Iturbe, I., and Rinaldo, A. 1997. Fractal river basins: chance and self-organization. New York: Cambridge University Press.

Schmidt-Nielsen, K. 1984. Scaling: why is animal size so important. Cambridge, UK: Cambridge University Press.

Sheldon, R.W., Prakash, A., and Sutcliffe, W.H.J. 1972. The size distribution of particles in the ocean. Limnol. Oceanogr. 17: 327–340.

Strahler, A.N. 1957. Quantitative analysis of watershed geomorphology. Am. Geophys. Union Trans. 38: 913–920.

Tatsumi, J.A., Yamauchi, Y., and Kono, Y. 1989. Fractal analysis of plant root systems. Ann. Botany 64: 499–503.

Vannote, R.L., Minshall, G.W., Cummins, K.W., Sedell, J.R., and Cushing, C.E. 1980. The river continuum concept. Can. J. Fisheries Aquatic Sci. 37: 130–137.

Virkkala, R. 1993. Ranges of northern forest passerines: a fractal analysis. Oikos 67: 218–226.

West, G.B., Brown, J.H., and Enquist, B.J. 1997. A general model for the origin of allometric scaling laws in biology. Science 276: 122–126.

West, G.B., Brown, J.H., and Enquist, B.J. 1999. The fourth dimension of life: fractal geometry and allometric scaling of organisms. Science 284: 1677–1679.

Wetzel, R.G. 1991. Land-water interfaces: metabolic and limnological regulators. Verhandlungen Internationale Vereinigung für Theoretische und Angewandte Limnologie 24: 6–24.

Whittaker, R.H. 1967. Gradient analysis of vegetation. Biol. Rev. 42: 207–264.

With, K.A., and Crist, T.O. 1996. Translating across scales: simulating species distributions as the aggregate response of individuals to heterogeneity. Ecol. Modelling 93: 125–137.

Yamamura, N. 1976. A mathematical approach to spatial distribution and temporal succession in plant communities. Bull. Math. Biol. 38: 517–526.

# 4
# Thoughts on the Generation and Importance of Spatial Heterogeneity in Ecosystems and Landscapes

JOHN PASTOR

## Abstract

Landscapes are spatially dynamic because materials and energy spread over them and change the distribution of ecosystem properties. This heterogeneity of the distribution of ecosystem properties can either be random or patterned. The landscape becomes patterned when the spread of materials and energy correlates an ecosystem property in one local neighborhood with that at another. When the spread of materials and energy does not correlate properties of different neighborhoods, then the landscape can still be heterogeneous but random. Various processes that result in spatial heterogeneity include physical disturbances (e.g., fire, erosion, etc.) that spread across neighborhoods and remove materials but whose spread is partly determined by previous disturbances; directional gradients in the flow of materials, energy, or information; and different diffusion rates of coupled ecosystem components combined with positive feedbacks, otherwise known as diffusive instability. Examples of these processes will be given from other papers in this conference and elsewhere.

## Introduction

The living world is not all green slime or a big leaf; things are different from place to place. This variety of the living world is what makes it a stunningly beautiful and interesting place to live. It is also what makes understanding ecological systems difficult.

Spatial heterogeneity of the distribution of ecosystem processes across the landscape can be random or patterned (or a combination of both). A heterogeneous spatial distribution of ecosystem properties is random if, given the value of an ecosystem property at a point, the value of that property at adjacent points cannot be predicted. In contrast, a heterogeneous spatial distribution is patterned if, given the value of an ecosystem property at a point, the value at adjacent points and possibly points further away can be predicted with some confidence. Because the spread of materials and energy across the landscape correlates values of an ecosystem property between adjacent local neighborhoods, this spread can therefore result in patterned heterogeneity.

For the most part, we know how to analyze spatially homogeneous distributions through analysis of variance and general linear statistical models. We know how to model their dynamics through coupled ordinary differential equations that depict energy and material flows between ecosystem components and whose parameters do not depend on position in space. In contrast, we are only beginning to learn how to describe the origin and dynamics of spatial heterogeneity. These require new mathematical, experimental, and observational tools for their description and analysis.

Physical disturbances create and sustain heterogeneities by removing materials from ecosystems or transferring materials from one ecosystem or ecosystem component to another. Physical disturbances often have a large random element, but they also may depend on underlying heterogeneity, which is often caused by previous disturbances. The spread of a disturbance correlates values of an ecosystem property at a given point with those at its neighbors and beyond to the boundary of the patch created by the disturbance.

Transport of energy and materials along a directional gradient, such as movement of water and suspended sediments or dissolved compounds downhill, also creates patterned heterogeneity. The transport of energy and materials along a directional gradient correlates ecosystem properties along the gradient. Ecosystem properties will therefore be similar for long distances along transects in the direction of the gradient but become less similar more rapidly along transects perpendicular to the gradient.

Spatial heterogeneities can also be generated by positive feedbacks between ecosystem components, such as soil, vegetation, and higher trophic levels (Meinders and van Breemen this volume). Such patterned heterogeneity can arise even in the absence of gradients and physical disturbances and can create patterned heterogeneity from homogeneity or random heterogeneity. This generation of pattern from homogeneity or randomness in the environment via positive feedbacks between ecosystem components is sometimes called "self-organized complexity" (Kauffman 1993; Bak 1997; Meinders and van Breemen this volume).

If two interacting ecosystem components also diffuse or spread across the landscape, new and surprising heterogeneities can arise even without any underlying heterogeneity in the physical environment (Okubo and Levin 2002). Under some circumstances, such heterogeneities could be stable. This seems to be especially prevalent in herbivore-vegetation systems where both the herbivore populations and the plant species that support them are diffusing across the landscape. For example, the spatial dynamics of balsam fir is coupled to the spatial dynamics of spruce budworm populations during an outbreak. In turn, the changes in the spatial distribution of balsam fir affect the fate of the outbreak (Holling 1978).

In this paper, I wish to explore how the spread of physical disturbances, the directional flows of materials down a gradient, the positive feedbacks between ecosystem components, and the diffusion of interacting components across the landscape all generate spatial heterogeneity. I will use the papers in this volume and additional ones from the literature as examples. My purpose is to seek some general principles of the sources and consequences of spatial heterogeneity and attempt to reach broad conclusions about similarities and differences between major ecosystem types in order to offer approaches for organizing future research.

## Physical Disturbances

The ecological literature on disturbance is vast, and it is not my intent to review it here. Instead, I wish to make a few remarks about some aspects of the nature of spatial heterogeneity caused by disturbances and why these might differ between terrestrial and aquatic ecosystems. By disturbance I mean some physical process that removes a fraction of an ecosystem component or adds to it. Thus, I exclude insect outbreaks, for example, because such biological processes (which are sometimes referred to as "disturbances") could be treated by other approaches involving diffusion of the population, which I discuss below. Physical disturbances, such as fire, erosion, landslides, avalanches, and so forth, are qualitatively different from "disturbances" initiated by growth of a population, because the physical disturbance itself is not a component or pool within an ecosystem but a process by which material is transferred spatially.

Disturbances have two aspects that are important for the generation of spatial heterogeneity. The first is where the disturbance is initiated, which has a large random component (e.g., where the lightening strikes) but also depends on the conditions in the initiation location (e.g., whether there is sufficient fuel of the right moisture content to ignite when struck by lightning). However, once initiated, the disturbance can and often does spread to adjacent locations whose conditions may not have been right for initiation but are sufficient for the spread (e.g., if your neighbor catches fire, you may burn, too). Thus, spatial heterogeneity caused by physical disturbances is partly random (through initiation) and partly patterned (through contagious spread).

Aquatic ecosystems, especially streams and rivers, are well mixed. Their components generally have rapid turnover because of short lifetimes of organisms and because currents break down structures by rolling and mixing of bedload and woody debris. Constant flux of water also dilutes the introduction of pollutants and contaminants at point sources. The spatial heterogeneities caused by many disturbances to streams, especially disturbances related to point-source pollution, are therefore quickly dissipated once the disturbance ends (Niemi et al. 1990).

In contrast, terrestrial ecosystems are not well mixed and often contain slow-growing perennial individuals. Therefore, the spatial pattern caused by a disturbance remains for long times. But if the recurrence interval of a disturbance is shorter than the recovery of a disturbed patch, then under some conditions the initiation and spread of any disturbance may partly depend on previous disturbances. How a disturbance moves through a landscape that previous disturbances have created is a major unanswered (and difficult) question of disturbance ecology.

From a modeling standpoint, this means that simple, first-order Markov chains, often used as a first approximation to modeling disturbance (see reviews by Baker 1989 and Pastor et al. 1993), will always be somewhat deficient. First-order Markov models assume that the probability of a transition in the system is constant and depends only on the current state of a system. But in fact, whether or not a disturbance happens at a point or propagates from it depends on disturbances back to some period in the past. Not only that, but the current state of a system (or local neighborhood) also includes the distribution of adjacent neighborhoods and their states (hence contagion). Cellular automata approaches are useful in dealing with these higher order effects because the change in a given cell depends in part on the state(s) of its neighbors (see review by Neuhauser 2001).

As Turetsky et al. (this volume) and Romme (this volume) show, disturbances are a particularly important source of heterogeneity in boreal regions and in coniferous forests of the arid West perhaps because of the slow recovery of vegetation owing to the slow growth rates of the species present (Chapin et al. 1986), because of the slow turnover rate of the soil N pool (Flanagan and Van Cleve 1983) that supplies the N required for plant growth (and hence recovery), and because of drought.

An excellent example of the importance of heterogeneity caused by a physical disturbance such as fire is the landscape of virgin forests of the Boundary Waters Canoe Area (BWCA) of northern Minnesota (Heinselman 1973). Virtually every stand in the BWCA originated from a fire, but the fire return intervals (which differ for different stands) are almost all less than the recovery time from the previous disturbance. Consequently, fires in the BWCA partly burn through previous burns. For example, some 44% of the BWCA burned during 1864, but only 20% of the current stands originated in the 1864 burns: the rest of the current landscape originated in fires that happened later but which spread to these burned areas from adjacent older stands that ignited first.

Romme (this volume) shows that the importance of spatial heterogeneity caused by previous burns in the arid West varies with forest type and climatic conditions. When the climate is dry and hot, everything burns and previous spatial heterogeneity is unimportant in fire spread. Thus, we get large-scale catastrophic fires as in Yellowstone during the late 1980s and in the southwest during 2001 and 2002. However, during more moderate years or in higher elevations where extended periods of hot and dry conditions are rare, the underlying spatial heterogeneity caused by previous burns is very important in determining initiation and spread of new fires.

These case studies raise several general questions. How do burns and other disturbances become overlaid on previous disturbances of the same type or of different types? Does it matter what the previous disturbance was, and if so, in what way does it matter? Is there a characteristic fractal or some other geometry of partly overlapping disturbances? If so, what determines it? Are these "geometries," if they exist, characteristic of a particular ecosystem or are there more general aspects common to two or more otherwise different ecosystems? These are some of the major questions, as I see them, which need to be answered to develop a more complete understanding of how disturbances produce and interact with spatial heterogeneities in any landscape.

## Directional Gradients

Both terrestrial and aquatic ecosystem ecologists have long dealt with gradients in the vertical dimension and its effects on ecosystem properties. The premier example of such vertical spatial heterogeneity is the extinction of light through a canopy and water column, the fundamental starting point of much of forest ecology and limnology. If we assume that leaves are randomly distributed through the canopy or that the water column is homogeneous, then this light gradient can adequately be treated by means of a linear model whereby the change in light through a given layer at some depth $d$ is some fraction $k$ of the light entering that layer, leading to the familiar exponential extinction curve:

$$I_d = I_0 e^{-kd}. \tag{4.1}$$

It is a relatively simple matter to incorporate heterogeneities in the distribution of leaves through the canopy or vertical changes in water column transparency simply by replacing $d$ with a function describing how leaf area or transparency change with depth and integrating down to depth $d$:

$$I_d = I_0 e^{-k\int_0^d LAI(\delta)d\delta}. \tag{4.2}$$

This vertical light gradient, $I_d$, often leads to a stratification of both terrestrial and aquatic communities according to the photosynthetic response curves of the constituent species (Shugart 1984; Tilman 1988). In terrestrial ecosystems, when the community is vertically stratified into shade-intolerant species above shade tolerants, then light-use efficiency and hence net primary production by the entire community may be maximized (Pastor and Bockheim 1984; Tilman 1988).

The most important horizontal directional gradient in landscape ecology may be topographic, causing water left after transpiration to flow transversely and downhill. Watershed studies have typically examined the mass balance of inputs to the watershed via precipitation and stream outputs, as demonstrated in an exemplary manner by the Hubbard Brook Ecosystem Study (Likens et al. 1970; Bormann et al. 1977). However, these watershed studies typically do not look in detail at the pathways and patterns of nutrients fluxes between stands within the watershed and how that affects the eventual transfer of nutrients to the streams (or lakes) at their base. Conversely, many detailed studies of nutrient cycles of ecosystems or stands within watersheds implicitly assume that the ecosystem sits on a flat table and leaching losses take place vertically rather than semihorizontally. To truly bridge watershed and stand-level approaches, we need to connect stands in the landscape by means of directional fluxes of nutrients down topographic gradients. Thus, the input-output balance of an ecosystem at a given point may depend as much on its position in the landscape and the delivery of nutrients to it from upslope as on the exchange of nutrients between components within it.

This has important consequences in both streams and the watersheds that surround them, perhaps especially so for the riparian zones. The riparian zone potentially receives nutrients from every stand above it, but the nutrients are delivered to it in sequence downslope. Therefore, the sequence of stands along a slope and their differing input-output balances may determine the loading of nutrients to the riparian zone.

Heterogeneity in riparian zones may also determine downstream flows of nutrients. Naiman et al. (this volume) review how sources of heterogeneity in riparian forests, such as coarse woody debris, denitrification hotspots, debris jams, formation of bars and side channels, and so forth, may mitigate large transfers of nutrients to aquatic ecosystems. The strong directional gradient that transfers nutrients along a topographic sequence may interact with fine-scale heterogeneity within the riparian zone to determine overall land-water material transfers. This fine-scale heterogeneity within the riparian zone may enhance nutrient retention if it increases the path length a molecule travels before it enters the stream channel, thus increasing its residence time within the riparian zone. The role of heterogeneity within the riparian zone must therefore be assessed in the context of the overall heterogeneity of the landscape and downslope transfers of nutrients to the riparian zone and how the heterogeneity of the riparian zone affects nutrient retention before the nutrient enters the stream channel. How heterogeneity of processes operating at different scales interacts to determine lateral transfers of material across landscapes is a difficult topic of great importance.

Urban ecosystems (Band et al. this volume) are distinguished partly by particular sorts of directional flows along the grid systems of streets. These directional flows can be parallel (one-way streets) or antiparallel (two-way streets) along two axes usually at 90 degrees to each other. This grid system is an attempt to impose some spatial order on travels of humans and commerce in a city, but it can have great consequences for the spatial dynamics of cities as landscapes. For example, city engineers must control downslope flows of water to prevent erosion and flooding of the roadbeds; these water diversions into storm sewers and along curbs have large effects on urban stream ecosystems (Band et al. this volume). Furthermore, pollutants from automobiles are dispersed from sources that move down streets and are dispersed further by wind tunnels or prevented from dispersing by wind-breaks caused by the buildings (Band et al. this volume). It would be interesting to learn how this grid system of directional flows of traffic, water, and wind disperses seeds of exotic plant species or diseases of boulevard trees.

When directional gradients of fluxes at boundaries of patches are very steep, the sign of the gradient can determine the degree of heterogeneity inside the boundary of a patch. Kratz and MacIntyre (this volume) remind us that there is a very important directional gradient at the surface of a lake, namely the heat flux gradient, which strongly determines the spatial heterogeneity within the lake. When the heat flux gradient at the lake surface is positive, heat flows out of the lake and the water column physically turns over, bringing nutrients from the sediment to the surface and oxygen from the surface to the lower depths. The lake is then also thermally homogeneous. But when the heat flux gradient at the lake surface is negative, heat flows into the lake and it becomes thermally stratified. This phenomenon, so important to aquatic ecosystems, depends on the fact that fluids such as water can be well mixed with fast time constants. Similar thermal stratification of the atmosphere over a city results in the formation of smog. Such spatial dynamics do not have any counterparts in terrestrial vegetation-soil systems because these systems cannot be well mixed over any reasonable ecological timescale.

Finally, positive feedbacks within ecosystems (Tongway this volume; Meinders and van Breemen this volume) can amplify the heterogeneity produced by directional gradients. Tongway shows how positive feedbacks between plants and soils in arid systems concentrate and retain soil moisture being delivered at a point such that water availability becomes raised above threshold levels required for plant growth, leading to the further development of patches of vegetation and high resource availability in a sea of low resource availability.

Such feedbacks and the spatial patterns that arise from them are not confined to arid systems. Peatlands are another excellent example of how plant-soil feedbacks lead to the formation of spatial patterns (Turetsky et al. this volume). Horizontal water flow patterns in peatlands are a result of microtopographic gradients and hydraulic permeability of the peat, both of which interact with the plant community. Broadly speaking, two different communities (bogs and fens) can be found in peatlands; these in turn appear to be related to hydrologic sources of nutrient inputs (Wright et al. 1992). In bogs, peat accumulation has raised the local water table above the regional water table; bogs therefore receive their exogeneous nutrient inputs solely from precipitation. Fens are in lower topographic positions or on the margins of peatlands and are not isolated from the regional groundwater table; they receive nutrient inputs from both precipitation and groundwater. *Sphagnum* mosses, ericaceous shrubs, and black spruce (*Picea mariana*) dominate the vegetation of bogs while sedges and other graminoids dominate fens.

These vegetation patterns are enhanced by positive feedbacks between the plant community and the type of peat formed from its litter (Glaser 1992). Sedges and other graminoids produce peat of high hydraulic permeability. Therefore, water preferentially flows through fens and maintains them. On the other hand, *Sphagnum*-derived peat has low permeability and water flow is diverted around it. *Sphagnum* mosses prefer these relatively drier conditions, and their continued dominance and production elevates the peat surface above the water table, leading to the formation of bogs (van Breemen 1995). These raised bogs shed precipitation to the surrounding wetter areas, further enhancing the dominance of graminoids there. Directional flows of water into peatlands from the upland is thus broken up into patterns of water tracks (occupied by fens) and raised bogs (occupied by *Sphagnum*), which are stabilized by these positive feedbacks between the plant community and the peat formed from it.

The positive feedbacks between peatland vegetation, peat formation, and hydrologic gradients and flows at local scales may have important implications for global carbon budgets. Although northern peatlands occupy less than 2% of the world's land surface (Post et al. 1982; Bridgham et al. 2001), they contain one third of the world's soil carbon and nitrogen pools (Post et al. 1982, 1985; Gorham 1991) and are the source for 6–9% of global methane emissions (Mathews and Fung 1987; Aselmann and Crutzen 1989; Bartlett and Harriss 1993). Carbon and nutrient budgets in bogs and fens are very different: bogs appear to accumulate more carbon and nutrients than fens (Glaser 1992; Bridgham et al. 1995, 2001). Therefore, the spatial distribution of bogs and fens and how that distribution arises from positive feedbacks between the plant community and water flow patterns may determine the pattern and degree of carbon balances of many northern regions.

# Diffusion, Diffusive Instability, and Pattern Formation and Destruction

Mahadevan (this volume) points out that diffusion of an ecosystem component or property destroys heterogeneity by dispersing the property or agent responsible for it across the landscape or seascape. Thus, a plume of nutrients or pollutants introduced at a point into a fluid, an insect outbreak at a spot, or the aggregation behavior of some zooplankton are all dispersed as these entities diffuse through the landscape or fluid. This dispersal can, to a first approximation, be described by random Brownian motion, otherwise known as Fickian diffusion. Thus, under some circumstances, random spatial motion destroys spatial heterogeneity (Murray 1989; Okubo and Levin 2003; Mahadevan this volume).

However, dispersing species also interact with each other (through predator-prey interactions, for example). This trophic interaction of two dispersing species can create, rather than destroy, spatial heterogeneity under certain conditions. If the growth of the lowest trophic level involves a positive feedback (autocatalysis) with itself (e.g., population growth) or with some underlying environmental condition (e.g., enhancement of nutrient availability through litter feedbacks) or is sustained by inputs from the surrounding environment and if the populations of species in different and interacting trophic levels spread or diffuse at different rates, then conditions are ripe for creation of a rich variety of spatial heterogeneities and patterns. In this case, the diffusion causes spatial heterogeneity by modifying the trophic interactions as the interacting populations away from points at different rates. This heterogeneity can, under certain circumstances, then be amplified by the interactions between trophic levels or between species and their resources. This phenomenon, known as reaction-diffusion or diffusive instability, was first mathematically described by Turing (1952) and is often called a Turing mechanism in his honor. Excellent reviews of this theory rich with ecological examples are given by Edelstein-Keshet (1988), Murray (1989), Holmes et al. (1994), Okubo and Levin (2002), and Levin (2003). This theoretical approach gives explicit conditions for when either spatial heterogeneity or homogeneity is stable and, through numerical solutions or simulations, it can also give some predictions about the pattern of heterogeneity. These explicit conditions and solutions can then be tested in experiments or observations.

To see the conditions under which such spatial heterogeneities arise, consider first a set of coupled equations for the interactions of two species in an otherwise homogeneous environment:

$$\begin{cases} \dfrac{dS_1}{dt} = F_1(S_1, S_2) \\[2mm] \dfrac{dS_2}{dt} = F_2(S_1, S_2), \end{cases} \tag{4.3}$$

where $S_1$ and $S_2$ are prey and predator, respectively, and $F_i$ are the differential equations (e.g., Lotka-Volterra predator-prey equations) describing their growth and interactions. "Predator" and "prey" are meant here in a general sense in that the predator "takes up" or consumes the prey. Thus, the "predator" can be a carnivore consuming an herbivore, an herbivore consuming a plant, or a plant species taking up a nutrient "prey" (an example of this will be given in a moment). For what follows, it is important to keep in mind that the growth of the "prey" population at the lowermost trophic level is either self-generating by means of autocatalysis, enhanced by positive feedbacks with some underlying environmental variable, or sustained by input from the outside environment.

Assume there is a spatially uniform (homogeneous) equilibrium in the absence of diffusion such that:

$$F_i(S_1^*, S_2^*) = 0,\tag{4.4}$$

where $S_1^*$ and $S_2^*$ represent equilibrium densities of $S_1$ and $S_2$. This equilibrium is spatially homogeneous and stable if small disturbances of size $\Delta S_i$ decay exponentially when the system is near equilibrium. Examples of disturbances of size $\Delta S_i$ could be harvesting or stocking of a population or enhancement of local nutrient availability by fertilization. The rates by which disturbances decay or grow are given by the eigenvalues of the Jacobian matrix $J$ of partial derivatives (sometimes called the "community matrix" by ecologists):

$$J = \begin{bmatrix} a_{11} & a_{12} \\ a_{21} & a_{22} \end{bmatrix},\tag{4.5}$$

where $a_{ij} = \partial F_i / \partial S_j$ and $J$ is evaluated at the equilibrium points $S_1^*$ and $S_2^*$. Analytical solutions of the eigenvalues near equilibrium are in terms of the parameters of the dynamical equations $F_i$; these parameters are usually the rate constants of fluxes between trophic levels of the system or the input-output terms.

The eigenvalues, $\lambda_n$, of $J$ give the rates of growth or decay of the perturbations in $n$ dimensions (where $n$ is the number of compartments of the system):

$$\begin{bmatrix} S_1(t) \\ S_2(t) \end{bmatrix} = \sum_n \begin{bmatrix} S_1(t_0) - S_1^* \\ S_2(t_0) - S_2^* \end{bmatrix} c_n \omega_n e^{\lambda_n t},\tag{4.6}$$

where $S_i(t_0) - S_i^* = \Delta S_i$ is the initial size of the perturbation to $S_i$, $\omega_n$ is the corresponding normalized eigenvector to $\lambda_n$, and $c_n$ are constants that depend on initial conditions. Clearly, if all $\lambda_n < 0$, then the perturbation $S_i(t_0) - S_i^*$ decays exponentially, and the system returns to its homogeneous equilibrium state of $S_1^*$ and $S_2^*$. Spatial homogeneity is then stable under these conditions. This happens when the trace of $J$ is negative and the determinant is positive, or

$$tr(J) = a_{11} + a_{22} < 0$$

and

$$\det(J) = a_{11} a_{22} - a_{12} a_{21} > 0.\tag{4.7}$$

Recall that $a_{11}$ represents the growth of $S_1$ with respect to itself, or the auto-catalytic/positive feedback in the system, and $a_{22}$ represents mortality of the predator ($S_2$) with respect to itself.

Now add diffusion terms to each equation (for simplicity, we will consider diffusion in only one lateral direction):

$$\begin{cases} \dfrac{\partial S_1}{\partial t} = F_1(S_1, S_2) + D_1 \dfrac{\partial^2 S_1}{\partial x^2} \\[2mm] \dfrac{\partial S_2}{\partial t} = F_2(S_1, S_2) + D_2 \dfrac{\partial^2 S_2}{\partial x^2}, \end{cases}\tag{4.8}$$

where $D_1$ and $D_2$ are rates of random spread or Fickian diffusion across space $(x)$, and the partial derivatives with respect to $x$ represent density or concentration gradients of $S_1$ and $S_2$ across space. (In much of the literature on reaction-diffusion equations, the prey is termed the "activator" because of the positive feedback, and the predator is termed the "inhibitor" because it consumes the prey, but I will continue to use the terms prey and predator in the general sense as defined above).

Perturbations to this spatially explicitly model (such as changing the population density of either species, corresponding, e.g., to an outbreak, an irruption, stocking, or harvesting) are introduced not simply at a point in time but at a point in both space and time. Furthermore, the perturbation propagates in space because the diffusion terms "spread" the perturbed population out in the $x$ direction. The perturbation is further modified by the interactions between the two species who spread or diffuse at different rates. The Jacobian now becomes:

$$J_{spatial} = \begin{bmatrix} a_{11} - D_1\sigma^2 & a_{12} \\ a_{21} & a_{22} - D_2\sigma^2 \end{bmatrix}, \tag{4.9}$$

where $\sigma$ is the wavenumber, or the number of a peak in population density assigned in increasing order away from the initial peak that was the perturbation. $\sigma$ is proportional to $2\pi/$distance between the peaks. The decay or growth of these perturbations is then necessarily a function of both space and time and is approximated by:

$$\begin{bmatrix} S_1(x,t) \\ S_2(x,t) \end{bmatrix} = \sum_n \begin{bmatrix} S_1(x_0,t_0) - S_1^* \\ S_2(x_0,t_0) - S_2^* \end{bmatrix} c_n\omega_n e^{\lambda_n t} \cos\sigma x. \tag{4.10}$$

Note the addition of the term $\cos\sigma x$ in comparison with Equation (4.6); this ensures that the fate of the disturbance depends both on space and on time. Again, the coexistence between $S_1$ and $S_2$ is stable and spatially homogeneous when the trace of $J_{spatial}$ is negative, the determinant is positive, and hence the real parts of $\lambda$ are all negative. These conditions obviously depend on the relative sizes of $D_1$ and $D_2$.

Assume that a perturbation is introduced at a point $x_0, t_0$. If $D_1 = D_2$, then some simple algebra shows that the heterogeneity introduced by the disturbance decays. Consequently, the spatially homogeneous distribution is stable with equilibrium values $S_1^*$ and $S_2^*$ [see Okubo and Levin (2002) for mathematical details and proofs].

But when $D_2 > D_1$ and $D_2/D_1$ is greater than some crucial value $C$, then the homogeneous steady-state distribution is not stable (the determinant becomes negative), and diffusive instability sets in. Spatial heterogeneity, rather than homogeneity, becomes the stable state of the system, and the disturbance propagates across space. Under these conditions, the two coexisting species are distributed heterogeneously across the landscape. Eventually, their distribution approaches a stable patterned heterogeneity [see Okubo and Levin (2002) for mathematical details and proofs].

The crucial value by which $C$ must be exceeded for patterned heterogeneities to develop varies with functions $F_1$ and $F_2$, but in general $C = f(a_{22}/a_{11})$. Therefore, if

$$\frac{D_2}{D_1} > f\left(\frac{a_{22}}{a_{11}}\right) > 1, \tag{4.11}$$

then spatial homogeneity of two interacting populations of different trophic levels is unstable, and spatial heterogeneity of the two interacting populations is stable. In other words, for patterned heterogeneity to be stable: (1) the diffusion rate of the predator must be greater than that of the prey and greater than some function of the ratio of per capita mortality of the predator to per capita growth of the prey; (2) the growth of the prey (at least at low population densities) must involve a positive feedback within its own population or with some underlying ecosystem property [plant litter-nutrient availabilities discussed by Meinders and van Breemen (this volume) could be one such feedback]; and (3) an increase in predator densities decreases prey density through consumption, and therefore eventually predator densities as well.

To see how this works, first consider a stable predator-prey system without diffusion. A random increase in prey density at a point in a landscape results in a further increase in both its density and that of the prey, but increased predator density at the point of random increase in prey density reduces the prey and is also self-limiting through mortality. The system is thereby stabilized, and the perturbation in prey density at the point of the disturbance dies away exponentially.

Introduction of diffusion terms dissipates the negative effect of the predator. If the diffusion rate of the predator is sufficiently greater than that of the prey ($D_2 > D_1 C$), then a local randomly introduced peak in prey density can grow because of autocatalysis or positive feedbacks to its population. The predator will be able to track the peak in prey density, causing "dents" to appear and separating the initial peak into two, which then grow by autocatalysis and the process repeats. Depending on the magnitudes of $D_i$ in both $x$ and $y$ directions and the exact form and magnitude of $C$, a rich variety of patterned spatial heterogeneities can develop (Okubo 1978; Murray 1989; Holmes et al. 1994).

The most surprising aspect of this theory is that these spatial heterogeneities are due entirely to the interactions of the two components diffusing randomly at different rates and not necessarily due to any persistent heterogeneity in the underlying environment or preferred directional flow of one or both species. If there are positive feedbacks in the growth of the prey population and greater rates of diffusion of predator than prey, then spatial patterns (heterogeneities) are almost inevitable. Thus, neither non-random foraging of a predator nor underlying environmental heterogeneity is required to produce spatial patterns in generalized predator-prey systems. This is not to say that predators necessarily forage at random nor does it deny the existence of underlying environmental heterogeneities. Rather, such underlying heterogeneities, if present, can modify the patterns further, and the mere presence alone of a pattern is not sufficient to invoke them.

Further theoretical explorations of this mechanism of generating spatial heterogeneity have been developed. As opposed to predator-prey models, Levin (1974) showed that diffusive instability cannot occur in simple two-species Lotka-Volterra competition models with diffusion, but Evans (1980) showed that it happens in three-species Lotka-Volterra competition models. Powell and Richerson (1985) showed that diffusive instability and pattern formation can happen between two species competing for two resources if the dynamics of both species and their resources are all modeled.

This mathematical approach has found applications in various ecological settings, beginning with ocean systems. Malchow (2000) gives an extensive review of recent developments in the theory of pattern formation in aquatic systems. Diffusive instability was first proposed to explain fine-scale spatial heterogeneities of herbivorous zooplankton and phytoplankton in the oceans by Segal and Jackson (1972) and independently by Steele (1974) and developed further by DuBois (1975) and Levin and Segal (1976). Later observations showed that both fine- and coarse-scale patchiness of zoo-plankton and phytoplankton require not only diffusive instability but also directional gradients caused by currents and gyres (Weber et al. 1986; Mahadevan this volume).

Levin (1977) extended the development of this approach by showing that a positive feedback in the prey is not necessary if the predator consumes prey according to a saturating function, such as a Michaelis-Menten func-tion. Okubo (1978) then showed that diffusive instability can occur between phytoplankton and the concentrations of limiting nutrients in the water col-umn if one assumes that the phytoplankton take up nutrients in a Michaelis-Menten function and herbivores are a constant sink for the phytoplankton.

This mechanism of generating patterned spatial heterogeneity is proba-bly not confined to aquatic systems, even though it has been more exten-sively investigated in such systems. One aspect of the above examples to notice is that a herbivore is present in all of them. Some recent studies also indicate that pattern formation through diffusive instability can arise in ter-restrial systems with herbivores. Maron and Harrison (1997) showed that tussock moths attain stable, locally high densities even though they disperse faster than their host plants because of the even faster dispersal of a more mobile parasitoid, thus introducing the possibility of diffusive instability in a plant-herbivore-parasatoid system. Pastor et al. (1999) showed that foraging by mobile model moose in a model landscape that was initially random eventually produced spatial patterns characteristic of diffusive instability. These theoretical patterns also conformed to field measurements made on Isle Royale (Pastor et al. 1998). Because the patterns that develop affect the energy balance of the mobile moose, only certain foraging strategies produced landscapes in which food was distributed in such a pattern that the moose sustained positive energy balances and thereby survived. Therefore, diffusive instability can produce spatially heterogeneous landscapes that can either be detrimental or crucial to the energy balance of foraging animals and thus the survival of their populations.

Terrestrial herbivore populations almost always disperse faster than their forage species disperse seeds or propagules. If it is also common that a forage species is part of a positive feedback with soil properties (Meinders and van Breemen this volume), then spatial heterogeneity would seem to be common in terrestrial ecosystems where herbivores have strong effects on plant community composition and nutrient cycles. If one is working in an ecosystem in which herbivores exert strong control over species composition, nutrient cycling rates, or both, then one should immediately suspect diffusive instability as a possible source of any patterns one finds.

Diffusive instability and spatial pattern formation through trophic interactions is currently an area of theoretical research rich with nontrivial predictions that can be tested experimentally. Some of these experiments may involve long-term observations to determine the scales over which spatial patterns arise (e.g., Grünbaum 1992; Pastor et al. 1998) or to determine if spatial heterogeneities change with time (e.g., Pastor et al. 1999). Long Term Ecological Research (LTER) sites, the Joint Global Ocean Flux Study (JGOFS) sites, and other sites with repeatedly monitored observation grids are possible sites to gather data to refine and test these theories.

## When Is Spatial Heterogeneity Important?

The above considerations beg the questions that the organizers of this conference have explicitly posed: When is spatial heterogeneity important? When is it not important?

These are difficult questions. In part, the answers depend on what is meant by "important." For example, to a moose walking across a landscape, the conditions in the next step may be important (e.g., whether or not there is edible food there). They may also be important to a behavioral ecologist trying to construct individual-based models of moose foraging. But whether or not they become important at population, ecosystem, and landscape levels depends on positive and negative feedbacks between the moose and plant growth and whether the recovery time of the browsed plant is longer or shorter than the average return time of a moose to each plant (Moen et al. 1998). Thus, the importance of spatial heterogeneity depends on the scale of the question being asked, a point made numerous times in the recent ecological literature and at this conference as well.

Disturbances create spatial heterogeneity in all systems almost by definition. To a crude first approximation, disturbances can be considered random losses of a certain percentage of biomass, easily modeled through stochastic linear processes such as Markov chains. Even when the dynamics are random and linear and therefore simple, they can be "important." Certainly, the loss of 75% of the biomass of an ecosystem over some mean recurrence interval and distributed more or less randomly over the landscape has large effects on ecosystem properties. We have gained a great deal of understanding of disturbances in ecosystems and landscapes through the application of linear data analysis and modeling techniques. Nonetheless, perhaps the more interesting and fruitful avenues for further exploration involve higher order effects of disturbances on landscape patterns, taking into account how and when heterogeneities created by one disturbance influence the spread and nature of future disturbances (see Romme this volume). Do these effects differ for different ecosystems? Do they depend on mean turnover rate of biomass or nutrient capital within the ecosystem, the rate of dispersal of component species, or the degree of mixing of materials, climatic conditions, or other forcing functions?

The spatial heterogeneity created by directional flows appears to be important when it affects the mass balance of materials in a local neighborhood: the position of the local neighborhood with respect to surrounding neighborhoods that deliver or receive materials from it must then be taken into account. This is particularly important when the materials limit growth, such as water, nitrogen, or photons of light, and especially when they are amplified by positive feedbacks within the local neighborhood, such as the formation of patterned communities in peatlands and arid lands. But we have much more to learn about this. When do lateral transfers become important and for what property or process? Is there a particular ratio of lateral inputs to internal rates of cycling above which we must consider position in the landscape and below which these lateral inputs can be ignored? Are there particular positions in the landscape such as riparian zones for which these lateral flows cannot be ignored? Do the importance of lateral flows increase "down gradient"?

The patterned heterogeneities created when positive feedbacks are coupled with different rates of diffusion between interacting trophic levels are important when they modify the success of individuals or populations of each trophic level in obtaining needed resources. This has obvious evolutionary implications, because it means that the landscape of selection pressures is dynamic precisely because of the interactions of individuals searching for food. Such dynamics may particularly be important in ecosystems in which herbivores control plant species composition and the cycling of nutrients and energy, but again we need to refine further these considerations. Does it matter how much the herbivore consumes? Or does the rate of recovery of plants from herbivore consumption matter even more? Or do both matter?

If nothing else, the papers of this conference show that we are only at the outset of being able to define the questions of how spatial heterogeneity is created in ecosystems and what are the consequences of it. Making cross-system comparisons will depend to what extent such questions can be more precisely defined so that experimental approaches can be brought to bear on them. The rich array of theoretical approaches to heterogeneity discussed above may prove useful in helping to define these questions.

## *References*

Aselmann, I., and Crutzen, P.J. 1989. Global distribution of natural freshwater wetlands, their net primary productivity, seasonality, and possible methane emissions. J. Atmospheric Chem. 8: 307–359.

Bak, P. 1997. How nature works. New York: Springer-Verlag.

Baker, W. 1989. A review of models of landscape change. Landscape Ecol. 2: 111–133.

Bartlett, K.B., and Harriss, R.C. 1993. Review and assessment of methane emissions from wetlands. Chemosphere 26: 261–320.

Bormann, F.H., Likens, G.E., and Melillo, J.M. 1977. Nitrogen budget for an aggrading northern hardwood forest ecosystem. Science 196: 981–983.

Bridgham, S.D., Johnston, C.A., Pastor, J., and Updegraff, K. 1995. Potential feedbacks of northern wetlands on climate change. BioScience 45: 262–274.

Bridgham, S.D., Ping, C.-L., Richardson, J.L., and Updegraff, K. 2001. Soils of northern peatlands: histosols and gelisols. In Wetland soils: genesis, hydrology, landscapes, and classification, eds. J.L. Richardson and M.J. Vepraskas, pp. 343–370. Boca Raton, FL: CRC Press.

Chapin, F.S., Vitousek, P.M., and Van Cleve, K. 1986. The nature of nutrient limitation in plant communities. Am. Naturalist 127: 48–58.

DuBois, D.M. 1975. A model of patchiness for prey-predator phytoplankton populations. Ecol. Modelling 1: 67–80.

Edelstein-Keshet, L. 1988. Mathematical models in biology. New York: Random House.

Evans, G.T. 1980. Diffusive structure: counter examples to any explanation. J. Theor. Biol. 82: 313–315.

Flanagan, P.W., and Van Cleve, K. 1983. Nutrient cycling in relation to decomposition and organic-matter quality in taiga ecosystems. Can. J. Forest Res. 13: 795–817.

Glaser, P. 1992. Ecological development of patterned peatlands. In Patterned peatlands of minnesota, eds. H.E. Wright, B.A. Coffin, and N.E. Aaseng, pp. 27–42. Minneapolis: University of Minnesota Press.

Gorham, E. 1991. Northern peatlands: role in the global carbon cycle and possible responses to climatic warming. Ecol. Applications 1: 182–195.

Grünbaum, D. 1992. Local processes and global patterns: biomathematical models of bryozoan feeding currents and density dependent aggregation in Antarctic krill. Ph.D. Thesis, Ithaca, NY: Cornell University.

Heinselmann, M.L. 1973. Fire in the virgin forests of the Boundary Waters Canoe Area, Minnesota. Quaternary Res. 3: 329–382.

Holling, C.S., ed. 1978. Adaptive environmental assessment and management. New York: John Wiley & Sons.

Holmes, E.E., Lewis, M.A., Banks, J.E., and Veit, R.R. 1994. Partial differential equations in ecology: spatial interactions and population dynamics. Ecology 75: 17–29.

Kauffman, S. 1993. The origins of order. Oxford: Oxford University Press.

Levin, S.A. 1974. Dispersion and population interactions. Am. Naturalist 108: 207–228.

Levin, S.A. 1977. A more functional response to predator-prey stability. Am. Naturalist 111: 381–383.

Levin, S.A. 2003. Complex adaptive systems: exploring the known, the unknown, and the unknowable. Bull. Am. Math. Soc. 40: 3–20.

Levin, S.A., and Segal, L.A. 1976. An hypothesis for the origin of planktonic patchiness. Nature 259: 659.

Likens, G.E., Bormann, F.H., Johnson, N.M., Fisher, D.W., and Pierce, R.S. 1970. Effects of forest cutting and herbicide treatment on nutrient budgets in the Hubbard Brook Watershed-Ecosystem. Ecol. Monogr. 40: 23–47.

Malchow, H. 2000. Non-equilibrium spatio-temporal patterns in models of non-linear plankton dynamics. Freshwater Biol. 45: 239–251.

Maron, J.L., and Harrison, S. 1997. Spatial pattern formation in an insect host-parasitoid system. Science 278: 1619–1621.

Mathews, E., and Fung, I. 1987. Methane emissions from natural wetlands: global distribution, area, and environmental characteristics of sources. Global Biogeochem. Cycles 1: 61–86.

Moen, R., Cohen, Y., and Pastor, J. 1998. Evaluating foraging strategies with a moose energetics model. Ecosystems 1: 52–63.

Murray, J.D. 1989. Mathematical Biology. New York: Springer-Verlag.

Neuhauser, C. 2001. Mathematical challenges in spatial ecology. Notices Am. Math. Soc. 48: 1304–1314.

Niemi, G.J., DeVore, P., Detenbeck, N., Taylor, D., Yount, J.D., Lima, A., Pastor, J., and Naiman, R.J. 1990. Overview of case studies on recovery of aquatic ecosystems from disturbance. Environ. Manage. 14: 571–588.

Okubo, A. 1978. Horizontal dispersion and critical scales for phytoplankton patches. In Spatial pattern in plankton communities, ed. J.H. Steele, pp. 21–42. New York: Plenum.

Okubo, A., and Levin, S.A. 2002. Diffusion and ecological problems. New York: Springer-Verlag.

Pastor, J., and Bockheim, J.G. 1984. Distribution and cycling of nutrients in an aspen-mixed hardwood-spodosol ecosystem in northern Wisconsin. Ecology 65: 339–353.

Pastor, J., Bonde, J., Johnston, C.A., and Naiman, R.J. 1993. A Markovian analysis of the spatially dependent dynamics of beaver ponds. In Theoretical approaches for predicting spatial effects in ecological systems pp. 5–27 ed. R.H. Gardner. Lectures on Mathematics in the Life Sciences, Vol. 23. American Mathematical Society. Providence, RI.

Pastor, J., Dewey, B., Moen, R., White, M., Mladenoff, D., and Cohen. Y. 1998. Spatial patterns in the moose-forest-soil ecosystem on Isle Royale, Michigan, USA. Ecol. Applications 8: 411–424.

Pastor, J., Cohen, Y., and Moen, R. 1999. The generation of spatial patterns in boreal landscapes. Ecosystems 2: 439–450.

Post, W.M., Emanuel, W.R., Zinke, P.J., and Stangenberger, A.G. 1982. Soil carbon pools and world life zones. Nature 298: 156–159.

Post, W.M., Pastor, J., Zinke, P., and Stangenberger, A. 1985. Global patterns of soil nitrogen storage. Nature 317: 613–616.

Powell, T.M., and Richerson, P.J. 1985. Temporal variation, spatial heterogeneity, and competition for resources in plankton systems: a theoretical model. Am. Naturalist 125: 431–464.

Segal, L.A., and Jackson, J.L. 1972. Dissipative structure: an explanation and an ecological example. J. Theor. Biol. 37: 545–559.

Segal, L.A., and Levin, S.A. 1976. Applications of nonlinear stability theory to the study of the effects of diffusion on predator-prey interactions. In Topics in statistical mechanics and biophysics: a memorial to Julius L. Jackson, pp. 123–152. New York: American Institute of Physics.

Shugart, H.H. 1984. A theory of forest dynamics. New York: Springer-Verlag.

Steele, J.H. 1974. Spatial heterogeneity and population stability. Nature 248: 83.

Tilman, D. 1988. Dynamics and structure of plant communities. Princeton, NJ: Princeton University Press.

Turing, A.M. 1952. The chemical basis of morphogenesis. Philos. Trans. R. Soc. London Ser. B 237: 37–52.

van Breemen, N. 1995. How *Sphagnum* bogs down other plants. TREE 10: 270–275.

Weber, L.H., El-Sayed, S.Z., and Hampton, I. 1986. The variance spectra of phytoplankton, krill and water temperature in the Antarctic Ocean south of Africa. Deep-Sea Res. 33: 1327–1343.

Wright, H.E., Coffin, B.A., and Aaseng, N.E., eds. 1992. Patterned peatlands of Minnesota. Minneapolis: University of Minnesota Press.

# 5
# Reciprocal Cause and Effect Between Environmental Heterogeneity and Transport Processes

William A. Reiners

## Abstract

The objective of this paper is to explore the relationships between environmental heterogeneities and the flows and movements that suffuse through all environments. Flows and movements are treated as propagations of ecological influence through environmental space. Propagations are composed of four elements: (1) initiating events or conditions, (2) transport vectors, (3) transported entities, and (4) deposition or impact processes. All four elements have multiple dimensions in type and scale, but vectors are the most convenient means of discussing these phenomena. At a medial level of causation, 10 major vectors are convenient descriptors. These vectors are molecular diffusion; transport by fluvial, colluvial, or glacial modes, gravitational sedimentation, currents (tidal and extratidal), wind (with fire as a special case) agencies; and by electromagnetic radiation, sound, and animal locomotion. Obviously, each of these vector types has different behavior. Propagations can be initiated, or modified by, environmental heterogeneities. But also, propagations can create, maintain, and destroy heterogeneities. Thus, reciprocal cause and effect relationships exist between propagations and environmental heterogeneities. Analysis and understanding of these reciprocal interactions between propagations and heterogeneities requires some understanding of the mechanics of propagations, whether they involve wind, waves, or wallabies. In the same sense, analysis and understanding of how environmental heterogeneities alter propagations requires an appreciation for the global range of heterogeneity types, whether they are ripples, runnels, or run-on patches. Spatially explicit two- and three-dimensional models of propagations in heterogeneous environments are useful ways to develop understanding and, with caveats, to predict how processes and patterns interact. Some of the representational issues of building such models are reviewed in this paper, and three model examples are described.

## Introduction

Although ecology has always been a geographically based science, for many decades basic ecological research tended to have a point-model focus. With some important exceptions (e.g., Watt 1947), this was reflected in the emphasis on putatively homogeneous sites, whether stands or watersheds, as the appropriate representation of nature (Wiens 2000; Reiners and Driese 2004). This was not true in applied areas of ecology such as forestry, wildlife, fisheries, and range sciences where spatially distributed representations of nature were imperative. Point models were of little use for predicting habitat usage by deer or the dispersal of white pine blister rust. This perspective has changed for basic ecology in the past two decades, however, as the point-model view of nature has largely given way to a spatially heterogeneous representation of nature (Turner et al. 2001; Chapin et al. 2002; Reiners and Driese 2004). With the advent of new foci such as landscape ecology, conservation biology, and earth system science, and with the practical application of tools for acquiring and managing spatial data, the conceptualization of nature and the practice of basic ecology have made the heterogeneous domain the primary focus (Turner et al. 2001).

A benefit of adopting a spatially distributed view of nature is an easier incorporation of flows and movements into our visualization and treatment of a spatially heterogeneous environment. Transport processes—so intrinsic to the way nature operates—underlie many of the more interesting and important aspects of ecology. Personal experiences tell us that transport processes are influenced by environmental heterogeneity. By stepping around the corner of a building on a windy day, for example, we notice significant changes in our bodily comfort. A spatial approach to ecology now allows us to appreciate, analyze, and model how spatial heterogeneity and transport phenomena are reciprocally related.

The objective of this paper is to review how flows and movements of different kinds affect, and are affected by, environmental heterogeneity. This paper is organized into six sections: (1) how transport phenomena act as propagations of ecological influence, (2) how transport processes are affected by environmental heterogeneity, (3) how propagations may produce, maintain, and destroy environmental heterogeneity, (4) issues in the spatial representation and modeling of propagations, (5) three examples of propagation modeling in heterogeneous environments, and (6) how a propagation perspective might influence our conceptualization of nature and ecology.

## Transport Phenomena as Propagations of Ecological Influence

Flows and movements can be generalized as propagation phenomena entailing four components: (1) initiating events or chronic conditions, (2) a

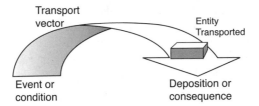

FIGURE 5.1. Diagram of the four components of propagation phenomena. Redrawn from Reiners and Driese (2004).

conveyance mechanism or vector operating through one or more media, (3) a conveyed entity, and (4) a locus of deposition or consequence (Figure 5.1) (Reiners and Driese 2001, 2003, 2004). There are analogies between ecologically relevant propagations and information transfer; indeed, some propagations primarily involve the transfer of information, such as the displays and sounds of many kinds of animals (Bradbury and Vehrencamp 1998). Propagations also involve transport of matter, such as slope-wash, or of energy, such as momentum of wind. Some, but not all, propagations are viewed as "fluxes" because with propagation, some quantity of an entity must move through some space or point at some rate. However, quantity and rate are not the essence of some propagations so that "flux" is too narrow as a general descriptor.

## Initiating Events or Conditions

Events or conditions initiating propagation can range from spatially discrete and brief phenomena, such as the crack of a twig under a predator's paw, to something as large and pervasive as an earthquake resetting slope angles and stream grades. In fact, initiating causes can vary in at least seven distinct ways. Initiating causes (1) can be characterized by the kind of environment in which the events or conditions occur, (2) may be of abiotic or biotic origin, (3) can emanate from a natural process or an anthropogenic action, (4) may be discrete events or chronic conditions, (5) have a spatial extent, (6) can vary in duration of the action, if they are discrete events, and (7) may have a periodic character, if they are discrete phenomena.

Properties (4) through (7) in the list above are relativistic problems, requiring explicit definitions of the scalar context for the immediate case in question (Peterson and Parker 1998). Determining the origin of cause is, itself, relativistic as illustrated by the familiar *butterfly effect* (Gleick 1987). Definition of causation at a distal level, however, may be philosophically satisfying but mechanistically frustrating. It really is not useful to know that the stroke of a butterfly's wing will ultimately lead to a tornado in Topeka. It is more useful to seek causation at a more proximal level (*sensu* Robertson 1989) such as meteorological dynamics over the Central Plains.

The extent of the initiating cause should not be confused with the extent of environmental space that is impacted. The extent of crustal displacement along fault scarps related to earthquakes may only be centimeters to meters, but the extent of the earth's surface disturbed by these displacements may be thousands of square kilometers. Similarly, a chronic condition like an acid mine seep may occupy square meters, but its outflows may alter stream chemistry for many kilometers.

## Entities

Propagations require that something be transported from the site of initiation to the locus of deposition (Figure 5.1). A neutral word for the item transported is *entity*. An initially simple approach is to classify entities into parcels of energy, matter, or information. Further thought reveals, however, that many entities one might consider to be energy also involve matter, such as the energy of atmospheric momentum. Likewise, many forms of matter bear with them some measure of energy, such as free energy of organic matter or reduced inorganic matter, or the momentum of transport itself. Finally, it may be neither the energy nor the material content itself but the information content that may impact the target destination. This is particularly true of biological targets. The transport of coded light signals generated by insects or fish, of programmed sounds such as mating calls of birds, or of genetic information bound in transported spores, pollen, and seeds are important from an informational point of view, not for incorporated energy and material content.

The definition of entities is dependent on their locus of deposition and on the point of view of the observer. Waves beating on the base of sea cliffs may be viewed as products of wind energy transformed to hydraulic energy eroding the cliff base through mechanical action. In this sense, the individual waves are energetic entities. But the waves also consist of water with dissolved and suspended substances, so that matter as well as energy is transported via wave motion. Whether waves moving onshore are material or energetic entities depends on the observer's phenomenological interest.

## Vectors

Propagations require a transport mechanism to move entities from places of origin to loci of deposition (Figure 5.1). *Vector* is a term for any agent providing transmission of an entity across space (Weins 1992). As with initiating causes, an operational level has to be selected for determining vectors. It is possible to generalize broadly and attribute a multitude of vectors under the category of gravity (e.g., surface and groundwater flows, tides, and mass wasting events). Obviously, this level of causation does not provide much useful information. At the other extreme, we can describe vectors in utmost detail that extends to particular cases, such as exactly the kind of mass wasting process (e.g., Summerfield 1991). An intermediate position between most distal and most proximal definitions of causation underlying vectors is used in the following discussions.

## Consequences

Eventually, transported entities are deposited, resulting in consequences somewhere in environmental space (Figure 5.1). Deposition may involve dissipation of heat, the triggering of an epidemic, absorption of sound waves, or the insertion of a new genetic variant. As emphasized earlier for causation and entities, definition depends on the viewpoint of the observer.

## Propagations as Space-Time Phenomena

In some cases, it is acceptable and appropriate to view propagations at their terminus of action, or as instantaneous phenomena, so that their trajectories through time can be ignored. But, the fact that propagations take place over finite periods of time must be kept in mind. Propagations are both spatial and temporal phenomena (Kelmelis 1998; Reiners and Driese 2003, 2004). The areal or volumetric extent of the zone of deposition, impact, or consequence changes over time, whether it is the spatial extent of a snow avalanche rollout during the fractional seconds of its passing, the expanding seepage zone of a pollutant leak as its plume flows outward over days and years, or the nearly continuous flux of trade winds. Obviously, the spatial extent of a propagation is related to the viewer's temporal scale; the longer the time, the greater the extent in many, but not all, cases. Extended further, propagation time-awareness implies that heterogeneities in the environment, such as depressions and mounds found on forest floors left by tree tip-ups, are legacies of propagations past. Thus, a local environment, however defined, is a product of ongoing propagations of varying types, frequencies, periodicities, and intensities, as well as of propagations of the past.

# How Transport Processes Are Affected by Environmental Heterogeneity

## Effects of Heterogeneous Media on Propagation Initiation

Environmental heterogeneities not only influence the transport of entities but also may be the immediate initiators of propagations (Figure 5.2). A riverine flood plain meandering through a grass- or shrubland can be a sand source for dune systems that may stretch for hundreds of kilometers downwind of the flood plain (Knight 1994). Analogously, an acidic spring can alter stream chemistry for many kilometers downstream (Schnoor 1996). Initiations caused by heterogeneities may be probabilistic as well as

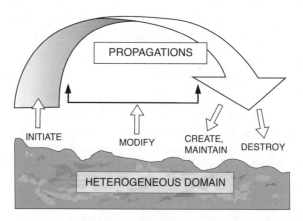

FIGURE 5.2. Diagrammatic relationships between environmental heterogeneities and propagations. Heterogeneities can initiate and very frequently modify the flow paths and intensities of propagations. Propagations, on the other hand, may be essential to create, maintain, or destroy environmental heterogeneities.

deterministic. There is a higher probability of lightning strikes on high ground than on low in terrain with relief. There is a higher probability of high nitrate fluxes from cultivated source areas than forested source areas on third-order watersheds (Herlihy et al. 1998).

## Effects of Heterogeneous Media on Propagation Flow Paths

On a perfectly homogeneous and infinite plane or in a homogeneous volume, a physically driven propagation would move across the plane or through the volume simply as a function of the underlying physics of the driving vector. In fact, such perfect propagations rarely occur because of environmental heterogeneity. In a finite world composed of a layered atmosphere overlying continents of variable roughness interspersed, in turn, among oceans, the media through which propagations are transmitted vary in space and time. Further, media interfaces like the land-atmosphere interface or the sea-atmosphere interface modify transport processes. Although much transport modeling has concerned itself with ideal cases involving putatively homogeneous media, for ecologists, the more interesting cases involve cases featuring heterogeneous environments (Kelmelis 1998; Reiners and Driese 2001, 2004).

The influence of heterogeneity on transport processes obviously depends on the physics or biology of the entities and vectors involved. For example, molecular diffusion must move from areas of high concentration to low. Stronger effluxes of biogenic gases emanate from soils where the dog is buried. Of course, burying the dog was yet another propagation event, in this context an accident of history. Once initiated, diffusion rates will be pathway-controlled, as described in the example in the next section. It should be noted that the term *diffusion* is used here strictly in the physical sense of heat or molecular movement in response to concentration gradients. Diffusion is also used by others as an analogy for other dispersal processes, especially those occurring over extents of meters to kilometers, involving composite transport processes, or even cultural transfers (e.g., Banks 1994; Barrell and Pain 1999; Nakicenovic and Grubler 1991; Okubo and Levin 2001; Turchin 1998).

For all vectors driven by gravity, surface topography or variation in subsurface properties are the primary source of heterogeneity. Water, rocks, ice, and particles move downward in response to gravity in cases of transport by fluvial, colluvial, or glacial modes and with gravitational sedimentation. In both subaerial and submarine environments, flows tend to initiate at higher points in the topography, but then flow directions and flux fields are directed by topographic variability. For groundwater flows, the effective topography is subsurface variability in conductivity. To an extent, some currents are also gravity driven by either astronomical forces or by the earth's gravity on density gradients. Although it is ultimately true that tidal currents are initiated by gravitational pull of the sun, the moon, and other planets, the better explanation at a global level is that irregularities of the sea surface caused by these attractions lead to water running "downhill" from high areas to low on the oceanic surfaces (Mellor 1996; Pinet 1998). In this sense, topographically directed gravitational fields are both initiators and modifiers of fluid flows.

Both aquatic and marine nontidal currents and wind are ultimately derived from differential heating over space but are then directionally altered by Coreolis force, boundary constraints (bottom surfaces, islands and continental interfaces), surface roughness, and by density gradients established by temperature and salinity differences. Thus, at all scales, currents and wind are initiated, or powered by, environmental heterogeneities. Whether these causes are initiating or modifying factors depends on one's scale and vantage point. Nocturnal, downslope winds originate at high elevations and dissipate at low elevations. At the scale of a mountain-valley complex, an observer would describe such winds as differential air densities associated with altitude. At the scale of a mountain slope segment, an observer would contend that topography constrained the velocity and direction of the propagation.

Electromagnetic radiation, including all wavelengths occurring on the earth, encompasses radiation from the sun and the moon, from all substances on and in the earth, and biologically generated light (Bradbury and Vehrencamp 1998). Obviously, radiation originates from sources, sometimes diffuse, like the atmosphere; sometimes from specific points, as from a bacterial cell. Once emitted, radiation can be refracted at media interfaces, absorbed, scattered, or reflected by substances suspended in any translucent material. Some of these substances can be large, like plant leaves, producing complicated direct light environments (Endler 1993). These large and small objects then become elements of environmental heterogeneity for radiation transfer.

Sound is in many ways analogous with light and diffusion. Sound may come from extensive sources like surf along a beach or from highly concentrated sources like a rasping cicada. In either case, sources are discrete elements of environmental heterogeneity. Sound dispersal is then very much controlled by heterogeneities in the transporting medium like thermal layering in air and water or by reflective barriers like hills and sedimentary plumes in water.

There are many ways in which environmental heterogeneity underlies the initiating condition for transport by animal locomotion. Pollen transport by insects from plants in one wetland site to another or forage transformation of grass to feces could be initiated in an upland grazing site, then transported to, and deposited in, a lowland watering site. As Aldo Leopold observed years ago, animals tend to feed high and transport altered materials to low points on the landscape (Leopold 1949).

The previous paragraphs illustrate how many propagations are initiated and directionally modified by environmental heterogeneities. As with many things ecological, definition of initiation is scale-dependent. To say that the Gulf Stream is part of a global transport complex receiving its name from the Gulf of Mexico is true and useful from a global perspective, but for those living on Bermuda, the Gulf Stream is a continuous flux that controls local water and air temperature. Also, when one views propagations over time, the importance of differential stochasticity becomes apparent. Some insect outbreaks leading to out-migrations are more likely to occur in old-growth forest stands than younger stands in heterogeneous, forested environments (Holling 1987). Similarly, mass wasting events are more likely to occur on slopes oversteepened by lateral cutting by streams below than on slopes above aggrading portions of the flood plain. Fluvially transported nitrate is more likely to originate from agricultural fields than from forest plots on a multiple-use watershed (Herlihy et al. 1998). Depending on the entity and vector, it is possible to map the probability of initiation and subsequent flow direction based on patterns of relevant environmental heterogeneities.

# How Propagations Produce and Destroy Environmental Heterogeneity

## The Role of Propagations in Creating and Maintaining Environmental Heterogeneity

The foregoing sections have emphasized the influence of environmental heterogeneity on propagations, both in their initiation and modification. In fact, there is a marked reciprocity between propagations and environmental heterogeneities inasmuch as the latter are often generated by, and subsequently maintained, modified, or obliterated by transport processes. In fact, one could make an argument that all environmental heterogeneity is caused by propagations. To take the extreme case, two of the most common and dominant forms of heterogeneity—topography and surface geology—are created on one hand by tectonics and volcanism (mass movements of the earth's crust), and, on the other hand, by erosion and deposition by various forms of gravity- or wind-driven transport. Although this extreme case may be true, such a broad view will be set aside because it defeats the usefulness of considering propagations as an active part of nature on shorter timescales typically used by ecologists.

More practically, let us consider the kinds of heterogeneity found in the environment at less than the scale of landscape evolution. For the purpose of this discussion, we can divide heterogeneities into those that are anthropogenic, such as road networks and land-use patterns, and those that are products of more or less natural processes. The latter class includes oceanic currents, gyres, and eddies (Barber 1988); stream networks (Harmon and Doe 2001; Smith et al. 1997); intrastream bars and banks (Fisher and Welter 2004); dune systems (Yaalon 1982); forest gap mosaics (Bormann and Likens 1979; Pastor et al. 1998); fire patches (Romme 1982); and various kinds of linear, wave-like structures observed in oceans (Mellor 1996), lakes (Kratz et al. this volume), and on land (Billings 1969; Sprugel 1976; Klausmeier 1999; Hiemstra et al. 2002; Wu et al. 2000; Tongway and Ludwig 2004). Of course, there can be interesting interplay between human-caused *versus* naturally caused patterns—an interesting topic in itself.

Within this range of examples, it is difficult to discern a case in which transport processes at an ecological time frame are not involved in production or maintenance of patterns. However, there surely are such cases, and they must carefully be sought out (Butler et al. 2003). For example, it is possible that some of Watt's classic cases of pattern and process of tussock or clonal patterning are totally autogenic and independent of resource flows (Watt 1947). A careful review of the large number of cases of environmental heterogeneities would be necessary to characterize patterns of causation. Nevertheless, it seems that many heterogeneities are created by either episodic propagations like glacial advances and retreats, wind storms, fires (a special case of wind), and other large-scale extremes, or by the interactions between biological damage by propagated physical stressors and resource sequestration provided by material fluxes.

The latter class of heterogeneities, those caused by interactions between propagated stresses and resource fluxes, is of particular interest to ecologists because of the seemingly self-organizing nature of such patterns. Patterned heterogeneities of this type were first described by Watt (1947) under the title "pattern and process," the meaning of which is how pattern reveals, and is caused by, process. Pattern and process has since been described numerous times. It was reviewed by White in 1979 and by Turner in 1989 and is the dominant theme of a recent landscape ecology book (Turner et al. 2001). "Process" in pattern and process actually has dual meanings: the processes underlying construction and maintenance of the physical pattern (*sensu* Watts 1947), and collective processes resulting from the pattern (e.g., Schlesinger et al. 1996). In fact, causative processes on one hand, and resulting processes on the other, may be restatements of the same phenomena. That tussocks capture water, organic matter, and nutrients transported downslope by sheet-wash describes the concentration of resources and extraordinary plant growth in islands or stripes and explains the result—the existence of those plants. The coincidences of reproductive mode (or plant life span) and crucial lengths between resource collection areas *versus* accumulated stressors leading to plant demise and the existence of vegetated patches are just interesting details crucial to the local example (Ludwig et al. 2000). The principal point here is that these kinds of self-organizing phenomena often depend on transport processes, so that there is a constructive relationship between environmental heterogeneity and propagations.

## The Role of Propagations in Destroying Environmental Heterogeneities

The intimate relationship between propagations and environmental heterogeneities is enhanced further by the fact that episodic propagations like tsunamis, hurricanes and tornadoes, ice storms, landslides, floods, fire, and lightning strikes also obliterate heterogeneities and possibly create new ones. If the intensity of a propagation event is sufficient and its footprint larger than the grain of the heterogeneous pattern, destruction of the antecedent heterogeneity, patterned or not, will result. Of course, a subsequent heterogeneity will then be established. Scaling relationships between destructive disturbances and heterogeneities probably exist for individual environments and episodic propagations characteristic of that environment. For example, there may be a scaling relationship between tree age and windstorm strength for a given vegetation type that will, most of the time, maintain a gap-phase mosaic but beyond which will occasionally destroy enough forest to eliminate the original, finer grained mosaic pattern (Foster et al. 1998).

## Spatial Representation and Modeling Propagations

To this point, the discussion of propagations and heterogeneities has been general and abstract. What about measurement and prediction in realistic situations? How are propagations through spatially varying media and over irregular surfaces actually measured in nature? Examples are found in several environmental science disciplines such as geomorphology (earth surface processes), hydrology, atmospheric sciences, epidemiology, animal behavior, oceanic hydrodynamics, fire science, and aerobiology (Reiners and Driese 2004). In many, if not the majority of cases, propagations are estimated by scaling up from a few point measurements. Scaling up may be a simple statistical process, such as kriging, but it usually involves joining observations with representations of the spatial domain with or without a Geographic Information System (GIS) (Fischer 2000; Fotheringham 2000) through some kind of modeling. Large-scale examples are global circulation models that assist in weather forecasting. These are highly mature three-dimensional models operating in a spherical geometry and incorporating (assimilating) point measurements from around the globe to update climate dynamics in order to estimate fluxes of energy and matter throughout the atmosphere (Henderson-Sellers and McGuffie 1987).

Modeling propagations over and through heterogeneous environments introduces two kinds of issues. The first is about environmental representation with spatial data; the second about simulating transport processes themselves in variable environmental fields. Discussion of these vital, methodological topics goes beyond this paper. Portals to this voluminous literature are Longley et al. (1999, 2001), Clarke et al. (2000), Varma (2002), and Reiners and Driese (2004).

## Producing Areal Estimates from Point Models of Flux Through Spatially Distributed Modeling

A commonly desired estimate is for vertical fluxes of energy or matter from sediment to water column, from water column to atmosphere, from atmosphere to soil, and so forth, extrapolated over a heterogeneous spatial domain. These are usually derived from point models (zero-order models), the outputs of which are varied according to heterogeneity of the spatial domain. Variation in the spatial domain can be represented in either vector (discretized map units based on aggregated environmental features) or raster (regular or irregular tessellations like rectangular raster) format (Burrough and McDonnell 1998). Outputs from all of the representative areas are then summed to give domain-level estimates of flux.

One assumption in such operations is that there are no lateral transfers between the representative areas within the time frame of the modeled phenomenon. If lateral fluxes do occur, they are parameterized or subsumed within site properties of the areas represented by the point models. For example, Reiners et al. (2002) estimated trace gas fluxes over a region using 1-ha cell rasters for six environmental variables. Lateral drainage transfers probably occur between the 1-ha cells but were assumed to be negligible over the time frame of the estimates (days to a year). Had the modeling time frame been extended to decades or centuries, estimates of lateral transfers between map units would have been required.

If some measure of variance with the estimated flux from the entire modeled domain is desired, it becomes necessary to account for covariance relationships for the multiple environmental drivers represented by multiple, overlain data sets. In fact, this is rarely done. In the same example cited above (Reiners et al. 2002), some of the spatially distributed data, such as soil texture, had statistical distributions rather than singly determined, categorical values (e.g., landcover type for each raster cell). Iterated model runs using random values drawn from these distributions served to produce replications of output from which means and variances, including covariances, could be calculated. Means from cohorts (tuples) covering the domain were added, and variances pooled, to gain summations of regional, vertical propagation with estimates of variance properly incorporating covariance. Regional estimates were also calculated with the typical method of simply summing singly determined values for cohorts. This latter, more commonly used method led to an underestimate of 8% for one gas and 18% for another.

## Modeling Propagations Moving Laterally Across Heterogeneous Environmental Fields

Perhaps more interesting are lateral propagations across heterogeneous environments. As wind blows across, or animals move through, terrain with variable vegetation cover, environmental heterogeneity influences the transport process itself. In other words, there are explicit interactions between points on the domain. These cases require two-dimensional spatial modeling, and in some cases, demand three-dimensional approaches. Two-dimensional modeling can involve vertically oriented as well as horizontally oriented planes. Glacial movement and oceanic currents are frequently modeled as vertical, two-dimensional planes (Holland 1986; Mellor 1996; Konrad et al. 1999). Of course, vertical plane, two-dimensional modeling can be combined with representations of environmental variation on orthogonal, horizontal planes to produce a pseudo-three-dimensional system. Two examples follow in a later section. Others are reviewed in Reiners and Driese (2004).

## Modeling Three-Dimensional Processes in Two Dimensions

Although most propagation phenomena are actually three-dimensional in physical character, many, if not most, are treated in two dimensions. This is managed by parameterizing the third dimension as functions of features represented in the two-dimensional map units. For example, transport by wind involves eddy formation and turbulent transfer between the atmosphere and land and water surface. These interactions are three-dimensional but are "flattened" to two dimensions by parameterizing roughness length and effective surface element height for the two-dimensional plane (Garratt 1992). Similarly, subsurface flows in hydrology models are handled in two dimensions by parameterizing estimated porosities and saturation values of watershed spatial units (Beven 2001).

What is lost by flattening propagation processes to two dimensions? Depending on the objectives, this flattening may be perfectly acceptable. In fact, given all the additional data, modeling and computations needed to treat explicitly the third dimension, a two-dimensional approach may be the more intelligent one. As with admonitions about using the proper data structure and scale, however, it is essential that investigators be aware that adoption of widely used practices may be inadequate for the question being addressed. If, for example, dry deposition to various layers of a three-dimensional forest canopy must be known, a three-dimensional approach is necessary. Similarly, if detailed subsurface conditions in the hyporheic zone are essential to predicting biogeochemical processes (Hedin et al. 1998; Hill et al. 1998; Schindler and Krabbenhoft 1998; Fisher and Welter, this volume), and these must be known over a horizontally variable domain, then three-dimensional modeling will be necessary.

## Three-Dimensional Modeling

It would seem that true, three-dimensional propagation modeling would be important to ecology. The foraging of martens on the ground and up trees, the spatially distributed deposition of nitric oxide within forest canopies, the changing redox state of soil aggregates with rainfall events, are all important phenomena that might best be dealt with in three-dimensional framework. There has been little development in this area in ecology, partially because of the enormous computational and parameterization demands of three-dimensional modeling, but also because of the lack of conventional software packages equivalent to GIS. Some of the conceptual potentials and problems in this area are reviewed by Couclelis (1999), Rogowski and Goyne (2002), and Peuquet (2002). More such work has been done in climate modeling, groundwater pollution, and oil and gas exploration and "production." Three-dimensional vector and voxel (cubic "pixels") methods are both available in these fields, and ecologists might be advised to investigate these possibilities for appropriate ecological problems.

Yet another technological frontier is the addition of the fourth dimension—time—to these problems. Ecologists usually regard nature in four dimensions, and the time will come when they will want to model in four dimensions as well. For thoughtful treatments on four-dimensional representations, see chapters in Egenhofer and Golledge (1998) and Longely et al. (1999).

# Three Examples of Propagation Modeling in Heterogeneous Environments

To better describe how propagations are modeled for heterogeneous environments, this section demonstrates how three vectors have been modeled to incorporate environmental heterogeneity (from Reiners and Driese

2004). The three examples are wind transport, molecular diffusion, and animal locomotion. Particular attention is paid to choice of environmental variables incorporated into the environmental representation, choice of environmental data structure, extent of modeling domain, modeling grain size, time steps used, how three-dimensional processes were handled, and the data platform and modeling languages typically used.

## Wind Transport

Wind is the motion of air relative to objects. It is one of the more pervasive transport vectors in the environment and features high variability in its direction and velocity. Wind entrains, transports, and deposits sensible heat, latent heat, hydrometeors, gases, and aerosols. Aerosols include condensation products of atmospheric .chemistry, soot, soil dust, salt spray, hydrometeors, and biological products. Biological aerosols include detritus, pollen, spores, seeds, fruits, and living invertebrates (Isard and Gage 2001). The relative importance of wind transport in environmental space varies locally depending on source strengths, wind trajectories and velocities, and surface properties (Reiners and Driese 2004).

A model for transport of snow by wind was adapted by Reiners and Driese (2004) from Hiemstra et al. (2002) for treeline in the Medicine Bow Mountains, WY. This model was adapted, in turn, from Liston and Sturm (1998). Static (in the time frame of the model operation) spatial data over the domain—the elements of environmental heterogeneity—are elevation, slope and aspect, patches of trees and krummholz, and the snow-holding capacity of vegetation types (Figure 5.3A). Temporally varying inputs— other aspects of environmental heterogeneity—include wind speed and direction, precipitation rate, temperature, and humidity. Model mechanics are based on calculated wind velocity at the surface and on the shear strength of the snow. All spatial data are represented in a 5-m rectangular raster. The entrainment, transport, and deposition of snow are parameterized with respect to topography and boundary layer surfaces. Thus, the third dimension is parameterized in terms of the surface plane so that this is a pseudo-three-dimensional model distributed over a two-dimensional surface. (See Figure 5.3B for results of one model run.) This example shows how a pseudo-three-dimensional approach is adequate for propagation processes in which the flux or deposition is expressed in terms of area.

## Molecular Diffusion

Diffusion is used here in the original sense of heat and mass transfer by the movement of molecules, or very small particles, due to their kinetic energy (Harris 1979; Monteith and Unsworth 1995). This is in contrast to the usage of diffusion described above as a default model for complex phenomena that are difficult to parameterize at the scale of the actual processes (e.g.,

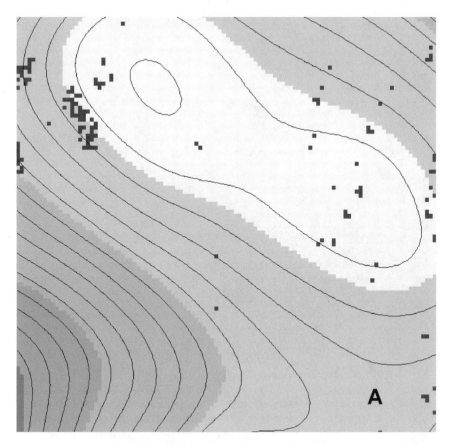

FIGURE 5.3. (A) Topography and tree vegetation of the Libby Flats treeline area of the Medicine Bow Mountains, WY. The area is 500 × 500 m in extent, and the elevation ranges from 3224 to 3239 m. Lighter shades are associated with higher elevations. Elevation contours are in black, and the areas occupied by trees are represented as dark pixels. (B) Modeled snow depths with darker shades indicating more snow. Black contours are elevation; white contours are snow depth. Snow depth ranges from 51 to 121 cm. Higher, windward (wind flows from left to right) locations tend to have less snow; positions leeward of topographic highs and trees tend to have more snow.

Pastor et al. 1998; Turchin 1998; Choy and Reible 2000; Hemond and Fechner-Levy 2000; Okubo and Levin 2001). Molecular diffusion involves transport lengths of only millimeters to centimeters but occurs over enormous surface areas ranging from the aggregate surface areas of bacterial cell walls to the ocean-atmosphere interface.

Reiners and Driese (2004) modified SNOWDIFF, a model originally formulated to simulate one-dimensional gas diffusion from soil through the snow pack to the atmosphere at one place on the landscape (Massman et al. 1997), to a two-dimensional mode. Diffusion is based on Fick's law with the

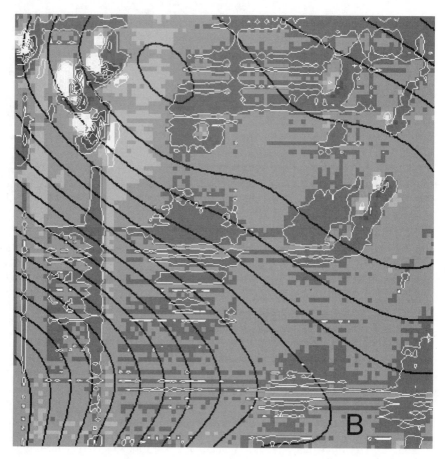

Figure 5.3. (*Continued*)

concentration gradient based on measured soil and atmosphere concentrations and resistances estimated from individual layers by thickness, porosity,
tortuosity, and temperature. Environmental variables used in the point
model—the elements of environmental heterogeneity—are snow depth and
porosity by layers. There is no estimate of lateral diffusion between points
(cells). $CO_2$ diffusion is extrapolated over the same $500 \times 500$ m treeline
domain described previously for the wind model by running the model for
each 5-m cell in a raster representation of landscape for which snow properties are known. Because flux is entirely a property of diffusion of gas
through snow and largely controlled by snow depth, estimates of gas flux
(Figure 5.4) are very similar to estimates of snow depth (Figure 5.3B). This
is actually a one-dimensional model run repeatedly over a two-dimensional
grid of snow profile properties to estimate a vertical flux in a pseudo-three-
dimensional space without lateral transport.

FIGURE 5.4. $CO_2$ efflux rates as a function of snow distribution (Figure 5.3B) resulting, in turn, from factors represented in Figure 5.3A. Black contours represent elevation; white contours indicate snow depth (see legend for Figure 5.3B). Lighter shades indicate higher $CO_2$ flux rates. Flux rates in this model output range from 3.9 to 9.1 $Mg\ m^{-2}\ s^{-1}$.

## Animal Locomotion

Locomotion is found in some stage of all members of the earth's biota but is particularly marked in animals and protists. As animals disperse, forage, flee, and mate, they move through environmental space—both aquatic and terrestrial, fluid and solid. In their movement, animals act as vectors by transporting their own biomass and leaving a trail of their influences, whether it is foraged materials, mechanical alterations of the medium, or exuvia.

Reiners and Driese (2004) produced a pine marten movement model to illustrate how some simple rules and environmental representations could lead to movements similar to those recorded in the field. The landscape

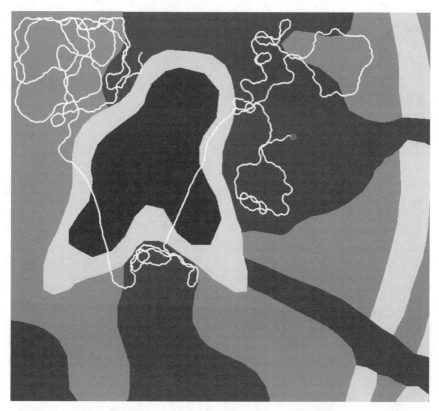

FIGURE 5.5. Map of simulated pine marten travel trajectory in a heterogeneous Rocky Mountain subalpine environment. The dark black patch in the center of the figure is a lake, a habitat type that is crossed quickly due to exposure to predators. The dark gray is the best habitat, medium gray is moderate quality, and light gray is poorest quality habitat. The line trace indicates 500 simulated movements of 5 m each.

configuration is based on actual U.S. Forest Service land cover data for an area of $1 \times 1$ km in the Medicine Bow National Forest. It is a two-dimensional vector-based representation of three levels of habitat suitability (high, medium, and low), plus water (Figure 5.5). In this case, environmental heterogeneity is represented as relatively large polygons relative to scale of animal movement. Rules for marten behavior were derived from observations on tracks in the snow. These were converted to probabilities of martens crossing habitat boundaries and turning angles made in transit through the environment as functions of the habitat type. Field data showed that angles were more acute in favorable habitat and obtuse in unfavorable habitat like water. Animal location and movement in the model is vector-based in the two-dimensional environment represented by the vector habitat layer (Figure 5.5). For heuristic purposes, users can vary turning angles associated with habitat types, number of time steps, and distance traveled per time step. This example illustrates how individual animal movement modeling may be done and subsequently coupled with animal impacts such as predation or habitat alteration.

# How Might a Propagation Perspective Influence Our Conceptualization of Nature and Ecology?

How we individually create mental frameworks for environmental heterogeneity and propagations is conditioned by our personal experiences, our intellectual predilections toward what we see, and our methods for representing nature. This is as true in natural science as it is in philosophy, religion, and art. The ways that ecologists view a landscape or seascape are influenced by what they know and how they individually represent nature based on their personal and disciplinary experiences. Interviews with several ecologists examining a common scene can reveal disparate "visions" of the scene. Some ecologists instinctively seize upon common color or textural "blocks" in the landscape (patch mosaic or patch matrix) as a means of mentally organizing variation in the domain in question regardless of what the eyes see. Others force a landscape image into raster cells similar to a remotely sensed image by ignoring unifying elements (continuous variation). Some ecologists will "see" environmental gradients, whereas others will "see" the imprints of historical events. Some see clues of ongoing change, whereas others see static patterns. Outside of ecology, many atmospheric scientists and oceanographers "see" their realms in terms of wave spectra. Our environmental cognitions vary in surprising ways and to considerable degrees.

How might a sense of flows and movements in all their variety influence our views of nature? Combined with sensitivity for heterogeneity at all scales, how would this alter our views of the environment? In the extreme, we might envisage the world as composed of temporary structures having more or less heterogeneity at given scales and bathed in a range of flows of variable intermittency and influence that alternatively create and destroy the heterogeneous features. Such a vision would return ecology to the spatial and geographic science that it once was.

*Acknowledgments.* Preparation for this review was made possible by a grant from the Andrew W. Mellon Foundation and support from the National Center for Ecological Analysis and Synthesis at the University of California, Santa Barbara. Much of the material on transport model adaptation was derived from Reiners and Driese (2004) for which Kenneth L. Driese performed the model adaptations described in the text. The author thanks Gary M. Lovett for helpful discussions that shaped this paper and the constructive remarks of two anonymous reviewers.

# References

Banks, R.B. 1994. Growth and diffusion phenomena: mathematical frameworks and applications. New York: Springer-Verlag.

Barber, R.T. 1998. Ocean basin ecosystems. In Concepts of ecosystem ecology, eds. L.L. Pomeroy and J.J. Alberts, pp. 171–193. New York: Springer-Verlag.

Barrell, R., and Pain, N. eds. 1999. Innovation, investment and the diffusion of technology in Europe. Cambridge, UK: Cambridge University Press.

Beven, K.J. 2001. Rainfall-runoff modelling. The primer. Chichester, UK: John Wiley & Sons.

Billings, W.D. 1969. Vegetational pattern near alpine timberline as affected by fire-snowdrift interactions. Vegetation 19: 192–207.

Bormann, F.H., and Likens, G.E. 1979. Catastrophic disturbance and the steady state in northern hardwood forests. Am. Scientist 67: 660–669.

Bradbury, J.W., and Vehrencamp, S.L. 1998. Principles of animal communication, 1st edition. Sunderland, MA: Sinauer Associates.

Burrough, P.A., and McDonnell, R.A. 1998. Principles of geographical information systems. Oxford, UK: Oxford University Press.

Butler, D.R., Malanson, G.P., Bekker, M.F., and Resler, L.M. 2003. Lithologic, structural, and geomorphic controls on ribbon forest patterns in glaciated mountain environment. Geomorphology 1388: 1–15.

Chapin III, F.S., Matson, P.A., and Mooney, H.A. 2002. Principles of terrestrial ecosystem ecology. New York: Springer-Verlag.

Choy, B., and Reible, D.D. 2000. Diffusion models of environmental transport. Boca Raton, FL: Lewis Publishers.

Clarke, K., Parks, B., and Crane, M. 2000. Integrating geographic information systems (GIS) and environmental modeling. J. Environ. Manage. 59: 229–233.

Couclelis, H. 1999. Space, time, geography. In Geographical information systems, 2nd edition, eds. P. Longley, M.F. Goodchild, D.J. Maquire, and D.W. Rhind, pp. 29–38. New York: John Wiley & Sons.

Egenhofer, M.J., and Golledge, R.G. 1998. Spatial and temporal reasoning in geographic information systems. New York: Oxford University Press.

Endler, J.A. 1993. The color of light in forests and is implications. Ecol. Monogr. 63: 1–27.

Fischer M.M. 2000. Spatial interaction models and the role of geographic information systems. In Spatial models and GIS. New potential and new models, eds. A.S. Fotheringham and M. Wegener, pp 33–43. London: Taylor & Francis.

Fisher, S.G., and Welter, J.R. 2004. Flow paths as integrators of heterogeneity in streams and landscapes. In Ecosystem function in heterogeneous landscapes, eds. G.M. Lovett., C.G. Jones, M.G. Turner and K.C. Weathers, pp. 1–4. New York: Springer.

Foster, D.R., Knight, D.H., and Franklin, J.F. 1998. Landscape patterns and legacies resulting from large, infrequent forest disturbances. Ecosystems 1: 497–510.

Fotheringham, A.S. 2000. GIS-based spatial modelling: a step forward or a step backwards? In Spatial models and GIS. New potential and new models. eds. A.S. Fotheringham and M. Wegener, pp. 21–30. London: Taylor & Francis.

Garratt, J.R. 1992. The atmospheric boundary layer. Cambridge, UK: Cambridge University Press.

Gleick, J. 1987. Chaos. Making a new science. New York: Viking Press.

Harmon, R.S., and Doe III, W.W. (editors). 2001. Landscape erosion and evolution modeling. New York: Kluwer Academic Publishers.

Harris, C.J. 1979. Mathematical modelling of turbulent diffusion in the environment. London: Academic Press.

Hedin, L.O., von Fischer, J.C., Ostrum, N.E., Kennedy, B.P., Brown, M.G., and Robertson, G.P. 1998. Thermodynamic constraints on nitrogen transformations and other biogeochemical processes at soil-stream interfaces. Ecology 79: 684–703.

Hemond, H., and Fechner-Levy, E.J. 2000. Chemical fate and transport in the environment (2nd edition). San Diego, CA: Academic Press.

Henderson-Sellers, A., and McGuffie, K. 1987. A climate modelling primer: Research and developments in climate and climatology. New York: John Wiley and Sons.

Herlihy, A.T., Stoddard, J.L., and Johnson, C.B. 1998. The relationship between stream chemistry and watershed land cover data in the Mid-Atlantic Region, U.S. In Biogeochemical investigations at watershed, landscape, and regional scales, eds. R.K. Wieder, M. Novak and J. Cerny, pp. 377–386. Dordrecht, The Netherlands: Kluwer Academic Publishers.

Hiemstra, C.A., Liston, G.E., and Reiners, W.A. 2002. Snow redistribution by wind and interactions with vegetation at upper treeline in the Medicine Bow Mountains, Wyoming. Arctic Antarctic Alpine Res. 34: 262–73.

Hill, A.R., Labadia, C.F., and Sanmugadas, K. 1998. Hyporheic zone hydrology and nitrogen dynamics in relation to the streambed topography of a N-rich stream. Biogeochemistry 42: 285–310.

Holland, W.R. 1986. Quasigeostrophic modelling of eddy-resolved ocean circulation. In Advanced physical oceanographic numerical modelling, ed. J.J O'Brien, pp. 203–231. NATO ASI Series. v. Series C: Mathematics and Physical Sciences, No. 186. Dordrecht, The Netherlands: D. Reidel Publishing Company.

Holling, C.S. 1987. The resilience of terrestrial ecosystems: local surprise and global change. In Sustainable development of the biosphere, eds. W.C. Clark, and R.E. Nunn, pp. 292–320. Cambridge, UK: Cambridge University Press.

Isard, S.A., and Gage, S.H. 2001. Flow of life in the atmosphere. An airscape approach to understanding invasive organisms. East Lansing, MI: Michigan State University Press.

Kelmelis, J.A. 1998. Process dynamics, temporal extent, and causal propagation as the basis for linking space and time. In Spatial and temporal reasoning in geographic information systems, eds. M.J. Egenhofer and R.G. Golledge, pp. 94–103. New York: Oxford University Press.

Klausmeier, C.A. 1999. Regular and irregular patterns in semiarid vegetation. Science 284: 1826–1828.

Knight, D.H. 1994. Mountains and plains. The ecology of Wyoming landscapes. New Haven, CT: Yale University Press.

Konrad, S.K., Humphrey, N.F., Steig, E.J., Clark, D.H., Potter, Jr. N., and Pfeffer, W.T. 1999. Rock glacier dynamics and paleoclimatic implications. Geology 27: 1131–1134.

Leopold, A. 1949. A Sand County almanac. Oxford: Oxford University Press.

Liston, G.E., and Sturm, M. 1998. A snow-transport model for complex terrain. J. Glaciol. 44: 498–516.

Longley, P., Goodchild, M.F., Maquire, D.J., and Rhind, D.W., Eds. 1999. Geographical information systems. Principles and technical issues. (2nd edition). New York: John Wiley & Sons.

Longley, P.A., Goodchild, M.F., Maguire, D.J., and Rhind, D.W., editors. 2001. Geographic information systems and science. Chichester, UK: John Wiley & Sons, Ltd.

Ludwig, J.A., Wiens, J.A., and Tongway, D.J. 2000. A scaling rule for landscape patches and how it applies to conserving soil resources in savannas. Ecosystems 3: 84–97.

Massman W., Sommerfeld, R.A., Mosier, A.R., Zeller, K.F., Hehn, T.J., and Rochelle, S.G. 1997. A model investigation of turbulence-driven pressure pumping effects on the rate of diffusion of $CO_2$, $N_2O$ and $CH_4$ through layered snow packs. J. Geophys. Res. 102: 18,851–18,863.

Mellor, G.L., 1996. Introduction to physical oceanography. New York, USA: Springer-Verlag.

Monteith, J. L., and Unsworth, M. 1995. Principles of environmental physics, 2nd edition. London:Arnold.

Nakicenovic, N., and Grubler, A., editors. 1991. Diffusion of technologies and social behavior. New York: Springer-Verlag.

Okubo, A., and Levin, S.A. 2001. Diffusion and ecological problems: modern perspectives (2nd edition). New York: Springer-Verlag.

Pastor, J., Dewey, B., Moen, R., Mladenoff, D.J., White, M., and Cohen, Y. 1998. Spatial patterns in the moose-forest-soil ecosystem on Isle Royale, Michigan, USA. Ecol. Applications 8: 411–424.

Peterson, D.L., and Parker, V.T. 1998. Ecological scale. Theory and applications. New York: Columbia University Press.

Peuquet, D. 2002. Representations of space and time. New York, USA: Guilford Press.

Pinet, P.R. 1998. Invitation to oceanography. Sudbury, MA: Jones and Bartlett Publishers.

Reiners, W.A., and Driese, K.L. 2001. The propagation of ecological influences across heterogeneous environmental space. BioScience 51: 939–950.

Reiners, W.A., and Driese, K.L. 2003. Transport of energy, information and material through the biosphere. Annu. Rev. Environ. Resources 28: 107–135.

Reiners, W.A., and Driese, K.L. 2004. Propagation of ecological influences through environmental space. Cambridge, UK: Cambridge University Press.

Reiners, W.A., Liu, S., Gerow, K.G., Keller, M., and Schimel, D.S. 2002. Historical and future land use effects on $N_2O$ and NO emissions using an ensemble modeling approach: Costa Rica's Caribbean Lowlands as an example. Global Biogeochem. Cycles 16: 223–240.

Robertson, G. P. 1989. Nitrification and denitrification in humid tropical ecosystems: Potential controls on nitrogen retention. In Mineral nutrients in tropical forest and savanna ecosystems, ed. J. Proctor, pp. 55–69. Oxford: Blackwell Scientific Publications.

Rogowski, A.S., and Goyne, J.L. 2002. Modeling dynamic systems and four-dimensional geographic information systems. In Geographic information systems and environmental modeling, eds. B. Parks, M. Crane and K. Clarke, pp. 122–159. Upper Saddle River, NJ: Prentice Hall.

Romme, W.H. 1982. Fire and landscape diversity in Yellowstone National Park. Ecol. Monogr. 52: 199–221.

Schindler, J.E., and Krabbenhoft, D.P. 1998. The hyporheic zone as a source of dissolved organic carbon and carbon gases to a temperate forested stream. Biogeochemistry 43: 157–174.

Schlesinger, W.H., Raikes, J.A., Hartley, A.E., and Cross, A.F. 1996. On the spatial pattern of soil nutrients in desert ecosystems. Ecology 77: 364–374.

Schnoor, J.L. 1996. Environmental modeling. Fate and transport of pollutants in water, air, and soil. New York: John Wiley & Sons.

Smith, T.R., Birnir, B., and Merchant, G.E. 1997. Towards an elementary theory of drainage basin evolution: I. The theoretical basis. Computers Geosciences 23: 811–822.

Sprugel, D.G. 1976. Dynamic structure of wave-generated *Abies balsamea* forests in the northeastern United States. J. Ecol. 64: 889–891.

Summerfield, M.A. 1991. Global geomorphology. Harlow, UK: Longman.

Tongway, D., and Ludwig, J. 2004. Heterogeneity in arid and semi-arid lands. In Ecosystem function in heterogeneous landscapes, eds. G.M. Lovett, C.G. Jones, M.G. Turner and K.C. Weathers, pp. 1–4. New York: Springer.

Turchin, P. 1998. Quantitative analysis of movement. Measuring and modeling population redistribution in animals and plants. Sunderland, MA: Sinauer Associates.

Turner, M.G. 1989. Landscape ecology: the effect of pattern on process. Annu. Rev. Ecol. Systematics 20: 171–197.

Turner, M.G., Gardner, R.H., and O'Neill, R.V. 2001. Landscape ecology in theory and practice. Pattern and process. New York: Springer-Verlag.

Varma, A. 2002. Data sources and measurement technologies for modeling. In Geographic information systems and environmental modeling. eds. K. Clarke, B.O. Parks and M.P. Crane, pp. 67–99. Upper Saddle River, NJ: Prentice Hall.

Watt, A.S. 1947. Pattern and process in the plant community. J. Ecol. 35: 1–22.

White, P.S. 1979. Pattern, process and natural disturbance in vegetation. Botanical Rev. 45: 229–299.

Wiens, J.A. 1992. Ecological flows across landscape boundaries: a conceptual overview. In Landscape boundaries. Consequences for biotic diversity and ecological flows, eds. A.J. Hansen and F. diCastri, pp. 217–235. New York: Springer-Verlag.

Wiens, J.A. 2000. Ecological heterogeneity: an ontogeny of concepts and approaches. In The ecological consequences of habitat heterogeneity, eds.M.J. Hutchings, E.A. John and A.J.A. Stewart, pp. 9–31. Oxford: Blackwell Science.

Wu, X.B., Thurow, T.L., and Whisenant, S.G. 2000. Fragmentation and changes in hydrologic function of tiger bush landscapes, south-west Niger. J. Ecol. 88: 790–800.

Yaalon, D.H. (editor.) 1982. Aridic soils and geomorphic processes. Cremlingin, Germany: Catena Verlag.

# Section II

# Perspectives from Different Disciplines

# Editors' Introduction to Section II: Perspectives from Different Disciplines

Many different scientific disciplines have to deal with spatial heterogeneity as a normal part of their systems of study. In this section, we asked representatives of four different disciplines to discuss how spatial heterogeneity is treated in their discipline, particularly in conceptual and mathematical models. The disciplines we chose are all tangentially related to ecosystem science—close enough to be relevant, but distant enough to be instructive. The result is a series of four distinct papers, each illuminating in its own way.

Lenore Fahrig and Bill Nuttle (Chapter 6) discuss the role of spatial heterogeneity in population ecology, beginning by tracing developments that led from a nonspatial approach to a spatially explicit perspective in this field. They emphasize the importance of separating the effects of compositional and configurational heterogeneity of the landscape and hypothesize that the effects of composition will generally be more important in determining population persistence. They reason that the effects of landscape configuration will be mediated primarily through influence on organismal movement and suggest conditions under which this influence may be important for population persistence. They then extend these ideas to consider when landscape configuration may influence ecosystem processes.

Christina Tague (Chapter 7) reviews the importance of spatial heterogeneity in hydrological models, including heterogeneity in inputs and parameters as well as the heterogeneity in underlying physical processes. She provides an overview of the different approaches used to represent spatial heterogeneity in hydrologic models, including spatial averaging and the use of "effective parameters," probabilistic distributions of parameters, and aggregation and partitioning strategies. These modeling techniques should be very useful to ecosystem scientists, who have to deal with similar problems: enormous variation across multiple scales of interest and insufficient data to characterize the fine-scale variation. As Tague points out, development of coupled ecological-hydrological models is complicated but is likely to advance both disciplines.

David Smith (Chapter 8) points out some interesting parallels between epidemiology and ecology; for instance, from the point of view of an infectious agent, a host organism is a habitat patch, and determining who comes in contact with whom is analogous to configurational heterogeneity in a landscape. Smith discusses the overwhelming complexity of disease transmission and argues for parsimony in modeling it. As he says, "... heterogeneity should be weighed and ignored, whenever possible." Nonetheless, he notes that heterogeneity usually does matter in disease transmission, and understanding the influence of spatial processes on the nonlinear aspects of epidemics often requires a model. He summarizes various modeling approaches used in epidemiology and gives examples of case studies in which heterogeneity was found to be important

The final in this paper in this section (Chapter 9) is by an oceanographer, Amala Mahadevan. This paper could just as easily have been included in Section III (Illustrations of Heterogeneity and Ecosystem Function), but we chose to include it here because of its unique perspective in integrating temporal and spatial heterogeneity. Mahadevan views heterogeneity as a dynamic entity, constantly created and dissipated by processes in the upper ocean. She mathematically describes the balance between generation and dissipation of heterogeneity as a function of the scale of the process being considered and describes how heterogeneity can be shifted from one scale to another. She also discusses how nonlinearities in processes can make it difficult to scale up.

Taken together, these four papers present a broad range of techniques and perspectives that can be used to conceptualize and model spatial heterogeneity in ecosystem processes. Each paper emphasizes the complexity of fully incorporating spatial processes into conceptual and mathematical models, and each discusses approaches to simplification that make the problem tractable.

# 6
# Population Ecology in Spatially Heterogeneous Environments

Lenore Fahrig and William K. Nuttle

## Abstract

Historically, population ecologists have equated environmental spatial heterogeneity with habitat spatial structure. Early models represented habitat spatial structure simply as population subdivision into habitat patches. Later models included at first partially and then fully explicit representation of the spatial relationships among habitat patches. More recently, landscape population ecologists have broadened the view of spatial heterogeneity to include the composition and configuration of the whole landscape. A change in landscape composition refers to a change in the cover types in the landscape, the proportions of each, or both. A change in landscape configuration refers to a change in the spatial pattern of cover types, independent of any change in landscape composition. We hypothesize that changes in landscape composition generally have much larger effects on population persistence than changes in landscape configuration. Landscape configuration should have a large effect on population persistence when both (i) configuration has a large effect on among-patch movement of the organism and (ii) among-patch movement has a large effect on population persistence. The first condition should hold for species whose movement direction is constrained, and the second condition should hold either (i) when colonization of empty habitat is important for persistence or (ii) for species that require more than one type of habitat. We discuss extensions of these ideas to the effects of landscape configuration on ecosystem processes.

## Introduction

The potential effects of environmental spatial heterogeneity on population dynamics and interactions have been of concern to population ecologists for decades. In this chapter, we review the ways in which spatial heterogeneity of the environment has been incorporated in models of population dynamics and interactions. We then discuss the current view of spatial heterogeneity in landscape population ecology, and we review the evidence for effects of *compositional heterogeneity* and *configurational heterogeneity* on population ecology. Finally, we present a hypothesis that predicts the circumstances in which a change in landscape configuration should have a large effect on population ecology, and we discuss possible extensions of the hypothesis to effects of landscape configuration on ecosystem processes.

## History of Environmental Spatial Heterogeneity in Population Ecology

The ways in which population ecologists incorporate environmental heterogeneity into population models have changed markedly over time. In this section, we review the implicit and explicit representation of environmental spatial heterogeneity in models of population dynamics and population interactions. Our review is limited to models in which the underlying environment is spatially heterogeneous in some way. We do not include the many spatially explicit population models in which the underlying environment is assumed to be homogeneous, such as reaction-diffusion models of population spread in a homogeneous environment (e.g., Lewis 1997), cellular automata models of disease spread in a homogeneous environment (e.g., Holmes 1997), or models exploring the generation of population spatial pattern in a homogeneous environment (e.g., Pacala and Levin 1997). Note that this review is not exhaustive; we have selected representative examples for each method of incorporating environmental spatial heterogeneity into models. In each case, we focus on the earliest examples that we know of, even though all the views of spatial heterogeneity persist simultaneously in the current literature.

### *Population Subdivision*

The first theories of population ecology assumed spatial homogeneity of the environment (e.g., Verhulst 1838; Lotka 1925; Volterra 1926; Nicholson and Bailey 1935). However, with Gause's classic experiments in 1934, population biologists began to understand that population theories based on spatial homogeneity are likely to fail in the real world. Gause showed that a predator-prey relationship was "inherently self-annihilative"; it could persist only when a portion of the prey population was protected by a "privileged sanctuary," or when reintroductions of prey occurred at intervals. This implied that persistence of natural populations depends on environmental patchiness or spatial heterogeneity. Laboratory experiments by Huffaker (1958) and Pimentel (1963) supported this conclusion.

Theoretical examination of the influence of environmental spatial heterogeneity on populations began with models that represented spatial heterogeneity as habitat subdivision, resulting in separation of the population into a number of subpopulations inhabiting habitat patches (e.g., Levins 1969, 1970; Reddingius and den Boer 1970; Hassell and May 1973; Roff 1974a,b; Vandermeer 1973; Levin 1974; Slatkin 1974; Hastings 1977; den Boer 1981; Shmida and Ellner 1984; Chesson 1985). *Metapopulation* or *patch occupancy* models predicted the proportion of patches that were occupied, based on rates of local extinction and colonization (Figure 6.1). Local population dynamics were not included in these models; the patches were either occupied or not occupied. The rate of colonization of empty patches was assumed to be independent of the spatial location of the patch (i.e., the models were not spatially explicit). In patchy population models, the population was divided into a number of subpopulations within which population dynamics and interactions occurred. Dispersal between subpopulations

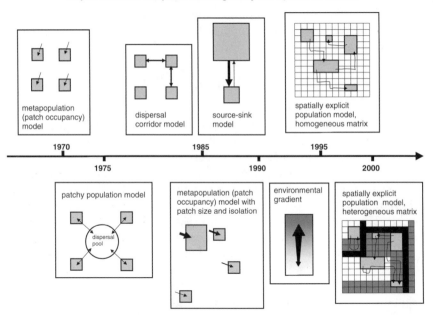

FIGURE 6.1. Representation of the progression of spatial heterogeneity in population models over four decades of ecological research. Time period for each model type represents the period over which it was established. Research using all model types continues to the present. Light gray rectangles represent habitat patches. In the metapopulation, or patch occupancy models, arrows represent colonization of patches; arrows only enter patches, to indicate that these models do not include emigration from patches. In the source-sink model, the patch sizes represent relative patch quality. In the spatially explicit models, the arrows represent movement paths of individuals. In the spatially explicit model with heterogeneous matrix, white, dark gray, and black areas represent matrix cover types; the black cover type represents a movement barrier (e.g., roads).

was "global," meaning that all patches were equally accessible to dispersers. Conceptually, this can be thought of as a *dispersal pool* into which a proportion of each subpopulation entered, and from which the dispersers were then redistributed among the subpopulations; again, these models were not spatially explicit (Figure 6.1).

This early theory suggested that under specific conditions, habitat subdivision could stabilize single-species population dynamics and species interactions. When local disturbances are asynchronous, population subdivision was predicted to stabilize single species dynamics by reducing the probability of simultaneous extinction of the whole population. Environmental patchiness was predicted to enhance the persistence of a predator-prey system if the prey species dispersed more readily than the predator species. Patchiness was also predicted to enhance two-species coexistence if there was a trade-off between dispersal rate and competitive ability. This trade-off, along with asynchronous disturbances that locally removed the superior competitor, would allow the inferior competitor (but superior disperser) to colonize the empty patches first, before being later displaced by the superior competitor.

## *Spatially Explicit Habitat Pattern*

Although the early theory did examine the effect of spatial heterogeneity *per se* (habitat subdivision or patchiness), it was not spatially explicit; the spatial relationships among subpopulations were not modeled. The first approaches to including such spatial relationships in a heterogeneous environment were intermediate between patch occupancy or patchy population models and fully spatially explicit models. In dispersal corridor models, the population was again assumed to be composed of several subpopulations in patches. However, dispersal was only possible between a pair of subpopulations if they were spatially connected (Figure 6.1). Spatial connection could represent patches that were close enough to each other for dispersal to occur or patches that were connected by a dispersal route or dispersal corridor. Lefkovitch and Fahrig (1985) used this type of model to predict that population persistence depends on the number of patches and how they are interconnected. The source-sink model (Pulliam 1988) was a version of the dispersal corridor model for a population divided into two linked subpopulations in patches of unequal quality. Dispersal between the subpopulations was asymmetric, with a higher dispersal rate from the high-quality patch to the low-quality patch (Figure 6.1). Source-sink models were conceptually the end-points of one-dimensional models in which habitat quality was represented as a continuous environmental gradient that influences dispersal rate (Thomas and Kunin 1999). Such models predicted that the interaction between organism movement and an environmental gradient can alter predator-prey dynamics (McLaughlin and Roughgarden 1991; Benson et al. 1993; Pascual and Caswell 1997).

Models in which immigration or colonization depended on patch size and isolation (e.g., Fahrig and Paloheimo 1988; Hanski 1991, 1994) also represented an intermediate approach between non-spatially-explicit patchy or metapopulation models and the truly spatially explicit models discussed below. They generally predicted that population persistence increases with increasing patch size and decreasing patch isolation. Particular patches (large, nonisolated ones) were predicted to be important for metapopulation persistence and persistence of systems of interacting species (e.g., Moilanen and Hanski 1995). Hanski (2001) labeled this type of model *spatially realistic*.

Although metapopulation models and patchy population models have continued to be used and developed in population ecology, fully spatially explicit population models, called *grid* or *lattice* models, have been used in population ecology since about the late 1980s (e.g., Nachman 1987; Fahrig 1991; Perry and Gonzalez-Andujar 1993; Dytham 1995; Wilson et al. 1998; Bonsall and Hassell 2000; Schiegg et al. 2002). These models represent the landscape as a spatial grid, in which each grid *cell* is either habitat or nonhabitat. Individuals or portions of the patch or cell populations move through the grid, according to movement parameters that determine movement distance and direction. Shugart (1998) labelled this type of model *interactive*.

A few grid models represent habitat quality as a continuous variable rather than the usual two-state variable (habitat or nonhabitat). For example, Colasanti and Grime (1993) assigned different resource levels to cells on a grid, arranged in a resource gradient. Engen et al. (2002) presented a model in which habitat quality varied continuously over the landscape and spatial heterogeneity was represented as spatial autocorrelation in local carrying capacities. Thomas and Kunin (1999) proposed representing habitat spatial heterogeneity in grid models by assigning a *neighborhood* value to each cell, which is a function of the distances to and qualities of all other cells on the grid.

Several studies have shown that the predictions of spatially explicit population models can be very different from the predictions of analogous non-spatially-explicit models (Adler and Nuernberger 1994; Bascompte and Solé 1994; Durrett and Levin 1994; Swihart et al. 2001; Buttel et al. 2002; Higgins and Cain 2002). For example, Swihart et al. (2001) compared predator-prey interactions in a patchy population model with global dispersal versus a spatially explicit model. They found large differences between the models in the predicted equilibrium levels of the predator and prey populations. The spatially explicit model predicted much higher abundances of the predator, and much lower sensitivity of the predator to habitat removal than did the non-spatially-explicit model. Higgins and Cain (2002) compared two-species competition in a metapopulation model and a spatially explicit model. They found that coexistence in the metapopulation model depended on a trade-off between competitive and dispersal abilities, whereas this trade-off was not necessary for coexistence to occur in the spatially explicit model.

What do spatially explicit population models predict regarding the effect of habitat spatial heterogeneity on population persistence? Spatial heterogeneity is typically highest at intermediate levels of habitat amount. Heterogeneity increases with increasing fragmentation of habitat, where fragmentation is defined as the breaking apart of habitat, independent of habitat loss (Figure 6.2). Spatially explicit population models predict that population persistence increases with increasing amount of habitat on the landscape and decreases with increasing fragmentation of the habitat (Henein et al. 1998; Hill and Caswell 1999; With and King 1999; Fahrig 2001; Flather and Bevers 2002). These models therefore predict that (i) a reduction in habitat from a high to a moderate amount (A/B to C/D in Figure 6.2) should produce a negative effect of increasing heterogeneity on population persistence, (ii) an increase in habitat from a low to a moderate amount (E/F to C/D in Figure 6.2) should produce a positive effect of increasing heterogeneity on population persistence, and (iii) a shift from low to high fragmentation (A/C/E to B/D/F in Figure 6.2) should produce a negative effect of increasing heterogeneity on population persistence.

Empirical studies confirm the predicted positive effect of habitat amount but do not generally confirm the predicted negative effect of habitat fragmentation (breaking apart of habitat; reviewed in Fahrig 2003). We are aware of 13 empirical studies of the effects of habitat fragmentation (independent of habitat amount) on the abundance and/or distribution of individual species (McGarigal and McComb 1995; Collins and Barrett 1997; Wolff et al. 1997; Collinge and Forman 1998; Meyer et al. 1998; Rosenberg et al. 1999; Trzcinski et al. 1999; Drolet et al. 1999; Flather et al. 1999; Villard et al. 1999; Caley et al. 2001; Langlois et al. 2001; Hovel and Lipcius 2001; reviewed in Fahrig 2003). In general, these studies indicate that habitat loss has a much larger effect than habitat fragmentation on population abundance and/or distribution. Of the species that were found to be affected by fragmentation, 9 showed declines and 17 showed increases in abundance or distribution with increasing fragmentation. Note that the observed positive effects of fragmentation cannot simply be explained as responses by "weedy," habitat generalist species. For example, McGarigal and McComb (1995) studied abundances of bird species that nest only in mature forest, in response to forest amount and fragmentation. They found that of the seven species that responded to fragmentation, six responded positively. Therefore, the direction of the relationship between habitat heterogeneity and population persistence is not consistently positive or negative. Possible explanations for both positive and negative effects of fragmentation are reviewed in Fahrig (2003).

For predator-prey or host-parasite interactions, increasing habitat heterogeneity by reducing habitat amount and/or increasing habitat fragmentation can result in outbreaks or persistently higher levels of the of the prey/host (Kareiva 1987; Roland 1993). It is hypothesized that habitat loss and fragmentation disrupt the ability of the predator or parasite to

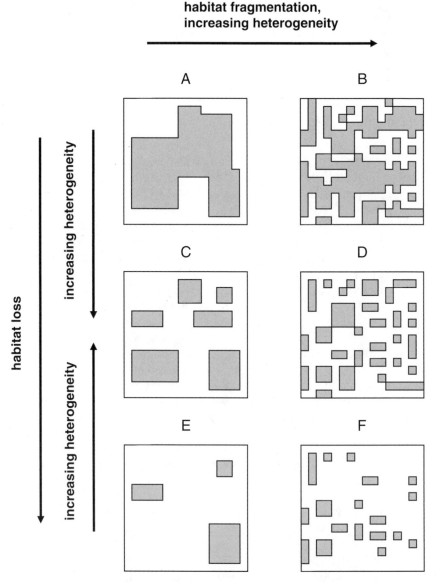

FIGURE 6.2. Effects of habitat fragmentation and habitat loss on habitat heterogeneity. Habitat heterogeneity increases from high to moderate habitat amount (from A to C, or B to D) and from low to moderate habitat amount (from E to C, or F to D), and increases with increasing habitat fragmentation (from A to B, or C to D, or E to F).

find and control the prey/host populations in time to avoid outbreaks. With et al. (2002) demonstrated the plausibility of this mechanism in an experimental study of the effects of habitat loss and fragmentation on patterns of aggregation of an insect predator-prey system. Some researchers have modeled effects of spatial heterogeneity of habitat on predator-prey interactions using reaction-diffusion equations, where diffusion occurs along an environmental gradient (McLaughlin and Roughgarden 1991; Benson et al. 1993; Pascual and Caswell 1997). These models predict that the environmental gradient interacts with organism movement to determine predator-prey dynamics, which can include outbreaks and chaos.

## Landscape Composition and Configuration

Most of the literature discussed so far deals implicitly or explicitly with only one kind of habitat, the habitat used by the species in question. Within this framework, consideration of spatial heterogeneity has increased over time from homogeneity to patchiness with global dispersal, to variation in patch sizes and connectedness, and finally to explicit spatial representation of the habitat on the landscape (Figures 6.1 and 6.2). The vast majority of current studies of the effect of environmental spatial pattern on population ecology still describe the landscape in terms of habitat and nonhabitat (nonhabitat is also called *matrix*).

In real landscapes, the matrix is not homogeneous but is composed of various cover types (final panel in Figure 6.1). Some of the cover types will represent habitat for the species in question. These may include different habitat cover types representing habitats that vary in quality resulting in, for example, different reproductive rates. Different cover types may also provide different types of resources that are needed at different times during the organism's life history (e.g., feeding habitat, mating habitat). Other cover types represent nonhabitat, which, again, may differ in quality, for example, in the probability of mortality of the organism while it is in the cover type.

What effect does taking account of this additional spatial heterogeneity have on our understanding of population ecology? Landscape ecologists describe landscape structure in terms of two main components: landscape composition and landscape configuration (Dunning et al. 1992; McGarigal and McComb 1995). Landscape composition refers to the different cover types present in the landscape and the proportions of each. Compositional landscape heterogeneity increases as the number of different cover types increases (Figure 6.3, from A to B or C to D), and if they occur in more similar proportions (Figure 6.3, from A to C or B to D). Compositional heterogeneity can be measured using, for example, the Shannon-Wiener diversity index applied to the number and proportions of cover types in the landscape (e.g., Jonsen and Fahrig 1997).

A change in landscape configuration refers to a change in the spatial pattern of cover types independent of any change in landscape composition (Figure 6.4). Configurational landscape heterogeneity increases with increasing interspersion of the different cover types, accompanied by increasing edge density in the landscape (Figure 6.4, from A to B or from C

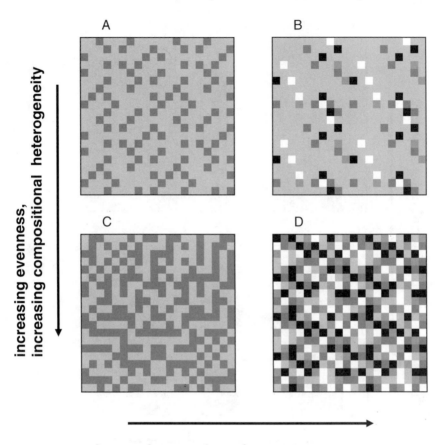

FIGURE 6.3. Illustration of the two components of compositional heterogeneity. Compositional heterogeneity increases with increasing number of cover types (from A to B, or C to D) and with increasing degree of evenness of representation of the cover types (from A to C, or B to D).

to D). Configurational heterogeneity can be measured using, for example, indices of edge density, shape complexity, edge contrast, and landscape subdivision (McGarigal 2002). Note that it is possible for landscape configuration to change without a change in landscape composition (Figure 6.4). Similarly, a change in the cover types while maintaining patch locations represents a change in landscape composition, with no change in landscape configuration. However, landscape composition and configuration are not completely independent; in particular, it is not possible to change the proportions of the different cover types (a change in composition) without changing landscape configuration.

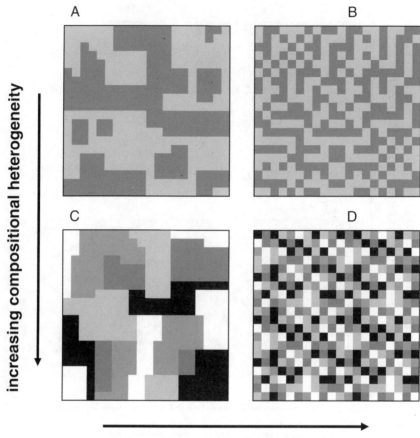

FIGURE 6.4. Illustration of configurational heterogeneity in comparison to compositional heterogeneity. (A) and (B) have the same compositional heterogeneity (50% of each of two cover types), but (B) has higher configurational heterogeneity than (A) because there is more interdigitation of the cover types. Similarly, (C) and (D) have the same compositional heterogeneity (20% of each of five cover types), but (D) has higher configurational heterogeneity than (C).

## Effects of Compositional and Configurational Heterogeneity on Population Ecology

There has to date been very little study of the effects of landscape heterogeneity on population ecology, so the following arguments represent mainly inference and conjecture. If the species relies on one kind of habitat only, then, as discussed above, an increase in compositional heterogeneity can imply a reduction in the amount of preferred habitat, which will cause a reduction in population persistence probability. For example, in Figure 6.3, landscape D contains less dark gray habitat than does landscape C, and D is more heterogeneous than C. If dark gray represents wetland and the species of interest relies solely on wetland habitat, this increase in heterogeneity will result in a reduction in the persistence probability of the population.

However, if the species relies on more than one kind of habitat, an increase in compositional heterogeneity may permit the species to persist in a landscape in which it otherwise would not. For example, Figure 6.3C contains no white habitat. If the species requires both dark gray and white habitats for persistence, it will not occur in landscape C but may occur in landscape D. This represents a positive effect of compositional heterogeneity on population persistence. The cooccurrence of two or more required habitat types within a landscape was labeled *landscape complementation* by Dunning et al. (1992). The importance of landscape complementation was demonstrated by Pope et al. (2000), who showed that leopard frog populations were more likely to occur in landscapes containing both large numbers of breeding ponds and accessible terrestrial habitat for foraging during the summer. Similarly, Thies and Tscharntke (2002) found that heterogeneous landscapes were more likely to harbor populations of parasites of an agricultural pest species than were homogeneous landscapes, presumably because the heterogeneous landscapes provided habitats containing resources for the parasites in addition to those offered by the habitats containing the pests.

Species that require landscape complementation may also be positively affected by increasing configurational heterogeneity. For the same landscape composition, a more heterogeneous landscape will have more interdigitation of different habitat types (Figure 6.4: from A to B, and from C to D). This should increase landscape complementation (Law and Dickman 1998; Tscharntke et al. 2002).

# Relative Effects of Landscape Composition and Configuration on Population Persistence: A Hypothesis

The only empirical studies of which we are aware that have examined the relative effects of landscape composition and landscape configuration are the studies discussed above (and reviewed in Fahrig 2003) on the effects of habitat fragmentation (a component of landscape configuration) and habitat loss (a component of landscape composition). These studies indicate that effects of fragmentation are generally much weaker than the effects of habitat loss. Does this conclusion hold for landscape composition and configuration in general? Are there situations in which the effect of landscape configuration is expected to be large? In this section, we propose a hypothesis for the conditions under which configuration should have a large effect on population persistence.

Before presenting the hypothesis, we emphasize that landscape spatial structure must be described from the point of view of the particular species of interest. For example, if a species of bird is susceptible to nest predation and nest predators occur preferentially in forest edges, then a configurational change to the forest that results in more forest edge (e.g., forest fragmentation) will have a negative effect on the bird species. However, because forest edges are of lower quality for the species than is the interior of forested areas, forest fragmentation also represents a compositional change to the landscape (i.e., a decrease in amount of high-quality habitat and an increase in amount of low-quality habitat). Therefore, to avoid potential ambiguity between composition and configuration effects, for this species it would be important to map forest edge as a separate cover type of lower quality. Conversely, if a species prefers habitat edges or shows higher growth rates in edges (e.g., Bowers and Dooley 1999), then edges should be mapped as a separate cover type of higher quality. The question can then be asked: Is there an effect of a change in landscape configuration (i.e., fragmentation) over and above the effect of changing landscape composition (i.e., increase in the amount of edge cover type)? As another example, for some species, very small patches of forest are of very low quality (Burke and Nol 2000), and patches smaller than some minimum patch size will not be occupied at all (Huhta et al. 1998). Such small patches should not be mapped as breeding habitat. In all of the discussion below, we are assuming that the landscape maps represent the landscape cover types correctly from the perspective of the particular species of interest.

Landscape composition has large, direct effects on population dynamics and persistence through its direct effects on reproduction and mortality. Landscape configuration, on the other hand, generally affects population dynamics indirectly through its effect on among-patch movement. To see this, imagine a species that does not move at all. Assume we begin with 100 individuals in each of landscapes A and B in Figure 6.4, equally divided between the light gray and dark gray areas (50 individuals in each). The overall reproductive rate and mortality rate will be exactly the same in the two landscapes, even though their configurations are very different. The only way that the difference in landscape configuration can affect population dynamics is if it affects among-patch movement and if among-patch movement affects population dynamics.

The effect of configuration on population persistence could also occur indirectly through its effect on among-patch movement of any mass, energy, or information that can influence the population in question. For example, if landscape configuration affects movement of a predator species, and predation by that species has a large effect on a prey species, this could produce an indirect effect of landscape configuration on the prey population dynamics. In this case, even though configuration is not affecting the movement of the prey, it affects the prey through its effect on the movement of the predator. Similarly, a population of denitrifying bacteria may indirectly be affected by landscape configuration if landscape configuration affects among-patch movement of nitrate. This leads to the interesting conclusion that landscape configuration could indirectly affect a population through its effect on an ecosystem process. In the section "Application to Ecosystem Processes" below, we present some examples of how landscape configuration might affect ecosystem processes. In the following, we present our hypothesis ignoring these indirect effects; we limit our consideration of movement to the movement of the organism in question.

We hypothesize that the effect of landscape configuration on population persistence is through its effect on (organism) movement, either facilitating or hindering habitat accessibility. Landscape configurations that facilitate habitat accessibility can indirectly increase the number of births and decrease the number of deaths in the population. This can occur through two processes, "landscape complementation" and "landscape supplementation" (Dunning et al. 1992). As discussed above, landscape complementation occurs when all required cover types are accessible to an organism that needs more than one landscape cover type to complete its life history. Landscape supplementation occurs when the organism can move among several resource patches of the same type to obtain sufficient resources for survival and reproduction. In either case, landscape configuration may facilitate or limit the ability of the organism to move about and obtain the resources required to avoid mortality and to reproduce successfully. For example, if roads represent a barrier to movement of the organism, then the particular placement of roads on a landscape may affect the ability of the organism to obtain crucial resources, which will affect the reproduction and/or mortality rate of the population, ultimately affecting its persistence.

Landscape configuration affects among-patch movement within the landscape when movement direction is highly constrained by the landscape. For example, some species are very reluctant to cross certain types of boundaries in the landscape (Tischendorf 2001). If the probability of crossing a boundary into a particular cover type (e.g., road surface) is low, this cover type represents a movement barrier in the landscape. If an organism is very reluctant to cross the boundary of its habitat into matrix, the configuration of habitat can have a large effect on population persistence. In this case, each habitat patch is isolated, so the persistence of the population in the landscape depends on the size of the largest piece of habitat (Figure 6.5). Movement within a stream network represents another example of highly constrained movement; Cumming (2002) showed that the form of the stream network can have a large effect on overall movement rate through the network.

Recent simulation studies suggest that strong effects of boundary type on boundary-crossing rates leads to a large effect of landscape configuration on among-patch movement rate through the landscape. Goodwin and Fahrig (2003) conducted simulations of animal movement on a grid containing habitat and two matrix cover types. They assumed that animal movement

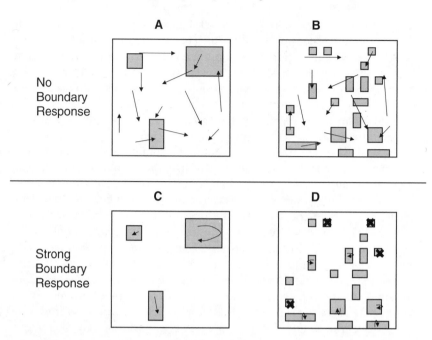

FIGURE 6.5. Comparison of the effect of landscape configuration on a species that does not respond to boundaries (top) *versus* a species that will not cross the habitat-matrix boundary (bottom). Landscape configuration has no effect on persistence of the species with no boundary response (A *vs.* B). For the species with strong boundary response, population persistence will be higher in the less fragmented configuration (C) than the more fragmented configuration (D).

rates and directionalities differed between the matrix cover types but that the animal showed no boundary responses. In contrast, Bender (2002) conducted simulations of animal movement in which different matrix cover types elicited different boundary-crossing probabilities by the simulated organism. Goodwin and Fahrig found no effect of matrix configuration on among-patch movement rate, whereas Bender found a very large effect of matrix configuration on among-patch movement rate. This suggests that landscape configuration is likely to have a large effect on movement rate for organisms that show strong behavioral responses to boundaries in the landscape.

A second way in which constrained movement can create an effect of landscape configuration on population persistence is when movement has an overall direction within the landscape. For example, if larval fish can only move downstream, the relative position of larval habitats and spawning habitats within the stream system can have a large effect on population persistence. Stream systems with larval habitat downstream relative to spawning habitat are more likely to contain viable populations than stream systems with larval habitat upstream of spawning habitat, even if they have the same amount of habitat (M. Power, personal communication).

It is important to note here that a large effect of landscape configuration on among-patch movement does not necessarily imply a large effect of landscape configuration on population persistence, relative to the effect of landscape composition, for several reasons. First, landscape composition also affects among-patch movement. Two empirical studies have examined the independent effects of landscape composition (habitat amount) and configuration (habitat fragmentation) on animal movement (Bélisle et al. 2001; With et al. 2002). Both found much larger effects of composition than configuration on movement. Second, population persistence at the landscape scale is ultimately determined by numbers of births and deaths, not movement within the landscape. Movement of an individual from one location to another within the landscape does not by itself affect overall population size. It can only affect population size indirectly if, by entering a new location, the individual changes its chance of reproducing or surviving. Landscape composition affects births and deaths directly, as well as indirectly through its effect on animal movement. Landscape configuration, on the other hand, only affects births and deaths indirectly through its effect on movement. Finally, some theoretical studies predict an interaction effect between composition and configuration, in which configuration affects population persistence only below certain threshold composition values (Fahrig 1998; Flather and Bevers 2002).

If landscape configuration affects population persistence through its effect on among-patch movement, then landscape configuration should have a large effect on population persistence when both (i) configuration has a large effect on among-patch movement and (ii) among-patch movement has a large effect on population persistence.

The hypothesis can be summarized as follows. (1) Landscape composition generally has a much larger effect than landscape configuration on population persistence, because composition directly affects births and deaths, whereas configuration only affects births and deaths indirectly through its effect on movement. (2) Landscape configuration has a large effect on among-patch movement for species whose movement direction is highly constrained. (3) Among-patch movement has a large effect on population persistence (i) when colonization of empty habitat is important for persistence or (ii) for species that require more than one type of habitat (landscape complementation). Note that condition (i) will occur when the probability of local extinction is high (e.g., high seasonal mortality) and following habitat restoration (Huxel and Hastings 1999). (4) Finally, landscape configuration has a large effect on population persistence when conditions under both (2) and (3) hold simultaneously. We emphasize that this is a general hypothesis; it is not limited to any particular sorts of species or landscapes, but it does depend on the assumption that the landscapes are correctly mapped from the point of view of the species of interest (see above). Interestingly, this hypothesis is not consistent with metapopulation theory, which predicts large effects of landscape configuration on population persistence, even for randomly moving organisms (Hill and Caswell 1999); a possible explanation for this difference is discussed in Fahrig (2002).

Recall here that the effect of landscape configuration on a population can also occur indirectly through its effect on movement of an interacting species or other mass, energy, or information that affects the species of interest. Our hypothesis can therefore be extended to state that landscape configuration can have a large effect on a population when both (i) movement direction of the interacting species, other mass, energy, or information is highly constrained, and (ii) the interacting species, other mass, energy, or information has a large effect on population persistence.

At this point, our hypothesis is supported only through the verbal arguments above; to date it has not been directly tested. Empirical testing will require comparisons across species and across landscapes. For example, we may know from previous studies that small mammal populations undergo frequent local extinctions (Merriam and Wegner 1992), which implies that movement is important for small mammal population persistence. From our hypothesis, we could then predict that the abundance of a small mammal species that shows strong avoidance of roads (i.e., its movement is highly constrained by roads) should be affected by a change in landscape configuration, whereas the abundance of a small mammal species that shows no behavioral response to roads should not be affected by a change in landscape configuration. To test this prediction, we would need first to study small mammal movement responses to roads to identify species that avoid roads and species that do not respond to roads. We would then compare the abundances of these small mammal species across a set of landscapes that vary in configuration (e.g., degree of habitat fragmentation). It will be important to select the landscapes in such a way that we can control for the effects on small mammal abundance of both road density and the amount of small mammal habitat (e.g., forest) in the landscapes. In particular, it is important that neither of these landscape composition variables is correlated with landscape configuration across the set of landscapes chosen for the study. There are clearly several challenges inherent in this type of research; these are discussed in Brennan et al. (2002).

## Application to Ecosystem Processes

We suggest that our hypothesis can be extended to the effects of landscape composition and configuration on ecosystem processes. A few examples illustrate parallels among the movement and persistence of organisms on the landscape and the flux and transformation of energy and nutrients that control ecosystem processes. Here, too, the composition of the landscape exerts the most direct influence over the net ecosystem functions of the entire landscape. Indeed, the trophic state of an ecosystem (i.e., oligotrophic, mesotrophic, or eutrophic) is often defined in terms of the type and biomass density of the primary producers (i.e., composition).

Landscape configuration may influence ecosystem function, depending on the distribution of areas of production and uptake relative to the pattern of movement of nutrients and energy, which may be affected by barriers that impede these fluxes. For example, consider the net processing of nutrients that move through an ecosystem in surface water and groundwater. Clearly, there will be a different effect on nutrient processing of wetlands taking up nutrients mobilized from farm fields, depending on whether the wetlands are located generally upstream or downstream of the farm fields. Where the conformation of the landscape generally acts to retard flow, nutrients are more subject to uptake and transformation by vegetation and microbial processes or to sequestration by sorption and sedimentation (Vollenweider 1975; Seitzinger 2000; Mitsch et al. 2001). Indeed, reengineering the landscape to promote these processes constitutes one of the principal tools used in environmental remediation and restoration (Mitsch et al. 2001; NRC 2002; Toth et al. 2002). Barriers to the movement of organisms can also affect ecosystem function. For example, fencing to exclude direct access by livestock to natural water bodies is an effective strategy for reducing the flux of nutrients into these surface water bodies and consequent effects on water quality and ecosystem processes downstream (NRC 2002; Mitsch et al. 2001). In the ocean, the seasonal development of strong, thermal stratification constrains phytoplankton from moving below the photic zone, and this triggers the increase in primary productivity observed as the spring plankton bloom (Sverdrup 1953). All these examples suggest that landscape configuration has its largest effect on ecosystem processes in situations where movement is constrained and/or directional.

The hypothesis may also apply to the propagation of disturbances across a landscape. For example, it appears that landscape configuration affects the spread of forest fire only when the fire is strongly limited by forest boundaries. In this situation, the total amount of forest burned is lower in landscapes where the forest is fragmented into small patches than in landscapes where the forest occurs in large tracts. Fires that start in small patches are unlikely to spread to the rest of the forest because there is low fuel availability between forest patches (Weir et al. 2000; Ricotta et al. 2001; Pitkänen et al. 2003). However, this boundary response can be reduced or eliminated in high winds, in very dry weather conditions, and in landscapes where there is a small difference in fuel load across the edge (Hargrove et al. 2000; Bessie and Johnson 1995; Moritz 2003). In these conditions, landscape configuration is unlikely to affect fire spread (Ricotta et al. 2001). Thus, landscape configuration only affects fire spread in situations where fire movement is constrained by forest boundaries (Figure 6.5, where the arrows now represent movement of fire rather than movement of organisms).

## Temporal Heterogeneity

In this paper, we have discussed the effects of spatial heterogeneity on population ecology. We do not mean to imply that temporal heterogeneity is unimportant. A small number of studies (all theoretical) have examined the combined effects of spatial heterogeneity and temporal heterogeneity on population persistence and population interactions. In general, these studies find that the rate and frequency of change of the landscape is extremely important. Fahrig (1992) and Bhar and Fahrig (1998) predicted that the rate of change of the landscape is much more important than habitat configuration in affecting population persistence. Keymer et al. (2000) predicted that the rate of landscape change has a large effect on the extinction threshold (i.e., the minimum amount of habitat required for population persistence). Finally, Bowers and Harris (1994) and Gourbiere and Gourbiere (2002) predicted that the outcome of interspecific competition depends strongly on the rate of environmental change.

## Conclusions

The incorporation of environmental spatial heterogeneity into population ecology has been a gradual process over a period of several decades. The concept itself has evolved from simple population subdivision, to effects of patch size and isolation, to spatially explicit representations of habitat, to spatially explicit representations of landscapes. At each level of heterogeneity, there are important predicted effects on population ecology. The overall picture is quite complicated, however, because although the way that population ecologists view environmental spatial heterogeneity has changed over time (Figure 6.1), all these views persist simultaneously in the current literature. In addition, the characteristics of the species studied (e.g., movement behavior) influence how the different aspects of spatial heterogeneity affect a population. Successful generalization will depend on cross-study comparisons, which, in turn, will depend on clear delineation of the important aspects of heterogeneity and species attributes. For example, it will be important to differentiate clearly between compositional heterogeneity and configurational heterogeneity, as these two aspects can have different predicted effects (Fahrig 2003).

Nevertheless, some tentative generalizations are possible. First, where increasing compositional heterogeneity reduces the amount of habitat available for a species, this will have a negative effect on population persistence. Second, information on organism movement behavior, in particular the responses of organisms to boundaries, will be important for predicting the likely effect of configurational heterogeneity on population ecology. Third, species that require landscape complementation can benefit from increases in both compositional and configurational heterogeneity. Landscape complementation will therefore be a central issue in developing a general understanding of the effects of spatial heterogeneity on population ecology.

*Acknowledgments.* We thank the Landscape Ecology Laboratory at Carleton University for comments on an earlier draft, particularly Julie Brennan, Neil Charbonneau, Stéphanie Duguay, Kringen Henein, Jeff Holland, Jochen Jaeger, Rachelle McGregor, Lutz Tischendorf, and Rebecca Tittler. Discussions with Gary Lovett, Hugh Possingham, and Mary Power and the comments of two anonymous reviewers were also very helpful. This work was funded by a grant from the Natural Sciences and Engineering Research Council of Canada.

## References

Adler, F.R., and Nuernberger, B. 1994. Persistence in patchy irregular landscapes. Theor. Popul. Biol. 45: 41–75.

Bascompte, J., and Sole, R.V. 1994. Spatially induced bifurcations in single-species population dynamics. J. Anim. Ecol. 63: 256–264.

Bélisle, M., Desrochers, A., and Fortin, M.-J. 2001. Influence of forest cover on the movements of forest birds: a homing experiment. Ecology 82: 1893–1904.

Bender, D.J. 2000. Wildlife movement in fragmented habitats: the influence of landscape complexity. Dissertation. Ottawa, Canada: Carleton University.

Benson, D.L., Sherratt, J.A., and Maini, P.K. 1993. Diffusion driven instability in an inhomogeneous domain. Bull. Math. Biol. 55: 365–384.

Bessie, W.C., and Johnson, E.A. 1995. The relative importance of fuels and weather on fire behaviour in subalpine forests. Ecology 76: 747–762.

Bhar, R., and Fahrig, L. 1998. Local vs. landscape effects of woody field borders as barriers to crop pest movement. Conservation Ecol. 2: 3. Available at http://www.consecol.org/vol2/iss2/art3.

Bonsall, M.B., and Hassell, M.P. 2000. The effects of metapopulation structure on indirect interactions in host-parasitoid assemblages. Proc. R. Soc. London, Ser. B: Biol. Sci. 267: 2207–2212.

Bowers, M.A., and Dooley, J.L. 1999. A controlled, hierarchical study of habitat fragmentation: responses at the individual, patch, and landscape scale. Landscape Ecol. 14: 381–389.

Bowers, M.A., and Harris, L.C. 1994. A large-scale metapopulation model of interspecific competition and environmental change. Ecol. Modelling 72: 251–273.

Brennan, J.M., Bender, D.J., Contreras, T.A., and Fahrig, L. 2002. Focal patch landscape studies for wildlife management: Optimizing sampling effort across scales. In Integrating landscape ecology into natural resource management, eds. J. Liu and W.W. Taylor, pp. 68–91. Cambridge, MA: Cambridge University Press.

Burke, D.M., and Nol, E. 2000. Landscape and fragment size effects on reproductive success of forest-breeding birds in Ontario. Ecol. Applications 10: 1749–1761. 2000.

Buttel, L.A., Durrett, R., and Levin, S.A. 2002. Competition and species packing in patchy environments. Theor. Popul. Biol. 61: 265–276.

Caley, M.J., Buckley, K.A., and Jones, G.P. 2001. Separating ecological effects of habitat fragmentation, degradation, and loss on coral commensals. Ecology 82: 3435–3448.

Chesson, P.L. 1985. Coexistence of competitors in spatially and temporally varying environments: a look at the combined effects of different sorts of variability. Theor. Popul. Biol. 28: 263–87.

Colasanti, R.L., and Grime, J.P. 1993. Resource dynamics and vegetation processes: A deterministic model using two-dimensional cellular automata. Functional Ecol. 7: 169–176.

Collinge, S.K., and Forman, R.T.T. 1998. A conceptual model of land conversion processes: predictions and evidence from a microlandscape experiment with grassland insects. Oikos 82: 66–84.

Collins, R.J., and Barrett, G.W. 1997. Effects of habitat fragmentation on meadow vole (Microtus pennsylvanicus) population dynamics in experiment landscape patches. Landscape Ecol. 12: 63–76.

Cumming, G.S. 2002. Habitat shape, species invasions, and reserve design: insights from simple models. Conservation Ecology 6: 3. Available at http://www.consecol.org/vol6/iss1/art3.

den Boer P.J. 1981. On the survival of populations in a heterogeneous and variable environment. Oecologia 50: 39–53.

Drolet, B., Desrochers, A., and Fortin, M.-J. 1999. Effects of landscape structure on nesting songbird distribution in a harvested boreal forest. Condor 101: 699–704.

Dunning, J.B., Danielson, B.J., and Pulliam, H.R. 1992. Ecological processes that affect populations in complex landscapes. Oikos 65: 169–175.

Durrett, R., and Levin, S. 1994. The importance of being discrete (and spatial). Theor. Popul. Biol. 46: 363–394.

Dytham, C. 1995. The effect of habitat destruction pattern on species persistence: A cellular model. Oikos 74: 340–344.

Engen, S., Lande, R., and Sæther, B.-E. 2002. Migration and spatiotemporal variation in population dynamics in a heterogeneous environment. Ecology 83: 570–579.

Fahrig, L. 1991. Simulation methods for developing general landscape-level hypotheses of single species dynamics. Pp. 417–442 in: Turner, M.G., and R.H. Gardner (eds.). Quantitative methods in landscape ecology. New York, NY: Springer Verlag.

Fahrig, L. 1992. Relative importance of spatial and temporal scales in a patchy environment. Theor. Popul. Biol. 41: 300–314.

Fahrig, L. 1998. When does fragmentation of breeding habitat affect population survival? Ecol. Modelling 105: 273–292.

Fahrig. L. 2001. How much habitat is enough? Biol. Conservation 100: 65–74.

Fahrig, L. 2002. Effect of habitat fragmentation on the extinction threshold: a synthesis. Ecol. Applications 12: 346–353.

Fahrig, L. 2003. Effects of habitat fragmentation on biodiversity. Annu. Rev. Ecol. Systematics 34: 487–515.

Fahrig, L., and Paloheimo, J.E. 1988. Determinants of local population size in patchy habitats. Theor. Popul. Biol. 34: 194–213.

Flather, C.H., and Bevers, M. 2002. Patchy reaction-diffusion and population abundance: the relative importance of habitat amount and arrangement. Am. Naturalist 159: 40–56.

Flather, C.H., Bevers, M., Cam, E., Nichols, J., and Sauer, J. 1999. Habitat arrangement and extinction thresholds: do forest birds conform to model predictions?

in: Wiens, J.A., and Moss, M.R., eds. International Association for Landscape Ecology 5th World Congress, Issues in Landscape Ecology, Snowmass Village, Colorado, July 29-August 3, 1999, Proceedings: Guelph, Ontario, Canada, The International Association for Landscape Ecology, p. 44–45.

Goodwin, B.J., and Fahrig, L. 2002. How does landscape structure influence landscape connectivity? Oikos 99: 552–570.

Gourbiere, S., and Gourbiere, F. 2002. Competition between unit-restricted fungi: a metapopulation model. J. Theor. Biol. 217: 351–368.

Hanski, I. 1991. Single-species metapopulation dynamics: concepts, models and observations. Biol. J. Linn. Soc. 42: 17–38.

Hanski, I. 1994. A practical model of metapopulation dynamics. J. Anim. Ecol. 63: 151–162.

Hanski, I. 2001. Spatially realistic theory of metapopulation ecology. Naturwissenschaften 88: 372–381.

Hargrove, W.W., Gardner, R.H., Turner, M.G., Romme, W.H., and Despain, D.G. 2000. Simulating fire patterns in heterogeneous landscapes. Ecol. Modelling 135: 243–263.

Hassell, M.P., and May, R.M. 1973. Stability in insect host-parasite models. J. Anim. Ecol. 42: 693–726.

Hastings, A. 1977. Spatial heterogeneity and the stability of predator-prey systems. Theor. Popul. Biol. 12: 37–48.

Henein, K., Wegner, J., and Merriam, G. 1998. Population effects of landscape model manipulation on two behaviourally different woodland small mammals. Oikos 81: 168–186.

Hill, M.F., and Caswell, H. 1999. Habitat fragmentation and extinction thresholds on fractal landscapes. Ecol. Lett. 2: 121–127.

Higgins, S.I., and Cain, M.L. 2002. Spatially realistic plant metapopulation models and the colonization—competition trade-off. J. Ecol. 90: 616–626.

Holmes, E.E. 1997. Basic epidemiological concepts in a spatial context. In Spatial ecology, eds. D. Tilman and P. Kareiva, pp. 11–136. Princeton, N.J: Princeton University Press.

Hovel, K.A., and Lipcius, R.N. 2001. Habitat fragmentation in a seagrass landscape: patch size and complexity control blue crab survival. Ecology 82: 1814–1829.

Huffaker, C.B. 1958. Experimental studies on predation: dispersion factors and predator-prey oscillations. Hilgardia 27: 343–83.

Huhta, E., Jokimaeki, J., and Rahko, P. 1998. Distribution and reproductive success of the Pied Flycatcher *Ficedula hypoleuca* in relation to forest patch size and vegetation characteristics; the effect of scale. Ibis 140: 214–222.

Huxel, G.R., and Hastings, A. 1999. Habitat loss, fragmentation, and restoration. Restoration Ecol. 7: 309–315.

Jonsen, I.D., and Fahrig, L. 1997. Response of generalist and specialist insect herbivores to landscape spatial structure. Landscape Ecol. 12: 187–195.

Kareiva, P. 1987. Habitat fragmentation and the stability of predator-prey interactions. Nature 326: 388–390.

Keymer, J.E., Marquet, P.A., Velasco-Hernandez, J.X., and Levin, S.A. 2000. Extinction thresholds and metapopulation persistence in dynamic landscapes. Am. Naturalist 156: 478–494.

Langlois, J.P., Fahrig, L., Merriam, G., and Artsob, H. 2001. Landscape structure influences continental distribution of hantavirus in deer mice. Landscape Ecol. 16: 255–266.

Law, B.S., and Dickman, C.R. 1998. The use of habitat mosaics by terrestrial vertebrate fauna: implications for conservation and management. Biodiversity Conservation 7: 323–33.

Lefkovitch, L.P., and Fahrig, L. 1985. Spatial characteristics of habitat patches and population survival. Ecol. Modelling 30: 297–308.

Levin, S.A. 1974. Dispersion and population interactions. Am. Naturalist 108: 207–28.

Levins, R. 1969. Some demographic and genetic consequences of environmental heterogeneity for biological control Bull. Entomol. Soc. Am. 15: 237–240.

Levins, R. 1970. Extinction. Lecture notes on mathematics in the life sciences 2, ed. M. Gerstenhaber, pp. 77–107. Providence, RI: American Mathematics Society.

Lewis, M.A. 1997. Variability, patchiness, and jump dispersal in the spread of an invading population. In Spatial ecology, eds. D. Tilman and P. Kareiva, pp. 46–69. Princeton, NJ: Princeton University Press.

Lotka, A.J. 1925. Elements of physical biology. Baltimore: Williams and Wilkins.

McGarigal, K. 2002. FRAGSTATS conceptual background: landscape pattern metrics. Available at http://www.umass.edu/landeco/research/fragstats/documents/ Conceptual%20Background/Background%20TOC.htm.

McGarigal, K., and McComb, W.C. 1995. Relationships between landscape structure and breeding birds in the Oregon coast range. Ecol. Monogr. 65: 235–260.

McLaughlin, J.F., and Roughgarden, J. 1991. Pattern and stability in predator-prey communities: How diffusion in spatially variable environments affects the Lotka-Volterra model. Theor. Popul. Biol. 40: 148–172.

Merriam, G., and Wegner, J. 1992. Local extinctions, habitat fragmentation, and ecotones. In Landscape boundaries: consequences for biotic diversity and ecological flows, eds. A.J. Hansen and F. di Castri, pp. 150–159. New York: Springer-Verlag.

Meyer, J.S., Irwin, L.L., and Boyce, M.S. 1998. Influence of habitat abundance and fragmentation on Northern Spotted Owls in Western Oregon. Wildlife Monogr. 139: 1–51.

Mitsch, W.J., Day, J.W., Gilliam, J.W., Groffman, P.M., Hey, D.L., Randall, G.W., and Wang, N. 2001. Reducing nitrogen loading to the Gulf of Mexico from the Mississippi River basin: strategies to counter a persistent ecological problem. BioScience 51: 373–388.

Moilanen, A., and Hanski, I. 1995. Habitat destruction and coexistence of competitors in a spatially realistic metapopulation model. J. Anim. Ecol. 64: 141–144.

Moritz, M.A. 2003. Spatiotemporal analysis of controls on shrubland fire regimes: age dependency and fire hazard. Ecology 84: 351–361.

Morris, J.T. 2000. Effects of sea-level anomalies on estuarine processes. In Estuarine science: a synthetic approach to research and practice, ed. J.E. Hobbie, pp. 107–128. Washington, DC: Island Press.

Nachman, G. 1987. Systems analysis of acarine predator-prey interactions. I. A stochastic simulation model of spatial processes. J. Anim. Ecol. 56: 247–265.

National Research Council (NRC). 2002. Riparian areas: functions and strategies for management. Washington, DC: National Academy Press.

Nicholson, A.J., and Bailey, V.A. 1935. The balance of animal populations. Proc. Zool. Soc. London. 1: 551–598.

Pacala, S.W., and Levin, S.A. 1997. Biologically generated spatial pattern and the coexistence of competing species. In Spatial ecology, eds. D. Tilman and P. Kareiva, pp. 204–232. Princeton, NJ: Princeton University Press.

Pascual, M., and Caswell, H. 1997. Environmental heterogeneity and biological pattern in a chaotic predator-prey system. J. Theor. Biol. 185: 1–13.

Perry, J.N., and Gonzalez-Andujar, J.L. 1993. Dispersal in a metapopulation neighbourhood model of an annual plant with a seedbank. J. Ecol. 81: 453–463.

Pimentel, D., Nagel, W.P., and Madden, J.L. 1963. Space-time structure of the environment and the survival of parasite-host systems. Am. Naturalist 97: 141–167.

Pitkänen, A., Huttunen, P., Tolonen, K., and Jungner, H. 2003. Long-term fire frequency in the spruce-dominated forests of the Ulvinsalo strict nature reserve, Finland. Forest Ecol. Manage. 176: 305–319.

Pope, S.E., Fahrig, L., and Merriam, H.G. 2000. Landscape complementation and metapopulation effects on leopard frog populations. Ecology 81: 2498–2508.

Pulliam, H.R. 1988. Sources, sinks, and population regulation. Am. Naturalist 132: 652–661.

Reddingius J., and den Boer, P.J. 1970. Simulation experiments illustrating stabilization of animal numbers by spreading of risk. Oecologia 5: 240–84.

Ricotta, C., Arianoutsou, M., Díaz-Delgado, R., Duguy, B., Lloret, F., Maroudi, E., Mazzoleni, S., Moreno, J.M., Rambal, S., Vallejo, R., and Vázquez, A. 2001. Self-organized criticality of wildfires ecologically revisited. Ecol. Modelling 141: 307–311.

Roff, D.A. 1974a. Spatial heterogeneity and the persistence of populations. Oecologia 15: 245–58.

Roff, D.A. 1974b. The analysis of a population model demonstrating the importance of dispersal in a heterogeneous environment. Oecologia 15: 259–75.

Roland, J. 1993. Large-scale forest fragmentation increases the duration of tent caterpillar outbreak. Oecologia 93: 25–30.

Rosenberg, K.V., Lowe, J.D., and Dhondt, A.A. 1999. Effects of forest fragmentation on breeding tanagers: a continental perspective. Conservation Biol. 13: 568–583.

Schiegg, K., Walters, J.R., and Priddy, J.A. 2002. The consequences of disrupted dispersal in fragmented red-cockaded woodpecker Picoides borealis populations. J. Anim. Ecol. 71: 710–721.

Seitzinger, S.P. 2000. Scaling up: Site-specific measurements to global-scale estimates of denitrification. In Estuarine science: a synthetic approach to research and practice, ed. J.E. Hobbie, pp. 211–240. Washington, DC: Island Press.

Shmida, A., and Ellner, S. 1984. Coexistence of plant species with similar niches. Vegetatio 58: 29–55.

Shugart, H.H. 1998. Terrestrial ecosystems in changing environments. New York: Cambridge University Press. 537 pp.

Slatkin, M. 1974. Competition and regional coexistence. Ecology 55: 128–34

Sverdrup, H.V. 1953. On the conditions for the vernal blooming of phytoplankton. Journal du Conseil. Conseil International pour l'Exploration de la Mer 18: 287–295.

Swihart, R.K., Feng, Z., Slade, N.A., Mason, D.M., and Gehring, T.M. 2001. Effects of Habitat Destruction and Resource Supplementation in a Predator-Prey Metapopulation Model. J. Theor. Biol. 210: 287–303.

Thies, C., and Tscharntke, T. 1999. Landscape structure and biological control in agroecosystems. Science 285: 893–895.

Thomas, C.D., and Kunin, W.E. 1999. The spatial structure of populations. J. Anim. Ecol. 68: 647–657.

Tischendorf, L. 2001. Can landscape indices predict ecological processes consistently? Landscape Ecol. 16: 235–254.

Toth, L.A., Koebel, J.W. Jr., Warne, A.G., and Chamberlain, J. 2002. In Flood pulsing in wetlands: Restoring the natural hydrological balance, ed. B.A. Middleton, pp. 191–222. New York: John Wiley & Sons.

Trzcinski, M.K., Fahrig, L., and Merriam, G. 1999. Independent effects of forest cover and fragmentation on the distribution of forest breeding birds. Ecol. Applications 9: 586–593.

Tscharntke T., Steffan-Dewenter, I., Kruess, A., and Thies, C. 2002. Contribution of small habitat fragments to conservation of insect communities of grassland-cropland landscapes. Ecol. Applications 12: 354–63.

Vandermeer, J.H. 1973. On the regional stabilization of locally unstable predator-prey relationships. J. Theor. Biol. 41: 161–70.

Verhulst, P.F. 1838. Notice sur la loi que la population suit dans son accroissement. Correspondence Mathématique et Physique, 10: 113–121.

Villard, M.-A., Trzcinski, M.K., and Merriam, G. 1999. Fragmentation effects on forest birds: relative influence of woodland cover and configuration on landscape occupancy. Conservation Biol. 13: 774–783.

Vollenweider, R.A. 1975. Input-output models with special reference to the phosphorous loading concept in limnology. Schweizerische Zeitschrift für Hydrologie 37: 53–84.

Volterra, V. 1926. Variations and fluctuations of the number of individuals of animal species living together. In Animal ecology, ed. R.N. Chapman, pp. 409–448. New York: McGraw-Hill.

Weir, J.M.H., Johnson, E.A., and Miyanishi, K. 2000. Fire frequency and the spatial age mosaic of the mixed-wood boreal forest in western Canada. Ecol. Applications 10: 1162–1177.

Wilson, H.B., Hassell, M.P., and Holt, R.D. 1998. Persistence and area effects in a stochastic tritrophic model. Am. Naturalist 151: 587–595.

With, K.A., and King, A.W. 1999. Extinction thresholds for species in fractal landscapes. Conservation Biol. 13: 314–326.

With K.A., Pavuk, D.M., Worchuck, J.L., Oates, R.K., and Fisher, J.L. 2002. Threshold effects of landscape structure on biological control in agroecosystems. Ecol. Applications 12: 52–65.

Wolff, J.O., Schauber, E.M., and Edge, W.D. 1997. Effects of habitat loss and fragmentation on the behavior and demography of gray-tailed voles. Conservation Biol. 11: 945–956.

# 7
# Heterogeneity in Hydrologic Processes: A Terrestrial Hydrologic Modeling Perspective

CHRISTINA TAGUE

## Abstract

Heterogeneity of land surface and atmospheric processes contributes to all aspects of the hydrologic cycle. Understanding the types and sources of this heterogeneity is a fundamental component of both theoretical and applied hydrology. Observations of heterogeneity occur at multiple scales ranging from within-canopy variation in water-holding capacity of a single leaf to spatial variation in precipitation at continental to global scales. Consequently, strategies for addressing heterogeneity in hydrologic modeling depend on the scale and type of process being modeled. Further, hydrologic models must address heterogeneity in both inputs and parameters as well as the representation of underlying physical processes. This paper provides an overview of heterogeneity and its implications for hydrologic modeling. Crucial examples of heterogeneity in inputs, parameters, and underlying physical processes are described, and approaches used to deal with heterogeneity within hydrologic modeling are discussed. In particular, the use of effective parameters, probabilistic approaches, and landscape tessellation are described as strategies to address heterogeneity in parameters and inputs. Explicit consideration of process heterogeneity is also considered from the perspective of physically based hydrologic modeling, and the implications for the coupling between hydrologic and ecological process models is discussed.

## Introduction

Analysis of heterogeneity in hydrology, as in other sciences, seeks to characterize and ultimately to explain spatial and temporal patterns of water in all of its forms—solid, liquid, and gas—and the pathways by which water is transported and stored on the surface of the earth. Observation of heterogeneity depends both on the spatial temporal scale of observation and the

groundwater) as well as measures of quantity, quality, and/or timing. Understanding and quantifying heterogeneity in these different variables across a range of scales and exploring how heterogeneity changes across scales and between measures can be viewed as one of the basic challenges in hydrologic science.

Many of the fundamental research areas as well as practical applications of hydrology must deal with heterogeneity. In theoretical studies, analysis of heterogeneity with respect to different components of the hydrologic cycle often provides insight into the underlying controlling mechanisms. In applied studies, prediction of system behavior and its sensitivity to change often depends on estimates of heterogeneity. In both these arenas, heterogeneity must be considered both as a cause and as an effect. Heterogeneity of variables of interest (i.e., streamflow, soil moisture, groundwater storage, etc.) is linked to heterogeneity in other related variables (soil hydraulic conductivity, land cover) that describe underlying controlling processes or characteristics of the system. Thus, hydrologic analysis must deal both with the characterization, explanation, and prediction of heterogeneity of hydrologic measures of interest and with assessing the role that heterogeneity in related measures plays in shaping these patterns. Hydrologic modeling attempts both to capture relevant heterogeneity in outputs and to represent crucial heterogeneity in inputs, parameters, and processes.

Hydrologic models are used to address a variety of basic and applied research questions. The extent to which heterogeneity matters depends on the research question being asked. This is true both in terms of the ability of models to represent heterogeneity of response and the extent to which models must incorporate information about heterogeneity in the underlying system in order to capture relevant dynamics. Models designed to estimate flood conditions in urban environments, for example, might not need to capture spatial-temporal heterogeneity in low flow volumes (response) nor incorporate heterogeneity in deeper soil hydraulic properties (parameters). Nonetheless, for many hydrologic models, there are commonalities both in terms of key inputs, parameters, and processes for which heterogeneity is often an issue and in terms of the techniques used to incorporate heterogeneity within a modeling framework. This paper will provide an overview of common sources of heterogeneity in hydrologic systems and then discuss some of the approaches used to account for heterogeneity at different scales within hydrologic models. It is important at this point to distinguish between heterogeneity and variability. Heterogeneity typically implies a difference in type or class (i.e., differences in soil texture classes). Variability can denote a difference in amount or degree, often within a type or class (i.e., differences in values for hydraulic conductivity within a soil class). How the type or class is defined can determine whether observed variation might be called heterogeneity. For example, if different soil structures result in variation in hydraulic conductivity, it might be reasonable to examine heterogeneity in hydraulic conductivity. Given this semantic problem, I will consider both heterogeneity and variability that likely arises from underlying structural differences of the property in question.

# Observations of Heterogeneity in Hydrology

In hydrology, the basic unit of analysis can range from a block of soil or the surface of a leaf at small scales, to hillslopes and watersheds at local to regional scales, and to the full hydrologic cycle at global scales. All of these systems, however, can be examined from the perspective of inputs and outputs of water and the internal state variables/parameters and processes that transform inputs to outputs. Heterogeneity of outputs at any scale may reflect heterogeneity in inputs, internal system parameters, and/or the processes involved.

## Heterogeneity in Inputs

One of the most important factors contributing to spatial heterogeneity in hydrologic response variables, including soil moisture, evapotranspiration, and streamflow, is spatial-temporal variation in precipitation inputs. At the continental scale, heterogeneity in all hydrologic processes can be explained based on the annual amount and seasonal variation in precipitation. Thus, annual differences in the amount and timing of streamflow in the northeastern *versus* southwestern United States can clearly be attributed to differences in the amount and timing of precipitation.

Most hydrologic models are constrained by an energy or mass balance equation where (Inputs − Outputs = ΔStorage). For mass-balance models in hydrology, precipitation is a fundamental input; thus, heterogeneity in precipitation can be seen as the starting point for heterogeneity of all hydrologic processes within the system. Quantifying heterogeneity in precipitation and incorporation of this heterogeneity into models, particularly at more local scales, is often confounded by limited rain gauge density. Smith et al. (1996) found that even a high density of rainfall gauges resulted in a significant underestimation of storm event precipitation when compared to radar estimates. Advances in rainfall observations through radar have contributed to mapping the heterogeneity in precipitation; however, data availability and error assessment remain issues (Krajewski and Smith 2002).

Irrigation and interbasin transfers of water can confound analysis of heterogeneity where precipitation is assumed to be the only input. In areas where interbasin transfers of water are significant, monitoring of these additional inputs can be essential for accurate modeling of streamflow and evapotranspiration. In the South Platte Basin of Colorado, for example, it is

in baseflow and annual flow patterns of subbasins within the South Platte can often be attributed to differences in irrigation regimes (Strange et al. 1999).

At the watershed scale, the temporal scale of interest often determines the extent of relevant heterogeneity in precipitation. Spatial heterogeneity at the timescale of individual storm events is often, but not always, greater than that of longer term (seasonal-annual) patterns. The mechanisms that generate precipitation events are important controls on the associated spatial length scales and their relationship with temporal scale. For a given storm event, convective rainfall, for example, varies at length scales of $< 1$ km, whereas frontal cyclonic storms may be organized over hundreds of kilometers (Bloschl and Sivapalan 1995). Thus, modeling runoff for individual storms for a first-order watershed may need to account for spatial variability in precipitation inputs, particularly in regions dominated by convective rainfall. Modeling runoff response to a flood producing storm event in Fort Collins, Colorado, for example, would need to account for a doubling of precipitation input within less than a kilometer (Ogden et al. 2000). For storm-events modeling at larger space scales, such as the Colorado Front Range, interpolation of rain gauge data for input into hydrologic models must account for both typical length scales of storm events and the stochastic nature of individual events.

At longer-term (i.e., annual) timescales, heterogeneity in precipitation within a given climatic region may often show a consistent spatial pattern. Precipitation, for example, is often dominated by topographic controls such that there is a significant relationship between mean annual precipitation and elevation across climatic regions of North America (Dingman 1994). Human modifications to the land surface may also contribute to a consistent long-term spatial variation of precipitation at relatively local scales. Urban heat island contributions to the frequency and intensity of convective rainfall, for example, can generate heterogeneity at storm event to annual timescales (Changnon 1992). In these cases, where heterogeneity in precipitation is temporally consistent, these patterns must be considered in longer term models of continuous streamflow, evapotranspiration, and so forth. Inputs, in this case, are often derived from atmospheric climate models such as Regional Atmospheric Modeling System (RAMs) (Walko et al. 2000) or models such as Parameter-elevation Regressions on Independent Slopes Model (PRISM) (Daly et al. 1994) that provide spatial estimates of precipitation by interpolating rain gauge data using topographic, wind direction, and other controls on spatial patterns.

Finally, in addition to precipitation inputs, energy balance approaches in hydrology must consider energy inputs or solar insolation as a key control on heterogeneity in response characteristics. Energy inputs often vary in structured predictable ways following topography (slope, aspect) and, at larger scales, latitude. As with precipitation, capturing this heterogeneity in input often requires going beyond available measured data and using models, such as Mtn-Clim (Running et al. 1987), to estimate spatial variation in radiation input.

## *Heterogeneity in System Characteristics or Parameters*

Distinctions between heterogeneity of system characteristics or parameters (i.e., variation in soil hydraulic conductivity) and heterogeneity of processes (i.e., saturation excess *vs* infiltration excess as runoff production mechanisms) depend on both the scale and the model being employed. Coefficient-based models in hydrology estimate runoff volumes as a function of precipitation using parameters related to land surface characteristics. The curve number approach developed by the U.S. Soil Conservation Service, for example, compiled data to determine standardized precipitation-runoff relationships for a variety soil (i.e., sandy loam, clay, silt) and land-use characteristics (i.e., high density urban, commercial, forest ). In these models, spatial heterogeneity in runoff coefficients can represent both a change in parameters or in the strength of relationships (i.e., an increase/decrease in infiltration capacity) and/or a mechanistic shift between dominant runoff production mechanisms (i.e., from subsurface to surface overland flow). In more process-based models, processes are explicitly represented, and parameters tend to reflect measurable characteristics that control the rates of these processes. In both types of models, however, several commonly used, physically based parameters are often the main drivers of heterogeneity in hydrologic responses. Key parameters include various measures that describe soil, vegetation/land cover, and topography as well as several measures of channel characteristics including channel geometry and surface roughness.

Soil parameters such as depth, texture, hydraulic conductivity, and porosity are often key inputs into hydrologic models. Significant efforts have been made in recent years to develop national databases (e.g., SSURGO; http://www.ncgc.nrcs.usda.gov/branch/ssb/products/ssurgo) that provide data on soil properties at scales ranging from 1:12,000 to 1:63,360. Nonetheless, significant uncertainty around the impact of soil properties on hydrologic behavior often remains, particularly at smaller (first order) watershed scales. For example, heterogeneity in soil characteristics is often represented by aggregate measures of hydraulic conductivity and has been shown to vary across multiple scales. Variation in hydraulic conductivity is often tied to soil type (i.e., fraction of sand, silt and clay; (e.g., Clapp and Hornberger 1978); however, site-specific variation within soil types can be significant. In particular, macropores—generated by roots, soil structure, and so forth—can result in significantly higher effective hydraulic conductivities than implied by the soil matrix (McDonnell 1990). Similarly, the role played by bedrock fractures, soil crusting, and so forth, can confound attempts to map heterogeneity in soil hydraulic characteristics based on typically available soil classification information. Given these uncertainties, soil hydraulic conductivity is often left as a calibrated parameter in hydrologic modeling (Beven and Binley 1992).

Heterogeneity in land cover characteristics often drives spatial hetero-

incorporation into hydrologic models has greatly been improved by remote sensing and, in particular, remote sensing estimates of leaf area index, which is a key parameter in many physically based hydrologic models (Waring and Running 1998). In more urban environments, land cover characteristics are typically derived from land use maps (i.e., Moglen and Casey 1998; Rose et al. 2001), although there is a potential for incorporating much finer and potentially more hydrologically relevant characteristics (i.e., impervious/pervious area) using remote sensing data. In both of these applications, scale becomes a crucial issue and is tied to the resolution of available sensors and/or mapping information.

It is important to consider that human activities, both agriculture and urbanization, can have a significant impact on heterogeneity of not only land cover but of other hydrologic parameters as well. Agricultural practices (such as tile drainage and plowing) can alter effective soil properties (i.e., infilitration rates, hydraulic conductivity) and even topography. More than 20.6 million acres within the U.S. Midwest can be classified as under agricultural drainage. The hydrologic impact of these agricultural drainage practices typically include both impacts on streamflow (i.e., increases peak runoff rates) and soil hydrologic conditions (i.e., reduction of swamp and wetland area) (Fausey et al. 1995). In these watersheds, human design often overwhelms natural controls on heterogeneity, and differences in agricultural practices can play a crucial role in defining hydrologic properties across a range of scales (Skaggs et al. 1994). Similarly, urbanization can increase watershed scale drainage efficiency through the development of storm sewer networks and impervious surfaces (Chester and Gibbons 1996). As discussed in Chapter 13 (Band et al. this volume), the net impact of urban design can alter heterogeneity in parameters and ultimately hydrologic behavior, although there is evidence of both increases and decreases in heterogeneity of response depending on the scale, location, and specific process of interest.

Heterogeneity in topography (slope, aspect, elevation) is probably the most accurate and readily available parameter used in hydrologic modeling. The geomorphic unit hydrograph (Rodrigues-Iturbe and Valdes 1979), for example, illustrates how topographic relationships readily derived from a digital elevation model (DEM) can account for spatial differences in stormflow behavior. Many simple coefficient-based rainfall-runoff models (i.e., Soil Conservation Service Curve number approach) use variation in slope to adjust or select coefficients that determine the relationship between rainfall and runoff for particular land-use types. Other models such as TOPMODEL (Beven and Kirkby 1979), which also consider within-watershed hydrologic conditions, use topographic indices to account for heterogeneity in soil moisture patterns as well as streamflow. Heterogeneity in topography occurs at multiple scales, and its impacts on hydrologic processes vary with these scales. At the plot scale, topographic heterogeneity might be expressed as surface irregularities that account for a surface detention storage capacity. At the hillslope scale, slope varies such that in particular regions, characteristic profiles emerge; for instance, Piedmont hillslopes are characterized by broad, gently sloping uplands, steep side slopes, and flat bottomlands, whereas the western Cascade mountains are characterized by steep slopes and narrow riparian zones. These characteristic profiles contribute to explanations for the rate that water moves through the landscape and within hillslope spatial variation in soil moisture. At these scales, differences in mean hillslope topographic characteristics (slope, aspect, elevation) account for heterogeneity in hydrologic responses.

In addition to topographic control on the rate of flow, topographic parameters can be used to indicate heterogeneity due to the magnitude and timing of latent and sensible heat fluxes. Variation in insolation follows both slope and aspect and contributes to spatial patterns of evapotranspiration and soil moisture, particularly in water-limited environments (Moore et al. 1988). Variation in air temperature associated with a change in elevation can explain heterogeneity in soil moisture due to differences in the timing and rate of snow melt. At larger, regional to continental scales, topographic variation reflects dominant geologic controls. However, at these large scales, the impact of topography on variation in hydrologic response is often secondary to differences in climatic regime.

Finally, it is worth noting that at all scales, the relationship between topographic parameters and processes and associated responses such as streamflow and spatial patterns of soil moisture can be complex. For example, Western et al. (1999) found topographic indices were highly correlated to measurements of soil moisture patterns during wetting and drying periods for the Tarrawarra catchment in Western Australia. During very dry periods, however, this relationship breaks down. The dynamic relationship between topography and soil moisture reflects a shift in the dominant control on heterogeneity—from topography, in a hydrologically connected landscape, to local soil properties in a drier, hydrologically disconnected landscape. Similar limitations to using topographic parameters as surrogates for other hydrologic properties occur in areas where the underlying bedrock topography does not follow surface topography and acts as the main control for the redistribution of soil moisture.

## Heterogeneity in Process

Ultimately, heterogeneity in hydrologic systems behavior may reflect heterogeneity in process. From a modeling perspective, spatial or temporal heterogeneity cannot always be easily represented by variation in parameters such as hydraulic conductivity, surface slope, or inputs such as the amount of rainfall. In these cases, heterogeneity is best explained by variation in space and time in the type of underlying processes rather than the intensity of those processes. For example, heterogeneity associated with differences in climate reflects a shift in underlying controlling processes. Variation in temperature, for example, can result in a shift from rain to snowmelt dominated hydrology. Snowmelt dynamics can then become the dominant control on the shape of seasonal hydrographs. Similarly, a shift from a climate dominated by short duration, high-intensity convective rainfall to one dominated by lower intensity frontal systems is often associated with a shift in runoff generation mechanisms from overland flow to subsurface throughflow. Modeling climate change impacts on hydrology, therefore, must be sophisticated enough to incorporate not only changes in input but also potential change in dominant controlling processes.

## Incorporating Heterogeneity in Hydrologic Modeling: Approaches

Given ample evidence of significant heterogeneity in parameters and inputs typically associated with hydrologic models, strategies for incorporating this heterogeneity into hydrologic models are needed and have been the subject of considerable research. The particular approach used depends on the specific modeling objective and the response to the following questions: (a)When and where does heterogeneity matter?(b)What data are available to characterize this heterogeneity? (c)What are the costs (in terms of complexity, computation efficiency, etc.) of including this heterogeneity in a given model?

There are a variety of ways in which heterogeneity of parameters and/or inputs can be incorporated into models. Models range from lumped to quasidistributed to fully explicit representations (Watts 1997) where the transition from lumped to distributed type models is often evoked specifically to account for spatial heterogeneity. For example, representation of the expansion and contraction of saturated areas (and hence spatial heterogeneity in soil moisture and runoff production) can explicitly be represented in a spatially distributed model. In contrast, a lumped bucket model (i.e., a model that produces runoff in proportion to rainfall only after a single finite hillslope scale volume/store has been filled) might underestimate flow during the runoff period following a storm (recession period) because it ignores this heterogeneity.

### Subunit Heterogeneity

Both lumped and spatially distributed models require estimation of parameters and inputs at the scale of the fundamental modeling unit. For a given modeling unit, the simplest approach is to use an estimate of the mean value of the parameter. Error associated with using a mean value will depend on the degree of nonlinearity of the process dependent on this parameter or input. Many hydrologic processes show significant nonlinearities. Numerous researchers have shown that nonlinearities in the relationship among soil properties, soil moisture, and evapotranspiration can result in under- or overestimation of evapotranspiration based on mean soil conditions (Kabat et al. 1997; Lammers et al. 1997). Runoff, particularly saturated overland flow, can also be highly nonlinear, given the threshold nature of the response. Many studies (reviewed by Giorgi and Avissar 1997) use soil-vegetation-atmospheric transfer (SVAT) models to estimate the coupling of land surface hydrology to the atmosphere for global climate models (GCMs) and have shown nonlinearities in the relationship between land surface characterizations and associated energy and moisture fluxes. Further, these studies show that these nonlinearities can result in significant errors in estimating these fluxes based on parameters averaged at the scales typically used in GCMs (e.g., Famiglietti and Wood 1994; Giorgi and Avissar 1997).

Spatial or temporal averaging of parameters to account for heterogeneity can also lead to errors when the scale at which the parameter is measured does not match the scale of application. For example, hydraulic conductivity is measured in the field at scales of the order centimeters to meters. Hillslope hydrology models, however, often include hydraulic conductivity as a parameter at scales of the order meters to kilometers. At this scale, heterogeneity in soil structure such as macro-pores, cracks, and so forth often increase effective conductivity (McDonnell 1990). Thus, mean soil hydraulic conductivity no longer controls the rate of flow. Instead, shallow subsurface resistance to flow is a complex function of soil matrix characteristics and the organization of flowpaths that produce an effective hydraulic conductivity. An alternative in this case is to use secondary field data, such as streamflow or lysimeter data, to infer effective parameter values through calibration. Even with calibration, however, the issue of using a single effective parameter to represent a distribution of conditions remains a problem when there is significant nonlinearity in the relationship between parameter values and response. Thus, a calibrated value for mean hillslope hydraulic conductivity may still result in error if distribution of actual values of hydraulic conductivity within the hillslope result in a nonlinear relationship between soil moisture and runoff production.

## Parameter Distribution Approaches

One alternative to the use of a single averaged or effective parameter value is to run the model over a distribution of parameter values for each modeling unit. Avissar (1992) defined this approach as a statistical dynamical approach and has used it to incorporate heterogeneity in stomatal resistance, leaf area index, and albedo in SVAT models of land surface evapotranspiration (Avissar 1992; Avissar 1993). Hartman et al. (1999) illustrated an increase in correspondence between observed and predicted runoff when a distribution rather than mean value for snow accumulation was used. Use of a distribution in this case accounted for heterogeneity in within-grid cell snow cover due to significant wind driven redistribution of snow in alpine regions.

The well-known TOPMODEL (Beven and Kirkby 1979) also uses probability distributions of a wetness index (7.1) to incorporate the effect of topography and soil characteristics on soil moisture and runoff production.

$$w_i = \ln\left(\frac{aT_i}{T_o \tan \beta}\right), \tag{7.1}$$

where $T_i$ and $T_o$ are local and mean watershed saturated soil transmissivity, respectively, $\tan \beta$ is the tangent of the local slope, and $a$ is upslope contributing area. Soil transmissivity is calculated as:

$$T = \int_{s_i}^{\infty} K_o e^{(-s/m)} ds, \tag{7.2}$$

where $K_o$ is saturated hydraulic conductivity at surface, $s$ is a saturation deficit (or depth from the surface to the water table), $s_i$ is local saturation deficit, and $m$ is a soil parameter that scales hydraulic conductivity with depth.

In TOPMODEL, the wetness index distribution is used to compute the distribution of local saturation deficits and runoff production. One of the strengths of TOPMODEL is that the topographic component of the wetness index distribution is easily derived from a DEM. Estimation of the distributions of $K_o$ and $m$ (which define local soil characteristics), however, presents a greater challenge and is often cited as explanation for differences between observed and predicted saturation deficits (Blazkova et al. 2002).

In most applications, TOPMODEL is calibrated by adjusting a mean $m$ and $K_o$ to achieve a best fit between observed and modeled streamflow. Calibration in this case reflects a method to deal with uncertainty in some of the underlying parameters—including the extent to which macropore flow and other heterogeneities in soil parameters impact the response. Calibration can also compensate for errors in estimating the distribution of the wetness index. In particular, the estimation of the TOPMODEL index has been shown to be sensitive to the resolution of the underlying DEM where too coarse a resolution will truncate the tails of the distribution and change the corresponding estimate of streamflow. Consequently, calibrated values for parameters based on DEMs of differing resolution tend to vary (Saulnier et al. 1997).

Errors in TOPMODEL as well as the need for calibration illustrate the extent to which the estimation of the required probability density function can be problematic. For other parameters that are not easily measured, such as stomatal resistance or deeper groundwater conductivities, deriving a reasonable distribution may depend solely on ancillary data or another model. The use of probability density function can also be problematic in a more complex model, with multiple parameters, given that modeling over a distribution is considerably more computationally and mathematically intensive than the use of a single effective parameter.

Nonetheless, there are many cases where the probability distribution can readily be derived and may be important in terms of capturing significant nonlinearities in response. Representing land cover (particularly in urban environments, where the length scale of heterogeneity is small) by the use of a distribution may be very useful and help to avoid a situation where large areas (i.e., major drainage basins encompassed within urban areas) must be modeled at very fine scale resolutions (i.e., individual lawns, houses, streets). Even in cases where high resolution data may be available to delineate these objects, the associated computational and data storage costs would preclude spatially explicit modeling, except for small localized neighborhoods.

## Aggregation or Partitioning Strategies

In spatially distributed models, an alternative to representing heterogeneity of inputs/parameters as either a probability density function (pdf) or an effective value is to explicitly represent heterogeneity through landscape tessellation. Defining the basic spatial modeling unit to minimize within-unit heterogeneity, however, again requires key issues of parsimony to be addressed including (a) When does heterogeneity matter? and (b) How simply can this heterogeneity be adequately described? Further, the use of effective or averaged parameters must be considered in conjunction with the strategy used to partition the landscape.

Numerous researchers have endeavored to derive optimal modeling units for representing landscape heterogeneity, given a specific hydrologic modeling task (e.g., Lammers et al. 1997). For many inputs/parameters/processes, aggregation often reduces heterogeneity. Wood et al. (1988) developed the concept of a representative elementary area to explore this effect with respect to runoff production. Evidence from both rainfall-runoff models and observed streamflow data illustrates that variability between different catchments within the same region tends to decrease as catchment size increases, such that a representative elementary area (REA) where variability between samples is minimized can be obtained (Woods et al. 1995). This effect is generally attributed to averaging of soil and topographic variability. At larger scales, of course, variability often increases again as regional scale climatic and geologic controls become important. For rainfall-runoff modeling at the regional scale, the concept of a REA provides a useful construct for dealing with heterogeneity. It illustrates that as the scale of the response variable (in this case runoff) changes, the scale of important heterogeneity also changes. The REA is a method to characterize this for topographic control of streamflow. The concept of a REA and associated scale analysis could also be applied to other hydrologic properties, such as effective hydraulic conductivity. In hydrologic modeling, however, response variables of interest may not necessarily be at the scale of a REA or, further, the response variable of interest or relevant inputs/parameters may not show this kind of scaling relationship. For example, a model designed to provide hydrologic information for the purposes of characterizing aquatic habitat must address streamflow defined at the scale of habitat sensitivity rather than scale (such as a REA) that simplifies analysis of streamflow behavior.

Theoretically, in situations where heterogeneity of the parameter produces nonlinear responses, the issue of heterogeneity in parameter values can be dealt with by partitioning the landscape into units with minimal within-unit parameter variation. RHESSys (Band et al. 2000), for example, allows patch size and shape to vary based on available input data and associated parameter variability. Proposed partitioning strategies based on topographic indices (slope, aspect, accumulated area) and land cover have been shown to reduce errors associated with averaging of observed nonlinear parameters/inputs (e.g., Lammers et al. 1997). In practice, however, the minimum modeling unit is often constrained by (a) resolution of available data and (b) computation memory/time. For example, distributed representation of land cover characteristics is often limited by the resolution of remote sensing data. On the other hand, as higher resolution data become available, computational limitations emerge.

## Spatial Connectivity

Finally, it is important to recognize that even fully explicit representations aggregate or lump the landscape at the scale of the fundamental modeling unit (e.g., a 30-m grid cell). It is useful, therefore, to distinguish between a single lumped model that is replicated over an array of spatial units and a fully explicit representation. In the fully explicit representation, in addition to accounting for spatial variation in inputs and parameters, the connectivity between units and the spatial organization of the units is considered.

In SVAT modeling to support atmospheric modeling, Giorgio and Avissar (1997) note that spatial heterogeneity can in fact generate meteorological behavior due to gradients created by heterogeneity in land surface characteristics. In this case, the organization of heterogeneous patches and fluxes between them must be considered in addition to the distribution of different patch characteristics. Similarly, in hydrologic models of biogeochemical cycling, the potential for uptake of nutrients along hydrologic flowpaths means that spatial organization of heterogeneity cannot be ignored. Further, connectivity between heterogeneous areas and the potential for that connectivity to change must then be represented in accounting for the impact of heterogeneity on water quality.

TOPMODEL is an approach that represents connectivity between heterogeneous landscape units implicitly, rather than explicitly. The higher wetness afforded to units with higher upslope contributing areas [$a$ in Eq. (7.1)] implies a movement of water to lower areas. TOPMODEL, however, does not actually move the water from one cell to another; thus, it does not necessarily account for processes where explicit connection is important. For example, in an urbanizing watershed, some upslope cells may have higher water loads due to lawn watering, and downslope cells that are hydrologically connected to these upper cells should be wetter than those in similar topographic positions but whose upland areas have not yet been developed.

In addition to ignoring specific upslope/downslope linkage, implicit approaches such as TOPMODEL typically assume a constant connectivity. With respect to subsurface flow, field evidence has shown that under dry conditions, upland areas within a watershed may be disconnected from lower regions (Western et al. 1999). Similarly, in urban environments, sewers and roads may act to alter topographically based hydrologic connectivity and result in the bypass of lowland areas (Djokic and Maidment 1991; Tague and Band 2001). These examples serve to illustrate (a) the need in some cases to account for explicit connections between heterogeneous areas and (b) the potential for those connections to vary with time. Models such as DHSVM (Wigmosta et al. 1994), RHESSys (Tague and Band 2001), Topog (Vertessy et al. 1996), and EPA's SWIMM account for explicit connections, although the adequacy of submodels and parameters used to define the strength of connectivity is an area of continued research.

## Physically Based versus Empirical Coefficient Models

Classification of hydrologic models also distinguishes between empirical-coefficient driven and physically based or process-based models (Watts 1997). This distinction, however, is a loose one because, as argued by Beven (1992), all physically based models include parameters derived from empirical relationships. Nonetheless, physically based models are more explicit in their representation of process heterogeneity. For example, observed differences in evapotranspiration and snowmelt between north- and south-facing slopes can be estimated in a physically based model that drives submodels of snowmelt and evapotranspiration with solar radiation inputs across spatially variable terrain (e.g., Band et al. 1993; Wigmosta et al. 1994) The increasing complexity of a physically based model, however, also increases the sources for potential error.

Physically based models are generally sensitive to interactions between specific inputs and/or processes. Soil moisture at any given point will be a function of rainfall, parameters controlling drainage such as hydraulic conductivity, and the representation of processes such as subsurface through-flow and evapotranspiration, which are both in turn dependent on current soil moisture conditions. The ability of process-based models to account for spatial/temporal heterogeneity assumes that the significant controls on variability, as well as covariation between different controls, inputs, and param...

an important heuristic tool by explicitly representing the impact of dominant processes and landscape features on hydrologic response. In this sense, they are distinct from coefficient-based approaches to the extent to which they can be used as tools to assess the implications of different explanations for causes and consequences of heterogeneity.

For example, Pauwels and Wood (1999) illustrate that incorporation of freeze/thaw cycles and distinct overstory (forest) and understory (moss) layers into a physically based model has a significant impact on the estimation of evaporative fluxes in a high-latitude boreal forest landscape. These results suggest that spatial and temporal patterns of these processes may play a significant role in boreal forest hydrology. Similarly, Bonan (1995) showed that including a distinct lake surface submodel in a SVAT approach significantly altered estimates of evaporative fluxes. By altering model structure rather than parameters, adaptive physically based models can be used where the research focus is understanding rather than prediction. However, using models to address process heterogeneity requires that model design be flexible enough that alternative models and/or additional processes can easily be implemented (Leavesley et al. 2002).

## Conclusions

Figure 7.1 presents a framework that summarizes the multiple avenues through which heterogeneity becomes an important consideration in hydrologic modeling. From one perspective, hydrologic models can be used to predict heterogeneity in variables of interest. Characterizing heterogeneity in hydrology responses such as streamflow is often a prerequisite for environmental planning directed at managing water resources. Simply quantifying heterogeneity in space and time of hydrologic fluxes (streamflow, evaporation, precipitation, and so forth.) remains a challenge that is currently being addressed both by extension of monitoring networks and by hydrologic modeling. Limited spatial-temporal coverage of monitoring networks and the potential for error in inputs, parameters, and the structure of hydrologic models, however, must be recognized and evaluated as sources of uncertainty in this information.

Both resource managers and scientists need a more complete understanding of the controls on heterogeneity in hydrologic responses. At the same time, the complementary issue of how heterogeneity in particular land surface characteristics impacts the way in which water moves through the landscape must also be recognized and evaluated. Hydrologic models are key tools that explore and illustrate both of these scenarios. The testing of hydrologic models against empirical data, therefore, improves the understanding of the role that heterogeneity of inputs, parameters, and processes plays in hydrology. By exploring the conditions under which different representations of heterogeneity (i.e., through effective parameters,

| Observations | | | Modeling Strategies - Input/Parameter Heterogeneity | | |
|---|---|---|---|---|---|
| *Heterogeneity of Causes/Controls* → | *Heterogeneity in Relevant Responses* → | | *Within-unit heterogeneity* → | *Between-unit heterogeneity* → | *Spatial organization of heterogeneity* |
| Inputs | Key Examples | Spatial/temporal variation | Effective Values | Landscape Tessellation | Implicit connectivity |
| | •precipitation •radiation | • streamflow, • ET, • soil moisture | pdf | | |
| | | | Calibration | | Explicit connectivity |
| Parameters | •Soil – porosity, hydrologic conductivity •Topography – slope, aspect | | **Modeling Strategies – Process Heterogeneity** | | |
| Processes | •Snowmelt •Saturation overland flow •Saturation excess flow •macro-pore flow | | Calibrated Parameters  Explicit representation of process & conditions under which they dominate | | |

FIGURE 7.1. A framework for considering the role of heterogeneity in hydrologic modeling. The framework acknowledges the distinction between heterogeneity in inputs, parameters, and processes and summarizes different approaches commonly used in hydrologic modeling to account for effect on model predictions.

probability density functions, or process algorithms) can adequately capture observed responses, hydrologic models are improved along with a basic understanding of key landscape controls on relevant hydrologic processes.

Linking hydrology with ecology broadens the context in which hydrologic models are used. Coupled hydro-ecological models employ many of the same techniques used in more classic hydrologic approaches. In these models, additional controls and feedbacks can become important drivers of heterogeneity. For example, models that couple vegetation carbon and nutrient cycling with hydrology must consider feedbacks between soil moisture and vegetation productivity and thus consider heterogeneity in both. The added complexity of considering interactions between hydrology and ecology means that parsimony becomes a crucial issue in model design. Ecological considerations, however, also help to bound the precision over which heterogeneity is relevant. For instance, for many ecological predictions, a 10% difference in streamflow or soil moisture may not be important. Further work that extends both the technical advances in addressing heterogeneity in hydrologic modeling and provides an ecological context for interpreting and evaluating model results, will, likely make valuable contributions to both disciplines.

# References

Avissar, R. 1992. Conceptual aspects of a statistical-dynamical approach to represent landscape subgrid-scale heterogeneities in atmospheric models. J. Geophys. Res. 97: 2729–2742.

Avissar, R. 1993. Observations of leaf stomatal conductance at the canopy scale: an atmospheric modeling perspective. Boundary Layer Meteorol. 64: 127–148.

Band, L.E., Patterson, J.P., Nemani, R., and Running, S.W. 1993. Forest ecosystem processes at the watershed scale: incorporating hillslope hydrology. Ag. For. Met. 63: 93–126.

Band, L.E., Tague, C.L., Brun, S.E., Tenenbaum, D.E., and Fernandes, R.A. 2000. Modelling watersheds as spatial object hierarchies: structure and dynamics. Trans. GIS 4: 181–196.

Beven, K. 2002. Towards an alternative blueprint for a physically based digitally simulated hydrologic response modelling system. Hydrol. Processes 16: 189–206.

Beven, K., and Binley, A.M. 1992. The future of distributed models: model calibration and uncertainty prediction. Hydrol. Processes 6: 279–298.

Beven, K., and Kirkby, M. 1979. A physically-based variable contributing area model of basin hydrology. Hydrol. Sci. Bull. 24: 43–69.

Blazkova, S., Beven, K., Tacheci, P., and Kulasova, A. 2002. Testing the distributed water table predictions of TOPMODEL (allowing for uncertainty in model calibration): the death of TOPMODEL? Water Resources Res. 38: 1257–1290.

Bloschl, G., and Sivapalan, M. 1995. Scale issues in hydrologic modelling, Hydrol. Processes 9: 251–298.

Bonan, G.B. 1995. Sensitivity of a GCM simulation to inclusion of inland water surface. J. Climate 8: 2691–2704.

Changnon, S.A. 1992. Inadvertent weather modification in urban areas: lessons for global climate change. Bull. Am. Meteorol. Soc. 73: 619.

Chester, L.A. Jr., and Gibbons, C.J. 1996. Impervious surface coverage; the emergence of a key environmental indicator. J. Am. Planning Association 62: 243–244.

Clapp, R., and Hornberger, G. 1978. Empirical equations for some soil hydraulic properties. Water Resources Res. 14: 601–604.

Daly, C., Nielson, R.P., and Phillips, D.L. 1994. A digital topographic model for distributing precipitation over mountainous terrain. J. Appl. Meteorol. 33: 140–158.

Dennehy, K.F., Litke, D.W., Tate, C.M., and Heiny, J.S. 1993. South Platte River Basin—Colorado, Nebraska and Wyoming. Water Resources Bull. 29: 647–683.

Dingman, S.L. 1994. Physical hydrology. Englewood Cliffs, NJ: Prentice Hall.

Djokic, D., and Maidment, D.R. 1991. Terrain analysis for urban stormwater modelling. Hydrol. Processes 5: 115–124.

Famiglietti, J.S., and Wood, E.F. 1994. Multiscale modeling of spatially variable water and energy balance processes. Water Resources Res. 30: 3061–3078.

Fausey, N.R., Brown, L.C., Belcher, H.W., and Kanwar, R.S. 1995. Drainage and water quality in Great Lakes and Cornbelt States. J. Irrigation Drainage Eng. 121: 283–288.

Giorgi, F., and Avissar, R. 1997. Representation of heterogeneity effects in earth system modeling: Experience from land surface modeling. Rev. Geophy. 35: 413–438.

Hartman, M.D., Baron, J.S., Lammers, R.B., Cline, D.W., Band, L.E., Liston, G.E., and Tague, C. 1999. Simulations of snow distribution and hydrology in a mountain basin. Water Resources Res. 35: 1587–1603.

Kabat, P., Hutjes, R.W.A., and Feddes, R.A. 1997. The scaling characteristics of soil parameters: from plot scale heterogeneity to subgrid parameterization. J. Hydrol. 190: 363–396.

Krajewski, W.F., and Smith, J.A. 2002. Radar hydrology. Adv. Water Resources 25: 1387–1394.

Lammers, R.B., Band, L., and Tague, C. 1997. Scaling behaviour in watershed processes. In Scaling-up: from cell to landscape, eds. P.R. van Gardingen, G.M. Foody, and P.J. Curran, pp. 295–317. Cambridge, UK: Cambridge University Press.

Leavesley, G.H., Markstrom, S.L., Restrepo, P.J., and Veger, R.J. 2002. A modular approach to addressing model design, scale and parameters estimation issues in distributed hydrological modelling. Hydrol. Processes 16: 173–187.

McDonnell, J.J. 1990. A rationale for old water discharge through macropores in a steep humid catchment. Water Resources Res. 26: 2821–2832.

Moglen, G.E., and Casey, M.J. 1998. A perspective on the use of GIS in hydrologic and environmental analysis in Maryland. Infrastructure 3: 15–25.

Moore, I.D., Burch, G.J., and Mackenzie, D.H. 1988. Topographic effects on the distribution of surface soil water and the location of ephemeral gullies, Trans. ASAE 31: 1098–1107.

Ogden, F.L., Sharif, H.O., Senarath, S.U.S., Smith, J.A., Baeck, M.L., and Richardson, J.R. 2000. Hydrologic analysis of the Fort Collins, Colorado flash flood of 1997. J. Hydrol. 228: 82–100.

Pauwels, V.R., and Wood, E.F. 1999. A soil-vegetation-atmosphere transfer scheme for the modeling of water and energy balance processes in high latitudes. 2. Application and validation. J. Geophys. Res. 104: 27823–27840.

Rodriquez-Iturbe, I., and Valdez, J.B. 1979. The geomorphologic structure of hydrologic response. Water Resources Res. 15: 1409–1420.

Rose, S., Peters, N. 2001. Effects of urbanization on streamflow in the Atlanta area (Georgia, USA): a comparative hydrological approach. Hydrol. Processes. 15: 1441–1457.

Running, S.W., Nemani, R.R., and Hungerford, R.D. 1987. Extrapolation of synoptic meteorological data in mountainous terrain and its use for simulating forest evaporation and photosynthesis. Can. J. Forest Res. 17: 472–483.

Saulnier, G.M., Beven, K.J., and Obled, C. 1997. Digital elevation analysis for distributed hydrological modelling: reducing scale dependence in effective hydraulic conductivity values. Water Resources Res. 33: 2097–2101.

Skaggs, R.W., Breve, M.A., and Gillam, J.W. 1994. Hydrologic and water quality impacts of agricultural drainage. Crit. Rev. Environ. Sci. Technol. 24: 1–32.

Smith, J.A., Seo, D.J., Baeck, M.L., and Hudlow, M.D. 1996. An intercomparison study of NEXRAD precipitation estimates. Water Resources Res. 32: 2035–2045.

Strange, E.M., Kurt, D.F., and Covich, A.P. 1999. Sustaining Ecosystem Services in human dominated watersheds: Biohydrology and ecosystem processes in the South Plate River Basin. Environ. Manage. 24: 39–54.

Tague, C., and Band, L. 2001. Simulating the impact of road construction and forest harvesting on hydrologic response. Earth Surface Processes Landforms 26: 135–151.

Vertessy, R.A., Hatton, T.J., Benyon, R.G., and Dawes, W.R. 1996. Long-term growth and water balance predictions for a mountain ash (Eucalyptus regnans) forest catchment subject to clear-felling and regeneration. Tree Physiol. 16: 221–232.

Walko, R.L., Band, L.E., Baron, J., Kittel, T.G., Kittel, F., Lammers, R., Lee, T., Ojima, D., Pielke, R.A., Taylor, C., Tague, C., Tremback, C.J., and Vidale, P.L. 2000. Coupled atmosphere-biophysics-hydrology models for environmental modeling. J. Appl. Meteorol. 39: 931–944.

Waring, R.H., and Running, S.W. 1998. Forest ecosystems: analysis at multiple scales, 2nd ed. San Diego: Academic Press.

Watts, G. 1997. Hydrological modelling in practice. In Contemporary hydrology: towards holistic environmental science, ed. R.L. Wilby, pp. 152–193. Hoboken, NJ: John Wiley & Sons.

Western, A.W., Grayson, R.B., Bloschl, G., Willgoose, G., and McMahon, T.A. 1999. Observed spatial organization of soil moisture and its relation to terrain indices. Water Resources Res. 35: 797–810.

Wigmosta, M.S., Vail, L.W., and Lettenmaier, D.P. 1994. A distributed hydrology-vegetation model for complex terrain. Water Resources Res. 30: 1665–1679.

Wood E.F., Sivapalan, M., Beven, K.J., and Band, L. 1988. Effects of spatial variability and scale with implications to hydrologic modelling, J. Hydrol. 102: 29–47.

Woods R., Sivapalan, M., and Duncan, M. 1995. Investigating the representative elementary area concept: an approach based on field data. Hydrol. Processes 9: 291–312.

# 8
# Spatial Heterogeneity in Infectious Disease Epidemics

David L. Smith

## Abstract

Infectious disease epidemics in populations are inherently spatial—infectious agents are spread by contact from an infectious host to a susceptible host nearby. Among-host differences can determine which hosts suffer disease and the population dynamics of infectious disease epidemics. From the perspective of the infectious agent, a host is a habitat patch; among-host differences that are epidemiologically important are related to the concepts of compositional and configurational heterogeneity in landscape ecology. Heterogeneous mixing in epidemiology encompasses factors that determine who comes into contact with whom; it is analogous to configurational heterogeneity in landscape ecology. Other sorts of heterogeneity are analogous to compositional heterogeneity, including among-host differences in the duration of an infection, susceptibility to infection, or the amount of an infectious agent that is dispersed from an infected host. In real epidemics, compositional heterogeneity and configurational heterogeneity can introduce an overwhelming amount of complexity. Mathematical modeling provides a method for understanding epidemic processes and for taming the complexity. The idea of epidemic distance is introduced as a way of comparing and contrasting two different epidemic processes, and it is used to compare and contrast some of the mathematical models used to understand the role of space and spatial heterogeneity in epidemiology. In understanding real epidemics, the notion of parsimony is a guiding principle—heterogeneity should be weighed and ignored whenever possible. Several case studies are presented in which compositional and configurational heterogeneity are shown to be important.

## Introduction

Infectious disease epidemics in populations are inherently spatial. Infectious agents persist by spreading from an infectious host to a susceptible host nearby. Each host has a location, although "nearby" and "location" have a different meaning for each infectious agent. Infectious agents spread along a network of hosts characterized by the biology of the host population, the transmission mode of the infectious agent, and the course of an infection. Location in an epidemic network may be determined by geographical position, position in a social or sexual network, proximity to vector breeding sites, the movement of hosts or infectious agents through commerce, air travel, wind, or something else. For example, the influenza A virus is spread by airborne particles; airborne transmission requires that two people must be within a few meters of the same place within a few minutes. In contrast, an *Anopheles* mosquito becomes infectious 10 days or more after becoming infected with malaria; the next host infected may be several kilometers away. Thus, location may have a different meaning for each infectious agent in each host population.

Infectious disease epidemics are complex processes involving heterogeneous host populations and genetically diverse parasite populations. Heterogeneous host factors may include genetics, behavior, immune status, or any epidemiologically important trait that is spatially distributed among hosts. From the perspective of a parasite, a host is a habitat patch. Epidemiologically important differences among hosts fall into two categories. The first category includes any factor that affects the position of a host in a contact network or the configuration (topology) of the network. Collectively, these differences are called heterogeneous mixing. Heterogeneous mixing is analogous to configurational heterogeneity in landscape ecology. In contrast, compositional heterogeneity refers to other differences among hosts. Important kinds of compositional heterogeneity include differences in the duration of the infectious period, susceptibility to infection following exposure, or the amount of an infectious agent that is shed or dispersed into the environment from an infected host.

Infectious disease epidemics are complex, nonlinear processes. Understanding epidemics involves statistical analysis combined with mathematical modeling. Homogeneous population models—those that assume all individuals are alike—are a useful starting point in a hierarchical approach to model building and play a role similar to statistical null models. Heterogeneous population models modify the simple assumptions of homogeneous models to incorporate heterogeneity in the distribution of some epidemiologically significant trait, whether it is configurational or compositional. Heterogeneity is often manifested in unique ways in each system. Put another way, homogeneous populations are all alike, but each heterogeneous population is heterogeneous its own way, like unhappiness in the Karenina family in Tolstoy's *Anna Karenina*. Heterogeneity is not something that can or should be studied for its own sake; heterogeneity must be understood in context. Thus, understanding heterogeneity is often limited to case studies, although some important generalizations can be made.

Some general observations about the effects of heterogeneity may be best understood by considering *simple* departures from homogeneity, such as the variance in the number of contacts per unit time. Associating an effect with heterogeneity amounts to an analysis of structural stability or sensitivity analysis on higher order terms; this may require building suites of mathematical models and associating cause and effect by comparing models, either by elaborating simple models or simplifying complex ones (Mollison 1984; Black and Singer 1987). Heterogeneity may or may not be *biologically important* depending on the question being asked. Determining the biological importance of heterogeneity through model building and analysis, model fitting, and model selection is an important activity in science, especially if one regards science as a process of successive approximation (Burnham and Anderson 1998). Some models are intrinsically bad, but no good model serves every purpose. George Box (1979) famously said, "All models are wrong, some models are useful."

Hundreds of epidemic models have been developed and analyzed, thousands more are plausible, and an infinite number of models are possible (Hethcote 1994). To avoid the sheer tedium of analyzing model after model, it is necessary to ask what makes one model different from another and how much do the models differ. One way to measure the differences is to ask how the models generate different predictions about the time course of an epidemic and the distribution of time to infection. In some sense, the question of whether to incorporate space or spatial heterogeneity into an epidemic model is a question about the most parsimonious way to represent the mixing patterns or the distribution of important epidemiological traits in a population of hosts.

Heterogeneity should be weighed and ignored, unless it is biologically important. In the following essay, I will present my own view of spatial heterogeneity in epidemiology with a specific focus on those situations where it cannot or should not be ignored. In epidemiology, configurational heterogeneity is particularly confusing because there are two natural points of departure: random mixing and homogenous space. The two are opposites, in some sense. The mathematical assumption of homogeneous mixing models is equivalent to rapid and even stirring of chemicals in a chemostat, whereas homogeneous spatial models assume a uniform distribution of individuals on a landscape. To avoid oversimplifying, I will consider the well-mixed models and homogeneous spatial models as two different points of departure for understanding configurational heterogeneity.

## Epidemic Models and Spatial Heterogeneity

Mathematical models have played an important role in the population biology of infectious diseases as conceptual tools, for statistical inference, and for developing and evaluating policy. Bernoulli's smallpox model in 1760 explored the efficacy of variolation, a precursor to modern vaccination that involved blowing the scabs from surviving small pox patients into the nose to induce a mild case of smallpox. Hamer published the first deterministic model for measles dynamics in 1906, followed by a mathematical model for malaria by Ross in 1911. Mathematical epidemiology was firmly established in 1927 by Kermack and McKendrick, and it has been a very active area of research during the past 25 years (Anderson and May 1991; Hethcote 2000). Some important contributions of mathematical models to epidemiological theory have been the following:

1. To establish a deterministic epidemic threshold, the basic reproductive number, $R_0$.
2. To establish an endemic threshold, a minimum population size or population birth rate for an infectious agent to persist in a stochastic epidemic.
3. To describe the relationship between epidemiological parameters and the long-term average prevalence or the fraction infected during an epidemic.
4. To explain periodicity in epidemics.

Most of these developments have been done under the classical assumptions that the population is homogeneous and well mixed; in other words, all individuals are alike, and a population is mixing uniformly and rapidly enough to prevent any pattern formation. Heterogeneous models modify one or more of these classical theoretical assumptions. Despite their simplicity, even well-mixed models are spatial in a limited sense; hosts are separate from one another, but the assumptions about mixing guarantee that location is irrelevant (see below). This is sometimes called *pseudospace*. Indeed, Levin's metapopulation model in ecology, developed to illustrate the qualitative aspects of space, is mathematically identical to a simple epidemic model (Levins 1969). Epidemic models have incorporated more realistic representations of space, but one important question is how these representations of space differ from one another. Is qualitative space good enough or is some more complicated spatial model necessary?

## Comparing Epidemic Models

Qualitative representations of space may differ from the familiar definition of the Cartesian distance between two hosts, because distance depends on context. The earth is not really flat or homogeneous, so the shortest effective distance between two points is different for birds, antelopes, and earthworms moving about on the same landscape. In some sense, different notions of epidemic space are motivated by the different ways an infectious agent moves through a population. To pave the way for a more rigorous understanding, I will define epidemic distance, motivated by the formula: distance = rate × time. Intuitively, distance in an epidemic is related to the length of a transmission chain from one host to another.

The "average path" metric is the expected time for an epidemic to reach one individual from another. Time units are measured as the average disease

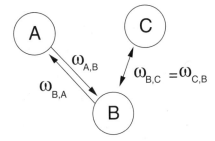

FIGURE 8.1. A simple network with three hosts, (A, B, and C). The infection rates are asymmetric between A and B ($\omega_{A,B} \neq \omega_{B,A}$) but they are symmetric between B and C ($\omega_{B,C} = \omega_{C,B}$).

generation, the average time elapsed between infection and transmission to another host; epidemic distance is the number of disease generations before an agent gets from one host to another. The distance between some pairs of hosts may be less than one because transmission rates between some hosts are higher than average. Thus, the epidemic distance concept is not simply analogous to the degrees of separation from network theory (popularized in the Kevin Bacon game and the movie *Six Degrees of Separation*), although the two are conceptual cousins (Watts 1999).

For an epidemic model defined on well-characterized set of discrete hosts, epidemic distance can be defined and computed. Graphs are one way of representing contact networks. A contact network can be represented as a weighted graph—each host is a vertex (Figure 8.1). Each pair of vertices is connected by two directed edges. The weight assigned to each weighted edge is the contact rate, $\omega_{i,j}$, the inverse of the expected waiting time for a new infection to be transmitted from one host to the other. In some pairs, $\omega_{i,j} = 0$, implying no *direct* connection. For an epidemic in a population of $N$ individuals, the distance is determined entirely by the direct pairwise connections, $\omega_{i,j}$, for each of the $N^2 - N$ ordered pairs of individuals. This formulation of distance is more accurately described as a directed distance concept—the distance from A to B may be longer than from B to A. For example, if a sexually transmitted disease is more easily transmitted from males to females than *vice versa*, the expected time to infection in a sexual partnership may be different from him to her than from her to him, even though transmission would occur through the same sexual acts. The degree of separation would be equivalent to epidemic distance if transmission were symmetric ($\omega_{i,j} = \omega_{j,i}$) and all direct contacts had weights of either zero or one.

## Distance in Simple Epidemics

In a well-mixed, closed, homogeneous population model, every host is alike and population density is a constant $N$. In the simplest epidemic models, individuals are either uninfected and susceptible to infection, infected and infectious, or recovered and immune; the density of individuals in each state is denoted S, I, and R respectively. These models typically assume that individuals recover at a constant per capita rate, $\alpha$, without respect to the time they became infected; recovery times are exponentially distributed with average infectious period of $1/\alpha$. In this model, the infectious period also defines a disease generation; the average time elapsed between consecutive infections. In an S-I-S model, individuals return to the susceptible state after infections, and $S + I = N$. In S-I-R models, recovered individuals become immune, and $S + I + R = N$.

The rate of infection for each susceptible in a population can be described by a general function $B(I)$; $1/B(I)$ gives an instantaneous estimate of the expected waiting time to infection. In density-dependent models, the rate of infection in the population is proportional to the density of infectious and of susceptible individuals, $B(I)S = \beta IS$. This assumption is also called mass-action in chemistry or a mean-field assumption in physics. This formula assumes that contact rates are proportional to the average crowding index, the squared density of hosts, $N^2$. Frequency dependent mixing assumes that mixing occurs at a constant rate, and the rate of infection is proportional to fraction of contacts that are infectious, $\beta IS/N$. Density and frequency dependent mixing models may be appropriate for different diseases.

For density dependent S-I-S and S-I-R models, the number of infected individuals over time is described by the equation $dI/dt = \beta SI - \alpha I$. In the S-I-R model, a second equation is necessary: $dR/dt = \alpha I$.

## The Basic Reproductive Number, $R_0$

A brief detour is necessary to introduce the concept of the basic reproductive number, $R_0$, a number that summarizes many important properties of an epidemic model. The term has its origins in demography where the basic reproductive number measures the lifetime reproductive output, the expected number of females born to an average female in a lifetime (Dietz 1993). For infectious agents, reproductive output could be measured either as the rate of reproduction within a host or as the rate of transmission among hosts. Because an infectious agent persists by maintaining an unbroken chain of infection from host to host, transmission is the more relevant measure of reproductive output. For epidemics, $R_0$ is defined as the expected number of new infectious cases caused by the first case introduced into an otherwise naïve population. At the beginning of the simple epidemic defined above, each individual infects others at the rate $\beta N$ and remains infectious for the $1/\alpha$ days; thus, $R_0 = \beta N/\alpha$. This formula applies to both S-I-S and S-I-R models; the development of immunity is irrelevant with respect to an infectious agent's ability to invade a naive population. The number of cases in the $n$th disease generation is approximately $R_0^n$. If $R_0 > 1$, the number of cases increases geometrically, at least until the epidemic depletes the number of susceptible individuals. Thus, $R_0$ plays a focal role in theory for infectious diseases by establishing an endemic threshold.

Shortest Path and Average-Path Metrics

Because every individual is alike and all individuals mix randomly, distance is one of two numbers; the distance from each individual to itself is zero, and the distance to any other individual is a positive number. This is the definition of pseudospace or qualitative space. The distance between any two individuals is the average or least number of disease generations elapsed for an infection to reach the individual, but how long does it take for the average person to become infected?

Assuming an epidemic occurs ($R_0 > 1$), the time to infection in the first generation is $1/\beta$; average time to infection in the first generation is $\alpha/\beta = N/R_0$. The "average path" metric depends on the time course of the whole epidemic, $I(t)$. In the S-I-S model, recovered individuals become susceptible again, so multiple infections are possible. All individuals become infected for the first time eventually, unless a stochastic epidemic fades out. In the S-I-R model, the epidemic depletes the susceptible population and eventually burns itself out; some fraction of the population remains uninfected. Using a definition that corrects for fade-out, epidemic distance is defined for $R_0 > 0$ (see the Appendix for details).

The distance from any host to itself is always zero. In homogeneous, well-mixed populations every host is equidistant from every other host. How far apart are two hosts? For the S-I-S and S-I-R models, average epidemic distance is very similar (Figure 8.2). The epidemic distance depends on $R_0$ and population size. Holding $\beta$ constant, the epidemic distance decreases with population size because $R_0$ is increasing (Figure 8.2a). Holding $R_0$ constant, the epidemic distance increases with population size because it takes more generations to reach all the population (Figure 8.2b).

## Heterogeneous Models

A heterogeneous model is one that departs from the assumptions of homogeneous well-mixed models. Simple departures in the composition of the host population, called compositional heterogeneity, tends to increase $R_0$ relative to a homogeneous model with the same average, all else being equal (Adler 1992; Dushoff and Levin 1995). Thus, an infectious disease is more likely to sustain a chain of transmission in a population with heterogeneous shedding, heterogeneous susceptibility, and heterogeneous duration of the infectious period relative to a homogeneous population with the same average.

Heterogeneity in the configuration of hosts, or heterogeneous mixing, implies a more complicated departure from homogeneous models. Since contact involves two individuals, mixing is inherently nonlinear (i.e., $\beta SI$). Thus, the mathematics of heterogeneous mixing is always more complicated. An important departure from homogenous mixing is positive assortative mixing (preferred mixing) in which the most active individuals mix preferentially with similar individuals, increasing $R_0$ (Dushoff and Levin 1995).

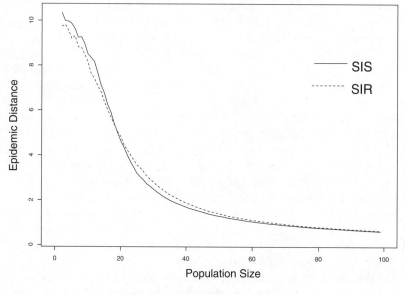

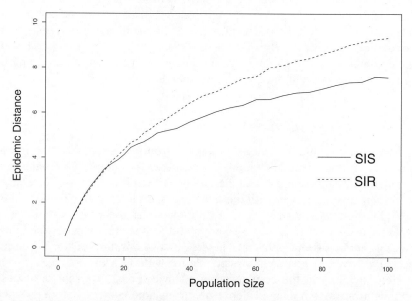

FIGURE 8.2. Epidemic distance varies with $R_0$ and population size. The average path metric for the S-I-S or S-I-R model with demographic stochasticity. The average path was estimated by simulating epidemics; the average was computed for an ensemble of 10,000 realizations. (a) Epidemic distance declines as population size increases, holding per capita transmission rates constant because $R_0$ increases. Here, $R_0$ ranges from 1 to more than 14. (b) Epidemic distance increases as a function of population size holding $R_0$ constant at 2.

These assumptions have been applied most commonly to the spread of sexually transmitted diseases (see Garnett and Anderson 1996, for example), but they may also reflect other kinds of population structure, such as age or grade in school, two factors that are important for childhood diseases. Spatial models are another kind of heterogeneous mixing; individuals are more likely to contact one another if they are close to the same place (Levin and Durrett 1996).

### Networks

Networks are a very general way of thinking about contact. Networks encompass a wide range of configurations, including random networks, in which the contacts are generated randomly, completely connected networks, and spatial networks (i.e., grids). Because a contact network on a finite set of hosts can be changed from a random network to a spatial network by making and breaking connections, networks provide a unified way of thinking about contact. From the network perspective, random mixing and spatial networks are special, limiting cases. An exhaustive study of epidemics on networks is beyond the scope of this paper (see Watts 1999 for a longer introduction).

In some cases, networks are the natural way of describing contact, such as sexual contact networks. Random mixing may be a reasonable approximation to many networks, as long as a given neighbor's neighbors are not importantly different from a random subset of all individuals. In practice, information about the structure of a contact network is extremely difficult to obtain. For many diseases, the spatial distribution of individuals is easier to characterize, the distance from an infected individual is a simple and reliable surrogate for the contact rate. Thus, space may be regarded as one useful way of understanding complicated contact networks.

## Homogeneous Spatial Models

The most common notion of space is the surface of the earth, although some habitats are roughly one-dimensional (e.g., a river) or three-dimensional (e.g., a lake). On a surface, the definition of a homogeneous spatial epidemic is ambivalent. A homogeneous spatial distribution of hosts is the uniform distribution, the points on a lattice or a uniform density in uncomplicated continuous space (Figure 8.3a), the variance is zero implying a departure from complete spatial randomness. A random distribution of points is drawn from a probability distribution function with a uniform *expectation*, the statistical definition of complete spatial randomness (Figure 8.3b). In a random distribution of hosts, the variance is equal to the mean. The major difference between these two distributions is the absence of pairs at very short distances in the uniform distribution (Figure 8.3c); in a uniform distribution (i.e., the trees in an orchard), hosts are arranged at regular intervals.

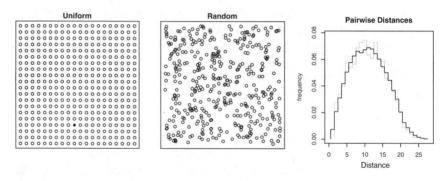

FIGURE 8.3. A homogeneous spatial epidemic is defined on a uniform distribution of points, not for random points drawn from a uniform distribution. The distribution of pairwise distances is similar for 400 uniform points or random points; the random distribution (black line) has slightly more pairs at very short distances compared with the uniform one (gray line).

In random distributions, some pairs are closer together than that interval. Very close pairs may be important for an infections agent to invade and persist in a population.

Aggregated distributions, in which the variance is greater than the mean, are clearly heterogeneous. Thus, we can establish a rule of thumb for three informal but meaningful classes of spatial epidemics. Homogeneous spatial epidemics occur in a completely uniform distribution of hosts. Heterogeneous distributions are those that are statistically indistinguishable from a uniform expectation (the variance is close to the mean), and those that are more aggregated (the variance is greater than the mean).

The most important property of a spatial distribution is that the majority of hosts are far away, but two hosts that have a close direct connection to a third host also have a fairly close direct connection to each other. This is not necessarily true for arbitrary networks. In space, if A and B are both close to C, then A and B must have some connection, even if C is removed. In contrast, A and B may be connected indirectly in an arbitrary contact network only through C; removing C may completely isolate A from B. Thus, when epidemics begin in geographic space, infected individuals are clustered. Epidemics spread as a front, and $R_0$ is proportional to the number of individuals in the neighborhood rather than the population density (Durrett and Levin 1994; Mollison and Levin 1995; Levin and Durrett 1996; Holmes 1997). Thus, spatial epidemics are characterized by a particular distribution of pairwise epidemic distances (Figure 8.3c). The same graph for pseudospace would have a spike at the one-distance.

Homogeneous spatial models can be classified by the representation of hosts as discrete or not and space as continuous or broken into patches (Durrett and Levin 1994). *Reaction-diffusion equations* (continuous space, no discrete individuals) are a straightforward extension of homogeneous mixing; nonspatial models can be transformed into spatial reaction-diffusion models by adding diffusion terms (Durrett and Levin 1994; Murray et al. 1986). *Spatially structured populations* (discrete space) subdivide a continuous landscape into a set of discrete patches. Each patch holds a population, and the local populations are well mixed. Epidemics spread by the movement of infected hosts or by dispersal of the infectious agent among patches. These models are typically deterministic with no discrete individuals, but directly analogous stochastic models can be built that incorporate demographic stochasticity and discrete individuals. For example, if patches are arranged on a grid and the net flux of individuals is proportional to the relative density, the model is a discrete analogue of the reaction-diffusion equations.

*Interacting particle systems* (discrete individuals on a lattice) are stochastic, grid-based models with at most one individual at each point on a lattice. On a grid, an individual has exactly four nearest neighbors, except possibly near the edge. In one common formulation of an epidemic on the grid, the probability of becoming infectious in some small interval of time is proportional to the number of infected neighbors. Alternatively, the infectious neighborhood can be expanded to the eight neighbors two-steps away, or the $4k$ neighbors $k$-steps away, ignoring edges. *Spatial point processes* (continuous space, discrete individuals) allow individuals to occupy any point in continuous space. Typically, the points are fixed representing the stem of a plant or the center of a territory or home range. The per capita probability of infection for an individual in a very small interval of time is approximately the sum of all infectious neighbors weighted by the probability of transmitting from some distance away, $\Sigma\beta(x)I(x)\,dx$.

## Heterogeneous Spatial Models

Spatial heterogeneity in epidemic models may include both configurational or compositional heterogeneity. A common type of configurational heterogeneity is a heterogeneous distribution of hosts because of variability in the quality of habitat. Host populations are usually distributed unevenly on a landscape, such as the distribution of humans in large cities and small towns. Alternatively, hosts may move faster through some areas, or the infectious agent may disperse among patches, effectively warping space. Heterogeneity may have been incorporated into interacting particle system models by eliminating some neighbors (Holmes 1997). Heterogeneity in spatial point processes may involve increasing the aggregation of the hosts.

Compositional heterogeneity includes heterogeneity in the distribution of an epidemiologically important trait. For example, heritable genetic traits that increase susceptibility to infection can be unevenly distributed among hosts on a landscape. Because epidemics can have demographic and/or selective effects on the host populations, compositional and configurational heterogeneity can be generated by an epidemic itself.

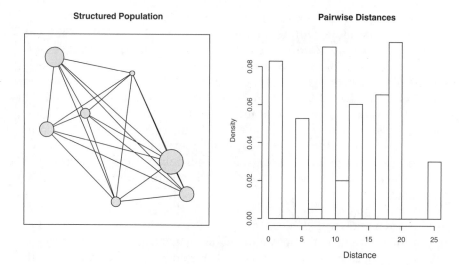

FIGURE 8.4. Structured epidemic models emphasize large-scale patchiness by ignoring local spatial structure (i.e., they assume that populations are locally well mixed). They are analogous to structured metapopulation models in ecology. Different subpopulations may have different sizes, represented here by the radius of the circle. For this distribution, the pairwise distances are much more heterogeneous than geographic space (compare with Figure 8.3).

## Structured Population Models

Structured population models (also called structured metapopulations) are a particularly useful hybrid between a well-mixed population model and a network model. A large population is subdivided into a set of local sub-populations connected by transmission or movement (diagram in Figure 8.4). Transmission in local subpopulations (patches) is well mixed, and local subpopulations are connected by transmission or the movement of hosts; the connectivity relationships among local subpopulations can be random hierarchical, spatial, or they can take any form specified by the network (Watts 1999). Adjacency is specified by a weighted graph describing the migration of individuals among local subpopulations (graph in Figure 8.4). Thus, structured population models emphasize spatial patchiness at large spatial scales by ignoring local spatial structure.

# Models and Data

To deal with the complexity of malaria, Sir Ronald Ross developed an approach to inference involving two complementary approaches that he called *a priori* and *a posteriori* (McKenzie 2000). *A priori* methods flow from cause to effect using mathematical models to understand the logical consequences of assumptions. A *posteriori* approaches involve reasoning from effect to cause by exploring patterns in data. Ideally, statistical inference proceeds within the general framework Ross described, although the *a posteriori* approach is far more familiar to most epidemiologists. Recent conceptual advances in likelihood, information theory, and complexity combined with the availability of inexpensive high-speed computers have made Ross's approach available to mainstream epidemiologists (Mollison 1984; Burnham and Anderson 1998).

More widespread use of *a priori* methods has been hampered by misperception and miscommunication. A model is a complicated hypothesis that explores the consequences of a set of assumptions. Mathematical models flow in the same direction of causation, from cause to effect, but the conclusions are only as good as the assumptions. In weighing the merits of a particular model or hypothesis, the standard is another model or hypothesis. Science proceeds by formulating different models and testing them against each other. Models must be judged critically by testing the assumptions or evaluating how well a model describes information in data. A perfect, but untested model may not have been formulated, a possibility that colors all conclusions with some skepticism. In the meantime, steady improvement in our approximating models provides a practical avenue to advance science. From this perspective, mathematical models are critical because they associate a biological mechanism with the pattern it generates. Moreover, because each effect has an associated effect size, mathematical models allow each effect to be quantified and interpreted.

Mathematical models are often criticized for being too simple, but population biology, like other academic disciplines, is guided by the principle of parsimony, a concept that has developed substantially since William of Occam's advice that we "shave away all that is unnecessary." Parsimony implies an appropriate level of complexity. Model building and inference for epidemics is a process of successive approximation, moving from simple to complex. In statistical inference, the best approximating model has an appropriate trade-off between bias and variance; explicit parsimony criteria have been derived from information theory as measures of information loss (Burnham and Anderson 1998). Formal measures of parsimony that are appropriate for scientific inference may not be appropriate for all uses. For example, models built for policymakers should robustly maximize an explicit policy outcome. Models that are too complex to be understood by policymakers are not useful, and those that are too simple to describe a complex system are not credible. The best model for making policy may not be the best model for inference (Ludwig and Walters 1985).

Deterministic models are easy to analyze and interpret, and the time course of an epidemic is entirely determined by the initial conditions. In stochastic models, an individual may be infected immediately in one realization but not in another. Stochastic epidemic models can be repeated as often as necessary. In contrast, real epidemics happen once, so everything about a complicated network must be inferred from the one and only epidemic. Moreover, real epidemics are extremely complicated and expensive to study. From this perspective, it should be obvious that even if everything about a real epidemic were perfectly observed, some information about the underlying process is lost (Anderson and May 1991)

Mathematical models help fill in the gaps; they are useful for developing concepts and quantitative intuition, synthesizing data from various sources, estimating and interpreting parameters, and integrating parameter estimates into a coherent picture of a whole process. Mathematical models are also a useful way to understand complex systems and evaluate the effects of heterogeneity. Epidemic distance may be a useful concept for comparing and contrasting models that are used for many purposes and selecting a simple one.

## Case Studies

The role of heterogeneity in infectious disease epidemiology has been demonstrated in a number of studies, especially those at the interface between basic and applied sciences. An important class of questions is the geographical spread of an invasive infectious disease, how fast does an epidemic wave move across a landscape? Spatial heterogeneity in the well-studied case of invasive rabies is discussed. Heterogeneity may also play a key role in the ability of an infectious agent to invade and persist in a local population. Examples include the role of spatial heterogeneity in an endemic plant disease, a vector-borne human disease, and the ongoing epidemic of antibiotic resistant bacteria. Finally, heterogeneity also has implications for the control of infectious diseases, including vaccination programs.

### Geographical Spread of Invasive Infectious Diseases

The invasive spread of infectious diseases in naïve populations are an interesting problem with potentially important management implications. For planning spatially oriented control measures, it is useful to identify areas where the spread is naturally inhibited by the landscape or by low population density (Murray et al. 1986; Shigesada et al. 1995). The rate of spread may be affected by configurational heterogeneity, such as differences in the spatial distribution of hosts or particular features of the landscape. Invasive infectious diseases, like invasive species, are prone to rare but important long-distance translocation events that establish nascent epidemic foci well in advance of an invading front (Mollison 1986).

One disease that has been well studied in both respects is rabies. The local movement of hosts tends to produce well defined and predictable traveling waves of infected hosts. In the case of rabies, geographical distance is a good proxy for epidemiological distance. Long-distance translocation tends to "homogenize" the landscape, effectively reducing the distance between any two hosts that are far apart.

## Fox Rabies

Epidemics of fox rabies in Europe at the end of World War II motivated the development of reaction-diffusion models for rabies epidemics in wildlife (Murray et al. 1986). On a homogeneous landscape with reaction-diffusion equations, an invasive disease spreads outward from the source in concentric circles. The speed of propagation of the epizootic wave-front depends on the local density of foxes. Heterogeneous distributions of foxes can speed up where hosts are most dense and slow down in areas where host density is too low to sustain an epidemic. High fox densities in the United Kingdom led to fears of a rabies epidemic there; the course of the epidemic was projected based on a population density map for foxes (Murray et al. 1986).

Other work on the fox rabies epidemic emphasizes the role of long-distance translocation in determining the speed of propagation and the shape of the invasive front (Mollison 1986). Diffusion implies rather strict conditions on the distribution of newly infected hosts; random movement or random dispersal may be modeled by many mathematical functions, including many fat-tailed distributions for which a relatively large fraction are dispersed long distances (Mollison 1991; Shigesada et al. 1995; Lewis and Pacala 2000).

## Raccoon Rabies

A recent epidemic of rabies in raccoons in the northeastern United States began in 1977 near the border between Virginia and West Virginia. Extensive testing of animals, including many that were behaving suspiciously, allowed the progress of the epidemic to be tracked over large spatial scales (Childs et al. 2000). Unfortunately, the location of raccoons that tested positive for rabies was recorded by county, or occasionally by township (usually a subdivision of a county). The political boundaries do not necessarily correspond to ecological boundaries, and the natural scale of the epidemic process is much smaller than the political unit. Thus, some spatial information about the epidemic was lost.

The incidence of raccoon rabies was recorded by township in Connecticut. Mathematical models for the epidemic in Connecticut considered the spread along a network of townships defined by geographic adjacency. The study sought to quantify spatial heterogeneity in the rate of spread and identify factors that would explain such heterogeneity. Models predicted that rabies would spread faster where host density was highest (Murray et al. 1986); raccoon density tends to be highest in suburban parkways (Hoffman and Gottschang 1977; Jones et al. 2003). Rivers were investigated as a possible barrier to dispersal. The effect of spatial heterogeneity was quantified and tested by fitting homogeneous and heterogeneous models to the data. Rivers were associated with a sevenfold slowing in the propagation of the epidemic

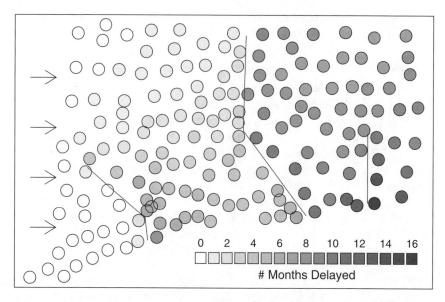

FIGURE 8.5. Large rivers can delay the spatial spread of an infectious disease. In Connecticut, rabies invaded from the east (arrows) but slowed at the rivers (lines) by a factor of seven; the delay is illustrated by degree of shading for each of the 169 townships plotted at the centroid of each township (circles). In the southeast corner, rivers would have delayed rabies by approximately 16 months. The configuration of the rivers determined the magnitude of the delay; for example, rabies can sometimes spread around the headwaters of the river faster than it would cross a river. In the actual epidemic, the rivers had a less important effect because of long-distance translocation. Rabies crossed the Connecticut River (middle line) early in the epidemic.

wave front (Figure 8.5), but no association was found between the speed of the traveling wave and human population density (Smith et al. 2002).

In Connecticut, rabies was detected in several townships in advance of the epidemic front, evidence that long-distance translocation of rabid raccoons was relatively frequent and important. At least one long-distance translocation event jumped the Connecticut river early in the epidemic. Such long-distance translocation events can minimize the slowing caused by heterogeneity, such as obstacles or areas of low population density, by occasionally leaping ahead of them (Smith et al. 2002).

## Spatial Heterogeneity and Endemicity

Measles epidemics have been instrumental in developing the concept of the critical community size (Anderson and May 1991). Threshold criteria describe the ability of an infectious agent to increase when rare, but the ability to persist depends on other aspects of the epidemiology. After the initial epidemic, infectious agents may fade-out and go extinct, especially if new susceptibles are generated too slowly to sustain a chain of infections. For example, childhood diseases with lifelong immunity tend to persist only in populations with high birth rates. Other aspects of persistence involve more ecological concepts, including metapopulations, source-sink and core-satellite relationships. The case studies that follow illustrate different concepts of epidemic distance.

A Foliar Pathogen

Configurational heterogeneity plays a particularly important role in the persistence of the fungal pathogen *Triphragmium ulmariae* and its host plant *Filipendula ulmaria* found growing on islands of the Skeppsvik Archipelago, northern Sweden (Burdon et al. 1995). The pathogen is host specific and virtually harmless. The host is a perennial plant that grows on the upper part of the shore. The host populations varied in size, and some islands had multiple host populations. The islands were created by glaciation, and channels of deep water arranged the islands into drumlin lines.

The complex life-cycle of the pathogen is mirrored by complex dispersal modes. As the host plant dies back to an underground rootstock during the winter, survival of the pathogen during this period is exclusively as teliospores found on dead leaf and stem fragments. *T. ulmariae* persists by reinfecting host plants as new stems grow through the spores in the detritus of the previous year's infected plant tissue, or possibly washed ashore as flotsam. New infections may also spread by windborne spores during the summer.

During the period 1990–2000, epidemiological patterns in the incidence, prevalence, and severity of disease were followed in this metapopulation. Model building and model selection were used to test different hypotheses about the role of configurational heterogeneity drawn from metapopulation and landscape theory. A suite of models incorporated different combinations of variables including the location and size of the host populations, hierarchical relationships among host populations imposed by the geology of the archipelago.

The study found strong evidence that the persistence of the parasite is determined by the configuration of the host populations. These are best described as core-satellite relationships; large host populations sustained stable populations of the parasites. A complex hierarchy in the spatial arrangement of host populations had a strong, secondary effect on persistence. Infection rates among populations on the same island were an order of magnitude greater than among islands in the same island chain. In turn, infection rates among populations on the same island chain were orders of magnitude higher than the baseline rate (Smith et al. 2003). Despite the obvious spatial distribution of the host populations, simpler population models that incorporated the natural spatial hierarchy performed as well as those that modeled the probability of infection as a function of distance. Structured population models were easier to understand, easier to interpret, and less computationally intensive.

## Antibiotic Resistance in Hospitals

Antibiotic resistance in nosocomial (hospital-acquired) infections is becoming increasingly frequent (NNIS 2001). As the name implies, transmission tends to be localized in hospitals, but compositional and configurational heterogeneity play an important role in understanding the spread of antibiotic resistance. Two of the most important bacterial pathogens are methicillin-resistant *Staphylococcus aureus* (MRSA) and vancomycin-resistant enterococci (VRE). A crucial feature of the epidemiology is the distinction between infection and colonization. Infections with VRE or MRSA, such as wound or bloodstream infections, are acute, serious clinical situations requiring treatment. In contrast, many people are colonized with VRE or MRSA; they carry bacteria in their gut, nasal passages, or skin without suffering illness. These carriers may shed resistant bacteria for years. Increased frequency of infection with antibiotic resistant bacteria in hospitals is a side effect of a largely silent epidemic of colonization with antibiotic resistant bacteria.

When bacteria persist for years, carriers move among institutions transmitting bacteria wherever they go. These individuals link all the institutions in a geographical region. Thus, antibiotic resistance epidemics are complicated spatial processes. Because the average frequency of resistance has been increasing, resistance must be spreading somewhere faster than it is lost. An important component of the public health response to epidemics of resistant bacteria is to identify where resistance is spreading.

Using structured population models, it is clear that some information about what kinds of institutions are responsible for transmission may be found by examining the time course of an epidemic (Smith et al. 2004b). Several simple models of well-mixed populations have focused on the spread primarily in hospital populations, but these models predict fairly fast epidemics; the average length of stay in a U.S. hospital is about 5 days.

Hospitals, long-term care facilities, and the community have different average lengths of stay, which are reflected in the dynamics of spread. As a rule, rapid increases in prevalence are driven by high transmission in institutions with fast turnover—usually hospitals (Figure 8.6a). Slower increases in the average prevalence of resistance in hospitals may occur because of epidemics sustained in places that have slower turnover, such as long-term care facilities or the community. Alternatively, in structured populations, resistant bacteria may increase in the catchment population of a hospital, even if no individual hospital or long-term care facility can sustain an epidemic. Increases in the frequency of resistant bacteria have had fast phases, including the epidemic of VRE in U.S. hospitals in the late 1980s and early 1990s, but these early epidemics have been followed by slow, steady increases since then (NNIS 2001).

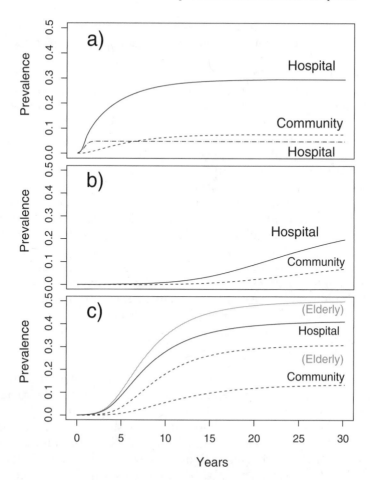

FIGURE 8.6. Configurational and compositional heterogeneity affect the time course of an epidemic of antibiotic-resistant bacteria; the epidemic, in this case, is a steady accumulation of antibiotic resistance carriers. (a) Population structure matters little for epidemics of hospital-acquired pathogens with short persistence times. Fast turnover of patients implies fast dynamics (dashed line). With persistent colonization, population structure is important. The rapid increase in the prevalence of resistance in the hospital (the proportion who are colonized by resistant bacteria) (solid line) is followed by a slow and steady rise in the number of carriers in the community, hence the proportion who return to the hospital still colonized (dashed line). (b–c) Compositional heterogeneity dramatically alters the time-course of an epidemic, compared to the homogeneous model. Daily transmission rates, per capita, are the same in the hospital and community in (b) and (c) and the *average* time to hospitalization is the same. In the bottom panel, heterogeneity in the frequency of hospitalization corresponds to the average hospitalization rates and composition of the elderly and nonelderly populations in the United States. Average prevalence in the hospital (black solid line) and in the community (black dashed line) increases more rapidly everywhere because of the elderly. Prevalence in the hospitalized elderly (gray solid line) and among the elderly in the community (gray dashed line) is always higher than average.

Heterogeneity in transmission rates associated with different health care institutions and the community, a kind of configurational heterogeneity, interact with other kinds of heterogeneity that are analogous to compositional heterogeneity. In the United States, there are about 600 people in the community for every occupied hospital bed, and the average period between hospital visits is about 8.7 years. These average measures of hospitalization ignore the fact that the elderly account for about 13% of the total population, but half the total days of care in hospitals. Correcting for this kind of heterogeneity, the elderly are hospitalized once every 2.2 years, on average, while nonelderly are hospitalized about once every 15 years (Figures 8.6b and 8.6c).

Frequently hospitalized populations are much more likely to be colonized from previous hospital visits and much more likely to remain colonized when they are readmitted to a hospital. These individuals represent a major challenge to hospital infection control programs. The elderly are one kind of population that is hospitalized much more frequently than average, but other populations may also play a role, including those on dialysis, cancer patients, and the mentally ill. In general, heterogeneity in the frequency of hospitalization interacts with heterogeneity in transmission making it much easier for antibiotic resistant bacteria to spread and persist.

## The Distribution of Risk in Mosquito-Borne Diseases

Malaria control programs have a long history with notable successes and failures, but malaria remains a leading cause of preventable death in children today, with the majority of deaths occurring in Africa (Killeen et al. 2003). Early models of Ross and MacDonald were instrumental in formulating malaria control policies (Ross 1911; Macdonald 1957). From these models, an estimate of $R_0$ was derived in terms of the basic parameters; in the control context, $R_0$ provides a measure of the factor by which transmission must be reduced to eliminate a disease. Early control focused on vector biting behavior and adult mosquito survivorship based in part on the analysis of these models; $R_0$ is most sensitive to these parameters (Macdonald 1957).

The Ross-MacDonald models assumed well-mixed, constant mosquito populations, but ignoring heterogeneity generates dramatic underestimates of $R_0$ and the difficulty of locally eliminating malaria. Mosquitoes prefer some hosts to others, for a variety of reasons (Takken and Knols 1999). Mosquito biting preferences generate heterogeneity in the human biting rate and large corresponding differences in $R_0$ (Dietz 1980; Dye and Hasibeder 1986). Spatial heterogeneity has increasingly been identified as an important issue because of small-scale spatial variability in the risk of disease (Staedke et al. 2003; van der Hoek et al., 2003). New technologies and high-speed computing have made it possible to develop continent-wide maps using remote sensing and GIS (Hay 1997; Rogers et al. 2002). Mathematical models can help guide studies that identify the distribution of risk at spatial scales ranging upwards from the daily flight distance of a mosquito.

Spatial variability in the distribution of larval habitat and humans can have similar effects. A common assumption about malaria, dengue and other mosquito-borne diseases is that the two main components of the risk of human infection—the rate at which people are bitten (human biting rate) and the prevalence of infection in mosquitoes—are positively correlated. In fact, these two risk factors are generated by different processes and may be negatively correlated in spatially heterogeneous environments. The uneven distribution of larval habitat creates a spatial mosaic of demographic sources and sinks. Mosquitoes seek blood meals; they tend to aggregate around areas where blood meals are readily available. Heterogeneous distribution of larval habitat and the populations that provide blood meals can generate complicated patterns in the distribution of risk for vector-borne diseases.

Models predict that the risk of human infection is highest near breeding sites where adult mosquitoes emerge or around aggregations of humans (Figure 8.7). In contrast, the prevalence of infection in mosquitoes reflects the age-structure of mosquito populations; it peaks where old mosquitoes are found, far from mosquito breeding habitat, and while mosquito density is declining (Aron and May 1982; Smith et al. 2004a).

## Measles

A century of measles reporting in the United Kingdom generated a long, detailed record of measles cases. Detailed analysis of the time series has

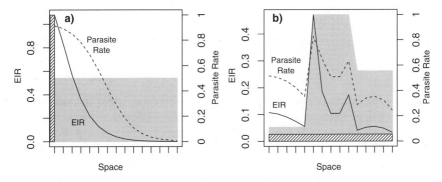

FIGURE 8.7. Vector searching behavior, the heterogeneous distribution of larval habitat, and heterogeneous distributions of humans can generate complicated patterns in the distribution of risk of mosquito-borne diseases. The prevalence of infection in humans (dashed line) is related to the entomological inoculation rate (EIR), the number of infectious bites, per human, per day (EIR, solid line). When larval habitat (dark hashed bars) and humans (gray background) are heterogeneously distributed, the distribution of risk is determined by (a) the distribution of larval habitat and (b) the behavior of adult female mosquitoes seeking a blood meal. These figures illustrate biting patterns that form on two different heterogeneous landscapes characterized by a) homogeneous hosts, but adult vector mosquitoes emerge from a point-source and b) homogeneous emergence of adult vector mosquitoes, but heterogeneously distributed hosts.

provided some important insights into the spatio-temporal persistence of childhood diseases (Bolker and Grenfell 1995). Despite complicated age structure and school structure, the spatio-temporal record of cases is explained well by a relatively simple model (Grenfell et al. 1995). The patterns are driven by stochastic fade-out of measles in small cities with net population birth rates too low to sustain epidemics and by persistence of measles in large cities. In large cities, such as London, measles is more or less always present. After stochastic fade-out in the smaller cities and the regeneration of the pool of susceptible individuals, the epidemics are reinitiated when an infectious individual moves from a city where measles has persisted. The spatio-temporal patterns across the United Kingdom reflect these stochastic dynamics; large cities synchronize the epidemics across the United Kingdom by initiating new epidemics in smaller cities where measles has gone locally extinct (Grenfell et al. 2001).

## Heterogeneity and Disease Control

Heterogeneity should be an important consideration in the planning and implementation of disease control. In simple models with vaccination, a disease may be eliminated if the population is susceptibles is reduced such that $R_0 < 1$. This is possible if the fraction protected exceeds $1 - 1/R_0$ (Anderson and May 1985, 1991). This is more difficult than expected in aggregated or otherwise heterogeneous populations, but the efficiency of vaccination programs can be improved by targeting those groups that are most susceptible or mixing at the highest rates (Anderson and May 1985). Uneven implementation of vaccine programs can generate heterogeneity, leading to their failure.

### Measles

Measles vaccination programs interact with heterogeneity in community size and many types of compositional or configurational heterogeneity. In Israel where regular vaccination was commonplace, a majority of measles cases occurred in the ethnic minority Bedouin population. In this case, the heterogeneity may have been generated by the uneven implementation of the vaccination program itself. One study concluded that improving vaccination coverage in the Bedouin population would have a disproportionate positive effect on measles elsewhere (Agur et al. 1993).

The special role played by large cities in sustaining measles may also provide an opportunity for vaccination. Measles may fade-out stochastically in small cities if vaccination is efficient enough in the larger cities (Anderson and May 1985). On the other hand, vaccination decreases the role that large cities play in synchronizing epidemics in small cities. Because epidemic peaks and troughs in the small cities tend to occur at different times, it is less likely that transmission will be interrupted everywhere all at once (Bolker and Grenfell 1996).

Foot-and-Mouth Disease

The virus causing foot-and-mouth disease is highly contagious, with $R_0$ estimated to be near 40 in the absence of control with frequent long-distance translocation (Haydon et al. 1997; Keeling et al. 2001). The elimination of foot-and-mouth disease in the United Kingdom focused on the farms that were near an infected farm. More recent studies of foot-and-mouth disease in England emphasized the role of large farms that were both more susceptible and more contagious than smaller farms; similar arguments apply to cattle farms compared with sheep farms (Keeling et al. 2001). This sort of compositional heterogeneity is important for management; vaccinating herds on the largest cattle farms would be substantially more efficient than a strategy that ignores heterogeneity (Keeling et al. 2003).

Polio

The eradication of poliomyelitis would mark a great achievement, but serious challenges remain (Dowdle et al. 2003). Wild-type polio virus has been eliminated from many places. The polio eradication initiative is now focused on interrupting transmission in a few populations, those with high birth rates, poor hygiene, and low vaccine coverage. Once transmission of the wild-type virus has ended, debate will turn to the "endgame issues." When and how will vaccination end? (Technical Consultative Group to the World Health Organization on the Global Eradication of Poliomyelitis 2002).

A serious obstacle to polio eradication may be the live oral poliovirus vaccine (OPV) itself. OPV is relatively inexpensive and easy to administer, and OPV viral strains are shed from vaccinated individuals, potentially exposing contacts who may also develop immunity. These advantages of OPV have made it the tool of choice for eliminating transmission of wild-type polio virus. On the other hand, a serious risk exists that the vaccine strain may revert to the wild-type in a vaccinated person or after transmission (Fine and Carneiro 1999; Nathanson and Fine 2002).

Mathematical modeling will be used to advise decision makers on the endgame strategies (Technical Consultative Group to the World Health Organization on the Global Eradication of Poliomyelitis 2002). The decision of how to weigh compositional and configurational heterogeneity will be an important feature of these models. After exposure to the vaccine strain, a small fraction of individuals become chronic shedders, a sort of compositional heterogeneity in the duration of infectiousness (Kew et al. 2004). In some places with high birth rates and poor hygiene, the population may be large enough to sustain a new epidemic in 2–5 years after the end of vaccination, especially in areas where polio has been difficult to eliminate. After the last wave of vaccination, polio may persist long enough to initiate a new epidemic through some combination of persistent shedding, transmission, and rapid regeneration of conditions suitable for transmission. The combination of configurational heterogeneity and compositional heterogeneity cast serious doubts on the prospects of ending the vaccination programs using only OPV. In the endgame, the polio eradication initiative may have to use inactivated polio vaccine (IPV) to eradicate OPV.

## Conclusions

Most epidemic theory for infectious diseases has been developed under the assumptions that all individuals are alike and that populations are well mixed. Epidemiological theory often stands in sharp contrast to studies that emphasize the role of compositional and configurational heterogeneity. Obviously, heterogeneity does affect who gets sick and the population dynamics of infectious diseases over time and space. Moreover, disease control programs are likely to create heterogeneity. Heterogeneity should be weighed and ignored, whenever possible, but in many cases, ignoring heterogeneity can mislead policymakers about the true nature of a problem and generate misguided policy.

Heterogeneity matters in epidemiology, but the details vary among diseases and populations. Understanding heterogeneity involves careful study, system by system. Mathematical epidemiologists must become familiar with the peculiar sorts of heterogeneity that are important in particular systems. To understand the interplay between heterogeneity and the nonlinear aspects of epidemics, modeling must become an integral part of infectious disease surveillance and control.

*Acknowledgments.* The author gratefully acknowledges support from the Department of Epidemiology and Preventive Medicine, University of Maryland School of Medicine and the Fogarty International Center, National Institutes of Health. Mark Miller introduced me to the problems associated with the polio eradication endgame. Jonathan Dushoff and F. Ellis McKenzie were especially supportive throughout, making important contributions to my thoughts about this essay.

## Appendix

The "average path" metric takes into account the expected time to infection considering all possible paths. For any finite population of hosts, we let epidemic distance between two individuals, A and B, denoted d(A,B), have following properties:

1. The epidemiological distance from an individual to itself is 0 [d (A,A) = 0].
2. The epidemic distance from A to B is the expected number of epidemic generations for an infection to reach B from A multiplied by the inverse of the probability that an infection starting at A reaches B. More precisely:

(a) p(A,B) denotes the probability that an epidemic started at A infects B before going extinct. The epidemiological distance for every pair is computed assuming that A is the index (i.e., first) case.

(b) $\tau(A,B)$ denotes the average time for an infection started at A to reach B, conditioned on the epidemic reaching B.

(c)  $\chi$ denotes the generation time of the infectious agent.

(d) $d(A,B) = \tau(A,B) / [\chi\, p(A,B)]$. This "average path" metric satisfies the mathematical requirements of a distance metric. In particular, the distance between any two different individuals is always positive [$d(A,B) > 0$ when $B = A$) and the distance satisfies the triangle inequality ($d(A,B) \leq d(A,C) + d(C,B)$], as the expected time from A to B is computed using all possible paths, including the one through C. As noted previously, it is not necessarily true that the distance from A to B is the same as the distance from B to A.

## References

Adler, F.R. 1992. The effects of averaging on the basic reproduction ratio. Math. Biosci. 111: 89–98.

Agur, Z., Danon, Y.L., Anderson, R.M., Cojocaru, L., and May, R.M. 1993. Measles immunization strategies for an epidemiologically heterogeneous population: the Israeli case study. Proc. R. Soc. London B Biol. Sci. 252: 81–84.

Anderson, R.M., and May, R.M. 1985. Vaccination and herd immunity to infectious diseases. Nature 318: 323.

Anderson, R.M., and May, R.M. 1991. Infectious diseases of humans. Oxford: Oxford University Press.

Aron, J.L., and May, R.M. 1982. The population dynamics of malaria. In Population dynamics and infectious disease, ed. R.M. Anderson, pp. 139–179. London: Chapman and Hall.

Black, F.L., and Singer, B. 1987. Elaboration versus simplification in refining mathematical models of infectious disease. Annu. Rev. Microbiol. 41: 677–701.

Bolker, B., and Grenfell, B. 1995. Space, persistence and dynamics of measles epidemics. Philos. Trans. R. Soc. London B Biol. Sci. 348: 309–20.

Bolker, B., and Grenfell, B. 1996. Impact of vaccination on the spatial correlation and persistence of measles dynamics. Proc. Natl. Acad. Sci. U S A 93: 12648–53.

Box, G.E.P. 1979. Robustness in the strategy of scientific model building. In R.L. Launer Robustness in statistics, eds. R.L. Launer and G.N. Wilkinson, pp. 201–236. New York: Academic Press.

Burdon, J.J., Ericson, L., and Müller, W.J. 1995. Temporal and spatial changes in a metapopulation of the rust pathogen *Triphragmium ulmariae* and its host, *Filipendula ulmaria*. J. Ecol. 83: 979–989.

Burnham, K.P., and Anderson, D.A. 1998. Model selection and inference: A. Practical information—theoretic approach. New York: Springer-Verlag.

Childs, J.E., Curns, A.T., Dey, M.E., Real, L.A., Feinstein, L., Bjornstad, O.N., and Krebs, J.W. 2000. Predicting the local dynamics of epizootic rabies among raccoons in the United States. Proc. Natl. Acad. Sci. U S A 97: 13666–13371.

Dietz, K. 1980. Models for vector-borne parasitic diseases. Lecture Notes Biomath. 39: 264–277.

Dietz, K. 1993. The estimation of the basic reproduction number for infectious diseases. Statistical Methods Med. Res. 2: 23–41.

Dowdle, W.R., Gourville, E. De, Kew, O.M., Pallansch, M.A., and Wood, D.J. 2003. Polio eradication: the OPV paradox. Rev. Med. Virol. 13: 277–291.

Durrett, R., and Levin, S.A. 1994. The importance of being discrete (and spatial). Theor. Popul. Biol. 46: 363–394.

Dushoff, J., and Levin, S. 1995. The effects of population heterogeneity on disease invasion. Math. Biosci. 128: 25–40.

Dye, C., and Hasibeder, G. 1986. Population dynamics of mosquito-borne disease: effects of flies which bite some people more frequently than others. Trans. R. Soc. Trop. Med. Hygiene 80: 69–77.

Fine, P.E., and Carneiro, I.A. 1999. Transmissibility and persistence of oral polio vaccine viruses: implications for the global poliomyelitis eradication initiative. Am. J. Epidemiol. 15: 1001–1021.

Garnett, G.P., and Anderson, R.M. 1996. Sexually transmitted diseases and sexual behavior: insights from mathematical models. J. Infect. Dis. 174: S150–S161.

Grenfell, B.T., Bjornstad, O.N., and Kappey, J. 2001. Traveling waves and spatial hierarchies in measles epidemics. Nature 414: 716–23.

Grenfell, B.T., Kleczkowski, A., Gilligan, C.A., and Bolker, B.M. 1995. Spatial heterogeneity, nonlinear dynamics and chaos in infectious diseases. Statistical Methods Med. Res. 4: 160–83.

Hamer, W.H. 1906. Epidemic disease in England. The Lancet i: 733–739.

Hay, S.I. 1997. Remote sensing and disease control: past, present and future. Trans. R. Soc. Trop. Med. Hygiene 91: 105–106.

Haydon, D.T., Woolhouse, M.E., and Kitching, R.P. 1997. An analysis of foot-and-mouth-disease epidemics in the UK. IMA J. Math. Appl. Med. Biol. 14: 1–9.

Hethcote, H.W. 1994. A thousand and one epidemic models. In Frontiers in mathematical biology, ed. S.A. Levin, pp. 504–515. Lecture Notes in Biomathematics Berlin: Springer.

Hethcote, H.W. 2000. The mathematics of infectious diseases. SIAM Rev. 42: 599–653.

Hoffman, C.O., and Gottschang, J.L. 1977. Numbers, distribution, and movements of a raccoon population in a suburban residential community. J. Mammol. 56: 623–636.

Holmes, E.H. 1997. Basic epidemiological concepts in a spatial context. In Spatial ecology, eds. D. Tilman and P. Kareiva, pp. 111–136. Princeton, NJ: Princeton University Press.

Jones, M.E., Curns, A.T., Krebs, J.W., and Childs, J.E. 2003. Environmental and human demographic features associated with epizootic raccoon rabies in Maryland, Pennsylvania, and Virginia. J. Wildlife Dis. 39: 869–874.

Keeling, M.J., Woolhouse, M.E., Shaw, D.J., Matthews, L., Chase-Topping, M., Haydon, D.T., Cornell, S.J., Kappey, J., Wilesmith, J., and Grenfell, B.T. 2001. Dynamics of the 2001 UK foot and mouth epidemic: stochastic dispersal in a heterogeneous landscape. Science 294: 813–7.

Keeling, M.J., Woolhouse, M.E., May, R.M., Davies, G., and Grenfell, B.T. 2003. Modeling vaccination strategies against foot-and-mouth disease. Nature 421: 136–142.

Kermack, W.O., and McKendrick, A.G. 1927. A contribution to the mathematical theory of epidemics. Proc. R. Soc. A115: 700–721.

Kew, O.M., Wright, P.F., Agol, V.I., Delpeyroux, F., Shimizu, H., Nathanson, N., and Pallansch, M.A. 2004. Circulating vaccine-derived polioviruses: current state of knowledge. Bull. WHO, 82: 16–23.

Killeen, G.F., Knols, B.G., and Gu, W. 2003. Taking malaria transmission out of the bottle: implications of mosquito dispersal for vector-control interventions. Lancet Infect. Dis. 3: 297–303.

Levins, R. 1969. Some demographic and genetic consequences of environmental heterogeneity for biological control. Bull. Entomol. Soc. Am. 15: 237–240.

Lewis, M.A., and Pacala, S. 2000. Modeling and analysis of stochastic invasion processes. J. Math. Biol. 41: 387–429.

Ludwig, D., and Walters, C.J. 1985. Are age-structured models appropriate for catch-effort data? Can. J. Fisheries Aquatic Sci. 42: 1066–1072.

Macdonald, G. 1957. The epidemiology and control of malaria. London: Oxford University Press.

McKenzie, F.E. 2000. Why model malaria? Parasitol. Today 16: 511–6.

Mollison, D. 1984. Simplifying simple epidemic models. Nature 310: 224–225.

Mollison, D. 1986. Modeling biological invasions: chance, explanation, prediction. Philosophical Trans. R. Soc. London B 314: 675–693.

Mollison, D. 1991. Dependence of epidemic and population velocities on basic parameters. Math. Biosci. 107: 255–287.

Mollison, D., and Levin, S.A. 1995. Spatial dynamics of parasitism. In Ecology of infectious diseases in natural populations, eds. B.T. Grenfell and A.P. Dobson, pp. 384–398. Cambridge, UK: Cambridge University Press.

Murray, J.D., Stanley, E.A., and Brown, D.L. 1986. On the spatial spread of rabies among foxes. Proc. R. Soc. London Ser. B Biol. Sci. 229: 111–150.

Nathanson, N., and Fine, P. 2002. Poliomyelitis eradication—a dangerous endgame. Science 296: 269–270.

NNIS. 2001. National Nosocomial Infections Surveillance (NNIS) system report, data summary from January 1992–June 2001, issued August 2001. Am J Infect Control 29: 404–421.

Rogers, D.J., Randolph, S.E., Snow, R.W., and Hay, S.I. 2002. Satellite imagery in the study and forecast of malaria. Nature 415: 710–715.

Ross, R. 1911. The prevention of malaria. London: Murray.

Shigesada, N., Kawasaki, K., and Takeda, Y. 1995. Modeling stratified diffusion in biological invasions. Am. Naturalist, 146: 229–251.

Smith, D.L., Lucey, B.T., Waller, L.A., Childs, J.E., and Real, L.A. 2002. Predicting the spatial dynamics of rabies epidemics on heterogeneous landscapes. Proc. Natl. Acad. Sci. U. S. A. 99: 3668–3672.

Smith, D.L., Ericson, L., and Burdon, J.J. 2003. Epidemiological patterns at multiple spatial scales; an 11-year study of a *Triphragmium ulmariae–Filipendula ulmaria* metapopulation. J. Ecol. 91: 890–903.

Smith, D.L., Dushoff, J., and McKenzie, F.E. 2004a. The risk of a mosquito-borne infection in heterogeneous environments. PLOS Biol. 2:e368.

Smith, D.L., Dushoff, J., Perencevich, E.N., Harris, A.D., and Levin, S.A. 2004b. Persistent colonization and the spread of antibiotic resistant nosocomial pathogens through health care systems: Resistance is a regional problem. Proc. Natl. Acad. Sci. U. S. A. 101: 3709–3714.

Staedke, S.G., Nottingham, E.W., Cox, J., Kamya, M.R., Rosenthal, P.J., and Dorsey, G. 2003. Short report: proximity to mosquito breeding sites as a risk factor for clinical malaria episodes in an urban cohort of ugandan children. Am. J. Trop. Med. Hygiene 69: 244–246.

Takken, W., and Knols, B.G.J. 1999. Odor-mediated behavior of Afrotropical malaria mosquitoes. Annu. Rev. Entomol. 44: 131–157.

Technical Consultative Group to the World Health Organization on the Global Eradication of Poliomyelitis. 2002. "Endgame" issues for the global polio eradication initiative. Clin. Infect. Dis. 34: 72–77.

van der Hoek, W., Konradsen, F., Amerasinghe, P.H., Perera, D., Piyaratne, M.K., and Amerasinghe, F.P. 2003. Towards a risk map of malaria for sri lanka: the importance of house location relative to vector breeding sites. Int. J. Epidemiol. 32: 280–285.

Watts, D.J. 1999. Small worlds. Princeton, NJ: Princeton University Press.

# 9
# Spatial Heterogeneity and Its Relation to Processes in the Upper Ocean

Amala Mahadevan

## Abstract

In the ocean, the spatial distribution of biogeochemical tracers is affected by their physical transport in the fluid medium. Many tracer distributions such as sea surface chlorophyll and temperature are highly correlated at length scales of 1–100 km on account of a commonality in the transport processes that affect them. We characterize and differentiate between the spatial heterogeneity of the tracers by using a variance-based measure for "patchiness." When we analyze the satellite-derived fields of surface chlorophyll and temperature, we find that chlorophyll is more patchy than temperature (i.e., a greater proportion of its variance occurs at small scales). We explain such differences in heterogeneity by taking the approach that the observed spatial heterogeneity of a tracer results from a balance between processes that generate variance and those that shift the variance from one length scale to another. The longevity of the tracer determines the extent to which the variance can be shifted to another scale. In the surface ocean, variance introduced at large scales due to geographic variations can be driven to smaller scales by the horizontal stirring and stretching of fluid filaments. On the other hand, small-scale vertical motion associated with fronts introduces small-scale variance that spreads to larger scales if the tracer anomalies are long lasting. For the latter case, we derive a quantitative relationship between a tracer's patchiness and the timescales of processes that modify its concentration in the upper ocean. This relationship links the observed spatial heterogeneity in the system to the processes that contribute to its generation. It lends hope to our being able to use quantitative measures of spatial heterogeneity, like the patchiness parameter defined here, to gain information about processes or, *vice versa*, to predict how the spatial heterogeneity might be modified as a result of a change in processes.

# Introduction

A key factor that influences spatial heterogeneity in the ocean and distinguishes it from heterogeneity in landscapes is that substances or properties in the ocean are transported within the fluid medium, which is in motion. Hence, the variability in the distribution of properties is largely linked to the dynamics of the fluid, which is complex, as it varies in both space and time. Spatial heterogeneity in the ocean is constantly evolving in time, in contrast to terrestrial systems, where the variability of the underlying medium (e.g., the geology or soil conditions) is more or less static on the timescales of relevance in the ocean. The fluid dynamical processes act over a wide range of time and length scales. In addition, there are a number of processes like warming or cooling at the surface, evaporation and precipitation, biological production of phytoplankton, and air-sea gas exchange that alter the properties of the ocean. Some processes generate variability and others annihilate it; our objective is to understand the net effect of these factors on the distribution of properties.

Transport in the ocean occurs through the process of advection, which carries properties along with the flow, and diffusion, due to which substances or energy can spread through the fluid. Diffusion is generally associated with small scales; it is important to individual plankton and bacteria and they may rely, for example, on the spatial variation in the concentration of a nutrient for its transport. A net diffusive flux occurs without the input of energy when the concentration gradient of a substance is spatially varying. The molecular diffusivity $\kappa$, of most substances is small and thus diffusive transport (quantified as $\kappa \nabla^2 c$, where $c$ is the tracer concentration and $\nabla$ the gradient operator) is relevant only at small length scales [the diffusive length scale $L_{diff} \sim (\kappa T)^{1/2} \approx 1$ mm, for $\kappa = 10^{-5}$ ms$^{-2}$ and a time interval $T = 10$ s]. At longer times $T$, and at length scales greater than a centimeter or so in the upper ocean where typical velocities $U$ are in the range of 0.01 to 1 ms$^{-1}$, advective transport by the fluid by far dominates diffusion. A net advective flux of tracer, $\mathbf{u} \cdot \nabla c$ occurs when there is a concentration gradient in the tracer $\nabla c$, in the direction of the fluid velocity $\mathbf{u}$. The length scale associated with advection, $L_{advec} \sim UT$, increases linearly with time $T$, as compared to the 1/2 power in the case of diffusion. Hence, with increasing time, advection affects larger length scales than diffusion. A process of considerable relevance in the ocean, and somewhat different from pure advection and diffusion, is mixing. It transfers energy and property gradients from larger to smaller scales and results in the homogenization of properties over the region on which it operates on relatively short timescales. Mixing, which is often induced by shear and convective instability, contributes much more to the flux of energy and tracers than molecular diffusivity, which is relatively negligible for length scales of more than a few centimeters. The effective flux of energy or tracer generated by random turbulent motions is often parameterized as a diffusion-like process by using an enhanced "eddy" diffusivity $\kappa_{eddy}$.

Another aspect that differentiates the ocean from the land surface is that it is a three-dimensional medium. However, the rotation of the earth, the small geometrical aspect ratio of depth to length scales in the ocean, and density stratification all contribute to the fact that motion in the horizontal plane dominates vertical motion. In fact, vertical velocities are several orders of magnitude smaller than horizontal velocities when one considers length scales of the order tens of meters or more. Hence, even though the ocean is three-dimensional, it is highly layered. The variation (i.e., the gradient) of properties is much stronger in the vertical than horizontal, but motion is much more rapid in the horizontal. Hence, the distributions of properties evolve more rapidly in the horizontal plane. The upper ocean, in particular, is more energetic and fast moving than the deep. Hence, the property distributions are rife with a highly transient variability that is influenced strongly by advection and mixing within the fluid, as well as forcing factors that modify the properties. Figure 9.1 displays the kind of spatial variability that results from the coupling between the biological and fluid transport processes.

In this chapter, we focus largely on upper ocean heterogeneity that has a transience time scale of the order weeks and a coherence length scale in the range 1–100 km. Because the fluid dynamical transport processes are common to the various substances or properties in the ocean, it is no surprise that the distributions of different tracers in the ocean are correlated on these length and time scales. Yet, the spatial heterogeneity of one property can vary from that of another to which it is closely linked. One challenge is therefore to quantitatively relate the heterogeneity of properties to the processes that affect them. Such an understanding might enable us to use one property as a proxy for the distribution of another and, secondly, to learn something about the processes at play from the characteristics of the distributions. Further, we might be able to anticipate a change in spatial heterogeneity resulting from a change in the processes or controlling parameters.

In general, the spatial and temporal heterogeneity of a system depends on the length and time scales of relevance and the process under consideration. Questions concerning the importance of heterogeneity, how we may model it, and how it might affect the ecosystem function are difficult to generalize and best viewed in a specific context. But, a somewhat general set of questions about heterogeneity and its link to processes that I would like to consider are:

1. What role does heterogeneity play for the question at hand? What is the effect of varying the heterogeneity (or some measure of it) on an ecosystem function, an ecosystem response, or on an integrated measure of interest?
2. How might one quantify the heterogeneity of the system?
3. What are the most important parameters/factors/processes behind the heterogeneity? How does the heterogeneity vary as a function of these parameters?

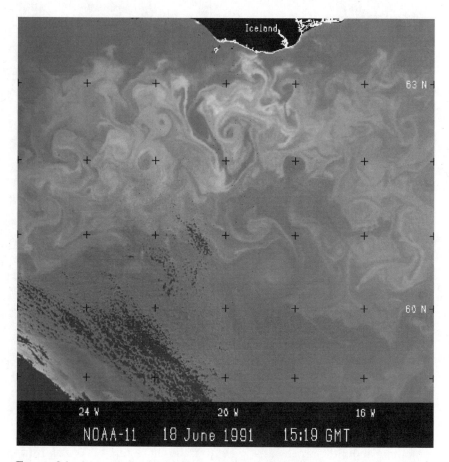

FIGURE 9.1.  An example of spatial heterogeneity in the ocean is seen in this Advanced Very High Resolution Radiometer (AVHRR) satellite image of a coccolithophore bloom south of Iceland in June 1991. Coccolithophores are a kind of phytoplankton that grows calcium carbonate plates that are shed, making the water appear milky in these images. The small crosses in the figure are 110-km apart. The image shows the strong coupling between physical flow fields and the biological distributions in the upper ocean. (AVHRR image received at the NERC Receiving Station Dundee and processed at the Plymouth Marine Laboratory. Courtesy Steve Groom.)

4. What is the relationship between the heterogeneity, the ecosystem response, and the parameters on which the heterogeneity depends?
5. Is it possible to account for the heterogeneity without explicitly modeling or measuring it? How might one sample a variable to correctly estimate an integrated measure of it?

Discussed below are some thoughts relating to these questions. Following this, I describe a study that attempts to explain and relate the spatial heterogeneity of different tracers at the sea surface over length scales in the range 1–500 km.

## The Relevance of Heterogeneity

Most oceanic processes are intermittent in space and time, and it is often important to account for variability to gain an accurate estimate of a process or flux. This is particularly true when a processes in nonlinear, in which case, even an integrated flux measure across the system's boundary requires a description of the heterogeneity. Let us consider, for example, the flux of carbon dioxide ($CO_2$) across the air-sea interface. We would like a time-integrated estimate of the flux of $CO_2$ into or out of the ocean, though the flux varies in time depending on the properties of the sea water, wind, and surface conditions. The air-sea gas flux is generally parameterized as the product of a gas exchange coefficient $k$, the solubility of the gas, and the difference in the partial pressure of the gas $\Delta pCO_2$ between sea and air. The gas exchange coefficient $k$ is estimated empirically and is typically a function of the wind speed raised to a power that varies between 1.6 and 3, depending on the formulation (Liss and Merlivat 1986; Wanninkhof 1992; Wanninkhof and McGillis 1999). This nonlinear dependence implies that short bursts and gusts of wind are more effective in fluxing $CO_2$ across the air-sea interface than a constant wind of the same mean intensity. Because the average of the instantaneous wind speed when squared is not the same as the square of the wind speed averaged in time, the averaging period and the frequency of sampling the wind speed become relevant to the estimate of air-sea gas transfer that one would obtain from such a relationship. An estimate for the global air-sea flux of $CO_2$ can vary by a factor of two depending on whether we use monthly averaged or 6-hourly winds to compute the fluxes. Any covariance between the variables $k$, $s$, and $\Delta pCO_2$ also affects the estimate and requires accounting for each of their variabilities independently over short timescales. Thus, the required resolution or the permissible period of averaging that is required to capture a process is highly dependent on the process and the distributions of the variables themselves. In this case, the integrated flux in and out of the system is dependent on the heterogeneity at the boundary, as the process has nonlinear dependencies.

As another example, consider the new production rate of phytoplankton in the subtropical gyres of oceans. New production (as opposed to the production that feeds off recently recycled organic matter) is derived from the supply of fresh nutrients from the subsurface, a processes that is highly episodic in time and space. A snapshot view of the ocean does not adequately represent this process, but the time-integrated effect of the process affects its state. Though the transport of nutrient by fluid advection may be considered a linear process, it is dependent on the spatial gradients in the nutrient. Quicker uptake of nutrient in the upper ocean and more efficient lateral transport at the surface ensures a steeper vertical gradient in the nutrient concentration and a greater net flux to the surface from below. Patchy upwelling generates a heterogeneous surface distribution of nutrient. This enhances the lateral nutrient transport and leads to stronger vertical nutrient gradients at the upwelling sites, consequently resulting in a greater supply of nutrients from the subsurface as compared to the situation where the upwelling is uniformly distributed in space (Martin et al. 2002). In this case, the transport of nutrient within various components of the system are linear processes, but the net productivity is nonetheless affected by the heterogeneity in the processes and distributions within a system. The net productivity in turn, affects the surface distribution of $pCO_2$ and the flux of $CO_2$ in and out of the system.

On much smaller scales, the transport of oxygen to a patch of decaying organic matter is dependent on the spatial heterogeneity in the oxygen distribution, as the diffusive flux is proportional to the second spatial derivative of the concentration. If the supply of oxygen is rate-limiting to the process, then the heterogeneity of the oxygen distribution that may be generated by the bacterial uptake itself is crucial for this activity. In such a case, the "patch" of decaying matter is not self-contained and depends on the spatial heterogeneity generated by itself or its neighbors for its survival. The turbulent diffusion of mechanical energy also occurs on a similar length scale to that at which diffusion operates because the molecular diffusivity of momentum is comparable to that of a trace substance in the fluid. However, turbulent dissipation is itself intermittent, and the intermittency in the turbulent energy dissipation rate is found to account for a decrease in zooplankton-phytoplankton encounter rates by 25–50%, an increase in the nitrogen flux to nonmotile phytoplankton cells by 6–62%, and a decrease in the coagulation and sedimentation of phytoplankton cells by 25–40% in experiments (Seuront 2001).

## What Causes Spatial Heterogeneity?

One way of thinking of spatial heterogeneity is that it results from competing processes: one set that tends to homogenize the distribution of a property and another that tends to introduce variance or heterogeneity in the system. If one considers, for example, temperature in the ocean, it is homogenized by mixing and diffusion at small scales, but unequal heating or cooling generates spatial heterogeneity in its distribution on very large scales. At the intermediate scale, one could think of long-wave radiation as relaxing the temperature to ambient atmospheric conditions and advective motions in the fluid as generating heterogeneity by stirring. The observed spatial distribution would be more homogeneous if the diffusion-like mixing processes were relatively vigorous or the relaxation to an ambient state were more rapid, and more heterogeneous if the unequally heated regions were stirred into fine scale structures more rapidly than can be homogenized. Processes like diffusion or relaxation to an ambient state tend to generate uniformity in the fields, whereas specific sources for the properties, like biological reproduction or generation by nucleation, create heterogeneity. In the case of phytoplankton, heterogeneity is generated by the variable response of phytoplankton to varying physical properties and the availability of light and nutrients but also by their reproduction, which is dependent on the presence of mature phytoplankton cells. They are, however, removed or reduced to an ambient state of low concentration by predation, death, and sinking. Their distributions are also homogenized by mixing and made more heterogeneous by advection, which can generate narrow filamentous structures by stirring.

## Relating Heterogeneity to Process TimeScales

The extent of the spatial heterogeneity in the distribution of a property results from the balance between the processes that homogenize and generate heterogeneity. These processes can be quantified in terms of the timescale on which they alter the concentration of the property. Hence, the rate of change of concentration $c$ of a property can be expressed as

$$\frac{\partial c}{\partial t} = \frac{c}{\tau_v} - \frac{c}{\tau_H}, \tag{9.1}$$

where $\tau_v$ is a timescale on which variance is increased in the system, and $\tau_H$ is a timescale on which the distribution is homogenized. In the case where $c$ describes the concentration of phytoplankton whose heterogeneity is considered over length scales ranging from 0.1 to 1 m, $\tau_v$ could be the timescale of net growth or reproduction, while $\tau_H$ might be the timescale of diffusion $\sim L^2/\kappa$, where $L$ is a length scale and $\kappa$ is the kinematic or eddy diffusivity. In the statistically steady-state, it is the balance between the right-hand-side terms in the above equation that determines the characteristics of the distribution. Hence, it is the ratio of timescales $\tau_H/\tau_v$ that determines the degree of spatial heterogeneity of the system. Later in this chapter, we will show how the patchiness or spatial heterogeneity varies with the ratio $\tau_H/\tau_v$, when variance is introduced at the small scales. It turns out the dependence is logarithmic, so that the distributions are more sensitive to the ratio $\tau_H/\tau_v$ when it is small.

## Accounting for Heterogeneity

We have earlier seen that the spatial heterogeneity of different properties can vary substantially. Thus, the grid resolution required in models and observation networks depends on the spatial heterogeneity of the property, given that one would like to observe the majority of its variance. Highly heterogeneous distributions require more resolution. Once again, the factor by which the resolution needs to be scaled up when going from one property to another more heterogeneous one can be related to the ratio $\tau_H/\tau_\nu$ for each of these. The question of how to account for heterogeneity is a more difficult one. It depends on the function or process that one wishes to account for (how it depends on the heterogeneous property) and also on the statistical characteristics of the property's distribution.

## Quantifying Heterogeneity

Several methods have been used to quantify heterogeneity in the oceans. The most common among these are spectral analysis (Platt and Denman 1975; Gower et al. 1980), semi-variogram analysis (Yoder et al. 1987; Yoder et al. 1993; Glover et al. 2002; Deschamps et al. 1981) and autocorrelation analysis (Campbell and Esaias 1985), probability density functions (pdf's), structure functions and multifractals (Seuront et al. 1999). In general, these methods analyze the variability of a distribution as a function of the length scale. A method that we have chosen to use in this presentation characterizes the variance as a function of the size of the region. When a greater proportion of the variance lies at smaller length scales, we tend to refer to the distribution as more heterogeneous, patchy, or intermittent. Once again, this depends on the range of length scales that one is considering.

In terms of processes, one may think of those that tend to shift the variance in a distribution to smaller scales or others that obliterate (smear) it. In a fluid, advection or stirring tends to drive variance to smaller scales because fluid filaments interleave and fold, generating finer scale filaments. Thus, stirring two fluids generates one in which variance moves downscale with time. Hence, the length scale at which the variance or heterogeneity is initially introduced is relevant. If it be at the large scale, then variance can increase with time due to advection by the fluid. But if a process introduces heterogeneity at the small scale, then it gets annihilated with time due to processes like diffusion that smear gradients and reduce variance.

## The Distribution of Biogeochemical Tracers at the Sea Surface

Sea surface temperature (SST) and chlorophyll (Chl) are two properties of the ocean that can be remotely measured from satellite platforms at a global resolution of approximately 1 km × 1 km. The distributions of these properties are highly correlated because both Chl and temperature are advected by the same underlying flow (Figure 9.2a). Oceanographic flow is largely

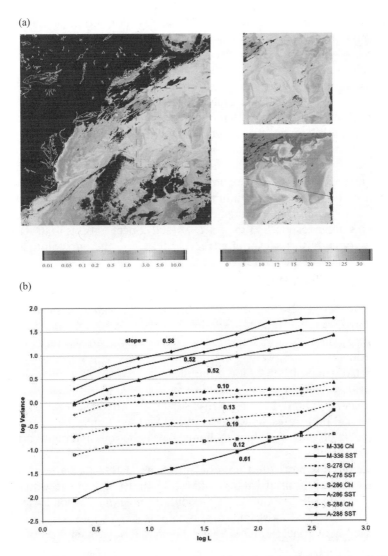

FIGURE 9.2. (a) Satellite image of sea surface chlorophyll (Chl) in the Atlantic Ocean acquired by the Moderate Resolution Imaging Spectroradiometer (MODIS). The boxes on the right show simultaneous views of the SST and Chl for the dashed region in the larger picture. This region is $512 \times 512$ km$^2$. The key is logarithmic for Chl but linear for SST. (b) The variance *versus* length scale (i.e., *V-L*) relationship for Chl and SST plotted on log-log axes for simultaneous satellite data of SST (bold) and Chl (dotted) as that shown above. The M-336 curves are based on MODIS data from the Arabian Sea. In addition, we show results for three concurrent AVHRR (A-278, A-286, and A-288) SST and SeaWiFS (S-278, S-286, and S-288) Chl images from the North Atlantic in October 2000. The length scales analyzed range from 2 to 512 km. The slopes indicated are estimated for lines fitted to the points between $L = 4$ km and 256 km. (Reproduced by permission of the American Geophysical Union from Mahadevan and Campbell 2002.)

two-dimensional; velocities in the vertical are several orders of magnitude smaller than in the horizontal when the length scales considered are such that the geometrical aspect (depth to length) ratio is small. Yet, these vertical velocities, which are of the order tens of meters per day at most, can introduce anomalous concentrations at the surface because the concentration gradient of most substances and properties in the ocean is very large in the vertical as compared to the horizontal. Upper ocean processes, such as air-sea exchange, heating, evaporation, and phytoplankton production, modify the concentrations of substances, whereas the slow rate of mixing across the thermocline (which extends to a depth of 500 or so meters beneath the surface mixed layer and is a region over which temperature and density change rapidly with depth) maintains a strong concentration gradient in the vertical. This common characteristic in the distribution of various properties ensures a similar response in their sea surface concentrations to upwelling, and consequently to horizontal advection that stirs the anomalous signatures introduced at the sea surface by upwelling.

In the following sections, I describe a study in collaboration with J.W. Campbell (Mahadevan and Campbell 2002, 2003) in which we relate the spatial heterogeneity of tracers at the sea surface to the characteristic response time of processes that modify them in the upper ocean. Our interest is in quantifying and understanding processes such as the air-sea flux of $CO_2$, new production, the upwelling rate of nitrate (a key nutrient for phytoplankton), organic carbon export from the surface ocean, and rates of remineralization of detrital organic matter that are related to the carbon cycle in the ocean. Spatial heterogeneity seems to affect several of these processes and their rates; hence, we would like to examine the reasons for the spatial heterogeneity in biogeochemical distributions and relate them to the underlying processes.

This study was motivated largely by the need to make a connection between remotely sensed variables like SST and Chl and those that are more difficult to observe but play an essential role in the carbon cycle, such as the total dissolved inorganic carbon content ($TCO_2$) and oxygen ($O_2$). The questions of concern are: Why do the spatial heterogeneities of these substances differ and how may they be related? How may we account for this when modeling or observing different variables?

When we analyze satellite data for the sea surface distributions of Chl and temperature in the pelagic ocean, we find they are correlated, but Chl has a greater percentage of its variance at smaller length scales as compared to temperature. This analysis is done by computing the variance in two concurrent views of SST and Chl in a region covered by $256 \times 256$ pixels of data, where each pixel is approximately 1 $km^2$ in area. The variance corresponding to a length scale $L$ is computed as $V(L) = (N-1)^{-1} \sum_{i=1}^{N} (x_i - \bar{x})^2$, where $N$ is the number of pixels, $x_i$ is the value of the variable at the $i$th pixel, and $\bar{x}$ is the mean over all $N$ pixels. $L$ is the dimension of the box over which the variance is computed. Having computed the variance $V_1$ over the whole domain of size $L_1$, we divide the domain into boxes of consequently smaller sizes $L_1/2, L_1/4, L_1/8$ ... and so forth and compute the average (over all the boxes in the domain) variance associated with each box of size $L_1/2$, $L_1/4, L_1/8$ .... We normalize the variance $V(L)$ by the total variance in the domain $V_1$ so that we can compare the correspondence between variance and length scale for different variables. In Figure 9.2b, we show $V$ plotted against $L$ on log-log axes for Chl and SST from different regions of the ocean. There are two notable features in these plots: (a) The plots of log $V$ versus log $L$ are more or less linear, suggesting that

$$V \sim L^p, \tag{9.2}$$

for Chl and SST over this range of length scales, and (b) $p$(Chl) is consistently less than $p$(SST). The slope $p$ of the log $V$ versus log $L$ plots is a measure of the spatial heterogeneity of the distribution. Smaller $p$ implies that a greater percentage of the total variance is contained in small scales and we think of the distribution as more "patchy." Hence, $p$ can be considered a patchiness index or a measure of the spatial heterogeneity, smaller $p$ corresponding to greater patchiness. Sea surface Chl is found to be more patchy than SST for length scales ranging from 1 to 500 km. The reason for this is explored in the next few paragraphs where $p$ is related to the properties of the tracer.

In an earlier study (Mahadavan and Archer 2000), we examined the variability of different biogeochemical tracers within a model for a region of the subtropical gyres of the ocean. This model is initialized with nutrients (for phytoplankton) absent from the surface waters but increasing with depth. When upwelling brings nutrients to the surface sunlit layers, they are converted to organic matter by the new production of phytoplankton. The phytoplankton production takes up $TCO_2$ in the surface layer, but $TCO_2$ is also modified by the air-sea exchange of $CO_2$ gas on a much longer timescale (several months to a year because $CO_2$ in the ocean equilibrates very slowly with the atmosphere). Dissolved $O_2$ is taken up by the remineralization of organic matter produced in the model, released by phytoplankton production, and also subject to air-sea exchange, but with a shorter equilibration time than $CO_2$. We include two idealized tracers that are initialized as varying exponentially from 0 at the surface to 1 at depth over an e-folding distance of 100 m. Their concentration changes due to the flow, but they are restored to their initial concentration profiles on timescales of 60 days and 3 days. The slowly restored tracer is meant to mimic dissolved organic carbon (DOC), and the fast-acting tracer resembles hydrogen peroxide $H_2O_2$.

When we analyze the surface distribution of the various tracers in the model, we find that the faster responding tracers or properties like new production, $H_2O_2$, and $O_2$ develop smaller scales or more patchiness than their respective counterparts: temperature, DOC, and $CO_2$ (Figure 9.3). This dependence on the timescale can be understood by considering the balance between the processes that generate variance (at the small scale) and those

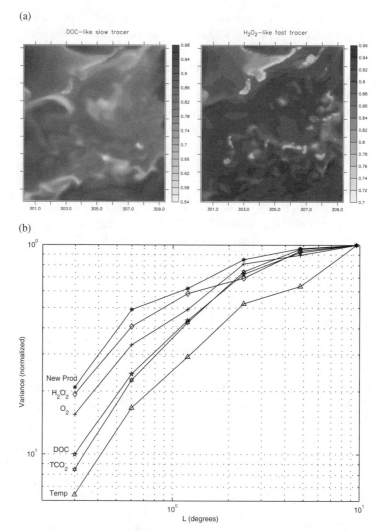

FIGURE 9.3. (a) Surface distributions of a fast and slowly responding tracer within a model of a $10^0 \times 10^0$ region of the Atlantic. The model was driven by time-dependent boundary conditions extracted from a global circulation model and run at a resolution of $0.1^0$ latitude-longitude. The tracers shown are the fast-acting $H_2O_2$-like tracer and the slower DOC-like tracer. They were initialized as being 1 at the surface and decreasing exponentially to 0 over an e-folding depth of 100 m. When the tracers deviated from this distribution due to advection, they were restored to it with an e-folding time of 3 and 60 days, respectively. The tracer with the shorter response time is seen to develop finer scale structure. (b) The log $V$ versus log $L$ curves demonstrate that the different tracers used in this model [temperature, dissolved inorganic carbon (DIC), oxygen ($O_2$), new production, and the $H_2O_2$-like fast and DOC-like slow tracers] develop different spatial heterogeneities. The faster acting tracers are more patchy and have smaller $p$ than the slower acting ones. Reproduced by permission of CRC Press from Mahadevan and Campbell 2003.

that obliterate it or shift it to larger scales. The range of length scales considered here (the sub-meso and meso scales covering 1–500 km) is rather energetic. This is because the natural length scale of frontal instabilities and eddies in the ocean, the so-called internal Rossby radius of deformation, lies within this range; it is typically between 10 and 100 km. Upwelling associated with fronts and eddies occurs along the edges of frontal meanders at sub-mesoscales (1–10 km), which are even finer than the internal Rossby radius. Because tracer concentration gradients are strong in the vertical, sub-mesoscale upwelling introduces an anomalous signature in the sea surface concentration of tracers, or can alternatively be thought of as introducing variance at the finest scales under consideration here. If the tracer considered is nutrient (nitrate or phosphate that are essential for phytoplankton production), then the upwelling results in the generation of small-scale phytoplankton patches. This is the situation when the surface of the pelagic ocean is depleted of nutrients, and phytoplankton production is limited by the supply of nutrients from the subsurface. Phytoplankton production takes up the nutrients from the surface, restoring it to its nutrient-depleted state. If the timescale for biological production in response to the nutrients is short, then the variance in phytoplankton is seen mostly at the small scales at which it was generated. If the response timescale is large, the nutrient patches generated by sub-mesoscale upwelling spread to larger regions due to the advection of the upwelling features in the flow. Consequently, the variance is distributed to larger scales.

A balance between the processes that generate small-scale variance (sub-mesoscale upwelling) and those that remove it (nutrient uptake) can be written as

$$w \frac{\partial c}{\partial z} \sim -\frac{c}{\tau}, \tag{9.3}$$

where $c$ is the anomalous tracer concentration at the surface (normalized by the mean), $w$ is the upwelling velocity, $z$ denotes the vertical coordinate direction, and $\tau$ is the response or removal timescale of the tracer (nutrient) in the surface layers. We scale this relation, by choosing $W$ to represent the typical upwelling velocity, $c$ and $c_\infty$ to be the normalized concentration of the tracer at the surface and depth, and $h$ to be the characteristic depth over which the upwelling occurs. Thus, $w \partial c / \partial z \sim W(c - c_\infty)/h \sim -c/\tau$. Taking the logarithm of both sides (terms 2 and 3), using the definition of the patchiness index $p \equiv \log V / \log L$, and considering the variance spread over the same range of length scales for different tracers with different $\tau$, we get the relation (Mahadevan and Campbell 2002)

$$p \sim \log \bar{\tau} + \log c_\infty \tag{9.4}$$

where $\bar{\tau}$ is the timescale $\tau$ normalized by the upwelling time scale $h/W$. The scaling suggests that the patchiness varies as the logarithm of the ratio of the timescales $\tau$ and $h/W$. The timescale $h/W$ may be difficult to ascertain in the field, but the spatial heterogeneity of various tracers can nonetheless be compared using this relation, because they are affected by the same flow field. Hence, their patchiness $p$ is expected to vary logarithmically with the characteristic timescale of the tracer's response to processes that modify it in the upper ocean. Although this scaling relation is derived using a rather simplistic balance between two processes (sub-mesoscale upwelling and nutrient uptake), tracers in the ocean may be affected by a number of competing processes with several timescales. It is possible to account for multiple timescales by calculating an effective nondimensional $\bar{\tau}$ using $\frac{1}{\tau} = \frac{1}{\tau_1} + \frac{1}{\tau_2} + \frac{1}{\tau_3} + \ldots$ .

This dependence of the spatial heterogeneity on the response timescale is tested with a three-dimensional fluid dynamical model of an ocean front that contains a tracer. The model is configured in an east-west periodic channel that is initialized with a north-south density gradient representative of an upper ocean front in the mid-latitudes. The front and the associated east-west jet are baroclinically unstable and form meanders and eddies, as are seen in the ocean. The tracer is initialized to resemble nitrate; it is absent from the surface and increases exponentially with depth. Any excursions of the tracer from this initial profile due to upwelling are restored to the initial state on a timescale $\tau$, where $\tau$ is chosen to be 2.5, 5, 10, 20, 40, and 80 days. The spatial heterogeneity of the surface layer of the model is analyzed for the tracers with different values of $\tau$ at different instants in the flow using the variance based analysis (Figure 9.4). The slope of the $V$-$L$ plot in log-log space gives the value of the patchiness index $p$. This index $p$ is found to vary with the log of the timescale $\tau$, suggesting that spatial heterogeneity of tracers with different response timescales that are affected by the same fluid dynamical motions is related by the scaling relationship we derived.

The above-discussed relationship between patchiness and response time applies to regions of the ocean where the phytoplankton growth is strongly limited by the upwelling of nutrients from the subsurface. Even though we neglected the effect of horizontal advection in generating small-scale variance while deriving the scaling relationship, the three-dimensional model results are fairly consistent. This is because the vertical motions make a dominant contribution to surface spatial distributions in the model, as is the case in many regions of the ocean. However, it is noteworthy that different results are obtained when the effects of horizontal advection are dominant as in the situation where phytoplankton blooms generate large-scale gradients in Chl between bloom regions and clear waters. In this case, variance is introduced at the large scale and stirred in to smaller scale filaments by advection (Martin 2003). The relationship between the longevity of the tracer and its patchiness is just the opposite of what is derived above, because with time, the tracers develop finer scale structure. Abraham (1998) uses the balance between advection and reaction in a two-dimensional flow to explain the difference in patchiness between phytoplankton and zooplankton populations in a model. In his

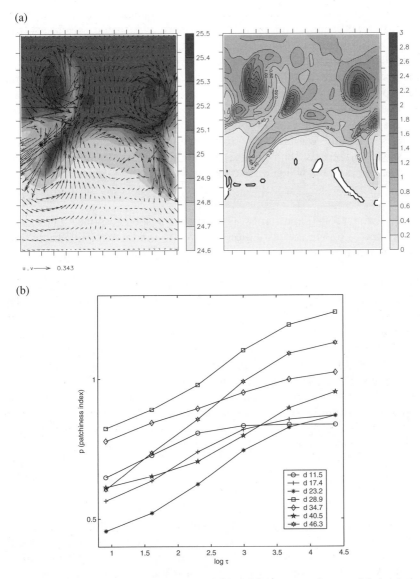

FIGURE 9.4. (a) Top left: Surface density field and flow vectors in a model simulation of a baroclinically unstable evolving front shown 17 days after initialization. The domain size is 258 km × 285 km and the horizontal resolution is approximately 4 km. Top right: Surface view of the tracer (nutrient) distribution in the upper 95 m. The tracer shown has a response time $\tau$ of 40 days. Tracers with a lesser value of $\tau$ show a similar distribution but are more patchy and less abundant. (b) The slope of the log $V$ *versus* log $L$ curves for the tracers in the model plotted *versus* the value of $\tau$ (on a logarithmic axis) to verify that $V \sim \log \tau$. Each curve corresponds to a different time in the simulation at which the spatial heterogeneity was analyzed. (Reproduced by permission of the American Geophysical Union from Mahadevan and Campbell 2002.)

modeling study, the longer-lived zooplankton generate finer scales than the phytoplankton. This demonstrates how relevant processes are to the spatial heterogeneity.

Finally, we address the issue of modeling or observing different tracers given what we have learned about their spatial heterogeneity and its dependence on the response time $\tau$. If our intention is to capture the bulk of the variance in the distribution of the tracer, then clearly, the tracer that is more patchy requires higher spatial resolution for modeling or observing. If, for example, we wish to capture 80% of the variance in a tracer distribution, then setting $V$, the nondimensional variance as 0.8 in Eq. (9.2) and taking $L$ to be the model resolution $\Delta$, we get $\log \Delta \sim 1/p$, which when combined with Eq. (9.4) gives

$$\Delta \sim \exp(-1/\log \bar{\tau}). \tag{9.5}$$

Thus, two tracers that differ by a factor of 10 in their response times $\tau$, differ by a factor of 4 in the resolution required to model or sample them. The faster acting tracer generates variance at finer scales and requires higher model and sampling resolution.

Phytoplankton reproduction and organic matter remineralization timescales are typically of the order a few days and are shorter than the timescale for equilibration of physical properties in the oceanic mixed layer with the atmosphere. (For example, the timescale on which the surface ocean temperature equilibrates with the atmosphere may be approximately a month, but the timescale for phytoplankton growth is only a few days.) This means that higher spatial resolution is needed to model and observe the phytoplankton ecology and carbon chemistry in the oceans than to model the distribution of temperature. Further, the logarithmic dependence of the spatial heterogeneity on $\tau$ implies that models are much more sensitive to the choice of the process timescale $\tau$, when $\tau$ is small.

## Conclusions

By analyzing the distribution variance $V$ of various properties in the surface ocean as a function of length scale $L$ and plotting $\log V$ *versus* $\log L$, we find that $V \sim L^p$ for satellite-derived fields of sea surface Chl and temperature on length scales ranging from 1 to 500 km. We use the slope $p$ of the $V$-$L$ curves as a measure of the spatial heterogeneity or patchiness; smaller $p$ corresponds to greater patchiness. We develop a relationship between the spatial heterogeneity of sea surface tracers and the timescale of processes that modify them by considering the balance between processes that generate variance at one scale and processes that shift the variance to another scale. The distributions of tracers at the sea surface are affected by submesoscale upwelling, which introduces variance at small scales, and also by processes like air-sea exchange, heat flux, evaporation, or biological production, which remove this variance and restore the surface ocean to an ambient state. The ratio between the timescale of upwelling and the timescale of the process that modifies the tracer in the upper ocean $\tau$ determines the patchiness of the system. For various tracers affected by the same flow field, the patchiness index $p \sim \log \bar{\tau}$. This relationship enables us to relate the spatial heterogeneity of different tracers with different response times. It also enables us to estimate the model resolution needed to model the variability of different substances in the ocean.

Finally, it is worth putting this oceanographic study in the context of terrestrial studies and asking whether our approach is generalizable to other systems. Distributions in the ocean tend to be transient and coupled to the dynamics of the fluid medium. Because properties in the ocean are evolving, we use a measure of spatial heterogeneity that is independent of the specifics of the distribution but characterizes it in terms of the distribution of variance over length scales. Such measures of heterogeneity can also be used for terrestrial systems if the appropriate variable (such as the concentration of a substance or property) can be measured. Assuming a steady state in the variance distribution, we then balance the factors that change the variance in the system in opposing ways. This approach could be applied to systems where the processes generating or changing variance are understood and quantifiable and when a steady state can be assumed in the variance characteristics. But many terrestrial systems evolve very slowly, and hence it is difficult to measure the processes that modify them and their rates of modification, particularly when the evolution of the system occurs in a highly erratic fashion. Nonetheless, we hope that this study would provoke thinking about what properties and parameters need to be measured in a system to explain its heterogeneity. Further, various measures for spatial heterogeneity might be conceived for different systems. If indeed relationships can be found between these measures and the processes that alter the system, then it is conceivable that the evolution of the spatial heterogeneity will be predictable from information about the processes, or inversely, the spatial heterogeneity of a system will provide information about the rates or strengths of the processes at play.

*Acknowledgments.* The SeaWiFS and MODIS data were obtained from the NASA Goddard DAAC. This work was sponsored by ONR (N00014-00-C-0079) and NASA (NAS5-96063 and NAG5-11258). Thanks to Steve Groom and Timothy Moore for Figures 9.1 and 9.2.

## References

Abraham, E. 1998. The generation of plankton patchiness by turbulent stirring. Nature 391: 577–580.

Campbell, J., and Esaias, W. 1985. Spatial patterns in temperature and chlorophyll on Nantucket Shoals from airborne remote sensing data, May 7–9, 1981. J. Marine Res. 43: 139–161.

Deschamps, P., Frouin, R., and Wald, L. 1981. Satellite determination of the mesoscale variability of the sea surface temperature. J. Phys. Oceanogr. 11: 864–870.

Glover, D.M., Doney, S.C., Mariano, A.J., Evans, R.H., and McCue, S.J. 2002. Mesoscale variability in time-series data: Satellite based estimates of the U.S. JGOFS Bermuda Atlantic Time-series Study (BATS) site. J. Geophys. Res. 107 (C8), 7–1 to 7–14, doi: 10.1029/2000JC000589.

Gower, J., Denman, K., and Holyer, R. 1980. Phytoplankton patchiness indicates the fluctuation spectrum of mesoscale oceanic structure. Nature 288: 157–159.

Liss, P., and Merlivat, L. 1986. Air-sea gas exchange rates: introduction and synthesis. In The role of air-sea exchange in geochemical cycling, ed. P. Buat-Menard, pp. 113–129. Hingham, MA: D. Reidel.

Mahadevan, A., and Archer, D. 2000. Modeling the impact of fronts and mesoscale circulation on the nutrient supply and biogeochemistry of the upper ocean. J. Geophys. Res. 105: 1209–1225.

Mahadevan, A., and Campbell, J. 2002. Biogeochemical patchiness at the sea surface. Geophys. Res. Lett. 29: 1926.

Mahadevan, A., and Campbell, J. 2003. Biogeochemical variability at the sea surface: How it is linked to process response times. In Handbook of scaling methods in aquatic ecology: Measurement, analysis, simulation, eds. L. Seuront and P.G. Strutton, pp. 215–227. Boca Raton, FL: CRC Press LLC.

Martin, A. 2003. Plankton patchiness: the role of lateral stirring and mixing. Progress Oceanogr. 57: 125–174.

Martin, A., Richards, K., Bracco, A., and Provenzale, A. 2002. Patchy productivity in the open ocean. Global Biogeochemical Cycles 16, 9–1 to 9–9, doi 10.1029/2001GB001449.

Platt, T., and Denman, K. 1975. Spectral analysis in ecology. Annu. Rev. Ecol. Systematics 6: 189–210.

Seuront, L., Schmitt, F., Lagadeuc, Y., Schertzer, D., and Lovejoy, S. 1999. Universal multifractal analysis as a tool to characterize multiscale intermittent patterns: example of phytoplankton distribution in turbulent coastal waters. J. Plankton Res. 21: 877–922.

Seuront, L. 2001. Microscale processes in the ocean: why are they so important for ecosystem functioning. La mer 39: 1–8.

Wanninkhof, R. 1992. Relationship between wind speed and gas exchange over the ocean. J. Geophys. Res. 97: 7373–7382.

Wanninkhof, R., and McGillis, W. 1999. A cubic relationship between air-sea $CO_2$ exchange and wind speed. Geophys. Res. Lett. 26: 1889–1892.

Yoder, J., McClain, C., Blanton, J., and Oey, L.-Y. 1987. Spatial scales in CZCS-chlorophyll imagery of the southeastern U.S. continental shelf. Limnol. Oceanogr. 32: 929–941.

Yoder, J., Aiken, J., Swift, R.N., Hoge, F., and Stegmann, P. 1993. Spatial variability in the near-surface chlorophyll a fluorescence measured by the Airborne Oceanographic Lidar (AOL). Deep Sea Res. II 40: 37–53.

# Section III

## Illustrations of Heterogeneity and Ecosystem Function

# Editors' Introduction to Section III: Illustrations of Heterogeneity and Ecosystem Function

Where does heterogeneity matter, and where does it not? We asked the authors of this section to consider these questions for specific ecosystems. The resulting chapters contain rich examples from a wide range of ecosystems, each detailing the drivers of heterogeneity. Many of the examples illustrate its importance—that it does matter—for specific ecosystem functions. The ecosystems described here range from those in which heterogeneity is quintessentially obvious and important and has been the focal point of research for decades, to those in which heterogeneity is more cryptic, and understanding its importance to ecosystem function is nascent. Though the examples are diverse, there are some common themes that emerge from these chapters.

First, real-world heterogeneity is often a result of a complex suite of abiotic controls upon which biology acts or to which it responds, and *vice versa*; an unsurprising but nonetheless recurring and interesting theme. The examples in this chapter fall along a continuum. At one end, there are strong abiotic controls that influence ecosystem function (e.g., arid and semiarid and riparian systems). At broad spatial and temporal scales, geology, geomorphology, and climate set the stage for heterogeneity. As Kratz et al. (Chapter 16) point out, "these factors set limits on physical properties, biogeochemistry, and biotic assemblages." At the other end of the continuum, strong biotic controls operate on an abiotic template (e.g., human influences on lakes, self- organization of terrestrial vegetation, and urban ecosystems). All along the continuum, however, the addition of biology makes it a challenge to model and predict both heterogeneity and ecosystem function: simple linear models are rarely sufficient. And to make life even more complex, heterogeneity is sometimes created, or maintained, by important feedbacks among biotic and abiotic processes (e.g., some aspects of riparian systems, *Sphagnum* bogs, semiarid systems). Second, the legacies of processes that created heterogeneity in the past can be important in structuring current heterogeneity in some ecosystems (e.g., boreal and riparian systems). Finally, all of the authors of this section acknowledge that links between spatial and temporal heterogeneity and ecosystem function are only beginning to be made and that a framework is (or frameworks are) needed. In regard to the latter, several of the authors of this section gave perspectives on conceptual frameworks for the generation and consideration of heterogeneity. For example, Meinders and van Breemen (Chapter 11) address the use of Turing's model, Naiman et al. (Chapter 14) and Band et al. (Chapter 13) apply patch dynamics to analyze structure and function of heterogeneity in the ecosystems on which they work, and Tongway and Ludwig (Chapter 10) offer a generic conceptual framework that links pattern to process for semiarid systems. All chapters are rich in detail and are food for thought about where and when heterogeneity matters to ecosystem function; highlights of the chapters follow.

Heterogeneity in vegetation pattern has long been observed, measured, modeled, and manipulated in arid landscapes where it *is* the name-of-the-game. Because resources are severely limited, coupling between crucial resources and vegetation response is tight, and thus they are excellent systems in which to develop (relatively) simple conceptual models. David Tongway and John Ludwig's paper (Chapter 10) gives an excellent overview of the history of inquiry and details heterogeneity and function in arid and semiarid ecosystems. They demonstrate that vegetation distribution and ecological function within these landscapes are coupled with the abiotic and biotic processes that control distribution of crucial resources: water, organic matter, propagules. The resulting redistribution and concentration of resources makes predictable the spatial distribution of vegetation. They offer conceptual models for determining the spatial distribution of crucial resources and for the processes through which the redistributions arise. Finally, they illustrate the necessity and utility of management goals in modeling ecosystem processes in these heterogeneous landscapes.

Using the examples of semiarid systems, peat bogs, and mull and mor-forming plants, Marcel Meinders and Nico van Breemen (Chapter 11) discuss the creation of landscape pattern as a result of soil-vegetation feedback interactions. Their examples show that there are, superimposed on geomorphology and climate, feedbacks between vegetation and soils that may create heterogeneity in an initially homogeneous landscape. First, they describe self-organized systems, based on Turing's model of self-organization, and note that, in regard to pattern formation, some systems comply, others do not (see also Chapter 4 by John Pastor for perspective on the Turing model). They then give compelling examples—*Sphagnum* bogs and mull and mor-forming plants—where ecosystem processes are influenced by the dominant vegetation in a self-serving manner: they alter soil to favor themselves, often creating competitive advantages.

Merritt Turetsky and colleagues (Chapter 12) describe how spatial heterogeneity in abiotic environmental factors controls carbon (C) storage in the boreal forest landscape. C accumulation varies markedly between different landscape units in the boreal forest, so knowledge of heterogeneity is crucial for scaling up. In boreal systems, heterogeneity is caused by physiographic controls on temperature and hydrologic regimes, which in turn control permafrost, peatland distribution, fire, and vegetation types. All of the latter factors influence C storage. This chapter offers an excellent example of how the environmental template controls vegetation and disturbance, which in turn influence both plant competition and ecosystem processes such as the cycling of nutrients.

Urban ecosystems are the focus of Chapter 13 by Larry Band and colleagues. They consider how engineered and natural features of the landscape, as well as human and institutional behavior, might affect heterogeneity and ecosystem processes (e.g., carbon, water, and nutrient cycling) in urban ecosystems. Band et al. make an excellent case for the use of existing frameworks that can be modified for urban landscapes. These frameworks include patch dynamics, distributed hydroecological modeling, and urban land-atmosphere interactions. Here, an urban patch dynamic framework is distinguished from classical ecological patch dynamics by the inclusion of people and their effects. Band et al. offer the general advice that mapping heterogeneity should be done in the context of knowing at what scale it matters to ecosystem function, and they offer some tools for patch aggregation. They also address the concern regarding configuration, noting that some models that take into account heterogeneity (e.g., the mosaic approach) will not detect the importance of configurational heterogeneity. This chapter offers innovative and straightforward ways of thinking about, and analyzing, heterogeneity in urban systems, with some nice examples from the Baltimore Ecosystem Study.

Robert Naiman and colleagues (Chapter 14) give a richly detailed description of heterogeneity and its controls for riparian ecosystems in the Pacific Coast region of North America. They emphasize the role of these dynamic systems in storage and dissipation of materials and energy across a region. Here, heterogeneity in the fluvial sorting and deposition of sediment has major effects on ecosystem processes, primarily because of the diverse physical properties of the sediment (e.g., variation in adsorption capacity, water and gas movement through pore spaces, and oxic and anoxic conditions). Large woody debris (LWD) is also featured as a driver for system heterogeneity. Naiman et al. use a conceptual model to consider riparian heterogeneity at multiple scales of space and time and their interactions. They, too, consider the ecological patch framework in their discussions of heterogeneity and ecosystem function, as do Band et al. (see above).

Stuart Fisher and Jill Welter (Chapter 15) offer a thoughtful presentation on functional heterogeneity in systems where aquatic and terrestrial perspectives are integrated. Building on many years of research on streams in desert landscapes, the authors provide an interesting and original view of how hydrologic flowpaths integrate the various patches present in stream ecosystems. The spatial heterogeneity is temporally dynamic as well, with disturbance events such as flooding and drying changing the patches, altering connectivity, and reinforcing spatial heterogeneity. They argue effectively for an integrated view of landscape function that is not separated into aquatic and terrestrial components.

Despite the superficial appearance of lakes as homogeneous entities (at least compared to arid ecosystems, for example), Tim Kratz et al. (Chapter 16) illustrate that, upon closer examination, most lakes are quite heterogeneous in structure and in function. They address heterogeneity from two different perspectives, within lake and within landscape. Within lakes, horizontal and vertical heterogeneity exists. This heterogeneity is strongly controlled by physical and chemical processes. In contrast, heterogeneity within landscapes (average characteristics and among-year dynamics) is largely controlled by the geomorphic setting of the lake, but humans can have a significant impact through changes in land cover. Humans can similarly affect significantly within-lake heterogeneity by simplifying the littoral zone. Both Kratz et al. and Band et al. (Chapter 14) consider the effects of humans and their structures (the "built" landscape) in destroying and creating heterogeneity and, as a result, controlling productivity and element cycling.

These seven chapters provide rich detail in illustrating how heterogeneity is important in different ecosystems. They emphasize how the interplay of abiotic and biotic factors creates heterogeneity and controls the interaction of heterogeneity and ecosystem function.

# 10
# Heterogeneity in Arid and Semiarid Lands

DAVID J TONGWAY AND JOHN A LUDWIG

## Abstract

Spatial heterogeneity is a hallmark of vegetation patterns in arid and semiarid landscapes. First observed in terms of the spatial array of vegetation patches, spatial heterogeneity is now more broadly interpreted as the cumulative outcome of the processes affecting the spatial and temporal distribution of vital resources such as water, topsoil organic matter, and propagules individually and collectively. Spatial resource redistribution is shown to be important at a variety of scales varying from millimetres to hundreds of metres and beyond and can be conveniently studied as a nested spatial hierarchy. The processes by which heterogeneous resource distribution arises are a mixture of physical and biological and can be represented by an information-structuring conceptual framework, which is described. Heterogeneity is crucial to the functioning of arid and semiarid lands, and changes in the scale of heterogeneity can be used to study and understand the processes underlying desertification and rehabilitation. Models of heterogeneous landscapes in semiarid landscapes have had two broad themes: a pragmatic approach, describing ecosystem function in landscapes under management, and a curiosity-driven approach, speculating about the *de novo* development of landscape heterogeneity.

## Introduction

The spatial heterogeneity of natural vegetation at broad scale in arid and semiarid lands began to be noted when people were able to view landscapes from aircraft (Gillett, 1941). Pattern had previously been difficult to identify on the ground in arid lands, due to the scale at which pattern elements were expressed and uncertainty as to whether heterogeneity was natural or due to adverse management. The great expansion of systematic aerial photographic surveys in the 1950s revealed the spatial extent, globally, of natural vegetation patterns of a distinctly geometric type (Macfayden 1950; Clos-Arceduc 1956; Greenwood 1957; Litchfield and Mabbutt 1962; Slatyer 1961;

Boaler and Hodge 1964; White 1969). These overtly patterned lands were a curiosity, and much of the early literature speculated about their origins and dynamics. Some writers saw these lands as the result of degradation from a formerly uniform cover of vegetation, due to adverse landscape use in historical times (Hemming 1965; Wickens and Collier 1971). Others suggested that the pattern was caused by climatic shifts during the Holocene (Clos-Arceduc 1956; Boaler and Hodge 1964). Biotic causes such as the slumping of termite mounds were suggested (Macfayden 1950). The most enduring proposals suggested geomorphic processes (Litchfield and Mabbutt 1962; Cornet et al. 1988).

Vegetation patchiness continues to have interest for ecologists in new locations and different spatial manifestations; for example, in the United States, Archer (1990); the Serengeti, Belsky (1995); in Mexico, Montana (1992); and Argentina, Aguiar and Sala (1999).

## Functional Heterogeneity: Linking Heterogeneity to Differential Soil Water Availability

The early publications were entirely descriptive and largely focused on speculation about the reasons behind the spatial disposition of vascular plants. However, it was not long before the role of rainfall in arid and semiarid lands was recognized as the primary driver of the pattern. Slatyer (1961) studied the overall water economy of a patterned landscape in central Australia and showed that water accession and capture into *Acacia* groves from bare soil intergroves was very high and, several days after rainfall, evaporation from the soil in the grove ceased and water loss thereafter was purely from transpiration, indicating high water use efficiency. This work was probably the first to propose the concept of the role of the temporal and spatial dynamics of water supply to patterned lands with empirical data. Slatyer (1961) measured the capture of rainfall by foliage and the channeling of water into the soil at the foot of the tree as well as measuring water runoff from bare, crusted soils upslope of the tree grove. This work needs to be seen as the foundation of the adoption of runoff/run-on processes as the primary explanatory tool in understanding landscape function in arid lands. More recent key work by Valentin and Bresson (1992) on the nature and formation of a variety of soil physical crusts on the interpatch zone has been crucial in explaining water runoff and run-on characteristics.

Rainfall in arid and semiarid lands is low by definition but in addition is typically unpredictable in timing and amount. Table 10.1 shows how skewed the quantity of rain is per event in a typical semiarid landscape. There are many, very small rainfall events that are ineffective for vascular plant growth, so for the survival and persistence of plants in these landscapes, we

TABLE 10.1. Rainfall Classification According to Jackson (1958) for Cobar, New South Wales, Australia

|  | VH | H | M | S | L | VL |
|---|---|---|---|---|---|---|
| Cumulative no. days in 20 years | 16 | 82 | 73 | 66 | 235 | 896 |
| Mean no. days year$^{-1}$ | 0.8 | 4.1 | 3.7 | 3.3 | 11.8 | 44.8 |

[a]VH, very heavy, >50 mm in 2 consecutive days; H, heavy, 23–50 mm in 2 consecutive days; M, moderate, 13–23 mm in 2 consecutive days: S, significant, 7.8–13 mm in a single day; L, light 2.8–7.8 mm in a single day; VL, very light, <2.8 mm in a single day.

need to understand the availability of soil water to plants in space and time. At the same time, cryptogams have been shown to respond to very small rain showers, facilitating brief bursts of biological activity in the surface few millimeters of soil (Eldridge et al. 2002). A vertical heterogeneity reflecting large differences in soil infiltration rate, nitrogen concentration, and aggregate stability consequently arises (Graetz and Tongway 1986; Belnap 2003; Warren 2003). Although vascular plants may have drought survival strategies, for example rhizosphere sheaths (Buckley 1982), there are also landscape-scale processes whereby water tends to be conserved in a spatially heterogeneous pattern. These are a mixture and interaction of biotic and abiotic processes.

Noy-Meir (1973, 1981) showed from a theoretical analysis that there would be higher biomass production in semiarid lands if rainfall were redistributed into run-on patches than if it were distributed evenly in the soil across the landscape. This is the reverse of the classic paradigm in higher rainfall climatic zones and is a cornerstone of the landscape ecology of arid lands. Note that three-dimensional water distribution was a clearly acknowledged feature of this proposal. The reasoning behind this goes back to the large number of small rainfall events and the small number of large ones referred to above. It also invokes the concept of a critical threshold in the availability of a scarce resource, in this case water. Many of the small rainfall events would be too small to elicit a response from perennial vascular plants, but if in some light rainfalls, water were to run off from one part in the landscape and run onto another part consistently over time, then the stored soil water would be enough to stimulate plant growth and support it for a period. The run-on zone would have a more mesic environment over time than average ambient rainfall would suggest. If this increased amount of stored soil water was higher than a critical threshold amount on an annual basis, then particular perennial vascular plants could persist and perhaps thrive. Different plant forms would of course have differing water requirements. The initial patterned vegetation observations had described grasslands, shrublands, and woodlands as comprising the characteristic vegetation in the patches. This suggests that different landscapes have different ratios of runoff to run-on, though the total amount of rain is also important. Lands with a smaller runoff ratio may support grasslands, whereas lands with larger run-on/runoff ratios might support a patterned shrubland or woodland.

Water capture by plant foliage and stem flow, thus augmenting water input to the patch, has also been demonstrated with different life forms (Slatyer 1961; Whitford et al. 1997).

## From Pattern to Process

By the mid-1980s, landscapes with much less overt pattern were being studied from a spatial resource availability point of view (Ludwig and Tongway 1995). Desertification studies often provided the impetus for framing questions about how landscapes worked and what happened under increasing stress and/or disturbance. The scale of natural heterogeneity was recognized as signifying the way landscapes worked as biogeochemical systems; however, spatial variation was more frequently studied in relation to deleterious land management (Graetz and Ludwig 1978). Tongway and Ludwig (1990) described the heterogeneity of an *Acacia* woodland as composed of a three-phase vegetation array (configurational heterogeneity) and also in terms of large differences in the chemical fertility (organic and inorganic) of the soil associated with the three distinct patch types. They identified the patches as "bare soil runoff," "grassy interception," and "*Acacia* grove" as characterizing the nature of the surface hydrology. The bare runoff patch type had a robust physical crust with a low infiltration rate and very sparse perennial grasses; the grassy interception zone was composed of a moderately dense perennial grassland with no soil physical crust, and the grove had a deep leaf litter floor with a very high infiltration rate. Greene (1992), working in the same landscape, showed that after a single rainfall event of 37.5 mm, the bare soil runoff zone infiltrated 15.7 mm or 42% of the incident rain, the "grassy interception" zone infiltrated 33.7 mm or 90% of the incident rain, and the *Acacia* grove infiltrated 51.5 mm or 138% of the incident rain. This data array strongly suggested that biophysical processes acting in space over time maintained the observed vegetation pattern. Subsequent work by Tongway et al. (1989) indicated that the role of soil fauna in facilitating organic matter decomposition, bioturbation, and nutrient cycling was a key process, augmenting the pattern established by the primary soil water redistribution process. Zaady and Bouskila (2002) and Zaady et al. (2003) also reported fauna as major agents of resource concentration. Ludwig and Tongway (1995) built on this landscape analysis process by examining two distinctively different landscape types, both of which had much more subtle expression of heterogeneity than the banded woodland. They extended the basic notion of economy of water redistribution that Slatyer and Noy-Meir pioneered to also include macro-organic matter transport, deposition, and decomposition, using the phrase "scarce, vital resources" as a omnibus term to indicate the scope of factors involved in ecosystem functioning in arid lands.

This study confirmed that the explanation of heterogeneous pattern needed an appreciation of the spatial and temporal dynamics of water, nutrients, and organic matter. Other roughly contemporaneous studies (in the United States, Schlesinger et al. 1990, Whitford et al. 1997; in Israel, Garner and Steinberger 1989; in Sahelian Africa, Seghieri et al. 1994) confirmed the concept of the need for fertile patches in the arid and semiarid. Interpatch (runoff) zones between the fertile (run-on) patches had distinctive properties, in particular soils with physical crusts with low infiltration, so that runoff was generated from quite small rainfall events (Valentin and Bresson 1992). Ludwig and Tongway (1997, 2000) proposed a generic conceptual framework that simply but comprehensively looks at the range of processes involved in a well-functioning landscape. This framework (Figure 10.1) treats landscapes as biophysical systems and focuses on processes that affect the way scarce limiting resources are used in landscapes. This simple but comprehensive framework facilitates the structuring of information derived from diverse studies and thus synthesizes more detailed information about ecosystem functioning. A *trigger*, such as rainfall, initiates *transfer* processes such as runoff and erosion that spatially relocates resources such as water, organic matter, and seeds across the landscape. Some of these resources may flow to waste and some may be stored in the soil (*reserve*). Some locations in the landscape absorb or capture more resources than other parts, due to differential runoff/run-on characteristics.

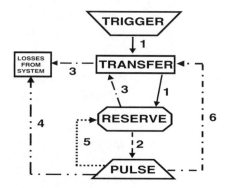

FIGURE 10.1. The trigger-transfer-reserve-pulse framework. This represents a sequence of resource mobilization or utilization processes explained in Table 10.1. key: 1, run-on, storage/capture, deposition, saltation capture; 2, plant germination, growth, nutrient mineralization, uptake processes; 3, run-off into streams, rill flow and erosion, sheet erosion out of system, wind erosion out of system; 4, herbivory, fire, harvesting, deep drainage; 5, seed pool replenishment, organic matter cycling/ decomposition processes, harvest/concentration by soil microfauna; 6, physical obstruction/absorption processes.

The *reserve* may be considered as a metaphorical "bank" dealing in many diverse "currencies"; water, nutrients, seeds, and soil fauna. A *pulse* of plant growth and of mineralized nutrients may ensue, the magnitude of which depends on the status of the reserve. Some of the growth may be *lost from the system* by fire or herbivory, but much is *cycled* back to the reserve. A growth pulse may also feed back to affect a subsequent spatial physical transfer process. Wind may also act as the trigger mobilizing loose soil and plant litter, either dispersing or concentrating the mobilized resources in patches.

## Linking Water and Nutrients

The close link between availability of water and the mineralization of nutrients in heterogeneous landscapes is not as well acknowledged as it should be. Charley and Cowling (1968) modeled the interacting dynamics of water and nutrient mineralization in a range of environmental circumstances that has scarcely been conceptually improved on, accounting for impressive pulses of plant growth after drought. Their model accommodates the small quanta and infrequency of rain received in arid and semiarid lands, building in small and/or short nitrogen mineralization events until a crucial amount of rain does fall. This both stimulates further nitrogen mineralization and provides enough water for the vegetation to put on a pulse of growth. Over time, the vegetation takes up the water and the available nitrogen, depleting each resource to near zero (provided no further rain falls). Crucially, the size of the mineralizable nitrogen pool is also depleted and needs time for potentially mineralizable nitrogen to build up again. This model clearly indicates the need for the synchrony of water and mineral nitrogen for an appropriate vegetation response. Gallardo and Schlesinger 1995, Whitford and Herrick 1996, and Guo and Brown 1997 have described the microbial processes involved in the mineralization of nitrogen alluded to in Charley and Cowling (1968) in some detail in heterogeneous semiarid landscapes in the southwestern United States. These studies all acknowledge the concept of the "fertile patch," a place where the soil chemical, physical, and biological properties are much more favorable for plant growth than in the sharply differentiated "interpatch" areas, which are relatively impoverished. Whitford et al. (1987) and Tongway et al. (1989) suggested that one form of landscape degradation disconnected the nexus between water and nitrogen availability. For example, perennial grass patches had maintained a positive nitrogen mineralization potential year-round, enabling them to use any rain that fell. The local extinction of perennial grass by grazing management and its replacement by ephemeral plants produced an initial large pulse of carbon from dead roots for microbes and fungi to metabolize, resulting in seasonal net nitrogen immobilization.

# Below-Ground Processes

Plant-induced soil chemistry patterns in general have had a long history of investigation (Roberts 1950; Ebersohn and Lucas 1965; Charley 1972), but in heterogeneous arid and semiarid lands it is particularly important for ongoing landscape function.

The role of soil macrofauna in maintaining environmental patches has been shown to be important, if not crucial. Soil fauna characterized as "ecosystem engineers" (Lavelle 1997) create stable tunnels and chambers that facilitate the rapid infiltration of water into the profile and facilitate gas exchange from roots and microbes. Greene (1992) showed that more than 90% of water infiltrating the soil in an *Acacia* patch was via "biopores" created by soil macrofauna and root channels. These soil macrofauna need to have their own environmental needs of food supply, habitat, and protection from predation properly satisfied so that they can provide the ecosystem services mentioned above. Biogeochemical cycling by vegetation in patches is markedly different from that in interpatches (Archer 1990). Woody species with more extensive root systems play a more prominent role in the cycling of mineral elements such as calcium, potassium, and manganese than do grassy interpatches (Tongway and Ludwig 1990).

# Scales of Heterogeneity

The importance of scale in ecology has had a lot of recent attention (e.g., Gardner et al. 1989; O'Neill 1989; Wiens 1997; Holling 1992; Pickett et al. 1997). Some of this work has been somewhat introspective, observing that changing the scale of observation makes a difference to the perception of heterogeneity, whereas there is also a need to understand scale-dependent processes to be able to understand and manage landscapes. The authors prefer to deal with heterogeneity in terms of nested scales of biophysical processes, connected or linked by biophysical processes, rather than simply the detection of boundaries between zones of differing properties.

Each nondegraded arid and semiarid landscape has a characteristic scale, over which resources are mobilized, transported, and deposited. The patterned or banded landscapes referred to above were coarse-scaled, their pattern visible in aerial photographs: resources moved and were recaptured at scales of many metres (Figure 10.2). Many other semiarid landscapes are quite fine-grained, and resources may only move some centimeters, for example grasslands in arid environments (Figure 10.3). Nevertheless, each of them is laterally heterogeneous in the spatial concentration of resources, often with sharp boundaries. We have noted above the sharp, vertical heterogeneity boundaries associated with microphytic crusts.

FIGURE 10.2. An example of heterogeneity in an *Acacia* woodland in western New South Wales, Australia. The landscape slope is directly toward the grove of trees; rain water runs off the bare crusted soil and is effectively infiltrated by the soil under the trees, which has very high infiltration capacity due to bioturbation by soil fauna. In this case, there is a pronounced ecotone characterized by a perennial palatable grassland. Nitrogen concentrations in the top 10 cm of the tree grove are about five times that in the bare soil zone. In the background can be seen another repetition of the pattern in heterogeneity.

Changes in the scale and nature of heterogeneity have been interpreted as desertification using evidence of species composition change and sand dune formation (Schlesinger et al. 1990). One of the hallmarks of desertification is lower productivity per millimeter of rainfall (le Houerou 1984). However, Huenneke (1996) showed that the primary productivity of the unpalatable shrubs was not lower than the displaced grassland, suggesting that there was no functional deterioration at landscape scale. However, range condition, the availability of forage for cattle, was markedly affected. Soil redistribution by wind had not resulted in a net loss of soil or other resources but had rearranged those resources at a coarser scale of spatial distribution so that the edaphic habitat better suited the colonizing shrub and greatly disadvantaged perennial grasses. This may well also be an example of another vertical expression of heterogeneity called the "inverse texture" effect (Noy-Meir 1981) where, after rainfall infiltrates deeply into sand dunes, the surface few centimeters dry out, breaking the capillary fringe, greatly reducing bare soil evaporation and thus "safely

FIGURE 10.3. A perennial grassland in a semiarid landscape. Although superficially homogeneous, this landscape is composed of grass tussocks with elevated infiltration capacity and nutrient pool sizes compared to the intertussock bare ground. Observations of the tussock base show accumulations of litter and sandy soil material that both contribute to the physical and chemical properties.

storing" water in the dune for plant use. In more mesic climates, fine textured soils tend to have higher productivity because of the higher soil water storage.

## Detecting Functional Heterogeneity at the Hillslope Scale

Recent studies of heterogeneity in arid and semiarid lands recognize landscape patches as regulators of the flux of vital resources such as water, topsoil particles, organic matter, and propagules. Landscapes may be characterized as "functional" if the loss of all of these resources is low and "leaky" if the rate and amount of loss of one or more of these resources beyond the boundary of the ecosystem can be observed and measured (Ludwig and Tongway 1997; Ludwig et al. 2002). A continuum exists between these conceptual extremes. All systems of course lose resources to some extent, but it is the rate of loss and the change in the balance caused by the loss of different resources that is the crucial factor in assessing dysfunction. For example, excessive loss of dissolved or adsorbed nutrients has consequences for stream eutrophication.

Many examples of heterogeneous landscape have been termed *mosaics*. This is a metaphor that is not fully accurate. In a classic mosaic, individual small pieces make up a picture that only makes sense at macroscale. In landscapes, pattern elements have a functional relationship that has functional relevance at both the fine and the coarse scales. The cover and spatial arrangement of perennial vegetation patches is an important indicator of landscape function. Bastin et al. (2002) have described four metrics by which functional heterogeneity can be assessed, covering both field traverse and remote sensing techniques. These were named (1) directional leakiness index, (2) weighted mean patch size, (3) lacunarity index, and (4) proximity index. Direct measurement of resource loss is extremely expensive, so remotely sensed data is an attractive way to detect change in patchiness. Methods 1 and 3 were able to assess the heterogeneity satisfactorily with a functional interpretation, but methods 2 and 4 did not. These latter methods are concerned only with spatial discrimination between pattern elements, whereas methods 1 and 3 are more strongly keyed into the flux of resources along an environmental gradient.

## Modeling Surface Processes

The work reported in this section largely conforms to a statement by Wiens (1984) as a goal for ecosystem studies: " to detect the patterns of natural systems, to explain them by discerning the causal processes that underlie them and to generalise these explanations as far as possible." The models developed for this purpose have a very applied objective: understanding landscape function for wise management. The understanding of spatial regulation of vital resources within the landscape boundaries is a key objective, but examining "internal" redistribution processes is the means of study. Essentially, this means looking at the nature of the processes in both the runoff and run-on sites in the landscape. This approach explicitly recognizes a continuum of the effectiveness of runoff/run-on systems ranging from as retentive as they can be to somewhat "leaky." Ludwig et al. (1994, 1999, 2002) have provided models with field validation data for woodlands at hillslope scale in Australia. In particular, Ludwig et al. (1999) were able to show how different proportions of runoff/run-on produce heterogeneous patterns of different types, which they characterized as "stripes, strands and stipples." This model type also shows that the loss of patchiness would greatly reduce productivity. Montana (1992) and Mauchamp et al. (1994) have modeled patch dynamics in the *mogote*, or shrubby grassland in Mexico. Both of these model types work with data collected explicitly in the direction of flow of resources across the landscape and use a minimum of two phases (runoff and run-on). Soil erosion and deposition are explicit: data are collected in real ecosystems. Most models use water as the main vector, but wind is important in sandy landscapes and in removing plant litter from unprotected sites. Animal dung is generally regarded as too dispersed in semiarid and arid ecosystems to be a major nutrient redistributor.

## Modeling Patch Initiation

Speculation about the initiation of heterogeneity and the spatial dynamics of patches in arid and semiarid lands has been a continuous thread since the earliest times (Macfayden 1950; Boaler and Hodge 1962) and continues today (Dunkerley 1997a, b, 1999; Orr 1996; Lefevre and Lejeune 1997). Upslope migration of patches is also a common modeling theme, though this has not been empirically observed with certainty at any field sites to date over the relatively short observation times of typical scientific experiments (Valentin et al. 2001). Thiéry et al. (2001) have reviewed a range of proposed models such as TIGREE, TIGRFLUX, and RUNOFF. Most of these models are based on different assumptions but have some common features imposed by the acceptance of heterogeneity as a "given." They are characterized by a level of abstraction that would suggest considerable use, but perhaps the simplicity of the underlying assumptions limits the uses to which the models can be put. This contrasts with the surface process models that explicitly acknowledge the complexity of the processes that are included. Thiéry et al. (2001) have reviewed a range of these models recently. Typically, long runs of simulated weather sequences are used to see to what extent patches form and change over time. For computational reasons, the models are low in complexity.

## Simulation Modeling for Management Optimization

Marsden (1998) compiled a series of simulation models based on the conceptual framework described in Ludwig et al. (1997), discussed here as the TTRP framework, Figure 10.1. These comprise five interacting models that simulate soil water, plant and soil nitrogen, plant production, sheep management, and wool production. This work takes the original scientific inquiry about ecological heterogeneity right through to management guidelines to ensure the continuance of ecological heterogeneity that is considered essential for long-term sustainable production in semiarid lands. The hydrology module uses water runoff and run-on factors that drive spatial vegetation heterogeneity so that inputs to all other modules rely on functional heterogeneity principles.

## Decline in Heterogeneity with Adverse Management

Landscape degradation may affect the capacity of patches to act as sinks for resources, so that the landscape as a whole becomes more leaky. Patches may become less competent, interpatches may discharge resources too rapidly, or patches might disappear. Wu et al. (2000) and Ludwig et al. (2002) have proposed monitoring procedures that look at the overall spatial array of patches to assess whether resources tend to be leaking away. Tongway and Hindley (2000) devised a monitoring procedure that rapidly assesses the capacity of patches and interpatches to regulate the flow of resources first at the hillslope scale and then at the individual patch scale. This procedure has been implemented extensively in Australia for assessing rangeland condition and mine-site rehabilitation success.

## Manipulating Heterogeneity to Rehabilitate Degraded Landscapes

If patches that regulate the movement and capture of vital resources are too few or too incompetent, active rehabilitation would involve reinstituting patches that would effectively capture those resources. An experiment that set out only to increase patchiness in landscapes after degradation by a century of sheep grazing was reported as having successfully done so (Ludwig and Tongway 1996; Tongway and Ludwig 1996). They created patches with tree branches stacked onto the ground, aligned to the contour to capture effectively resources being transported by surface flows of water over the soil surface. The experimental layout was such that the ratio of upslope interpatch to run-on experimental patch were similar to nondisturbed examples of the same landscape type. This was a design element to ensure that resource amount, mobilization, and transport was sufficient to enrich the experimental patch. After only 3 years, soil properties had markedly improved and palatable, perennial grasses has self-established in the treatment patches. Ringrose-Voase and Tongway (1996) later showed with micromorphological examination that soil physical crusts in the treatment patches had been completely perforated and dismantled by soil faunal bioturbation, and that tunnels and galleries were abundantly ramified in the surface soil, thus explaining an order of magnitude increase in water infiltration over the first 3 years of the experiment.

## Conclusions

Spatial heterogeneity in arid and semiarid lands when interpreted in terms of the availability of scarce, vital resources in space and time is the key to understanding how those landscapes work as biogeochemical systems. A full range of biophysical processes is facilitated by resource transfers from poor or low concentrations (interpatches) to rich or highly concentrated pools of resources (fertile patches). There appears to be no scale of resource processing where this proposition is invalid. Sequences of processes and processes nested within the scale of another process can be accommodated. This paradigm facilitates the better-informed study of complex natural ecosystems, their wise management, and the rehabilitation of lands degraded in the past.

## References

Aguiar, M.R., and Sala, O.E. 1999. Patch structure, dynamics and implications for functioning of arid ecosystems. Trends Ecol. Evolution 14: 273–277.

Archer, S. 1990. Development and stability of grass/woody mosaics in a subtropical savanna parkland, Texas, USA. J. Biogeogr. 17: 453–462.

Bastin, G.N., Ludwig, J.A., Eager, R.W., Chewings, V.H., and Liedloff, A.C. 2002. Indicators of landscape function: comparing patchiness metrics using remotely sensed data from rangelands. Ecol. Indicators 1: 247–260.

Belnap, J. 2003. Factors influencing nitrogen fixation and nitrogen release in biological soil crusts. In Biological soil crusts: structure function and management, eds. J. Belnap and O.L. Lange, pp. 241–262. New York: Springer.

Belsky, A.J. 1995. Spatial and temporal landscape patterns in arid and semi-arid African savannas. In Mosaic landscapes and ecological processes, eds. J. Hansson, L. Fahrig, and G. Meeriam, pp. 31–56. London: Chapman and Hall.

Boaler, S.B., and Hodge, C.A.H. 1962. Vegetation stripes in Somaliland. J. Ecol. 50: 465–474.

Boaler, S.B., and Hodge, C.A.H. 1964. Observations on vegetation arcs in the northern region, Somalia Republic. J. Ecol. 52: 511–544.

Buckley, R. 1982. Sand rhizosheath of an arid grass. Plant Soil 66: 417–421.

Charley, J.L., and Cowling, S.W. 1986. Changes in soil nutrient status resulting from overgrazing and their consequences in plant communities of semi-arid areas. Proc. Ecol. Soc. Australia. 3: 28–38.

Charley, J.L. 1972. The role of shrubs in nutrient cycling. In Wildland shrubs—their biology and utilisation, eds. C.M. McKell, J.P. Blaisdell, and J.R. Goodwin, pp. 182–203. USDA Forest Service General Technical Report INT-1. Ogden, Utah, USA.

Clos-Ardceduc, M. 1956. Etude sur photographies aeriennes d'une formation vegetale sahelienne; la Brousse Tigree. Bulletin l'Institute francais d'Afrique noire Serie. A 18: 677–684.

Cornet, A., Delhoume, J.P., and Montaňa, C. 1988. Dynamics of striped vegetation patterns and water balance in the Chihuahuan Desert. In Diversity and pattern in plant communities, eds. H.J. During, M.A. Werger, and H.J. Willems, pp. 221–231. The Hague: SPB Academic Publishing.

Dunkerley, D.L. 1997a. Banded vegetation: development under uniform rainfall from a simple cellular automaton model. Plant Ecol. 129: 103–111.

Dunkerley, D.L. 1997b. Banded vegetation: survival under drought and grazing pressure based on a simple cellular automaton model. J. Arid Environ. 35(3): 419–428.

Dunkerley, D.L. 1999. Banded chenopod shrublands of arid Australia: modelling responses to interannual rainfall variability with cellular automata. Ecol. Modelling 121: 127–138.

Ebersohn, J.P., and Lucas, P. 1965. Trees ands soil nutrients in south-western Queensland. Queensland J. Agric. Anim. Sci. 22: 431–435.

Eldridge, D.J., Zaady, M.E., and Shachak, M. 2002. Microphytic crusts, shrub patches and water harvesting in the Negev desert: the Shikim system. Landscape Ecol. 17: 587–597.

Gallardo, A., and Schlesinger, W.H. 1995. Factors determining soil microbial biomass and nutrient immobilisation in desert soils. Biogeochemistry 28: 55–68.

Gardner, R.H., O'Neill, R.V., Turner, M.G., and Dale, V.H. 1989. Quantifying scale-dependent effects of animal movement with simple percolation models. Landscape Ecol. 3: 217–227.

Garner, W., and Steinberger, Y. 1989. A proposed mechanism for the formation of "fertile islands" in the desert ecosystem. J. Arid Environ. 16: 257–62.

Gillett, J. 1941. The plant formations of western British Somaliland and the Harar province of Abyssinia. Kew Bulletin 2: 37–75.

Graetz, R.D., and Ludwig, J.A. 1978. A method for analysis of piosphere data applicable to range assessment. Australian Rangeland J. 1: 126–36.

Graetz, R.D., and Tongway, D.J. 1986. Influence of grazing management on vegetation, soil structure and nutrient distribution and the infiltration of applied rainfall in a semi-arid chenopod shrubland. Australian J. Ecol. 11: 347–360.

Greenwood, J.E.G.W. 1957. The development of vegetation patterns in Somaliland Protectorate. Geographical J. 123: 465–473.

Greene, R.S.B. 1992. Soil physical properties of three geomorphic zones in a semi-arid mulga woodland. Australian J. Soil Res. 30: 55–69.

Guo, Q., and Brown, J.H. 1997. Interactions between winter and summer annuals in the Chihuahuan desert. Oecologica 111: 123–28.

Hemming, C.F. 1965. Vegetation arcs in Somaliland. J. Ecol. 53: 57–67.

Holling, C.S. 1992. Cross-scale morphology, geometry, and dynamics of ecosystems. Ecol. Monogr. 63: 447–502.

Huenneke, L.F. 1996. Shrublands and grasslands of the Jornada long-term ecological research site: desertification and plant community structure in the northern Chihuahuan Desert, p. 48–50. In: Barrow, J.R., McArther, E.D., Sosebee, R.E., and Tausch, R.J. (eds.). Proceedings: Shrubland ecosystem dynamics in a changing environment. Gen. Tech. Rep. INT-GTR-338, Intermountain Res. Sta., U.S. for Serv., Ogden, Utah. USA. 275p.

Jackson, E.A. 1958. A study of the soils and some aspects of the hydrology of Yudnapinna Station, South Australia. CSIRO Division of Soils, Soils and Land Use Series No. 24. Adelaide, Australia.

Lavelle, P. 1997. Faunal activities and soil processes: adaptive strategies that determine ecosystem function. In Advances in Ecological Research, eds. M. Begon, and A.H. Fitter, pp. 93–132. San Diego: Academic Press.

Le Houerou, H.N. 1984. Rain use efficiency: A unifying concept in arid-land ecology. J. Arid Environ. 7: 1–35.

Litchfield, W.H., and Mabbutt, J.A. 1962. Hardpan soils of sem-iarid western Australia. J. Soil Sci. 13: 148–159.

Lefever, R., and Lejeune, O. 1997. On the origin of tiger bush. Bull. Math. Biol. 59: 263–294.

Ludwig, J.A., and Tongway, D.J. 1995. Spatial organisation and its function in semi-arid woodlands, Australia. Landscape Ecol. 10: 51–63.

Ludwig, J.A., and Tongway, D.J. 1996. Rehabilitation of semi-arid landscapes in Australia. II. Restoring vegetation patches. Restoration Ecol. 4: 398–406.

Ludwig, J.A., and Tongway, D.J. 1997. A landscape approach to rangeland ecology. In Landscape ecology function and management: principles from Australia's rangelands, eds. J. Ludwig, D. Tongway, D. Freudenberger, J. Noble, and K. Hodgkinson, pp. 1–12. Melbourne, Australia: CSIRO.

Ludwig, J.A., and Tongway, D.J. 2000. Viewing rangelands as landscape systems p. 39–52. In Rangeland desertification, eds. O. Arnalds and S. Archer. Dordrecht: Kluwer Academic.

Ludwig, J.A., Tongway, D.J., and Marsden, S.G. 1994. A flow-filter model for simulating the conservation of limited resources in spatially heterogeneous, semi-arid landscapes. Pacific Conservation Biol. 1: 209–215.

Ludwig, J., Tongway, D., Freudenberger, D., Noble, J., and Hodgkinson, K., eds. 1997. Landscape ecology, function and management: principles from Australia's rangelands. Melbourne, Australia: CSIRO Publishing.

Ludwig, J.A., Tongway, D.J., and Marsden, S.G. 1999. Stripes, strands or stipples: modelling the influence of three landscape banding patterns on resource capture and productivity in semi-arid woodlands, Australia. Catena 37: 257–273.

Ludwig, J.A., Eager, R.W., Bastin, G.N., Chewings, V.H., and Liedloff, A.C. 2002. A leakiness index for assessing landscape function using remote sensing. Landscape Ecol. 17: 157–171.

Macfayden, W.A. 1950. Vegetation patterns in British Somalilands. Nature 165: 121.

Marsden, S. 1998. The SEESAW Model: simulation of the ecology and economics of the semi-arid woodlands. Canberra, Australia: CSIRO Wildlife and Ecology.

Mauchamp, A., Rambal, S., and Lepart, J. 1994. Simulating the dynamics of a vegetation mosaic: a specialized functional model. Ecol. Modelling 71: 107–130.

Montana, C. 1992. The colonisation of bare areas in two-phase mosaics of an arid ecosystem. J. Ecol. 80: 315–27.

Noy-Meir, I. 1973. Desert ecosystems: environment and producers. Annu. Rev. Ecol. Systematics 4: 25–51.

Noy-Meir, I. 1981. Spatial effects in modelling of arid ecosystems. In Arid land ecosystems, eds. D.W. Goodall and R.A. Perry, pp. 411–432. Cambridge, UK: Cambridge University Press.

O'Neill, R.V. 1989. Perspectives in hierarchy and scale. In Perspectives in ecological theory, eds. J. Roughgarden, R.M. May, and S.A. Levin, pp. 140–146. Princeton, NJ: Princeton University Press.

Orr, B. 1996. Modeling Nigerian brousse tigrée. Photocopy discussion notes. Houghton, MI: Michigan Technological University.

Pickett, S.T.A., Ostfeld, R.S., Shachak, M., and Likens, G.E. eds. 1997. The Ecology basis of conservation: heterogeneity, ecosystems, and biodiversity. New York: Chapman and Hall.

Ringrose-Voase, A.J., and Tongway, D.J. 1997. Micromorphological effects of rehabilitating overgrazed, semi-arid mulga woodlands using log mounds. In Soil micromorphology: studies on soil diversity, diagnostics, dynamics (proceedings 10th international working meeting on soil micromorphology, Moscow, Russia, July 1996), eds. S. Shoba, M. Gerasimova, and R. Miedema. Moscow University, Moscow, Russia.

Roberts, R.C. 1950. Chemical effects of salt-tolerant shrubs on soils. Fourth Int. Congress Soil Sci. 1: 404–406.

Seghieri, J., Floret, C., and Ponanier, R. 1994. Development of an herbaceous cover in a Sudano-Sahelian savanna in North Cameroon in relation to available soil water. Vegetatio 114: 175–199.

Schlesinger, W.H., Reynolds, J.F., Cunningham, G.L., Huenneke, L.F., Jarrel, W.M., Virginia, R.A., and Whitford, W.G. 1990. Biological feedbacks in global desertification. Science 247: 1043–1048.

Slatyer, R.O. 1961. Methodology of a water balance study conducted on a desert woodland (*Acacia aneura* F.Muell.) community in central Australia. UNESCO Arid Zone Research. 16: 15–26.

Thiéry, J.M., Dunkerley, D.L., and Orr, B. 2001. Landscape models for banded genesis. In Banded vegetation patterning in arid and semiarid environments, eds. D. Tongway, C. Valentin, and J. Seghieri, pp. 167–197. New York: Springer-Verlag.

Tongway, D., and Hindley, N. 2000. Assessing and monitoring desertification with soil indicators. In Rangeland desertification, eds. O. Arnalds and S. Archer, pp. 89–98. Dordrecht: Kluwer Academic.

Tongway, D.J., and Ludwig, J.A. 1990. Vegetation and soil patterning in semi-arid mulga lands of eastern Australia. Australian J. Ecol. 15: 23–24.

Tongway, D.J., and Ludwig, J.A. 1996. Restoration of semi-arid landscapes in Australia. I. Restoring productive soil patches. Restoration Ecol. 4: 388–397.

Tongway, D.J., Ludwig, J.A., and Whitford, W.G. 1989. Mulga log mounds: fertile patches in the semi-arid woodlands of eastern Australia. Australian J. Ecol. 14: 263–268.

Valentin, C., and Bresson, L.M. 1992. Morphology, genesis and classification of surface crusts in loamy and sandy soils. Geoderma 55: 225–245.

Valentin, C., Tongway, D.J., and Seghieri, J. 2001. Banded landscapes: ecological developments and management consequences. In Banded vegetation patterning in arid and semiarid environments, eds. D. Tongway, C. Valentin, and J. Seghieri, pp. 228–243. New York: Springer-Verlag.

Warren, S.D. 2003. Synopsis: influence of biological soil crusts on arid land hydrology and soil stability. In Biological soil crusts: structure function and management, eds. J. Belnap and O.L. Lange, pp. 241–262. New York: Springer.

White, L.P. 1969. Vegetation arcs in Jordan. J. Ecol. 57: 461–464.

Whitford, W.G., Reynolds, J.F., and Cunningham, G.L. 1987. How desertification affects nitrogen limitation of primary production on Chihuahuan desert watersheds. General Technical Report, Rocky Mountain Forest and Range Experiment Station, USDA Forest Service. RM-150: 143–153.

Whitford, W.G., Anderson, J., and Rice, P.M. 1997. Stemflow contribution to the "fertile island" effect in creosotebush, *Larrea tridentata* J. Arid Environ. 35: 451–457.

Whitford, W.G., and Herrick, J.E. 1996. Maintaining soil processes for plant productivity and community dynamics. In Proceedings of Vth international rangelands congress (July 1995, Salt Lake City, Utah), ed. N.E. West, pp. 33–37. Denver: Society for Range Management.

Wickens, G.E., and Collier, F.W. 1971. Some vegetation patterns in the Republic of the Sudan. Geoderma 6: 43–59.

Wiens, J.A. 1984. On understanding a non-equilibrium world: myth and reality in community patterns and processes. In Ecological communities: conceptual issues and the evidence, eds. D.R. Strong, D. Simberloff, L.G. Abele, and A.B. Thistle, pp. 439–457. Princeton, NJ: Princeton University Press.

Wiens, J.A. 1997. The emerging role of patchiness in conservation biology. In The ecology basis of conservation: heterogeneity, ecosystems, and biodiversity, eds. S.T.A. Pickett, R.S. Ostfeld, M. Shachak, and G.E. Likens, pp. 93–107. New York: Chapman and Hall.

Wu, X.B., Thurrow, T.L., and Whisenant, S.G. 2000. Fragmentation and functional change of tiger bush landscapes in Niger. J. Ecol. 88: 790–800.

Zaady E., and Bouskila, A. 2002. Lizard burrows association with successional stages of biological soil crusts in an arid sandy region. J. Arid Environ. 50: 235–246.

Zaady E., Groffman, P.M., and Shachak, M. 2003. Consumption and release of nitrogen by the harvester termite *Anacanthotermes ubachi Navas* in the northern Negev desert, Israel. Soil Biol. Biochem. 35: 1299–1303.

# 11
# Formation of Soil-Vegetation Patterns

Marcel Meinders and Nico van Breemen

## Abstract

Vegetation patterns often resemble the pattern of the geological substratum. In some cases, however, correlations between soils and vegetation in patterned distributions appear to have developed in an initial homogeneous landscape. Here, soil-vegetation feedback processes appear to be responsible for the development of such patterns. In this paper, we discuss various systems and their feedbacks that may lead to formation of patterns. In semi-arid systems, soil-water-vegetation feedbacks might lead to Turing-like self-organized pattern formation, as indicated by previously published models. In other cases of patterned soil-vegetation systems, feedback mechanisms may be involved that locally enhance growth of one species and inhibit that of other species. These interactions do not fulfill the criteria of Turing for pattern formation. However, such strong competitive interactions may lead to patterned vegetation as is shown by a study of a competitive model including spreading of the species. This pattern is not due to self-organization but depends on the initial boundary conditions.

## Introduction

Many plant species are confined to sites with more or less particular soil properties. This reflects a better ability of such species to cope with certain adverse soil conditions, or to better use available soil resources, than competing species in the vicinity. As a result, spatial soil heterogeneity is almost invariably associated with differences in vegetation cover. Such heterogeneity is in fact the basis of most soil surveys using aerial photographs or other remotely sensed data. Usually, the observed vegetation pattern resembles the pattern of the geological substratum. An example is given in Figure 11.1, where sparsely vegetated sandy point bar deposits alternate with more densely vegetated clayey depressions in a recent floodplain, bordered by high forest on higher land with older, better-drained soils.

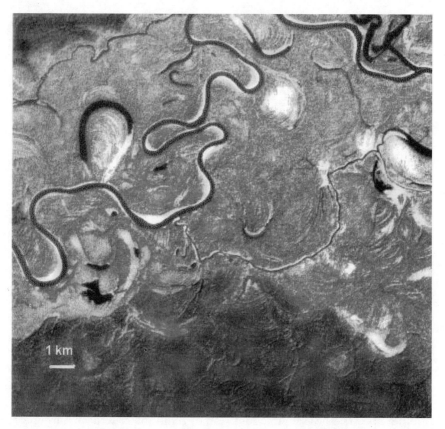

FIGURE 11.1. Satellite image from a river floodplain in Amazonia, Brazil (courtesy W.G. Sombroek, personal communication). Different grey scales show variable soil and hydrological conditions, as reflected by differences in plant cover under pristine conditions. In the active floodplain, sparsely vegetated sandy point bar deposits (light grey) alternate with more clayey filled depressions with a more luxuriant vegetation (medium grey) and open oxbow lakes (black). The high forest on either side of the active floodplain (dark grey) grows on older, better drained soils. Such patterns in plant cover that mirror underlying soil differences due to geological processes have traditionally been used in soil survey. Courtesy W. G. Sombruek.

Sometimes, however, heterogeneity in vegetation cover does not, or only partially, reflects differences in soil parent material (Figure 11.2). In such cases, ecosystem processes within an initially more or less homogenous landscape must have been responsible for the observed patterns. In this paper, we will investigate the hypothesis that soil-vegetation feedback processes can be responsible for such pattern formation.

This hypothesis builds on the well-known phenomenon that soil properties influence vegetation growth, and vegetation and associated soil-dwelling biota influence many chemical and physical soil properties (Hole

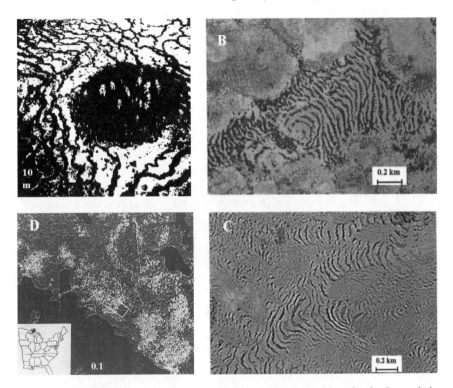

FIGURE 11.2. Aerial photographs of vegetation patterns, considered to be formed via soil-vegetation feedback processes. Scales are approximate only. (A) Aapa mire in northern Finland. *Sphagnum*-dominated bands (dark) tend to follow contours surrounding a tree island on mineral soil; light-colored areas are mainly graminoid-dominated fen vegetation (Metsäsaareke, Martimoaapa-Lumiaapa-Penikat, Simo-Keminmaa, photo by Aarno Torvinen). (B and C) Typical banded vegetation patterns in semiarid areas in (B) Niger and (C) Somalia. Vegetated ground is dark, bare soil is light shaded. (Reprinted from Valentin et al. 1999 with permission from Elsevíer.) (D) Clusters of hemlock and sugar maple in an old-growth stand, Sylvania Wilderness, Michigan, USA. (Reprinted by permission from Frelich et al. 1993.)

1982; van Breemen 1993). The effect of biota on soils was already recognized by Dokuchayev in 1879, who identified biota as one of the five soil-forming factors. These were later incorporated in Jenny's state factor approach (Jenny 1941), which says that the properties of any soil are a function of five state factors: climate, parent material, topography, biota, and time elapsed since the beginning of soil formation. The state factor approach still forms a major theoretical framework in pedology and has been used as the basis for ecological studies as well.

Biota-soil feedbacks were recognized long ago but were largely ignored by pedologists, perhaps because these feedbacks frustrated the state factor approach. Whereas the climate, parent material, topography, and time are practically independent of the soil formed under their influence, this is not true for biota. The state factor biota was seen as ". . . a real bugbear. Like everybody else I could see that the vegetation affects the soil and the soil affects the vegetation, the very circulus vitiosus that I was trying to avoid" (Jenny 1980). In an effort to break this vicious circle, Jenny took the immigration of individuals and input of propagules as the biotic factor.

However, in this way Jenny ignored the mutual feedback interaction between soil and vegetation. In general, the biota-soil system involves many interacting components, and the dynamic behavior is not easy to predict. When such interactions are nonlinear, they may lead to chaotic behavior and self-organization. Such phenomena have been observed in many areas, ranging from physics and earth system science to social science and economics, and have led to a comprehensive research in the dynamical behavior of nonlinear dynamical systems. Characteristic vegetation patterns in (semi-)arid systems have indeed been ascribed to self-organization, resulting from nonlinear feedback interactions between vegetation and water (Klausmeier 1999; Von Hardenberg et al. 2001; Rietkerk et al. 2002; Shnerb et al. 2003).

In this paper, we focus on feedback mechanisms in soil-vegetation systems and how they may lead to pattern formation. First, we summarize Turing's (1952) model of self-organized pattern formation; next, a model is described that includes competitive interaction between species. We will discuss applications of these models to the development of vegetation patterns observed in (semi-)arid systems and soil-vegetation systems showing interactions among species that enhance growth of similar species but inhibit that of others.

# Dynamical Models

## Turing's Mechanism for Self-Organized Pattern Formation

Studies of initially homogenous systems of interacting elements that self-organize into ordered patterns were pioneered by Prigogine (Nicolis 1977; Prigogine 1945) and Turing (1952). Turing considered a system of two reacting chemical species and with different diffusion constants:

$$\frac{\partial A}{\partial t} = f_A(A, B) + D_A \nabla^2 A$$

$$\frac{\partial B}{\partial t} = f_B(A, B) + D_B \nabla^2 B$$

(11.1)

where $A = A(\mathbf{x}, t)$ and $B = B(\mathbf{x}, t)$ denote the concentrations of the two species at time $t$ and position x. The functions $f_A(A, B)$ and $f_B(A, B)$ describe the reaction kinetics, and $D_A$ and $D_B$ are the diffusion constants of the corresponding components. Turing discovered that under certain conditions, a spatially uniform state that is stable in absence of diffusion could become unstable to small perturbations when the chemicals are allowed to diffuse. The conditions can be derived from a linear stability analysis, which gives

$$a_{11} + a_{22} < 0 \tag{11.2a}$$

$$a_{11}a_{22} - a_{12}a_{21} > 0 \tag{11.2b}$$

$$D_A a_{22} + D_B a_{11} > 0 \tag{11.2c}$$

where the coefficients $a_{ij}$ come from the Jacobian matrix J evaluated at the spatially uniform stationary state $S = (A_s, B_s)$:

$$J = \begin{pmatrix} \dfrac{\partial f_A}{\partial A}\bigg|_S & \dfrac{\partial f_A}{\partial B}\bigg|_S \\ \dfrac{\partial f_B}{\partial A}\bigg|_S & \dfrac{\partial f_B}{\partial B}\bigg|_S \end{pmatrix} = \begin{pmatrix} a_{11} & a_{12} \\ a_{21} & a_{22} \end{pmatrix} \tag{11.3}$$

The first criterion [Equation (11.2a)] in combination with the third [Equation (11.2c)] implies that the signs of $a_{11}$ and $a_{22}$ are opposite and (assuming $a_{11} > 0$) that component 1 enhances its own instability (positive feedback or activator), whereas component 2 decreases its own instability (negative feedback or inhibitor). Furthermore, if follows that also $a_{12}$ and $a_{21}$ must have opposite signs. The situation for $a_{12} > \phi$ and $a_{21} < \phi$ is schematically pictured in the feedback diagram shown in Figure 11.3A. The third criterion [Equation (11.2c)]

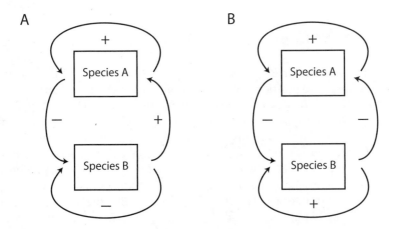

FIGURE 11.3. (A) Feedback diagram of the Turing mechanism. (B) Feedback mechanism for a competitive mechanism. An arrow from A to B with a + (−) sign means that A has a positive (negative) effect on B.

implies that the diffusion length of component 1 is smaller than that of component 2 ($l_A{}^2 = D_A/a_{11} < l_B{}^2 = D_B/a_{22}$). These necessary criteria for Turing instability are often referred to as "local activation with long range inhibition."

When only linear interactions are involved, a small perturbation would exponentially grow with time. However, nonlinear terms in the interaction could saturate the growth and bound the perturbation, leading to a spatially inhomogeneous steady state. The final pattern that will evolve is difficult, if not impossible, to calculate. The stable pattern, however, results from the internal dynamics of the system and is independent of the initial form of the perturbations. Such a system is said to be self-organizing.

Based on this reaction diffusion mechanism, various models have been studied in many research areas which show self-organized pattern formation. Especially in morphogenesis, which had the interest of Turing, this pattern-forming reaction-diffusion concept has been used to explain coat patterns of animals such as panthers, zebras, giraffes, and tigers (see, e.g., Murray 1989 or Edelstein-Keshet 1987 and references therein). In a following section, some examples of models are shown that describe vegetation patterns in semiarid regions.

## Competition Model Including Spreading

As will be seen below, feedback mechanisms that enhance growth of similar species but inhibit that of others occur commonly. A simple competition model, describing the interactions between two species or life forms $A$ and $B$, may describe these systems:

$$\frac{\partial A}{\partial t} = f_A(A,B) + D_A\nabla^2 A = A - A^2 - \alpha AB + D_A\nabla^2 A$$

$$\frac{\partial B}{\partial t} = f_B(A,B) + D_B\nabla^2 B = B - B^2 - \beta AB + D_B\nabla^2 B$$

$$(11.4)$$

where $A = A(x,t)$ and $B = B(x,t)$ stand for the biomass of the respective species at time $t$ and position x in arbitrary dimensionless units. The functions $f_A(A,B)$ and $f_B(A,B)$ describe the feedback interactions. The $+A$ and $+B$ terms correspond to the growth of the species, which is a positive feedback. The nonlinear quadratic terms $-A^2$ and $-B^2$ describe the limitation of growth due to, for example, a limitation in nutrients. The $-\alpha AB$ and $-\beta AB$ terms correspond to the negative feedback between species that compete for the same limiting resources or via their contrasting effects on soil properties. The diffusion terms $D_A\nabla^2 A$ and $D_B\nabla^2 B$ describe the lateral spread of the species. The parameters $\alpha$ and $\beta$, regulating the strength of the competitive interaction, and $D_A$ and $D_B$, regulating the spreading rates, are chosen to be positive and dimensionless. For simplicity, the constants that describe the strength of feedback interaction between similar species are taken to be 1.

The feedback diagram of this competition model is schematically drawn in Figure 11.3B, and from this it is expected that the competition model does not show Turing instability and self-organized pattern formation.

Analysis of the competition model (see, e.g., Murray 1989 or Edelstein-Keshet 1987) shows that in the absence of diffusion, four steady-states can be identified, viz. the bare state $S_1 = (A_{SI}, B_{SI}) = (0, 0)$, a state with only $A$ species $S_2 = (1, 0)$, a state with only $B$ species $S_3 = (0, 1)$, and a state where both species coexist $S_4 = ((1 - \alpha)/(1 - \alpha\beta), (1 - \beta)/(1 - \alpha\beta))$. From a linear stability analysis it follows that the bare state $S_1$ is unstable, $S_2$ is stable when $\alpha > 1$ and unstable when $\alpha > 1$, while $S_3$ is stable when $\beta > 1$ and unstable when $\beta > 1$. State $S_4$ is only stable when both species have little competition ($\alpha < 1$ and $\beta < 1$) so that both species can coexist. When the competition of one species is large and the other small, than the species with largest competition will survive. Thus, when $\alpha > 1$ and $\beta < 1$, $A$ will survive ($B$ dies) and when $\alpha < 1$ and $\beta > 1$, then $B$ will survive. When there is a strong competition between the species ($\alpha > 1$ and $\beta > 1$), then $S_4$ is unstable for small perturbations. In this case, it depends on the initial conditions whether the system will evolve toward a stable state consisting of only species $A$ or only species $B$.

When the species are allowed to diffuse, it is therefore expected that the system will evolve to a spatially homogeneous stable state (consisting of only $A$ or only $B$ or coexisting $A$ and $B$) except when there is strong competition for both species. To investigate whether or not vegetation patterns could evolve for strong competitive feedback interactions, the model was solved numerically using forward Euler integration on a square grid consisting of $100 \times 100$ cells. Calculations were started with an empty grid with 100 randomly chosen cells having species $A$ on it ($A = 1$) and another 100 randomly chosen cells having $B = 1$. Results are shown and discussed below.

## Self-Organization of Vegetation in (Semi-)Arid Regions

Dotted, striped (Figure 11.2B and 11.2C), and labyrinthic vegetation patterns in (semi-)arid regions have been explained by effects of biota on soils and hydrology (e.g., Aguiar and Sala 1999). Slatyer (1962) was among the first to point out that runoff from bare patches and infiltration of run-on water in vegetated patches plays a role in the formation of the spatially heterogeneous vegetation cover typical of most arid and semiarid lands. In addition to greater water availability, vegetated patches are characterized by higher concentrations of soil nutrients such as N, P, and K than adjacent barren areas, giving rise to the term *islands of fertility*. These islands dominate the structure of arid and semiarid landscapes worldwide (Reynolds et al. 1997).

There is a rich literature on how in dry regions these patches are formed by processes involving lateral redistribution of surface water and nutrients under the influence of the patch-forming plants, facilitating establishment of other species (e.g., Charley and West 1975; Pugnaire et al. 1996; Schlesinger et al. 1996; Aguiar and Sala 1999; Zaady et al. 1998; Maestre et al. 2001). The vegetated patches may be dominated by trees, shrubs, and perennial grasses. The bare areas in between are normally characterized by a microphytic soil crust, containing algae, cyanobacteria, bacteria, mosses, and lichens (Belnap and Lange 2001; www.soilcrust.org). The crusts are usually erosion-resistant and if smooth have a lower capacity for water infiltration than the same soil material without a crust (Eldridge et al. 2000). Accumulation of coarse organic debris and sediments, conveyed by rain splash, water runoff, or wind, may increase the nutrient pool in vegetated patches (Parsons et al. 1992; Wainwright et al. 1999). This helps formation of mound-shaped vegetation patches, which further increases capture of runoff water (Shachak and Lovett 1998).

In a search for the underlying (generic) mechanism that causes the observed vegetation patterns, dynamical models are developed that are based on Turing's reaction-diffusion mechanism and that incorporates the mechanisms described above. Various models have been published that could describe the observed vegetation patterns accurately (Klausmeier 1999; Von Hardenberg et al. 2001; Rietkerk et al. 2002; Shnerb et al. 2003). The self-organized pattern formation is essentially caused by nonlinear interactions between water and vegetation.

The models include a vegetation growth that depends positively on the available amount of vegetation and water; vegetation that spreads (diffuses) slowly with respect to water. Water is depleted by evaporation and root uptake. Furthermore, infiltration rates of water in vegetated areas is assumed to be higher than that in bare soil, reflecting the absence of microbiotic crust and differences in soil structure under the influence of vegetation. These mechanisms correspond to the Turing mechanism of pattern formation (Figure 11.3A).

For flat land, assuming only lateral water transport through soil (without runoff), these models produce spotted and labyrinthic patterns (see, e.g., Figure 11.4), dependent on the amount of precipitation. In sloping land, where runoff takes place, vegetated strips along the contours that move uphill are obtained in the model simulations. Similar pattern and pattern-forming mechanisms were found by Couteron and Lejeune (2001) for water-limited systems and by Lejeune et al. (2002) for strongly nutrient-limited systems. These workers used a phenomenological propagation-inhibition model that produces pattern by the interplay between short-range facilitative and long-range competitive interaction between the vegetation.

A number of processes presumably involved in the feedback interactions in these systems have been summarized in Figure 11.5 and 11.6. Figure 11.5 illustrates how cementation of dust particles by cyanobacteria via sticky polysaccharide sheaths around their cells aid the formation of a

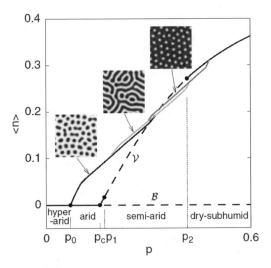

FIGURE 11.4. Modeled vegetation patterns of spatially averaged biomass $n$ (where a biomass density of 1.25 kg/m$^2$ occurs at $n = 1$) *versus* precipitation $p$ (where $p = 1$ corresponds to 800-mm annual rainfall). Curve $v$ designates the uniform vegetation state that is unstable if dashed (i.e., in the precipitation range $p_1 < p < p_2$). (Reprinted by permission from Von Hardenberg et al. 2001.)

crust of high stability in the presence of slaking and mechanical impact (Malam Issa et al. 2001) Low infiltration capacity of the crust increases runoff and helps to remove coarse particles. This in turn decreases roughness of the soil crust surface and increases supply of water and solids to the vegetated patch. Different feedback processes all contribute to the creation of a patchwork of microphytic and macrophytic domains with

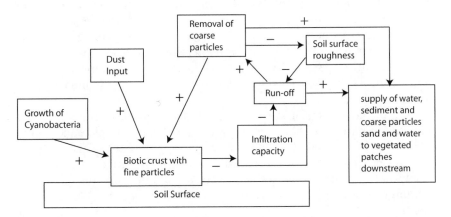

FIGURE 11.5. Feedback process involved in decreasing the permeability of biotic crusts and increasing surface runoff from patches with biotic crusts. An arrow from A to B with a + (−) sign means that A has a positive (negative) effect on B.

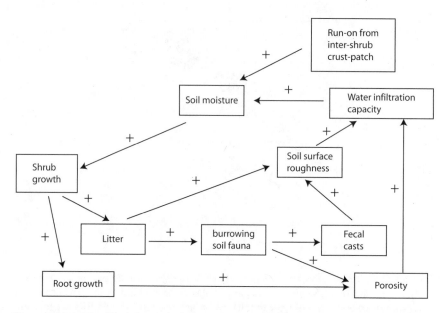

FIGURE 11.6. Feedback loops helping to increase surface roughness and soil porosity in vegetated patches. An arrow from A to B with a + (−) sign means that A has a positive (negative) effect on B.

different soil properties. Figure 11.6 illustrates a number of processes that further stimulate separation of vegetated and non-vegetated patches via increased soil porosity and surface roughness under the shrubs (Rostagno 1989). Furthermore, we expect a competitive interaction between the microphytes and the macrophytes: Development of microphytes is hindered in macrophytic patches by shading effects of the canopy and litter and by physical disturbance by higher activity of burrowing soil fauna. Macrophytes, on the other hand, will have more problems germinating and surviving as seedlings on the bare microphytic crust than in the sheltered vegetation patch with its more permeable and more fertile and moist soil. These relationships may be qualitatively summarized as depicted in Figure 11.3B.

The infiltration rates of water in vegetated areas is assumed to be higher than that in bare soil. Therefore, the feedback diagram depicted in Figure 11.3B may be simplified to that shown in Figure 11.3A. This feedback mechanism can lead to self-organization when a positive feedback acts locally and a negative feedback acts over a greater distance, resulting from a difference rate of lateral movement of the vegetation (slow) and water (fast).

The models described in this section strongly indicate that the underlying mechanism of pattern formation in (semi-)arid regions might be due to the nonlinear interaction between water and vegetation.

# Pattern Formation Involving Plants That Alter Soil to Favor Themselves

In this section, we will investigate the possibilities for pattern formation among species that strongly affect soil properties in direction that favor them *vis-à-vis* potential competitors. We will discuss *Sphagnum* and mull- and mor-forming plants.

## Peat Bogs

Peat moss is an ecological engineer (Jones et al. 1994) with tremendous effects on soils and landscapes. Well developed *Sphagnum* bogs are characterized by a nearly treeless landscape, with only few stunted xeromorphic trees on the better drained parts of the bog. Bogs are often characterized by complex spatial patterns. At fine (meter) scales, these involve hollows and hummocks with different species of peat moss and associated vascular plants. At coarser scales, these spatial patterns include large (up to a few kilometers in diameter and up to 6 m deep) raised bogs, with more or less concentric rings of hollows and hummocks and the shallower blanket bogs typical of the wettest climates. Patterned fens such as the Aapa mires in Finland and peat lands in Minnesota often have the low-lying flats dominated by graminiod species alternating with elongated strings of dense vegetation consisting mainly of *Sphagnum* and small trees (Figure 11.1A; Wright et al. 1992; Vasander 1996).

Van Breemen (1995) reviewed how *Sphagnum* creates peat bogs with their unfavorable properties for many vascular plants, including almost all trees. *Sphagnum* competes with vascular plants via specific organochemical, morphological, anatomical, and physiological properties that enable it to form acidic, slowly permeable and hence rain-water-dependent and nutrient-poor peat. "Almost nothing eats Sphagnum" (Clymo and Hayward 1982), in spite of its high content of polysaccharides and its lack of lignin. Its low palatability to herbivores and decomposers alike may be caused mainly by its high content of polyphenols, which partly form a polymeric network that is probably covalently linked to the cell wall polysaccharides (Van der Heijden 1994). Due to the pliable nature of the material and its many open, porous ("hyaline") cells, dead *Sphagnum* collapses to a finely porous material that holds large quantities of water yet is extremely slowly permeable to water. High nutrient-use efficiency in *Sphagnum,* the presence of polyuronic acids in plants cells, and anoxia contribute to a very slow rate of litter decomposition. As a result, *Sphagnum* peat bogs rise above the mineral soil base and the lithotrophic groundwater associated with it. The resulting bogs are increasingly dependent on nutrients deposited from the atmosphere (ombrotrophy) and are subject to even greater nutrient limitation.

Collectively, high acidity, low nutrient availability, and water saturation in bogs reduce the competitive ability of upland, vascular plants because of their requirement for the uptake of nutrients by roots and the difficulty in doing so. The absence of vascular plants in bogs increases light availability, decreases the rate of evapotranspiration, and increases water availability. High acidity, low availability of nutrients, water saturation, and a sparse cover of light-shading vascular plants favor the development of a positive feedback loop by promoting *Sphagnum* growth and peat formation. By contrast, the more easily decomposable litter of vascular plants tends to counteract peat formation, maintaining the cycling of appreciable amounts of lithotrophic elements such as Ca and Mg through the vegetation, which hampers growth of typical bog-forming *Sphagnum* species. Through their tendency to maintain a degree of lithotrophy and to depress light availability to mosses below their canopy, vascular plants compete with *Sphagnum* species (Figure 11.7). This is an example of litter decomposability as a component of plant fitness, as proposed by Berendse (1994). The interactions between *Sphagnum* species and typical fen-dwelling vascular plants via

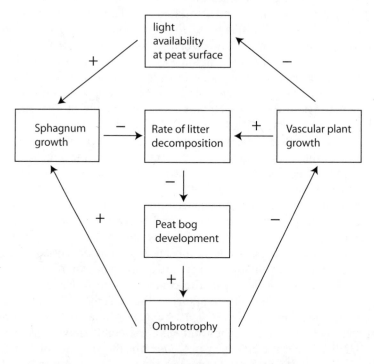

FIGURE 11.7. Effects of *Sphagnum* species and vascular wetland plants on the formation of ombrotrophic peat bogs. Vascular plants depress *Sphagnum* via their negative effect on light availability (negative feedback interaction times a positive one combines to a negative one, $- \times + = -$) and their tendency to reduce bog growth (with its positive effects on ombrotrophy and therefore on *Sphagnum*) by providing more relatively easily decomposable litter ($+ \times - \times + \times + = -$). *Sphagnum* has similar negative effects on vascular plants via stimulation of bog growth and resulting low nutrient supply ($- \times - \times + \times - = -$). An arrow from A to B with a $+ (-)$ sign means that A has a positive (negative) effect on B.

their contrasting effects on resource availability is probably mainly responsible for the highly varied vegetation patterns in fens.

*Sphagnum* and different species of vascular plants have a contrasting effect on peat accumulation and associated aspects of resource availability. These specific effects of contrasting plant types interact with geological and meteoric factors and disturbance (e.g., fire) to create the highly varied vegetation patterns of bogs and fens. Pastor et al. (2002) constructed and analyzed a process-based model of peat lands to account for the roles that vascular plants and mosses play in carbon and nutrient flux and storage and how they interact with nutrient supply. The model mimics the broad features of successional development of peat lands from fens to bogs often found in the paleorecords of peat cores. The model is not spatially explicit but does predict the presence of alternative stable states. By expanding the model to include diffusion terms, we expect it should be possible to explain the patterning typical in fens and hollow-hummock bogs.

## Mull- and Mor-Forming Plants

Northern temperate forests are composed of relatively few species and are usually found on young soils with largely unweathered, glacial deposits. Many soil properties in such forests are often closely correlated with the dominant tree species present. These properties include profile morphology (particularly with regard to the organic surface layer, or *forest floor* (Muller 1887; Ovington 1953), availability of N (Finzi et al. 1998a), and soil pH and exchangeable cation concentrations (Zinke 1962; Lefevre and Klemmedson 1980; Alban 1982; Klemmedson 1987; Boetcher and Kalisz 1990; Finzi et al. 1998b). In the surface soils, N availability, soil pH, and the quantity of exchangeable base cations (expressed per unit area) all tend to be lower when organic surface layers are thick ("mor" humus form) than when they are thin or absent ("mull" humus form). Foresters' experience and observations by pedologists from natural stands led to the notion that those differences are caused by differences in tree species. Binkley (1995) reviewed the evidence for and against that notion and concluded that it is often impossible to infer whether such soil-tree combinations resulted from prior differences in soils or from the influence of species. He called for replicated common garden experiments to test further the hypotheses that species cause those differences.

In fact, one such experiment of appreciable duration (about 30 years) is available and does demonstrate a dramatic species effect, in line with the traditional notion. The Siemianice common garden experiment of the Agricultural University of Poznan (Poland) involves triplicated 20 × 20 m plots planted in 1968 with 14 different tree species (6 conifers, 8 hardwoods) in uniform sandy brown podzolic soil that was earlier under *Pinus sylvestris*. Different nutrient cycling strategies and litter composition of these species have caused clear differences in soil pH, soil Ca levels, and organic matter content, correlated with the litter Ca contents of the different tree species (P.B. Reich and J. Oleksyn, personal communications).

Van Breemen et al. (1997) found that soil pH and exchangeable Ca levels were higher under sugar maple (*Acer saccharum*) than under eastern hemlock (*Tsuga canadensis*) growing in the same area on soils developed from glacial tills that did not differ significantly in mineral-bound Ca and Mg. This suggested that these two late-successional species influence exchangeable Ca and soil pH and available N in opposite directions. Dijkstra and Smits (2002) demonstrated that greater uptake of Ca from deeper soil layers under sugar maple than under hemlock could account for these differences and showed that exchangeable Ca levels could increase (under sugar maple) or decrease (under hemlock) several-fold within the lifetime of an individual tree (i.e., within 200 years). If the changes in soil chemistry and nutrient availability beneath these species would confer some competitive advantage or disadvantage to each of these species, this should lead to separate patches dominated by each of the species. Indeed, Frehlich et al. (1993) showed that old-growth sugar maple and hemlock occur in monospecific patches that are stable over long periods of time (>1000 years) in the Sylvania Wilderness in northern Michigan (Figure 11.2D). They concluded that seedbed effects (burial of hemlock seedlings beneath sugar maple leaf litter and low N availability to sugar maple beneath hemlock) explained the development and maintenance of the patches over long periods of time. We believe that the differential effects of trees on the soil's base status and pH and related effects of pH on the thickness of the litter layer and the N dynamics lay at the basis of the spatial segregation. Walters and Reich (1997) found that high nitrate availability to sugar maple seedlings increased growth under low light. Kobe et al. (1995) observed that deeply shaded sugar maple saplings survived better in calcareous than in noncalcareous soils, implying that the small size of the exchangeable pool beneath hemlock competitively displaces sugar maple. In a regional survey, Watmough (2002) showed that sugar maple grew better on soils with pH 6 and high levels of water-extractable Ca than on soils of pH 4.5 with lower contents of water-extractable soil Ca.

So, sugar maple, through its effect on soil chemistry and nutrient availability, creates habitats that favor self-replacement while inhibiting hemlock seedlings by burying them under leaf litter. Through its effect on soil, sugar maple probably also favours itself *vis-à-vis* other late successional species such as American beech (*Fagus grandifolia*) and red oak (*Quercus rubra*), which either prefer more acid sites or are outcompeted by sugar maple in more base-rich soils (Kobe et al. 1995). Hemlock appears to influence seedbed conditions favoring offspring success while hampering the establishment of sugar maple. The low soil pH beneath hemlock appears to be created by the production of highly acidic, slowly decomposing leaf litter because of high lignin and tannin contents (e.g., White 1986, 1991; Millen 1995). The low pH depresses the rate of net N mineralization and the rate of net nitrification, which could dramatically reduce the growth rate of sugar maple seedlings beneath hemlock. Although low N availability and low soil pH may not be an optimal growth environment for hemlock, hemlock's tolerance to such conditions (Godbold and Huttermann 1994 and references therein) may be sufficient to give hemlock seedlings a competitive edge over sugar maple (and other) seedlings that germinate beneath hemlock.

In the examples described above (trees and common fen plants in *Sphagnum* bogs; sugar maple in hemlock stands), individuals of a particular species decrease soil pH and nutrient availability to levels that are more unfavorable to competitors than to themselves. This mechanism is apparently more widespread. For example, in eastern U.S. forests, *Rhododendron maximum* appears to inhibit the colonization of broad-leaved trees that dominate on adjacent, relatively fertile, nonpodzolized soils (Orbell et al. 1980). Rhododendron litter is recalcitrant to the decomposition process and depresses earthworm activity, thus facilitating the development of a thick forest floor, which could confer a competitive advantage to their seedlings over other understorey vegetation (Boetcher and Kalisz 1990, 1991). Berendse (1994) described how slowly decomposing litter of ericacious plants tends to slow down the accumulation of plant available nutrients (especially N) in soil, so that the succession to more nutrient-demanding grasses is delayed. This illustrates how species that depress the growth of competitors via their adverse effect on soil fertility tend to stall succession [cf. Connell and Slatyer's (1977) "inhibition model"]. They follow Grime's (1977) stress-tolerant strategy, with traits characteristic of many slow-growing plants on nutrient-poor soils: low foliar nutrient concentrations and low palatability to herbivores (Grime et al. 1997). The results in the Grime et al. (1997) paper expand Grime's (1977) model to include an active component (i.e., creating stress) as part of the competitive strategy of certain mid- and late-successional species. This is consistent competition being defined as "the tendency of neighbouring plants to utilise the same quantum of light, ion of a mineral nutrient, molecule of water, or volume of space . . ." (Grime 1977).

Other early- to mid-successional plant species that stimulate the supply of soil resources fit Connell and Slatyer's (1977) "facilitation" model (e.g., *Molinia* in heathlands, desert shrubs, and islands of fertility). Dahlgren et al. (1997) showed that blue oak (*Quercus douglasii*) significantly increased organic C and N, cation exchange capacity, exchangeable bases, and pH in the soils under their canopies in California. Similarly, *Quercus robur* invading *Calluna* heathlands increased soil pH, decreased the thickness of the organic surface layer, and decreased the contrast between Albic and Spodic horizons in podzols in Denmark (Nielsen et al. 1987). In addition to physical and chemical changes in soils, balsam fir tree islands in dry, alpine areas influence microclimate—snow cover, water supply and weather-related stresses—that tend to increase plant growth (Van Miegroet and Hysell 1995).

## Competition Model Including Plant Spreading

In all these cases, soil effects that are likely to result from particular physical and physiological properties of particular plant species will feed back

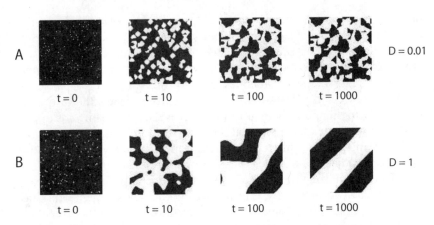

FIGURE 11.8. Spatial patterns observed for the competition model that allows for diffusion of two species and positive feedback between similar and negative feedback between different species [Equation (11.4)] for a strong competition between the species $A$ and $B$ ($\alpha = \beta = 100$) at different times $t$. In the upper panel results are shown for small diffusion constants of $A$ and $B$, $D_A = D_B = 0.01$ (slow spreading), and in the lower panels results are shown for $D_A = D_B = 1$. The white color corresponds to species $A$ ($A = 1, B = 0$), and black corresponds to species $B$ ($A = 0, B = 1$).

positively to individuals that caused them and possibly negatively to certain other species. The question is whether such processes do lead to spatial self-organization.

Therefore, we studied the competition [Equation (11.4)] model described in a previous section. It is found that the system evolves toward a spatially homogeneous state except for strong competition. For strong competitive interaction between different species and slow lateral diffusion (spreading), stable patterns develop with areas consisting of only $A$ and others of only $B$. This is shown in Figure 11.8A ($\alpha = \beta = 100$ and $D_A = D_B = 0.1$). Initially, the randomly distributed species $A$ and $B$ both grow and spread slowly. When $A$ and $B$ meet, inhibition by strong competition in combination with the slow diffusion prevents interference, apparently causing a stable distribution with (sharp) boundaries between monospecific patches. At a boundary between two species, there is a continuous intrusion of one species into the area of the other. How fast the "intruder" will die depends on the strength of the competition. The width of such a border region, which is not in equilibrium and where species coexist, will depend on interaction strength and diffusion constants.

The shape of the spatial distribution is determined by the initial (random) distribution of $A$ and $B$ and appears to be defined by the positions where species happen to meet. These patterns as well as the underlying mechanism seem to be similar to those found by Frelich et al. (1998) using a spatially referenced Markov model including similar feedback interactions among trees in the neighborhood. Such patchiness lacks the spatial periodicity of the self-organized patterns shown in Figure 11.3 and is not a result of Turing-like self-organization. However, when the diffusion constants are of the order one, a striped pattern develops, as can be seen in Figure 11.8B. The origin of the formation of these patterns is yet unclear and needs further exploration.

## Conclusions

Spatial soil heterogeneity is almost invariably associated with differences in vegetation cover due to differences in "preferences" of particular plant species for particular soil properties. Such heterogeneity is in fact the basis of most soil surveys using aerial photographs or other remotely sensed data.

Often, prior geologically determined soil differences form the basis of such soil-vegetation patterns. Where soil parent material differences cannot explain such soil-vegetation patterns, we propose that they are caused by soil-vegetation feedback processes.

Patterns observed in (semi-)arid regions in subhumid tropical grassland may be explained by self-organized pattern formation due to the nonlinear interaction between soil water and vegetation (Klausmeier 1999; Von Hardenberg et al. 2001; Rietkerk et al. 2002; Shnerb et al. 2003; Couteron and Lejeune 2001; Lejeune et al. 2002). During rain storms, faster infiltration of water at vegetation sites (where surface litter is more abundant and soil porosity is higher) and (much) faster lateral transport rates of water than of vegetation propagules form the basis of such spatial self-organization. Depending on the amount of precipitation, the models yield spotted and labyrinthic patterns on flat land and vegetated strips on sloping land. These patterns correspond well with those observed in nature. The models are based on the reaction-diffusion mechanism pioneered by Turing, who showed that under conditions of local activation and long-range inhibition, self-organized pattern formation could occur.

In other examples such as fen plants in *Sphagnum* bogs and patchy mixed stands of sugar maple and hemlock, competitive interactions involving contrasting effects of plants on soil properties appear to play a role in pattern formation. Stable monospecific patchiness is produced by a simple spatial model involving slow diffusion, strongly positive local feedback for similar species, and strong negative feedback between different species. Although the origin and nature of the patterns requires further study, it is clear that the interactions involved do not fulfill Turing's conditions for pattern formation. It might, however, be that lateral transport of water and/or nutrients under influence of one of the species can create a Turing-like situation.

We expect that pattern formation is not restricted to the examples given in this paper, but that soil-vegetation patchiness caused by contrasting effects of different species on soils are widespread. Further research is needed to indicate qualitatively and quantitatively the processes involved and to obtain a fundamental understanding of their nature.

## References

Aguiar, M.R., and Sala, O.E. 1999. Patch structure, dynamics and implications for the functioning of arid ecosystems. TREE 14: 263–277.

Alban, D.H. 1982. Effects of nutrient accumulation by aspen, spruce and pine on soil properties. Soil Sci. Soc. Am. J. 46: 853–861.

Belnap, J., and Lange, O.L. eds. 2001 Biological soil crusts: structure, function and management. Heidelberg: Springer Verlag. 503 pp.

Berendse, F. 1994. Litter decomposability: a neglected component of plant fitness. J. Ecol. 78: 413–427.

Binkley, D. 1995. The influence of tree species on forest soils: processes and patterns. In Proceedings of the Trees and Soil Workshop, eds. D.J. Mead and I.S. Cornforth, pp. 1–33. Canterbury, New Zealand: Lincoln University Press.

Boetcher, S.E., and Kalisz, P.J. 1990. Single-tree influence on soil properties in the mountains of eastern Kentucky. Ecology 71: 1365–1372.

Boetcher, S.E., and Kalisz, P.J. 1991. Single-tree influence on earthworms in Eastern Kentucky. Soil Sci. Soc. Am. J. 55: 862–865.

Charley, J.L., and West, N.E. 1975. Plant-induced chemical patterns in some shrub-dominated semidesert ecosystems of Utah. J Ecol. 63: 945–963.

Clymo, R.S., and Hayward, P.M. 1982. The ecology of *Sphagnum*. In Bryophyte ecology, ed. A.J.E. Smith, pp. 229–289. London: Chapman and Hall.

Connell, J.H., and Slatyer, R.O. 1977. Mechanisms of succession in natural communities and their role in community stability and organization. Am. Naturalist 111: 1119–1144.

Couteron, P., and Lejeune, O. 2001. Periodic spotted patterns in semi-arid vegetation explained by a propagation-inhibition model. J. Ecol. 89: 616–628.

Dahlgren, R.A., Singer, M.J., and Huang, X. 1997. Oak tree and grazing impacts on soil properties and nutrients in a California oak woodland. Biogeochemistry 39: 45–64.

Dokuchayev, V.V. 1879. Abridged historical account and critical examination of the principal soil classifications existing. Trans. Petersburg Soc. Nat. 1: 64–67 (in Russian).

Dijkstra, F.A., Smits, M.M. 2002. Tree species effects on calcium cycling: the role of calcium uptake in deep soils. Ecosystems 5: 385–398.

Edelstein-Keshet, L. 1987. Mathematical models in biology. Birkhauser Mathematics Series. New York: McGraw Hill.

Eldridge, D.J., Zaady, E., and Shachak, M. 2000. Infiltration through three contrasting biological soil crusts in patterned landscapes in the Negev, Israel. Catena 40: 323–336.

Finzi, A.C., van Breemen, N., and Canham, C.D. 1998a. Canopy tree-soil interactions within temperate forest: species effects on soil carbon and nitrogen. Ecol. Applications 8: 440–446.

Finzi, A.C., Canham, C.D., and van Breemen, N. 1998b. Canopy tree-soil interactions within temperate forests: species effects on pH and cations. Ecol. Applications 8: 447–454.

Frelich, L.E., Calcote, R.R., Davis, M.B., and Pastor, J. 1993. Patch formation and maintenance in an old-growth hemlock-hardwood forest. Ecology 74: 513–527.

Frelich, L.E., Sugita, S., Reich, P.B., Davis, M.B., and Friedman, S.K. 1998. Neighbourhood effects in forest: implications for within stand patch structure. J. Ecol. 86: 149–161.

Godbold, D.L., and Hutterman, A. eds. 1994. Effects of acid rain on forest processes. New York: Wiley-Liss.

Grime, J.P. 1977. Evidence for the existence of three primary strategies in plants and its relevance to ecological and evolutionary theory. Am. Naturalist 111: 1169–1194.

Grime, J.P., Thompson, K., Hunt, R., Hodgson, J.G., Cornelissen, J.H.C., Rorison, I.H., Hendry, G.A.F., Ashenden, T.W., Askew, A.P., and Band, S.R. et al. 1997. Integrated screening validates primary axes of specialisation in plants. Oikos 79: 259–281.

Hole, F.D. 1982. Effects of animals on soil. Geoderma 25: 75–112.

Jenny, H. 1941. Factors of soil formation. New York: McGraw-Hill.

Jenny, H. 1980. The soil resource. Origin and behaviour. New York: Springer Verlag.

Jones, C.G., Lawton, J.H., and Schachak, M. 1994. Organisms as ecosystem engineers. Oikos 69: 373–386.

Klausmeier, C.A. 1999. Regular and irregular patterns in semiarid vegetation. Science 284: 1826–1828.

Klemmedson, J.O. 1987. Influence of oak in pine forests of central Arizona on selected nutrients of forest floor and soil. Soil Sci. Soc. Am. J. 51: 1623–1628.

Kobe, R.K., Pacla, S.W., Silander Jr, J.A., and Canham, C.D. 1995. Juvenile tree survivorship as a component of shade tolerance. Ecol. Applications 5: 517–532.

Lefevre, R.E., and Klemmedson, J.O. 1980. Effect of Gambel oak on forest floor and soil of a Ponderosa pine forest. Soil Sci. Soc. Am. J. 44: 842–846.

Lejeune, O., Tlidi, M., Couteron, P. 2002. Localized vegetation patches: A self-organized response to resource scarcity. Phys. Rev. E 66: 010901/1–010901/4.

Malam Issa, O., Le Bissonnais, Y., Défarge, C., and Trichet, J. 2001. Role of a cyanobacterial cover on structural stability of sandy soils in the Sahelian part of western Niger. Geoderma 101: 15–30.

Maestre, F.T., Bautista, S., Cortina, J., and Bellot, J. 2001. Potential for using facilitation by grasses to establish shrubs on a semiarid degraded steppe. Ecol. Applications 11: 16541–1655.

Millen, P.E. 1995. Bare Trees. Zadock Pratt, Master Tanner & what happened to the Catskill mountain forests. Hensonville, NY: Black Dome Press Corp. 100 p.

Muller, P.E. 1887. Studien über die natürlichen Humusformen und deren Einwirkungen auf Vegetation und Boden. Berlin: Julius Springer. 324 pp.

Murray, J.D. 1989. Mathematical Biology. Berlin: Springer.

Nicolis, G., and Prigogine, I. 1977. Self Organization in Non-Equilibrium Systems. New York: Wiley and Sons.

Nielsen, K.E., Dalsgaard, K., and Nornberg, P. 1987. Effects on soils of an oak invasion of a Calluna heath, Denmark. I and II. Geoderma, 41: 79–106.

Odling-Smee, F.J., Laland, K.N., Feldman, M.W. 2003. Niche construction: the neglected process in evolution. Princeton, NJ: Princeton University Press.

Orbell, G.E., Parfitt, R.L., and Furkert, R.J. 1980. Guide Book for Tour 6—Specialist North Auckland. Field guide for the "Soils with Variable Charge" conference. Palmerston North, New Zealand. February 1981. Wellington, New Zealand: P.D. Hasselberg, Government Printer.

Ovington, J.D. 1953. Studies on the development of woodland conditions under different trees. I Soil pH. J. Ecol. 41: 13–34.

Parsons, A.J., Abrahams, A.D., and Simanton, J.R. 1992. Microtopography and soil-surface materials on semi-arid piedmont hillslopes, southern Arizona. J. Arid Environ. 22: 107–115.

Pastor, J., Peckham, B., Bridgham, S., Weltzin, J., and Jiquan Chen, J. 2002. Plant community dynamics, nutrient cycling, and alternative stable equilibria in peatlands. Am. Naturalist 160: 553–568.

Prigogine, I. 1945. Bull. Acad. Roy. Belg. Cl. Sci. 31: 600.

Pugnaire, F.I., Haase, P., and Puigdefabregas, J. 1996. Facilitation of higher plant species in a semi-arid environment. Ecology 77: 1420–1426.

Reynolds, J.F., Virginia, R.A., and Schlesinger, W.H. 1997. Defining functional types for models of desertification. In Functional Types, eds. T.M. Smith, H.H. Shugart, and F.I. Woodward. Cambridge, UK: Cambridge University Press. Pp. 195–216.

Rietkerk, M., Boerlijst, M.C., Van Langevelde, F., HilleRisLambers, R., Van de Koppel, J., Kumar, L., Prins, H.H.T., and de Roos, A.M. 2002. Self organization of vegetation in arid ecosystems. Am. Naturalist 160: 524–530.

Rostagno, C.M. 1989. Infiltration and sediment production affected by soil surface in a shrubland of Patagonia, Argentina. J. Range Manage. 42: 382–385.

Schlesinger, W.H., Raikes, J.A., Hartley, A.E., and Cross, A.F. 1996. On the spatial pattern of soil nutrients in desert ecosystems. Ecology 77: 364–374.

Shachak, M., and Lovett, G.M. 1998. Atmospheric deposition to a desert ecosystem and its implications for management Ecol. Applications 8: 455–463.

Shnerb, N.M., Sarah, P., Lavee, H., and Solomon, S. 2003. Reactive glass and vegetation patterns. Phys. Rev. Lett. 90: 038101/1–038101/4.

Slatyer, R.O. 1962. Methodology of a water balance study conducted on a desert woodland (*Acacia aneura* F. Muell.) community in central Australia. UNESCO Arid Zone Research 16: 15–26.

Turing, A.M. 1952. The chemical basis of morphogenesis. Phil. Trans. Roy. Soc. Lond. B. 237: 37–72.

Van Breemen, N. 1993. Soils as biotic constructs favouring net primary productivity. Geoderma 57: 183–211.

Van Breemen, N. 1995. How *Sphagnum* bogs down other plants. Trends Ecol. Evol. 10: 270–275.

Van Breemen, N., Finzi, A.C., and Canham, C.D. 1997. Canopy tree-soil interactions within temperate forests: effects of soil texture and elemental composition on species distributions. Can J. Forest Res. 27: 1110–1116.

Van der Heijden, E. 1994. A combined anatomical and pyrolysis mass spectrometric study of peatified plant tissues. Ph.D. Thesis, University of Amsterdam, The Netherlands, 157 pp.

Van der Putten, W.H., Van Dijk, C., and Peters, B.A.M. 1993. Plant-specific soil-borne diseases contribute to succession in vegetation. Nature 362: 53–55.

Vasander, H., ed. Peatlands in Finland. Helsinki: Finnish Peatland Society. 168 pp.

Van Miegroet, H., and Hysell, M.T. 1995. The effect of tree islands on soil properties in the spruce-fir zone of Northern Utah. p. 307. Agronomy abstracts. ASA, Madison, WI.

Von Hardenberg, J., Meron, E., Shachak, M., and Zarmi, Y. 2001. Diversity of vegetation patterns and desertification. Phys. Rev. Lett. 87: 1981011–1981014.

Wainwright, J., Parsons, A.J., and Abrahams, A.D. 1999. Rainfall energy under creosotebush. J. Arid Environ. 43: 111–120.

Walters, M.B., and Reich, P.B. 1997. Growth of Acer saccharum seedlings in deeply shaded understories of northern Wisconsin: effects of nitrogen and water availability. Can. J. Forest Res. 27: 237–247.

Watmough, S.A. 2002. A dendrochemical survey of sugar maple (Acer saccharum Marsh) in south-central Ontario, Canada. Water Air Soil Pollution 136: 165–187.

White, C.S. 1986. Volatile and water-soluble inhibitors of nitrogen mineralizatiuon and nitrification in a ponderosa pine ecosystem. Biol. Fert. Soils 2: 97–104.

White, C.S. 1991. The role of monoterpenes in soil nitrogen cycling processes in ponderosa pine. Biogeochemistry 12: 43–68.

Wright, H.E., Coffin, B., and Aaseng, N.E. 1992. Patterned peatlands of Minnesota. Minneapolis: University of Minnesota Press, 544 pp.

Zaady, E., Groffman, P., and Shachak, M. 1998. Nitrogen fixation in macro- and microphytic patches in the Negev desert. Soil Biol. Biochem. 30: 449–454.

Zinke, P.J. 1962. The pattern of influence of individual forest trees on soil properties. Ecology 43: 130–133.

# 12
# Spatial Patterning of Soil Carbon Storage Across Boreal Landscapes

MERRITT R. TURETSKY, MICHELLE C. MACK, JENNIFER W. HARDEN, and KRISTEN L. MANIES

## Abstract

The boreal forest covers 14% of the earth's vegetated surface but contains about 27% of the world's vegetation carbon and between 25% and 30% of the world's soil carbon. Unique features of this biome include cold climates, large areas of relatively flat topography, discontinuous permafrost, large and severe fire events, and the accumulation of peat. These characteristics are important in controlling energy and carbon cycling and either influence or are influenced by regional climate and hydrological regimes. Total carbon accumulation within an ecosystem reflects the balance between net primary production (NPP), decomposition, and nonrespiratory losses (dissolved carbon export, fire, and land-use changes). In this chapter, we use soil carbon storage as a long-term estimate of net ecosystem productivity (NEP; the balance between NPP and decomposition) and nonrespiratory losses that integrates annual variability in the ecosystem processes contributing to carbon balance. Our overall hypothesis is that a combination of regional and local physiography creates spatial heterogeneity in hydrology and soil temperatures. Hydrology and thermal regimes, in turn, influence distributions of fire, permafrost, peatlands, and vegetation and ultimately control long-term carbon storage in many boreal climatic zones. Soil carbon storage varies tremendously between boreal stand types or features and is particularly large in poorly drained peatland and permafrost ecosystems. Landscape composition, then, is important for scaling carbon storage in boreal regions. However, whether the configuration of upland and lowland ecosystems influences carbon processes has not been adequately explored but likely is important to variations in carbon emissions during fire. Biological controls such as herbivory and insect outbreaks are important to the distribution of plant species and nitrogen availability in forest ecosystems, but their influence on wetland systems or long-term carbon dynamics is not well understood.

# Introduction

The boreal biome covers 18.5 million km$^2$ across interior Alaska, Canada, Fennoscandia, Russia, and parts of Mongolia and China. This biome actually represents a number of ecoclimatic zones that support coniferous or mixed conifer-hardwood forests. Most boreal regions experience large annual changes in solar input, short growing seasons (3–4 months), and extremely cold winter temperatures (Eugster et al. 2000). Low precipitation and temperatures may limit plant productivity in boreal ecosystems (Baldocchi et al. 2000). However, throughout the Holocene, soils in boreal regions have served as an important reservoir for terrestrial carbon (C) (Harden et al. 1992). Today, boreal forests contain approximately 27% of the world's vegetative C and between 25% and 30% of the world's soil C, approaching 500 Gt C (Gorham 1991; Dixon et al. 1994). Carbon sinks in the boreal forest are relatively small, averaging between 0.3 and 0.5 Pg ($=10^{15}$g) C yr$^{-1}$, and the size of this sink varies spatially and temporally (Apps et al. 1993; Goodale et al. 2002). Thus, C sequestration in boreal regions is dictated by the small difference between larger C inputs and outputs, which makes it difficult to assess spatial or temporal controls on C balance. We argue that soil C storage is a long-term (decadal and longer) proxy for net ecosystem production (NEP; total ecosystem C storage) that integrates annual variation in processes such as net primary productivity (NPP) and decomposition (Randerson et al. 2002) and thus is useful for assessing spatial or temporal controls on NEP.

Identifying spatial controls on ecosystem-level processes is important for scaling current ecosystem dynamics and planning future responses to global change. An approach based in landscape ecology, or the study of how pattern effects process (Turner 1989), can help to identify the controls of landscape structure on ecosystem processes. However, a landscape approach requires an integrated investigation of patterns and processes at varying spatial and/or temporal scales. For example, large-scale patterns of deglaciation and sediment deposition exert major controls on hydrology and ecosystem development in boreal regions. Glacial till of varying thickness overlies bedrock across much of interior and eastern Canada, increasing in thickness to 20 m in low-lying areas. Glacial movement across Canada created relatively flat lake and outwash plains in the west and moraines in central Canada. Peatlands occur extensively across glacial plains in western and central Canada because of poor drainage. Rocky outcrops with increasing elevation are found farther east. Across Alaska and Siberia, large regions repeatedly have been covered by windblown loess (largely silt or finer particles) derived from river floodplains and glacial outwash plains (Pewe 1958; Van Cleve et al. 1993). Silt is transported to lowlands, mixed with organic debris, and incorporated into permafrost layers (see the section "Discontinuous Permafrost" below). Loess deposits generally are less well drained than other glacial deposits and thereby influence heterogeneity in surface hydrology.

Although postglacial topography and sediment deposition influence ecosystem development and hydrologic processes, there is considerable local heterogeneity that influences C storage. For example, landscape position, topography, soil texture/bedrock, and stand age create variation in soil hydrological and temperature regimes. Our goals here are to discuss (1) important spatial controls on soil C storage across boreal ecosystems and (2) how landscape composition (amount of different ecosystems or habitat types) and configuration (the spatial arrangement of ecosystems or habitat types) influence the spatial patterning of C storage. Water availability both within and between ecosystems or habitat types controls biological patterns (vegetation structure) and both biological (insect outbreaks) and abiotic (fire) processes. Thus, our overall hypothesis is that a combination of regional, landscape, and local physiography (i.e., bedrock, topography, soil texture) creates heterogeneity in state factors such as hydrology and soil temperatures, which manifests through distributions of fire, permafrost, peatlands, and dominant vegetation to control carbon storage (Figure 12.1).

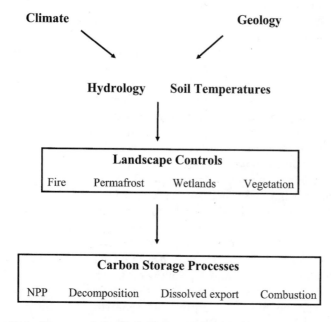

FIGURE 12.1. Conceptual framework for describing spatial heterogeneity in the boreal forest and its influence on long-term carbon storage. Climate as well as regional and local physiography influence soil hydrology and temperatures. Hydrologic processes control distributions of permafrost, dominant vegetation, peat, and fire behavior, which ultimately regulate the processes contributing to carbon storage.

## Spatial Controls on C Storage

Rather than provide a comprehensive list of factors that create spatial heterogeneity within each climatic zone in the boreal forest, we focus in more detail on the processes most important to NEP and long-term carbon storage. Specifically, we describe in detail what controls variation in the distributions of fire, permafrost, peatlands, and vegetation composition in boreal regions and how these factors influence soil C storage.

### Fires

Forest fires are a common disturbance in the boreal forest, occurring on average every 60 to 200 years in North America (Stocks and Kaufman 1997). An average of 2 million ha of Canadian forest has burned annually from 1959 to 1999, but more than 7 million ha of forest can burn during extreme fire years. Conard et al. (2002) estimate that in 1998 about 13.3 million ha burned in Siberia alone. Fire frequencies have increased in North America over the past several decades (Kasischke and Stocks 2000; Podur et al. 2002). About twice as much area burned in Canada during the 1980s and 1990s compared to the previous two decades (Stocks et al. 2003).

Fires in boreal regions of North America and Western Europe tend to occur as intense crown fires (Stocks and Kaufman 1997) that burn large areas and destroy the majority of forest floor litter (Stocks 1991; Kasischke et al. 1995). These fires also usually are stand replacing, initiating new stands of spruce, jack pine, and aspen. Because boreal regions contain large volumes of biomass that are susceptible to burning, boreal forest fires emit large quantities of $CO_2$, CO, $CH_4$, other trace gases, and particulates to the atmosphere. Emission ratios of $CH_3Br$ and $CH_3Cl$ from boreal burning are much higher than are measured from savanna and chaparral fires, likely due to lower combustion efficiencies during smoldering (Manö and Andreae 1994). Overall, boreal fire emissions may be equivalent to 14–20% of regional fossil fuel emissions in Canada and Siberia (Conard et al. 2002).

Fire influences terrestrial C stocks in boreal forests by releasing C to the atmosphere through combustion, determining age-class structure, and by altering nutrient stocks and availability as well as decomposition regimes (cf. Van Cleve et al. 1996; Bhatti et al. 2002; O'Neill et al. 2002). Plant production usually decreases initially following fire but increases over time with vegetation recovery. Plant biomass stored belowground in root structures also can be damaged during fire, contributing to decreases in autotrophic soil respiration (Weber 1990; Burke et al. 1997, Amiro et al. 2003). Fire activity influences decomposition by changing soil thermal and hydrologic environments and by altering nutrient availability through ash and runoff. However, microbial responses and potential changes in heterotrophic respiration in soils following fire remain unclear. Phenolic inhibition by charcoal (DeLuca et al. 2002), nutrient fertilization (Viereck et al.

1979; Dyrness and Norum 1983; Pietikainen and Fritze 1996; Van Cleve et al. 1996) and higher soil temperatures (O'Neill et al. 2002) may stimulate decomposition post-fire. However, the loss of microbial biomass (Ahlgren and Ahlgren 1965; Fritze et al. 1994), alteration of microbial communities (Bissett and Parkinson 1980), and/or removal of labile soil substrates may decrease decomposition rates. Hogg et al. (1992) found that ash additions to peat in the laboratory had mixed effects on $CO_2$ production but increased $CH_4$ production. In upland stands, Fritze et al. (1994) and Pietikainen and Fritze (1996) suggest that low soil moisture actually limits respiration rates in burned areas.

The accumulation of above- and belowground biomass also significantly affects fire behavior. Net primary production leads to the accumulation of fuels in aboveground biomass. Imbalances between NPP and organic matter losses over time lead to the accumulation of fuels on the forest floor. Fuel loading and continuity are important controls on the total area and severity of burning. Thus, greater NEP likely increases the susceptibility of terrestrial boreal ecosystems to burning. However, this is complicated by interactions with hydrology, substrate physiography, species composition, and soil thickness. For example, black spruce has a higher probability of burning than hardwood species because of greater flammability and ladder fuel structure (Hely et al. 2000). Yet, drainage-fire associations occur even within black spruce communities. Black spruce is associated with high burn frequency and larger fires in more well drained sites where feather mosses and lichens dominate the understories. Black spruce associations with *Sphagnum* may burn less frequently or less severely. In these ecosystems, poor drainage controls fire behavior and/or decomposition and lead to greater long-term NEP (Harden et al. 2000). However, drought conditions will increase fire activity in peatlands (Turetsky et al. 2004), and likely will lead to large losses of organic matter from poorly drained areas due to increase fire and decomposition.

## Peatland Distributions

The largest expanse of peatlands globally occurs in the boreal regions of North America and Eurasia ($350 \times 10^6$ ha; Gorham 1991; Botch et al. 1995; Rugo and Weiss 1996) because of glacial lake basins formed during Holocene deglaciation, low evaporation rates, and relatively low relief (Gignac and Vitt 1994; Scott 1995). Across continental, western Canada (Alberta, Saskatchewan, and Manitoba), peatlands cover about 365,000 km$^2$ and store 42 Pg of soil C (Vitt et al. 2000). In Russia's West Siberian Lowland (WSL), the largest peatland complex in the world, peatlands cover about 600,000 km$^2$ and store about 72 Pg C (Kremenetski et al. 2003; Smith et al. 2004). To our knowledge, there are no statewide estimates of peatland distributions or carbon stocks in Alaska. However, ombrotrophic bogs in Alaska's northern tundra and interior boreal forest generally have thicker organic soil layers and greater C storage than other ecosystems in Alaska (Ping et al. 1997).

Both climate and physiography control distributions of wetlands in boreal regions. In North America, the thermal seasonal aridity index (TSAI; total annual precipitation/mean growing season temperature) is an important climatic control on the southern limit of peatlands (Halsey et al. 1997). Physiographic controls at regional (i.e., glacial lake basins), landscape (i.e., bedrock characteristics), and local (i.e., soil texture) scales control hydrologic and vegetation processes in wetlands (Glaser 1992; Almquist-Jacobson 1995; Halsey et al. 1995; Halsey et al. 1997). For example, fine-grained mineral soil with low hydraulic conductivity promotes poor drainage and the development of bogs, whereas fens tend to develop on substrates with higher hydraulic conductivity and greater groundwater components. In Manitoba, Canada, bogs occur preferentially on acidic Precambrian rock, whereas fens (particularly calcareous-rich fens) tend to be found on calcareous bedrock (Halsey et al. 1997).

Although peat accumulation is influenced by the spatial patterning of hydrologic processes, dynamic feedbacks exist between peatland development, hydrology, and physiography (see also the description of permafrost aggradation, below). Peat initiation generally occurs with the stabilization of seasonal water levels, restriction of water flow through a landform, and leaching of salts from mineral layers, which favor the development of a moss ground layer (Zoltai and Vitt 1990; Kuhry et al. 1993). Mosses accumulate and retain nutrients and minerals in forms largely unavailable to vascular plants, while cation exchange by *Sphagnum* species contributes to local acidification (Clymo and Hayward 1982). The high thermal conductivity of wet and frozen peat in fall and winter months generally makes it a good conductor of heat. Dry peat has low thermal conductivity, allowing it to insulate underlying soil layers from warmer temperatures in summer months. As a result of dry *Sphagnum*'s low thermal conductivity, bogs have lower surface water temperatures than other surrounding organic and nonorganic soils (Vitt et al. 1994). Generally, vegetation in fens are less likely to influence the thermal dynamics of ground surfaces, as water circulation in fens increases heat flux. Sedge peat also tends to have a greater hydraulic conductivity than *Sphagnum* peat, which can raise water table levels. Bogs generally are characterized by a diplotelmic soil structure, where properties of the acrotelm (zone above the regional water table) control water table levels and supply to vegetation, particularly *Sphagnum* mosses (Ingram 1978). Removal of surface vegetation and soil layers with disturbance or land-use disrupts this feedback: increased oxidation of surface peat decreases pore size and reduces water storage capacity and saturated hydraulic conductivity (Price and Whitehead 2001). The complex feedbacks between climate, hydrology, and the autogenic nature of peatland development have been explored primarily through modeling (Clymo 1984; Hilbert et al. 2000; Belyea and Clymo 2001; Pastor et al. 2002). Collectively, these efforts show that nonlinear relationships between peatland hydrology, physiography, and ecosystem processes can create multiple stable states and that some systems could adjust quickly to changes in soil moisture.

Although peatlands cover only 3–5% of the earth's terrestrial surface, they may store up to 30% of the world's terrestrial soil C (Gorham 1991; Zoltai and Martikainen 1996; Moore et al. 1998). Currently, boreal peatlands are thought to function globally as a net sink for atmospheric $CO_2$, sequestering approximately 76 Tg ($10^{12}$g) C $yr^{-1}$ (Zoltai and Martikainen 1996) and as a net source of $CH_4$ (Gorham 1991, 1994; Wahlen 1993; Gorham 1995). Northern wetlands emit an estimated 65 Tg $CH_4$ $yr^{-1}$ (Walter et al. 2001), representing about 25% of $CH_4$ emissions from natural sources (Prather et al. 2001). However, recent work has shown that $CO_2$ exchange in peatlands can switch between sinks and sources between wet and dry years (Shurpali et al. 1995; Alm et al. 1999).

Generally, peat accumulates when C fixation through net primary production (NPP) exceeds losses from decomposition, leaching, and/or disturbance losses throughout the peat column. Soil C storage in peatlands generally is not controlled by fast NPP (Figure 12.2). For example, although NPP varied

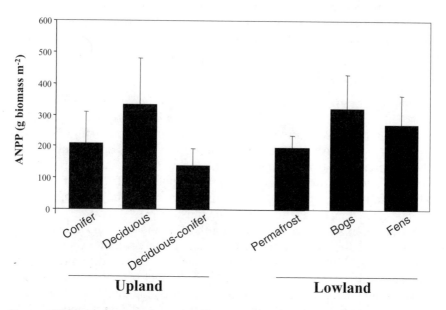

FIGURE 12.2. Aboveground net primary productivity (ANPP) of major types of upland stands and peatlands. Data are means ± one standard error. Data are from Reader and Stewart (1972), Grigal et al. (1985), Bartsch and Moore (1985), Oechel and Van Cleve (1986), Billings (1987), Bazilevich (1993), Hall (1997), Szumigalski and Bayley (1997), Thormann and Bayley (1997b), Linder (1998), Comeau and Kimmins (1999), Havas (1999), Gower et al. (2001), Reich et al. (2001), Rencz and Auclair (2001), Camill et al. (2001), Vitt et al. (2001), Camill (personal communication).

threefold among wetland and well drained sites in Manitoba, the turnover times for deep soil C was much faster in the upland stands (80–130 years) than in the wetland (>3000 years) (Trumbore and Harden 1997). Peatlands lose C to dissolved organic carbon (DOC) export and generally are considered important sources of DOC to aquatic ecosystems, though the influence of DOC export on net C balance in peatlands is not well quantified (but see Fraser et al. 2001). Water table position controls the area of peatland contributing to DOC export (Fraser et al. 2001). At the Experimental Lakes Area in Ontario, Canada, flooding of a catchment with peatlands increased pore water and pond DOC concentrations and led to greater amounts of hydrophilic neutral fractions in the DOC (Moore et al. 2003).

Although losses of C to DOC export might be greater from peatlands than from upland stands, depending on climate and catchment characteristics, losses of C to decomposition and fire likely are lower from peatlands than uplands over a variety of spatial and temporal scales. However, losses of C to disturbance may become more important to peatland NEP if water tables are reduced under future climate change. For example, fire frequency and severity may increase with lower water table positions and drier peat layers in the acrotelm. Turetsky et al. 2004 found a positive correlation between the area of peatland burned in western Canada and warmer, drier fire weather, suggesting that more peatlands will burn under the drier climate regimes predicted for some boreal regions.

## Discontinuous Permafrost

Permafrost is a ground-temperature phenomenon and occurs in earth materials with temperatures below 0 °C (cryotic) for 2 or more years (Thie 1974; Woo et al. 1992). Permafrost is both a product and determinant of climate, as it influences heat and water balance in surface ground and air layers. Permafrost creates a strong heat sink in the summer that reduces surface temperature and energy exchange with the atmosphere (cf. Eugster et al. 2000). Frozen soils can retain large amounts of precipitation due to impeded drainage, creating excess humidity (Gavrilova 1993).

Local factors such as snow cover (Smith 1975) and the distribution of heat by water bodies control the distribution of discontinuous permafrost (Halsey et al. 1995). Permafrost development is regulated by factors that influence surface energy balance and soil thermal properties, such as site physiography (slope, aspect, elevation). Soil organic matter thickness and moisture content also influence permafrost aggradation and stability (Swanson 1996). In summer months, thick soil organic matter insulates underlying permafrost from solar insolation, and in later months freezing and water saturation increase heat loss (Railton and Sparling 1973). Williams and Burn (1996) concluded that at least 11 cm of organic soil is required for the presence of discontinuous permafrost in the central Yukon Territory, Canada.

Feedbacks also occur between permafrost and the thermal properties of vegetation (i.e., shading of ground surfaces). Black spruce that colonize bogs and permafrost peatlands in North America intercept snowfall, leading to thinner snow pack depths that promote the cooling of peat (Zoltai 1995; Camill 1999; Beilman 2001). Also, by minimizing wind velocity near the soil surface, trees prevent high air-ground heat and humidity fluxes (Brown 1969). Establishment of *Sphagnum* cover in peatlands promotes the persistence of ice in soils (Brown 1969; Seppälä 1988; Kuhry 1998) and the high albedo of bryophytes and lichens generally can protect permafrost features from solar radiation in the summer (Railton and Sparling 1973).

Across the boreal forest, permafrost largely is discontinuous in its distribution, ranging from continuous coverage in the northern Continuous Permafrost Zone to localized permafrost landforms or frost mounds (cf. Beilman et al. 2001) in the Localized Permafrost Zone (Zoltai 1995). Localized permafrost features occur almost exclusively in peatlands. In Alaska, the driest soils lack permafrost and tend to occur on south-facing slopes, convex landforms, or upper slope positions (Swanson 1996; Harden et al. 2003). Dry stands without permafrost have thinner surface soils over gravelly material compared to stands with recent permafrost thaw (Swanson 1996). The coldest and wettest permafrost landscapes in Alaska occur on concave to planar features, lower slope positions, and north-facing slopes, whereas warmer permafrost environments occur on crests and shoulders, and east-, west-, or south-facing midslopes (Swanson 1996). Silt mixed with organic debris often is incorporated into permafrost layers. This inhibits water drainage in soils by decreasing permeability and supplying water through thaw of the active layer.

Localized permafrost peatlands near the southern limit of discontinuous permafrost accumulate C at rates similar to unfrozen peatlands, whereas permafrost features further north accumulate C more slowly than adjacent unfrozen peatlands (Robinson et al. 2003). Permafrost processes leading to colder and wetter soil conditions likely inhibit the NPP of some vascular plants while promoting bryophyte NPP (Skre and Oechel 1981). Permafrost layers may prevent dissolved organic compounds from penetrating into deeper soil layers where they could be mineralized. Thus, permafrost may reduce the export of DOC and nitrogen to streams during lateral flow (MacLean et al. 1999).

Recent warming across boreal regions has led to substantial warming and thawing of discontinuous permafrost features (Vitt et al. 1994; Osterkamp and Romanovsky 1999; Osterkamp et al. 2000; Vitt et al. 2000; Jorgenson et al. 2001). Changes in surface energy balance following fire also melts near-surface permafrost and increases active layer depths for several decades (Viereck 1983; Zoltai 1993; Robinson and Moore 1999; O'Neill et al. 2002). The influence of permafrost collapse on soil drainage and C balance depends on site physiography, permafrost conditions, and underlying substrates (Hinzman et al. in press). Soil drainage might improve following

the thaw of permafrost in black spruce stands underlain with gravelly outwash. Drier soil conditions following thaw will increase organic matter
decomposition (O'Neill et al. 2002) and susceptibility to fire. However,
permafrost degradation in peatlands or ecosystems underlain by ice-rich
permafrost leads to decreased soil drainage (Vitt et al. 1994, 2000; Hinzman
et al. in press). For example, permafrost thaw in peatlands creates wet
depressional fens called internal lawns (resulting from the thaw of localized
permafrost features) or collapse scars (resulting from thaw within peat
plateaus or palsas). These thaw features initially are colonized by semiaquatic *Sphagnum* mosses and *Carex* species (Zoltai 1993; Camill 1999;
Beilman 2001) and accumulate C faster than adjacent permafrost peatlands
(Robinson and Moore 2000; Turetsky et al. 2000; Camill et al. 2001). Until
peat accumulates well above the water table, saturated conditions may protect the C stored in internal lawns and collapse scars from burning.

## Dominant Vegetation

Generally, low precipitation and temperatures in boreal regions can limit NPP
(Baldocchi et al. 2000). Species composition, however, also is an important
control on the productivity and nutrient cycling of boreal ecosystems. Early
successional, deciduous species have greater nutrient requirements than coniferous species because of greater productivity and lower nutrient use efficiencies (cf. Bridgham et al. 1998; Harden et al. 2000). Early successional species
also tend to have more labile litter that rapidly decays (cf. Pastor et al. 1999;
Figure 12.3). Pastor et al. (1999) used spatial modeling to show that a combination of vascular plant population traits (seed dispersal) and litter decomposition rates can lead to the spatial patterning of soil nitrogen availability.
Species composition also controls the rate at which soluble compounds leach
from litter and soils in boreal regions (Neff and Hooper 2002). For example,
solute flushing is related to bryophyte desiccation tolerance, as carbohydrate
pools that accumulate during dehydration can be leached during rewetting
(Proctor 1982; Carleton and Read 1991; Wilson and Coxson 1999).

Bryophytes and lichens are important components of the boreal forest
understory, and NPP of these species can be 500% greater than NPP of black
spruce foliage (Van Cleve et al. 1983). Generally, bryophytes appear to have
large and persistent spore banks in forest and wetland soils (Jonsson 1993;
Sundberg and Rydin 2000) that may explain their widespread occurrence
across boreal regions (Gajewski et al. 2001; Vitt et al. 2001). Moss and lichen
species maintain low temperatures and higher moisture contents in soils
(Railton and Sparling 1973; Yu et al. 2002). The genus *Sphagnum* (peat
mosses) is an important component of aboveground biomass in areas of
poor drainage (Turetsky 2003) and increases NEP through its slow decomposition (Verhoeven and Toth 1995; Aerts et al. 1999). *Sphagnum* species
also wick water upwards through external capillary action and store large
volumes of water in dead, hyaline cell structures.

Complex interactions occur between vegetation composition and perturbations such as herbivory and fire. For example, Rupp et al. (2002) showed that the frequency and size of large fires increases positively with the cover of black spruce forest. Generally, conifer stands generate more flammable and continuous fuels than deciduous stands (Hely et al. 2000). Thus, stands are less susceptible to burning during deciduous revegetation until conifer species recolonize (cf. Viereck 1983). Wet or moist moss and other ground layer species may decrease combustion efficiency but become more efficient fuel during periods of dessication.

Herbivory influences competitive interactions between plant species and influences nutrient cycling through changes in vegetation structure, litter chemistry, and the proportion of nutrients returned as feces. Feeding patterns and trampling/disturbance caused by herbivores can lead to spatial patterning of vegetation and nutrient availability in northern ecosystems (McInnes et al. 1992; Bridgham et al. 1998; Olofsson and Oksanen 2002). Deciduous vegetation generally is preferred by herbivores such as moose (Pastor et al. 1988), and herbivory can control litter and soil quality by favoring particular species. For example, snowshoe hare browsing influences the spatial distribution of regenerating white spruce in the boreal forest (Dale and Zbigniewicz 1997). Moose browsing decreased total litter production, reduced sapling growth of deciduous species, and increased herb litter production in the Isle Royale National Park, Minnesota (McInnes et al. 1992). Browsing also can reduce N mineralization, with additional feedbacks to vegetation and soil quality (Pastor et al. 1993). Plants often respond to herbivory by increasing allocation to structural and/or chemical defenses (cf. Shaver and Aber 1996).

Generally, herbivory of bryophyte and lichen species tends to be low, likely because of low nutrient content and the production of defensive chemicals (Clymo and Hayward 1982; Davidson et al. 1990). However, certain species of insects graze on bryophytes. Caribou use lichens as a winter food source, preferentially using treed peatlands in Canada either for food sources and/or to avoid wolf predation (Bradshaw et al. 1995). The influence of caribou herbivory on litter quality in boreal peatlands has not been documented to our knowledge. However, lichens decompose preferentially in peatlands, and thus the removal of lichen biomass by caribou likely does not influence long-term C storage in peatland soils.

Outbreaks of insects such as the spruce budworm (*Choristoneura fumiferana*) and forest tent caterpillar (*Malacosoma disstria*) are important to the structure of boreal forest stands. For example, outbreaks of spruce budworm occur every 30–35 years and can be synchronous across large areas of the eastern United States and Canada (Weber and Schweingruber 1995; Williams and Liebhold 1995, 2000; Lussier et al. 2002). Oscillations of forest tent caterpillar populations occur every 10–12 years under natural conditions but tend to be asynchronous in boreal landscapes.

Insect outbreaks lead to the defoliation, reduced growth, and increased mortality of both deciduous and conifer species. Under extreme conditions, tree mortality will increase coarse woody debris and favor the growth of shade-intolerant species. Trees may also allocate more resources to the production of defensive compounds (Schultz and Baldwin 1982), potentially decreasing allocation to reproduction. During outbreaks, insects stimulate the conversion of foliar N into green leaf fall and throughfall, as well as insect feces and biomass. Some studies have found increased leaching export of nitrogen following insect outbreaks, with potential consequences for NPP in N-limited systems. However, Lovett et al. (2002) conclude that most of the nitrogen release caused by insect defoliation in temperature forests is redistributed within the ecosystem. Increasing areas of insect disturbance over the past several decades in eastern Canada has been linked to increased C emissions from upland ecosystems (Kurz and Apps 1999), though the influence of outbreaks on the ecosystem processes contributing to NEP is not well understood. Insect-caused mortality of trees also can increase vulnerability of secondary infections and the risk of burning (cf. McCullough et al. 1998). Conversely, Cappuccino et al. (1998) showed that lower rates of budworm-induced mortality occurred in boreal forest with frequent fire activity.

## Does Pattern Affect Process? Landscape Composition versus Configuration

### Composition

Landscape composition (the amount of different habitat or ecosystem types) is important to soil C storage in the boreal forest. The spatial patterning of soil C stocks across boreal landscapes is controlled largely by interactions between soil drainage and decomposition (Figures 12.3 and 12.4). To demonstrate this, we calculated potential C storage using mean aboveground net primary productivity (ANPP) and decomposition rates reported in the literature for deciduous stands, conifer stands, bogs, and fens (Table 12.1). This approach illustrates how correlations between ANPP and decomposition across types of boreal ecosystems can influence C storage. Generally, estimates of $k$ across study sites tend to be more variable than ANPP, with large coefficients of variance both within and between stand types (Table 12.1). Although ANPP is greater in deciduous than in conifer stands (Figure 12.2), potential C storage in deciduous stands is low due to rapid decomposition rates (Figure 12.3; Table 12.1). Differences between conifer and deciduous stands would be accentuated if litterfall rates were used to estimate potential C storage (see notes for Table 12.1).

Estimates of ANPP in boreal peatlands are comparable to those reported for upland sites (Figure 12.2; Table 12.1). Initial losses of C from litter decomposition in peatlands also are relatively high (Figure 12.3; Table 12.1). Using decay values obtained from litterbag measurements, potential C storage is not greater in peatlands than in upland stands (Table 12.1). However,

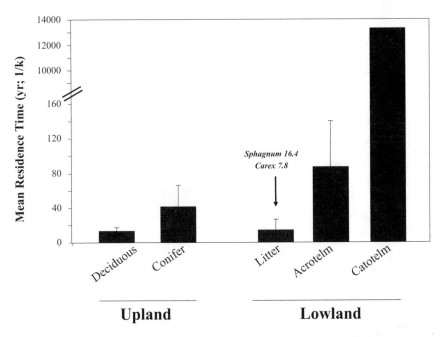

FIGURE 12.3. Mean residence time (yr; $1/k$) in upland and peatland boreal systems. Data are means ± one standard error. Litter decomposition in peatlands may be similar to or even faster than decomposition rates in some upland conifer stands. However, accumulation of peat in the acrotelm (obtained with [210]Pb-dating) or in the entire peat column (obtained with [14]C dating) suggests that decomposition in buried organic matter is much slower in peatlands than in upland stands. Data from Fox and Van Cleve (1983), Rochefort et al. (1990), Gorham (1991), Harden et al. (1997), Szumigalski and Bayley (1996), Thormann and Bayley (1997a), Clymo et al. (1998), Harmon et al. (2000), Yatskov (2000), Yu et al. (2001), Storaunet and Rolstad (2002), Bond-Lamberty et al. (2002).

the higher water tables in poorly drained systems lead to slower decomposition in deeper soil layers (cf. Trumbore and Harden 1997). As soil organic matter is transferred into deeper acrotelm and catotelm layers, turnover rates decline and lead to greater accumulation of C (Table 12.1).

Our calculations of potential C storage are meant to demonstrate how correlations between NPP and decomposition can influence C accumulation, and they ignore the important influences of belowground productivity or losses to leaching or disturbance. Rapid turnover rates in well drained deciduous stands (Figure 12.3) lead to little accumulation of ground fuels (Table 12.1), with consequences for fire behavior and combustion losses. Severe fires may reduce decomposition losses by removing organic matter substrates (Harden et al. 2000). Post-fire, however, decomposition rates will drive ecosystem C accumulation, and NPP will become a stronger determinant of C accumulation later in stand development (Zhuang et al. 2003).

TABLE 12.1. Calculations of potential C storage at steady-state in conifer stands, deciduous stands, bogs, and fens using data from Figures 12.2 and 12.3[a].

| | ANPP[b] $(\text{g m}^{-2}\text{ yr}^{-1})$ | $k$ $(\text{yr}^{-1})$ | $C^b$ $(\text{g m}^{-2})$ |
|---|---|---|---|
| Upland | | | |
| Conifer | 208.98 (0.48) | 0.02 (1.66) | 8634.38 (0.77) |
| Deciduous | 331.00 (0.45) | 0.08 (2.84) | 4316.24 (0.57) |
| Peatland | | | |
| Bog | 259.10 (0.39) | 0.08 (1.74)[c] | 3409.49 (0.69)[c] |
| | | 0.01 (1.83)[d] | 17,841.79 (0.67)[d] |
| | | 0.0001 (1.17)[e] | 2,139,152 (0.94)[e] |
| Fen | 270.42 (0.35) | 0.07 (0.97)[c] | 3936.79 (1.09)[c] |
| | | 0.01 (—)[d] | 43,616.19 (0.35)[d] |
| | | 0.0001 (0.92)[e] | 4,657,158.00 (1.14)[e] |

[a] Potential C storage is calculated as $C = l(ANPP)/k$, where $k$ is the fractional turnover rate of soil carbon and $l$ is the fraction of aboveground net primary productivity (ANPP) shed as litter. While average litterfall (as measured by leaf, moss, and twig biomass) was equivalent to 42% of ANPP in deciduous stands and 89% of ANPP in conifer stands (Mack unpublished data; Gower et al. 2001), woody debris is an important component of long-term soil C accumulation (Manies et al. 2005). Therefore, here we assume that 100% of ANPP eventually is incorporated into litter (see also Yu et al. 2001 for peatlands). These estimates of potential C storage ignore belowground productivity and losses of C to leaching or disturbance. Coefficients of variance are reported in parentheses and were compounded using Gaussian distributions. Error terms are dominated by differences between various study sites and likely reflect methodological differences as well as spatial heterogeneity among the ecosystem processes contributing to C storage.
[b] Assumes that 100% of ANPP is shed as litterfall.
[c] $k$ derived from litter decomposition.
[d] $k$ from $^{210}\text{Pb}$ curve fits through acrotelm peat.
[e] $k$ from $^{14}\text{C}$ curve fits through acrotelm and catotelm peat.

Long-term modeling suggests that wetlands lose less organic matter to burning than upland boreal stands (Harden et al. 2000), either because of decreased fire frequency and/or severity (Figure 12.4). Recent work, however, has shown that moderately to poorly drained areas of Alaska and Canada burn as frequently or more frequently than well drained ecosystems (Harden et al. 2003; Turetsky et al. unpublished manuscript). Peatlands across the boreal forest vary widely in moisture availability, which is important to the efficiency of fuel combustion during fire events. For example, whereas peat plateaus and other peatlands underlain by permafrost have low rates of C storage because of relatively high fire frequencies (Zoltai 1993; Robinson and Moore 2000), mires in western Siberia were found to have burned only 2–3 times over the past 8000 years, with no evidence of decreasing soil C with fire activity (Turunen et al. 2001).

Carbon storage at landscape or regional scales in the boreal forest region is influenced more by the spatial heterogeneity of fire and decomposition processes rather than spatial heterogeneity in NPP (Table 12.1; Frolking et al. 1998; Harden et al. 2000; Vitt et al. 2001). However, there still is considerable

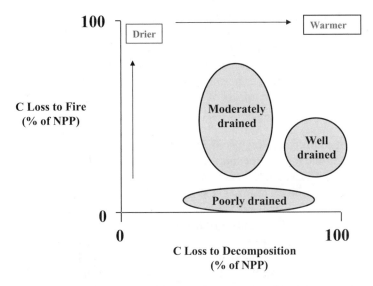

Figure 12.4. Conceptual framework for evaluating C losses to the atmosphere from boreal ecosystems of varied soil drainage. Losses of C to decomposition generally increase positively with soil drainage. Losses of C to fire also generally will increase with soil drainage due to more efficient combustion but also will depend on fuel loading. Due to large accumulations of above- and belowground biomass, moderately drained ecosystems may be most susceptible to C loss to fire. Across soil drainages, losses of C to fire will increase with drought, and losses of C to decomposition. Figure adapted from figure originally published in Harden et al. (2003).

variation in NPP across stand types in boreal systems (Figure 12.2). Stand age (time since last stand-replacing fire) is more important in driving aboveground productivity in Siberian Scots pine forest than forest characteristics such as vegetation structure, surface fire damage, age-related reduction of NPP, and interannual variability (Wirth et al. 2002). Thus, it may be possible to scale NPP within a similar stand-type with information on fire history alone. Across stands, however, factors associated with fire (severity, propagule availability) determine post-fire stand composition, stand structure, and successional trajectories in boreal systems and thus also will be important for scaling NPP.

## Configuration

The influence of landscape configuration (the spatial arrangement of different ecosystem or habitat types) on ecosystem processes in the boreal forest has not been well studied. The configuration of upland and lowland stands with varying soil and vegetation characteristics likely will influence biological processes such as insect outbreaks and herbivory. Insect outbreaks are one of the most common disturbances across the boreal forest of eastern

Canada and may stimulate C emissions from upland ecosystems (Kurz and Apps 1999). Outbreaks may become more common under increasing land-use and changing climatic regimes, and modeling the spatial patterning of vegetation, carbon, or nutrients in response to outbreaks will require modeling efforts with interaction among spatial patches to simulate dispersal. Similarly, modeling the impacts of herbivory requires the ability to transfer biomass and nutrients among patches with animal movement and habitat selection.

Landscape configuration also is important to the movement of water and sediments on the landscape and thus may be important in predicting C storage in some ecosystems. For example, in areas with strong topographic relief such as interior Alaska, landscape configuration may be important for water runoff into lowland ecosystems. Smaller and isolated wetland catchments have greater ratios of surrounding upland slope compared to larger wetland catchments. Increasing slope may increase the rate of water movement. Increased water flow may alter site-level hydrology but also could accelerate transport of nutrients, spores, DOC, and possibly microbial biomass into the catchment. Landscape configuration around geological sources of Aeolian deposition also may be important to ecosystem functions such as NEP, as stands closer to and downwind from floodplains will receive more Aeolian deposition than sites farther away or upwind. Silt layers impede soil drainage and consequently will influence vegetation distribution, decomposition rates, and fire combustion.

Landscape configuration likely will influence fire behavior because of relationships between drainage, fuel, and climate discussed throughout this chapter. Vegetation and soils influence fire through fuel loading and soil climates, climate influences both water table and fuel moisture, and fires influence terrestrial organic matter through combustion and nutrient redistribution. The configuration of boreal wetlands, which is dictated largely by topography, geology, and hydrology, likely is important to fire spread and severity. For example, combustion of fuels may be less severe if burning is initiated in wetter areas or if the fire front spreads quickly to an adjacent peatland margin. Smaller patches of isolated lowlands may burn more severely and/or frequently than landscapes with larger or more continuous wetland. Landscapes with greater amounts and continuity of well drained fuels will increase burn size and spread. Thus, determining how fires ignite and spread through boreal landscapes may require some interactive modeling, with feedbacks between landscape patches to determine the rate and severity of fire spread.

## Conclusions

Carbon storage in terrestrial ecosystems is influenced by a number of processes including net primary productivity, decomposition, leaching export, and disturbance losses. Each of these ecosystem processes can best be understood at a particular spatial and temporal scale. For example, Randerson et al. (2002) conclude that fire impacts on NEP or total carbon storage are best studied over relatively large spatial and long temporal scales relative to fire return intervals. The importance of individual ecosystem processes in contributing to NEP may vary across time-space scales. Decomposition is dependent on relatively short-term variations in weather, while fire combustion also is driven by annual and decadal cycles in fire behavior. Thus, climatic shifts will influence NEP not only by altering decomposition on short timescales but also by influencing hydrology and fire behavior over longer time frames.

At landscape and regional scales in the boreal forest region, spatial patterning in decomposition and fire losses with drainage control long-term soil C storage. Generally, soil drainage is influenced by water-holding capacity, hydraulic conductivity, and position of the seasonal water table. Thus, soil drainage classes condense a large amount of heterogeneity stemming from hydrologic controls on species composition, wetland cover, and permafrost distribution and may be useful for scaling ecosystem processes at large scales. Our hypothesis is that soil drainage classification will be a useful framework for scaling NEP across the boreal forest (Harden et al. 2003). However, detailed (i.e., the use of 5–7 drainage classes) or large-scale soil drainage classification requires rigorous testing at the plot and landscape scales. A predictive framework should test whether variability between drainage classes outweighs variability within each class. For example, as controlling factors on site level hydrology, the cover of wetlands, permafrost, and vegetation can contribute to within-drainage class variability. For soil drainage to be an effective scaling tool in ecosystem ecology/biogeochemistry, variability in ecosystem function caused by these various landscape factors must be explained by differences in drainage class terms.

Finally, we note that whereas large areas of Siberia and Alaska remain pristine, other boreal regions have experienced or currently are experiencing increased land-use and development. Land uses such as linear disturbances (roads, cutlines, seismic lines, pipelines), forestry, agriculture, and resource extraction increasingly are influencing hydrologic and vegetation patterns in northern ecosystems. For example, clearcutting in the boreal forest has been found to produce more landscape heterogeneity in vegetation and edaphic conditions than burning (Schroeder and Perera 2002). The influence of disturbance on peatland C storage has not received adequate attention, but regional C budgets suggest that peat harvesting, mining, and reservoir creation in the peatlands of western Canada lead to losses of 280 Gg ($10^9$g) C yr$^{-1}$, equivalent to about 4% of regional peatland C stocks (Turetsky et al. 2002). However, due to their impacts on vegetation, soils, and hydrology, increasing land-use and development in boreal regions adds complexity to scaling regional patterns of carbon sinks/sources at northern latitudes. Forest fragmentation, for example, decouples caterpillar growth from their classic growth regulators (parasitoids and viruses), leading to longer pest outbreaks

(Roland and Taylor 1997; Rothman and Roland 1998; Roland et al. 1998). Given the large store of terrestrial C at northern latitudes and the strong control of regional and local physiography on terrestrial C in northern regions, landscape approaches to scaling C storage will be valuable across varying timescales.

*Acknowledgments.* We thank Phil Camill for access to data. Jon Wynn, Jon Carrasco, John Pastor, and an anonymous reviewer provided helpful comments on this chapter. We also would like to thank John Pastor for ideas and suggestions regarding the calculation of potential C storage.

## References

Aerts, R., Verhoeven, J.T.A., and Whigham, D. 1999. Plant-mediated controls on nutrient cycling in temperate fens and bogs. Ecology 80: 2170–2181.

Ahlgren, I.F., and Ahlgren, C.E. 1965. Effects of prescribed burning on soil microorganisms in a Minnesota jack pine forest. Ecology 46: 304–310.

Alm, J., Schulman, L., Silvola, J., Walden, J., Nykänen, H., and Martikainen, P.J. 1999. Carbon balance of a boreal bog during a year with an exceptionally dry summer. Ecology 80: 161–174.

Almquist-Jacobson, H. Foster, D.R. 1995. Toward an integrated model for raised bog development: theory and field evidence. Ecology 76: 2503–2516.

Amiro, B.D., MacPherson, J.I., Desjardins, R.L., Chen, J.M., and Liu, J. 2003 Post-fire carbon dioxide fluxes in the western Canadian boreal forest: evidence from towers, aircraft and remote sensing. Agric. Forest Meteorol. 115: 91–107.

Apps, M.J., Kurz, W.A., and Price, D.T. 1993. Estimating carbon budgets of Canadian forest ecosystems using a national scale model. In Proceedings of the workshop on carbon cycling in boreal forests and subarctic ecosystems eds. T. Vinson and T. Kolchugina, pp. 241–250. Washington, DC: US EPA, Office of Research and Development.

Baldocchi, D., Kelliher, F.M., Black, T.A., and Jarvis, P. 2000. Climate and vegetation controls on boreal zone energy exchange. Global Change Biol. 6: 69–83.

Bartsch, I., and Moore, T.R. 1985. A preliminary investigation of primary production and decomposition in four peatlands near Schefferville, Quebec. Can. J. Botany 63: 1241–1248.

Bazilevich, N.I. 1993. Biological productivity of ecosystems of Northern Eurasia. Moscow: Nauka Publishers, 293 p.

Beilman, D.W. 2001. Plant community and diversity change due to localized permafrost dynamics in bogs of western Canada. Can. J. Botany 79: 983–993.

Beilman, D.W., Vitt, D.H., and Halsey, L.A. 2001. Localized permafrost peatlands in western Canada: definitions, distributions and degradation. Arctic Antarctic Alpine Res. 33: 70–77.

Belyea, L.R., and Clymo, R.S. 2001. Feedback control of the rate of peat formation. Proc. R. Soc. London Biol. Sci. 268: 1315–1321.

Bhatti, J.S., Apps, M.J., and Jiang, H. 2002. Influence of nutrients, disturbances, and site conditions on carbon stocks along a boreal forest transect in central Canada. Plant Soil 242: 1–14.

Billings, W.D. 1987. Carbon balance of Alaskan tundra and taiga ecosystems: past, present, and future. Quaternary Sci. Rev. 6: 165–177.

Bissett, J., and Parkinson, D. 1980. Long-term effects of fire on the composition and activity of the soil microflora of a subalpine, coniferous forest. Can. J. Botany 58: 1704–1721.

Bond-Lamberty, B.P., Wang, C., and Gower, S.T. 2002 Annual carbon flux from woody debris for a boreal black spruce fire chronosequence. J. Geophys. Res. No. D23, 8220, doi: 10.1029/2001JD000839.

Botch, M.S., Kobak, K.I., Vinson, T.S., and Kolchugina, T.P. 1995. Carbon pools and accumulation in peatlands of the former Soviet Union. Global Biogeochem. Cycles 9: 37–46.

Bradshaw, C.J.A., Hebert, D.M., Rippin, A.B., and Boutin, S. 1995. Winter peatland habitat selection by woodland caribou in northeastern Alberta. Can. J. Zool. 73: 1567–1574.

Bridgham, S.D., Updegraff, K., and Pastor, J. 1998. Carbon, nitrogen, and phosphorus mineralization in northern wetlands. Ecology 79: 1545–1561.

Brown, J. 1969. Soil properties developed on the complex tundra relief of northern Alaska. Biuletyn Peryglacjalny 18: 153–167.

Burke, R.A., Zepp, R.G., Tarr, M.A., Miller, W.L., and Stocks, B.J. 1997. Effect of fire on soil-atmosphere exchange of methane and carbon dioxide in Canadian boreal forest sites. J. Geophys. Res. 102: 29289–29300.

Camill, P. 1999. Patterns of boreal permafrost peatland vegetation across environmental gradients sensitive to climate warming. Can. J. Botany 77: 721–733.

Camill, P., and Clark, J.S. 1998. Climate change disequilibrium of boreal permafrost peatlands caused by local processes. Am. Naturalist 151: 207–222.

Camill, P., Lynch, J.A., Clark, J.S., Adam, J.B., and Jordan, B. 2001. Changes in biomass, aboveground net primary production, and peat accumulation following permafrost thaw in the boreal peatlands of Manitoba, Canada. Ecosystems 4: 461–478.

Cappuccino, N., Lavertu, D., Bergeron, Y., and Régnicre. J. 1998. Spruce budworm impact, abundance and parasitism rate in a patchy landscape. Oecologia 114: 236–242.

Carleton, T.J., and Read, D.J. 1991. Ectomycorrhizas and nutrient transfer in conifer feather moss ecosystems. Can. J. Botany 69: 778–785.

Clymo, R. 1984. The limits to peat bog growth. Philos. Trans. R. Soc. London Ser. B 303: 605–654.

Clymo, R.S., and Hayward, P.M. 1982. The ecology of *Sphagnum*. In Ecology of bryophytes, ed. A.J.E. Smith, pp. 229–289. London: Chapman and Hall.

Clymo, R.S., Turunen, J., and Tolonen, K. 1998. Carbon accumulation in peatland. Oikos 81: 368–388.

Comeau, P.G., and Kimmins, J.P. 1999. NPP Boreal Forest: Canal Flats, Canada, 1984. Data set. Available at http://www.daac.ornl.gov.

Conard, S.G., Sukhinin, A.I., Stocks, B.J., Cahoon, D.R., Davidenko, E.P., and Ivanova, G.A. 2002. Determining effects of area burned and fire severity on carbon cycling and emissions in Siberia. Climatic Change 55: 197–211.

Dale, M.R.T., and Zbigniewicz, M.W. 1997. Spatial pattern in boreal shrub communities: effects of a peak in herbivore densities. Can. J. Botany 75: 1342–1348.

Davidson, A.J., Harborne, J.B., and Longton, R.E. 1990. The acceptability of mosses as food for generalist herbivores, slugs in the Arionidea. Botanical J. Linnean Soc. 104: 99–113.

Dixon, R.K., Brown, S., Houghton, R.A., Solomon, A.M., Trexler, M.C., and Wisniewski, J. 1994. Carbon pools and flux of global forest ecosystems. Science 263: 185–190.

Dyrness, C.T., and Norum, R.A. 1983. The effects of experimental fires on black spruce forest floors in interior Alaska. Can. J. Forest Res. 12: 879–893.

Eugster, W., Rouse, W.R., Pielke, R.A., McFadden, J.P., Baldocchi, D.D., Kittel, T.G.F., Chapin, F.S., Liston, G.E., Vidale, P.L., Vaganov, E., and Chambers, S. 2000. Land-atmosphere energy exchange in Arctic tundra and boreal forest: available data and feedbacks to climate. Global Change Biol. 6: 84–115.

Fox, J.F., and Van Cleve, K. 1983. Relationships between cellulose decomposition, Jenny's k, forest-floor nitrogen, and soil temperature in Alaskan taiga forest. Can. J. Forest Res. 13: 789–794.

Fraser, C.J.D., Roulet, N.T., and Moore, T.R. 2001. Hydrology and dissolved organic carbon biogeochemistry in an ombrotrophic bog. Hydrol. Processes 15: 3151–3166.

Fritze, H., Smolander, A., Levula, T., Kitunen, V., and Malkonen, E. 1994. Wood-ash fertilization and fire treatments in a Scots pine forest stand: effects on the organic layer, microbial biomass, and microbial activity. Biol. Fertil. Soils 17: 57–63.

Frolking, S., Bubier, J.L., Moore, T.R., Ball, T., Bellisario, L.M., Bhardwaj, A., Carroll, P., Crill, P.M., Lafleur, P.M., McCaughey, J.H., et al. 1998. The relationship between ecosystem productivity and photosynthetically active radiation for northern peatlands. Global Biogeochem. Cycles 12: 115–126.

Gajewski, K., Viau, A., Sawada, M., Atkinson, D., and Wilson, S. 2001. Sphagnum peatland distribution in North America and Eurasia during the past 21,000 years. Global Biogeochem. Cycles 15: 297–310.

Gavrilova, M. 1993. Climate and permafrost. Permafrost Periglacial Processes 4: 99–111.

Gignac, L.D., and Vitt, D.H. 1994. Responses of northern peatlands to climate change: effects on bryophytes. J. Hattori Botanical Lab. 75: 119–132.

Glaser, P.H. 1992. Peat landforms. In The patterned peatlands of Minnesota, eds. H.E. Wright, B.A. Coffin, and N.E. Aaseng, pp. 3–14. Minneapolis, MN: The University of Minnesota Press.

Goodale, C.L., Apps, M.J., Birdsey, R.A., Field, C.B., Heath, L.S., Houghton, R.A., Jenkins, J.C., Kohlmaier, G.H., Kurz, W., Liu, S.R., et al. 2002. Forest carbon sinks in the Northern Hemisphere. Ecol. Applications 12: 891–899.

Gorham, E. 1991. Northern peatlands: role in the carbon cycle and probable responses to climatic warming. Ecol. Applications 1: 182–195 .

Gorham, E. 1994. The Future of research in Canadian peatlands: a brief survey with particular reference to global change. Wetlands 14: 206–215.

Gorham, E. 1995. The biogeochemistry of northern peatlands and its possible responses to global warming. In Biotic feedbacks in the global climatic system—will the warming feed the warming? eds. G.M. Woodwell and F.T. Mackenzie, pp. 169–187. New York: Oxford University Press.

Gower, S.T., Krankina, O., Olson, R.J., Apps, M., Linder, S., and Wang, C. 2001. Net primary production and carbon allocation patterns of boreal forest ecosystems. Ecol. Applications 11: 1395–1411.

Grigal, D.F., Buttleman, C.G., and Kernick, L.K. 1985. Biomass and productivity of the woody strata of forested bogs in northern Minnesota. Can. J. Botany 63: 2416–2424.

Hall, F.G. 1997. NPP Boreal Forest: Superior National Forest, U.S.A., 1983–1984. Data set. Available at http://www.daac.ornl.gov.

Halsey, L.A., Vitt, D.H., and Zoltai, S.C. 1995. Disequilibrium response of permafrost in boreal continental western Canada to climate change. Climatic Change 30: 57–73.

Halsey, L., Vitt, D., and Zoltai, S.C. 1997. Climatic and physiographic controls on wetland type and distribution in Manitoba, Canada. Wetlands 17: 243–262.

Harden, J.W., Meier, R., Darnel, C., Swanson, D.K., and McGuire, A.D. 2003. Soil drainage and its potential for influencing wildfire in Alaska. In Studies in Alaska by the U.S. Geological Survey, J. Galloway, ed. U.S. Geological Survey Professional Paper 1678.

Harden, J.W., Trumbore, S.E., Stocks, B.J., Hirsch, A., Gower, S.T., O'Neill, K.P., and Kasischke, E.S. 2000. The role of fire in the boreal carbon budget. Global Change Biol. 6: 174–184.

Harden, J.W., O'Neill, K.P., Trumbore, S.E., Veldhuis, H., and Stocks, B.J. 1997. Moss and soil contributions to the annual net carbon flux of a maturing boreal forest. J. Geophys. Res. 102: 28805–28816.

Harden, J.W., Sunquist, E.T., Stallard, R.F., and Mark, R.K. 1992. Dynamics of soiil carbon during deglaciation of the Laurentide Ice Sheet. Science 258: 1921–1924.

Harmon, M.E., Krankina, O.N., and Sexton, J. 2000. Decomposition vectors: a new approach to estimating woody detritus decomposition dynamics. Can. J. Forest Res. 30: 76–84.

Havas, P. 1999. NPP Boreal Forest: Kuusamo, Finland, 1967–1971. Data set. Available at http://www.daac.ornl.gov.

Hely, C., Bergeron, Y., and Flannigan, M.D. 2000. Coarse woody debris in the southeastern Canadian boreal forest: composition and load variations in relation to stand replacement. Can. J. Forest Res. 30: 674–687.

Hilbert, D., Roulet, N., and Moore, T.M. 2000. Modelling and analysis of peatlands as dynamical systems. J. Ecol. 88: 230–242.

Hinzman, L.D., Bettez, N.D., Bolton, W.R., Chapin, F.S., Dyurgerov, M.B., Fastie, C.L., Griffith, B., Hollister, R.D., Hope, A., Huntington, H.P., et al. In press. Evidence and implications of recent climate change in terrestrial regions of the arctic. Climatic Change (in press).

Hogg, E.H., Lieffers, V.J., and Wein, R.W. 1992. Potential carbon losses from peat profiles: effects of temperature, drought cycles, and fire. Ecol. Applications 2: 298–306.

Ingram, H.A.P. 1978. Size and shape in raised mire ecosystems: a geophysical model. Nature 297: 300–302.

Jonsson, B.G. 1993. The bryophyte diaspore bank and its role after small-scale disturbance in a boreal forest. J. Veg. Sci. 4: 819–826.

Jorgenson, M.T., Racine, C.H., Walter, J.C., and Osterkamp, T.E. 2001. Permafrost degradation and ecological changes associated with a warming climate in central Alaska. Climate Change 48: 551–579.

Kasischke, E., and Stocks, B.J. 2000. Introduction. In Fire, climate change, and carbon cycling in the boreal forest, eds. E. Kasischke and B.J. Stocks, pp. 1–6. Ecological Studies 138. New York: Springer-Verlag.

Kasischke, E.S., Christensen, N.L., and Stocks, B.J. 1995. Fire, global warming, and the carbon balance of boreal forests. Ecol. Applications 5: 437–451.

Kremenetski, K.V., Velichko, A.A., Borisova, O.K., MacDonald, G.M., Smith, L.C., Frey, K.E., and Orlova, L.A. 2003. Peatlands of the Western Siberian lowlands: current knowledge on zonation, carbon content, and Late Quaternary history. Quaternary Sci. Rev. 22: 703–723.

Kuhry, P. 1998. Late Holocene permafrost dynamics in two subarctic peatlands of the Hudson Bay Lowlands (Manitoba, Canada). Eurasian Soil Sci. 31: 529–534.

Kuhry, P., Nicholson, B.J., Gignac, L.D., Vitt, D.H., and Bayley, S.E. 1993. Development of Sphagnum-dominated peatlands in boreal continental Canada. Can. J. Botany 71: 10–22.

Kurz, W., and Apps, M.J. 1999. A 70-year retrospective analysis of carbon fluxes in the Canadian forest sector. Ecol. Applications 9: 526–547.

Linder, S. 1998. NPP Boreal Forest: Flakaliden, Sweden, 1986–1996. Data set. Available at http://www.daac.ornl.gov.

Lovett, G.M., Christenson, L.M., Groffman, P.M., Jones, C.G., Hart, J.E., and Mitchell, M.J. 2002. Insect defoliation and nitrogen cycling in forests. BioScience 52: 335–341.

Lussier, J.M., Morin, H., and Gagnon, R. 2002. Mortality in black spruce stands of fire or clear-cut origin. Can. J. Forest Res. 32: 539–547.

MacLean, R., Oswood, M.W., Irons, J.G., and McDowell, W.H. 1999. The effect of permafrost on stream biogeochemistry: a case study of two streams in Alaskan taiga. Biogeochemistry 47: 239–267.

Manies, K.L., Harden, J.W., Bond-Lamberty, B.P., O'Neill, K.P. 2005. Woody debris along an upland chronosequence in boreal Manitoba and its impact on long-term carbon storage. Can. J. For. Res. 35: 472–482.

Manö, S., and Andreae, M.O. 1994. Emission of methyl bromide from biomass burning. Science 263: 1255–1257.

McCullough, D.G., Werner, R.A., and Neumann, D. 1998. Fire and insects in northern and boreal forest ecosystems of North America. Annu. Rev. Entomol. 43: 107–127.

McInnes, P., Naimain, R.J., Pastor, J., and Cohen, Y. 1992. Effects of moose browsing on vegetation and litter of the boreal forest, Isle Royale, Michigan, USA. Ecology 73: 2059–2075.

Moore, T.R., Roulet, N.T., and Waddington, J.M. 1998. Uncertainty in predicting the effect of climatic change on the carbon cycling of Canadian peatlands. Climatic Change 40: 229–245.

Moore, T.R., Matos, L., and Roulet, N.T. 2003. Dynamics and chemistry of dissolved organic carbon in Precambrian Shield catchments and an impounded wetland. Can. J. Fisheries Aquatic Sci. 60: 612–623.

Neff, J.C., and Hooper, D.U. 2002. Vegetation and climate controls on potential $CO_2$, DOC, and DON production in northern latitude soils. Global Change Biol. 8: 872–884.

O'Neill, K.P., Kasischke, E.S., and Richter, D.D. 2002. Environmental controls on soil $CO_2$ flux following fire in black spruce, white spruce, and aspen stands of interior Canada. Can. J. Forest Res. 32: 1525–1541.

Oechel, W.C., and Van Cleve, K. 1986. The role of bryophytes in nutrient cycling in the taiga. In Forest ecosystems in the Alaskan taiga, eds. K. Van Cleve, F.S. Chapin, P.W. Flanagan, L.A. Viereck, and C.T. Dyrness, pp. 121–137. New York: Springer.

Olofsson, J., and Oksanen, L. 2002. Role of litter decomposition for the increased primary production in areas heavily grazed by reindeer: a litterbag experiment. Oikos 96: 507–515.

Osterkamp, T.E., and Romanovsky, V.E. 1999. Evidence for warming and thawing of discontinuous permafrost in Alaska. Permafrost Periglacial Processes 10: 17–37.

Osterkamp, T.E., Viereck, L., Shur, Y., Jorgenson, M.T., Racine, C., Doyle, A., and Boone, R.D. 2000. Observations of thermokarst and its impact on boreal forests in Alaska, U.S.A. Arctic Alpine Res. 32: 303–315.

Pastor, J., Naiman, R.J., Dewey, B., and McInnes, P. 1988. Moose, microbes, and the boreal forest. BioScience 38: 770–777.

Pastor, J., Dewey, B., Naiman, R.J., McInnes, P.F., and Cohen, Y. 1993. Moose browsing and soil fertility in the boreal forests of Isle Royale National Park. Ecology 74: 467–480.

Pastor, J., Cohen, Y., and Moen, R. 1999. Generation of spatial patterns in boreal forest landscapes. Ecosystems 2: 439–450.

Pastor, J., Peckham, B., Bridham, S., Weltzin, J., and Chen, J. 2002. Plant dynamics, nutrient cycling, and multiple stable equilibria in peatlands. Am. Naturalist 160: 553–568.

Pewe, T.L. 1958. Geology of the Fairbanks (D-2) Quadrangle, Alaska. Washington, DC: U.S. Department of the Interior, U.S. Geological Survey.

Pietikainen, J., and Fritze, H. 1996. Soil microbial biumass: determination and reaction to burning and ash fertilization. In Fire in ecosystems of boreal Eurasia, eds. J.G. Goldhammer and V.V. Furyaev, pp. 337–349. Dordrecht, The Netherlands: Kluwer Academic Publishers.

Ping, C.L., Michaelson, G.J., and Kimble, J.M. 1997. Carbon storage along a latitudinal transect in Alaska. Nutrient Cycling Agroecosystems 49: 235–242.

Podur, J., Martell, D.L., and Knight, K. 2002. Statistical quality control analysis of forest fire activity in Canada. Can. J. Forest Res. 32: 195–205.

Prather M. Ehhalt, D., Dentener, F., Derwent, R., Dlugokencky E., Holland, E., Esaksen, I., Katima, J., Kirchhoff, V., Latson, P., Midgley, P., Dang M. 2001. Atmospheric chemistry and greenhouse gases. In Climate change 2001: the scientific basis, eds. Houghton J.T., Ding, Y., Noguer, M., Vander Linder, P.J., Dai, X., Maskell, K, Johnson, C.A., pp. 241–280. UK and New York, NY, USA: IPCC, Cambridge University Press.

Price, J.S., and Whitehead, G.S. 2001. Developing hydrologic thresholds for Sphagnum recolonization on an abandoned cutover bog. Wetlands 21: 32–40.

Proctor, M.C.F. 1982. Carbon-14 experiments on the nutrition of liverwort sporophytes *Pellia epiphylla, Cephalozia bicuspidata*, and *Lophocolea heterophylla*. J. Bryol. 12: 279–284.

Railton, J.B., and Sparling, J.H. 1973. Preliminary studies on the ecology of palsa mounds in northern Ontario. Can. J. Botany 51: 1037–1044.

Randerson, J.T., Chapin, F.S.I., Harden, J.W., Neff, J.C., and Harmon. M.E. 2002. Net ecosystem production: a comprehensive measure of net carbon accumulation by ecosystems. Ecol. Applications 12: 937–947.

Raymond, P.A., and Bauer, J.E.A. 2001. DOC cycling in a temperate estuary: a mass balance approach using natural C-14 and C-13 isotopes. Limnol. Oceanogr. 46: 655–667.

Reader, R.J., and Stewart, J.M. 1972. The relationship between net primary production and accumulation for a peatland in southeastern Manitoba. Ecology 53: 1024–1037.

Rencz, A.N., and Auclair, A.N.D. 2001. NPP Boreal Forest: Schefferville, Canada, 1974. Data set. Available at http://www.daac.ornl.gov.

Robinson, S.D., and Moore, T.R. 1999. Carbon and peat accumulation over the past 1200 years in a landscape with discontinuous permafrost, northwestern Canada. Global Biogeochem. Cycles 13: 591–601.

Robinson, S.D., and Moore, T.R. 2000. The influence of permafrost and fire upon carbon accumulation in high boreal peatlands, Northwest Territories, Canada. Arctic Antarctic Alpine Res. 32: 155–166.

Robinson, S.D., Turetsky, M.R., Kettles, I.M., and Wieder, R.K. 2003. Permafrost and peatland carbon sink capacity with increasing latitude. In Proceedings of the 8th

International Conference on Permafrost, Vol. 2, pp.965–970. Zurich, Switzerland: Balkema Publishers.

Rochefort, L., Vitt, D.H., and Bayley, S.E. 1990. Growth, production, and decomposition dynamics of Sphagnum under natural and experimentally acidified conditions. Ecology 71: 1986–2000.

Roland, J., and Taylor, P.D. 1997. Insect parasitoid species respond to forest structure at different spatial scales. Nature 386: 710–713.

Roland, J., Mackey, B.G., and Cooke, B. 1998. Effects of climate and forest structure on duration of forest tent caterpillar outbreaks across central Ontario, Canada. Can. Entomol. 130: 1–12.

Rothman, L., and Roland, J. 1998. Relationships between forest fragmentation and colony performance in the forest tent caterpillar, Malacosoma disstria. Ecography 21: 383–391.

Rugo, O., and Weiss, A. 1996. Preserving Russia's carbon sink: strategies for improving carbon storage through boreal forest protection. Int. J. Environ. Pollution 6: 131–141.

Rupp, T.S., Starfield, A.M., Chapin, F.S., and Duffy, P. 2002. Modelling the impact of black spruce on the fire regime of Alaskan boreal forest. Climatic Change 55: 213–233.

Schroeder, D., and Perera. A.H. 2002. A comparison of large-scale spatial vegetation patterns following clearcuts and fires in Ontario's boreal forests. Forest Ecol. Manage. 159: 217–230.

Schultz, J.C., and Baldwin, I.T. 1982. Oak leaf quality declines in response to defoliation by gypsy moth larvae. Science 217: 149–151.

Scott, G.A.J. 1995. Canada's vegetation. A world's perspective. Montreal, PQ: McGill-Queens University Press.

Seppälä, M. 1988. Palsas and related forms. In Advances in Periglacial Geomorphology, ed. M.J. Clark, pp. 247–278. Chichester: Wiley.

Shaver, G.R., and Aber, J.D. 1996. Carbon and nutrient allocation in terrestrial ecosystems. In Global change: effects on coniferous forests and grasslands, eds. A. Melillo and A. Breymeyer, pp. 183–198. SCOPE Synthesis Series Chichester, UK: John Wiley.

Shurpali, N.J., Verma, S.B., and Kim, J. 1995. Carbon dioxide exchange in a peatland ecosystem. J. Geophys. Res. 100: 14319–14326.

Skre, O., and Oechel, W.C. 1981. Moss functioning in different taiga ecosystems in interior Alaska, USA 1. Seasonal phenotypic and drought effects of photosynthesis and response patterns. Oecologia 48: 50–59.

Smith, L.C., MacDonald, G.M., Velichko, A.A., Beilman, D.W., Borisova, O.K., Frey, K.E., Kremenetski, K.V., and Sheng, Y. 2004. Siberian peatlands a net carbon sink and global methane source since the early Holocene. Science 303: 353–356.

Smith, M. 1975. Microclimatic influences on ground temperatures and permafrost distribution, Mackenzie Delta, Northwest Territories. Can. J. Earth Sci. 12: 1421–1438.

Stocks, B.J. 1991. The extent and impact of forest fires in northern circumpolar countries. In Global biomass burning: atmospheric, climatic, and biospheric implications, ed. J. S. Levine, pp. 197–202. Cambridge, MA: MIT Press.

Stocks, B.J., and Kaufman, J.B. 1997. Biomass consumption and behaviour of wildland fires in boreal, temperate, and tropical ecosystems: parameters necessary to interpret historic fire regimes and future fire scenarios. In Sediment records of biomass burning and global change, eds. J. S. Clark, H. Cachier, J. G. Goldammer, and B. Stocks, pp. 169–188. NATO ASI Series no. I, Global Environmental Change, Berlin: Heidelberg, New York: Springer.

Stocks, B.J., Mason, J.A., Todd, J.B., Bosch, E.M., Wotton, B.M., Amiro, B.D., Flannigan, M.D., Hirsch, K.G., Logan, K.A., Martell, D.L., and Skinner, W.R. 2003. Large forest fires in Canada, 1959–1997. J. Geophys. Res. 108: 8149, doi: 10.109/2001JD000484.

Storaunet, K.O., and Rolstad, J. 2002. Time since death and fall of Norway spruce logs in old-growth and selectively cut boreal forest. Can. J. Forest Res. 32: 1801–1812.

Sundberg, S., and Rydin, H. 2000. Experimental evidence for a persistent spore bank in Sphagnum. New Phytologist 148: 105–116.

Swanson, D.K. 1996. Susceptibility of permafrost soils to deep thaw after forest fires in interior Alaska, U.S.A. and some ecologic implications. Arctic Alpine Res. 28: 217–227.

Szumigalski, A.R., and Bayley, S.E. 1996. Decomposition along a bog to rich fen gradient in central Alberta. Can. J. Botany 74: 573–581.

Szumigalski, A.R., and Bayley, S.E. 1997. Net above-ground primary production along a peatland gradient in central Alberta, Canada in relation to environmental factors. Écoscience 4: 385–393.

Thie, J. 1974. Distribution and thawing of permafrost in the southern part of the discontinuous permafrost zone in Manitoba. Arctic 27: 189–200.

Thormann, M.N., and Bayley, S.E. 1997a. Decomposition along a moderate-rich fen-marsh peatland gradient in boreal Alberta, Canada. Wetlands 17: 123–137.

Thormann, M.N., and Bayley, S.E. 1997b. Response of aboveground net primary plant production to nitrogen and phosphorus fertilization in peatlands in southern boreal Alberta, Canada. Wetlands 17: 502–512.

Trumbore, S.E., Chadwick, O.A., and Amundson, R. 1996. Rapid exchange between soil carbon and atmospheric carbon dioxide driven by temperature change. Science 276: 393–396.

Trumbore, S.E., and Harden, J.W. 1997. Accumulation and turnover of carbon in organic and mineral soils of the BOREAS northern study area. J. Geophys. Res. 102: 28817–28830.

Turetsky, M.R. 2003. Bryophytes in carbon and nitrogen cycling. The Bryologist 106: 395–409. 2003.

Turetsky, M.R., Wieder, R.K., Williams, C.J., and Vitt, D.H. 2000. Organic matter accumulation, peat chemistry, and permafrost melting in peatlands of boreal Alberta. Ecoscience 7: 379–392.

Turetsky, M.R., Wieder, R.K., Halsey, L.A., and Vitt, D.H. 2002. Current disturbance and the diminishing peatland carbon sink. Geophys. Res. Lett. 29: 10.1029/2001 GL014000.

Turetsky, M.R., Amiro, B.D., Bosch, E., Bhatti, J.S. 2004. Peatland burning and its relationship to fire weather indices in western Canada. Glob. Biogeochem. Cyc. 18: GB4014, doi: 10.1029/2004GB002222.

Turner, M.G. 1989. Landscape Ecology: the effect of pattern on process. Annu. Rev. Ecol. Systematics 20: 171–197.

Turunen, J., Tahvanainen, T., Tolonen, K., and Pitkanen, A. 2001. Carbon accumulation in west Siberian mires, Russia. Global Biogeochem. Cycles 15: 285–296.

Van Cleve, K., Dyrness, C.T., Viereck, L.A., Fox, J., Chapin III, F.S.F.S., and Oechel, W.C. 1983. Taiga ecosystems in interior Alaska. BioScience 41: 78–88.

Van Cleve, K., Viereck, L.A., and Marion, G.M. 1993. Introduction and overview of a study dealing with the role of salt-affected soils in primary succession on the Tanana River floodplain of interior Alaska. Can. J. Forest Res. 23: 879–888.

Van Cleve, K., Viereck, L.A., and Dryness, C.T. 1996. State factor control of soils and forest succession along the Tanana River in Interior Alaska, U.S.A. Arctic Alpine Res. 28: 388–400.

Verhoeven, J.T.A., and Toth, E. 1995. Decomposition of *Carex* and *Sphagnum* litter in fens: effect of litter quality and inhibition by living tissue homogenates. Soil Biol. Biochem. 27: 271–275.

Viereck, L.A. 1983. The effects of fire in black spruce ecosystems of Alaska and northern Canada. In The role of fire in northern circumpolar ecosystems, eds. R.W. Wein and D.A. MacLean, pp. 201–220. Wiley: New York.

Viereck, L.A., Foote, J., Dryness, C.T., Van Cleve, K., Kane, D., and Seifert, R. 1979. Preliminary results of experimental fires in the black spruce type of interior Alaska. PNW research note 332, U.S. Forest Service, Pacific Northwest Forest and Range Experiment Station: Portland, Oregon. 27 pp.

Vitt, D., Halsey, L., and Zoltai, S. 1994. The bog landforms of continental western Canada in relation to climate and permafrost patterns. Arctic Alpine Res. 26: 1–13.

Vitt, D.H., Halsey, L.A., and Zoltai, S.C. 2000. The changing landscape of Canada's western boreal forest: the current dynamics of permafrost. Can. J. Forest Res. 30: 283–287.

Vitt, D.H., Halsey, L.A., Campbell, C., Bayley, S.E., and Thormann, M.N. 2001. Spatial patterning of net primary production in wetlands of continental western Canada. Ecoscience 8: 499–505.

Wahlen, M. 1993. The global methane cycle. Annu. Rev. Earth Planetary Sci. 21: 407–426.

Walter, B.P., Heimann, M., and Matthews, E. 2001. Modeling modern methane emissions from natural wetlands: 1. Model description and results. J. Geophys. Res. 106: 34189–34206.

Weber, M.G. 1990. Forest soil respiration after cutting and burning in immature aspen ecosystems. Forest Ecol. Manage. 31: 1–14.

Weber, U.M., and Schweingruber, F.H. 1995. A dendroecological reconstruction of western budworm outbreaks (Choristoneura occidentalis) in the Front Range, Colorado, from 1720 to 1986. Trees 9: 204–213.

Williams, D.J., and Burn, C.R. 1996. Surficial characteristics associated with the occurrence of permafrost near Mayo, Central Yukon Territory, Canada. Permafrost Periglacial Processes 7: 193–206.

Williams, D.W., and Liebhold, A.M. 1995. Herbivorous insects and global change: potential changes in the spatial distribution of forest defoliator outbreaks. J. Biogeogr. 22: 665–671.

Williams, D.W., and Liebhold, A.M. 2000. Spatial scale and the detection of density dependence in spruce budworm outbreaks in eastern North America. Oecologia 124: 544–552.

Wilson, J.A., and Coxson, D.S. 1999. Carbon flux in a subalpine spruce fire-forest: pulse release from Hylocomium splendens feather-moss mats. Can. J. Botany 77: 564–569.

Wirth, C., Schulze, E.D., Kusznetova, V., Milyukova, I., Hardes, G., Siry, M., Schulze, B., and Vygodskaya, N.N. 2002. Comparing the influence of site quality, stand age, fire, and climate on aboveground tree production in Siberian Scots pine forests. Tree Physiol. 22: 537–552.

Woo, M., Lewkowicz, A.G., and Rouse, W.R. 1992. Response of the Canadian permafrost environment to climatic change. Phys. Geogr. 13: 287–317.

Yatskov, M.A. 2000. A chronosequence of wood decomposition in the boreal forests of Russia. M.S. Thesis, Oregon State University.

Yu, Z., Turetsky, M.R., Campbell, I.D., and Vitt, D.H. 2001. Modeling long-term peatland dynamics. II. Processes and rates as inferred from litter and peat-core data. Ecol. Modeling 145: 159–173.

Yu, Z.C., Apps, M.J., and Bhatti, J.S. 2002. Implications of floristic and environmental variation for carbon cycle dynamics in boreal forest ecosystems of central Canada. J. Vegetation Sci. 13: 327–340.

Zhuang, Q., McGuire, A.D., Melillo, J.M., Clein, J.S., Dargaville, R.J., Kicklighter, D.W., Myneni, R.B., Dong, J., Romanovsky, V.E., Harden, J.W., and Hobbie, J.E. 2003. Carbon cycling in extratropical ecosystems of the Northern Hemisphere during the 20th century: a modeling analysis of the influences of soil thermal dynamics. Tellus 55B: 751–776.

Zoltai, S. 1993. Cyclic development of permafrost in the peatlands of northwestern Alberta, Canada. Arctic Alpine Res. 25: 240–246.

Zoltai, S. 1995. Permafrost distribution in peatlands of west-central Canada during the Holocene warm period 6000 years B.P. Géographie physique et Quaternaire 49: 45–54.

Zoltai, S.C., and Tarnocai, C. 1971. Properties of a wooded palsa in northern Manitoba. Arctic Alpine Res. 3:115–129.

Zoltai, S.C., and Martikainen, P.J. 1996. The role of forested peatlands in the global carbon cycle. Nato ASI Series I 40: 47–58.

Zoltai, S.C., and Vitt, D.H. 1990. Holocene climatic change and the distribution of peatlands in the western interior of Canada. Quaternary Res. 33: 231–240.

# 13
# Heterogeneity in Urban Ecosystems: Patterns and Process

LARRY E. BAND, MARY L. CADENASSO, C. SUSAN GRIMMOND,
J. MORGAN GROVE, and STEWARD T.A. PICKETT

## Abstract

Heterogeneity in urban ecosystems derives from a combination of natural
and engineered landscape features, as well as behavior of human individuals and institutions. Modern urban regions in North America and elsewhere
are no longer uniformly compact and densely populated but have extended
into surrounding regions and include intricate mixes of residential, commercial, and residual agricultural, forest, and other managed and unmanaged
vegetated areas. Compared to less developed ecosystems, heterogeneity in
water, carbon, nutrient, and energy cycling may be enhanced, specifically over
the short distances associated with urban development patterns. We review
conceptual approaches to characterizing and representing heterogeneity in
urban ecosystems and illustrate some of the main sources of heterogeneity
resulting from interactions within and between urban patch networks, with
special reference to examples drawn from the Baltimore Ecosystem Study.

## Introduction

Cities are rapidly expanding in both area and population on a global scale.
For example, about 80% of the U.S. population currently lives in urban areas
as defined by the U.S. Bureau of the Census. In addition, increases in the area
of urban land use have typically exceeded increases in urban population,
resulting in urban sprawl characterized by lower density and more extensive
areas (Berry 1990). The shift in extent and density of cities has resulted in a
change in the morphology of urban areas. The formerly compact and uniformly developed coverage of North American cities has evolved into more
extensive and spatially heterogeneous patterns that contain significantly
contrasting land covers (Garreau 1991). The contemporary metropolis contains lower density urban development including an assemblage of residential,
commercial, and industrial land uses interspersed with residual agricultural
and forest land, and other unmanaged vegetation (Zipperer et al. 1997). The

current patch network may still maintain some memory of previous states because of slowly changing variables such as soil and groundwater conditions (Effland and Pouyat 1997; Pouyat et al. submitted). Spatial and temporal interactions between different patch types within these areas may be important regulators of aggregate system behavior (Cadenasso et al. 2003b). Distinct dynamics of individual patches can be dependent on local neighborhood connectivity, as well as successional history (Pickett et al. 2004).

This paper discusses the influence of heterogeneity on ecosystem processes in urban environments. We first present and then attempt to integrate a set of conceptual frameworks for representing heterogeneity and its effects in urban ecosystems. The frameworks we seek to integrate include (1) ecological patch dynamics, (2) distributed hydroecological modeling, and (3) urban land-atmosphere interactions.

Our main focuses are on (1) the forms of spatial heterogeneity in urban patch networks, (2) the interaction of human individual and institutional activity with landscape cycling of matter and energy, and (3) the impact and influence of heterogeneity on ecosystem behavior. Throughout the chapter, we maintain an approach that treats urban ecosystems as specific cases of ecosystems in general (Pickett and Cadenasso 2002), varying only in the degree of influence of specific ecosystem "agents," including human activity and the built environment. Illustrative material is drawn from work in several sites but focuses on examples from the Baltimore Ecosystem Study (BES) Long-Term Ecological Research program (http://beslter.org).

## Characterization of Urban Ecosystem Heterogeneity

Any ecosystem can be defined in terms of a set of state, flux, and transformation variables that are linked by a set of statements assuming conservation of mass and energy. Animal and plant populations are incorporated through trophic structures that both contribute and respond to the mass and energy regulation of the system. One approach to characterizing the spatial patterns and distributions of the ecosystem is to use patch networks, in which discrete areas, or patches, are interconnected with defined flows of material, energy or information (Kolasa and Pickett 1991; Wu and Loucks 1995). Ecological patch dynamics (Pickett and White 1985; Fahrig 1992; Fisher 1993; Wiens 1995) is a widely recognized approach in the study of landscape ecological patterns that incorporates interactions between patches but also considers the space-time dynamics of patch structures across scales by a set of slow (e.g., successional) and fast (disturbance) processes. Conservation statements are applied to each patch and to the fluxes between and within connected patches. The form of connections (e.g., downslope, first or higher order neighborhoods, land-atmosphere) is crucial to the behavior and properties of each patch and of the full ecosystem and provides crucial linkages between the three frameworks of patch dynamics, hydroecology, and land-atmosphere interactions. For example, hydrological connections are determined by the arrangement of source and sink patches for water along slopes (Black 1991), and land-atmosphere connections are determined by hydrologic state and the roughness and arrangement of patch structures.

In order to scale the behavior of individual patches to the structure and behavior of the urban ecosystem, it is necessary to characterize the patterns and interactions of the population of patches. This may be done by (1) explicitly representing patches in space through mapping, (2) by statistically describing the patch population, potentially including spatial relational or pattern information, or (3) by a hybrid approach of mapping areas to a particular scale (e.g., subwatersheds, hillslopes, or neighborhoods), then statistically describing the patch subpopulations below that scale. For example, spatially explicit mapping compellingly shows richness, shapes, and configuration of patches, whereas statistical approaches characterize the patch array in aggregate terms such as means, frequencies, and variation. A time dimension can be added to any of the analytical approaches above by incorporating information on patch history (Brush 2003) and the appearance and disappearance of different patches and patch types (Pickett et al. 2000).

The hybrid approach maintains spatially explicit information at the higher levels of patch aggregation, while subsuming the characteristics of lower, more detailed levels of patches in statistical terms. Band et al. (1991, 1993, 2000) have used this approach in spatial hydroecological modeling by the use of digital terrain and image analysis to map the set of hillslopes within a watershed and then statistically describing the variance and covariance in canopy and terrain conditions within each hillslope.

An essential question pertinent to ecosystem heterogeneity is to determine what form and scales of variance and pattern in the landscape need to be accounted for to determine ecosystem behavior. Assuming that there may be an irreducible level of variance at all scales, at what scale and in what circumstances do the effects of heterogeneity in the form of explicit spatial patterns, become significant relative to the ecosystem processes of interest? This fundamental question should guide the scale of "mapping" ecosystem patterns. This is in contrast to representation of heterogeneity as spatially inexplicit, or aspatial, variance, or the corollary of "lumping" or effectively ignoring ecosystem variance entirely. Ecologists have traditionally lumped system spatial variance, and more recently, have expressed variance in spatially inexplicit terms (Wiens 2000).

## Representation of Ecosystem Heterogeneity and Aggregation Approaches

Methods of incorporating heterogeneity into environmental process models have been developed and used extensively over the past two decades. Good examples in hydrology include Bresler and Dagan (1988), who considered the effects of soil heterogeneity within a field on rates of infiltration, soil moisture, and crop yield. Band et al. (1991, 1993) explored the effects of incorporating different levels of heterogeneity in canopy leaf area index, terrain and soil moisture conditions on catchment water and carbon cycling. Rastetter et al. (1991) summarized different methods of aggregating fine-scale ecosystem heterogeneity to coarse scale in process models. More recently, Strayer et al. (2003) discussed different types of heterogeneity in patch structures and approaches for representing their effects at the ecosystem level. The following discussion draws from a set of these sources.

We can characterize a population of patches as a functional set, **P**, representing a geographic distribution of simple or complex ecosystem processes such as mass and energy flux or transformation within discrete elements of a landscape. The ecosystem processes are determined by a set of factors including state variables and control parameters, denoted as a vector **x**. The dynamics or behavior of each patch may be represented by a simple rule, equation, or complex model at each patch, given as $p(\mathbf{x})$. The simplest model of the landscape (full patch network) is to ignore heterogeneity and treat the full landscape as a "metapatch" characterized by mean or modal parameter and variable values. This approach assumes

$$E[p(\mathbf{x})] = p[E(\mathbf{x})] \tag{13.1}$$

indicating spatial mean or expected values of **x** can reproduce the effects of a distribution function of the controlling parameters including biotic, abiotic, and human socioeconomic factors. In this case, heterogeneity in the form of distributions or patterns in **x** do not need to be known. This may occur if the form of $p$ is linear with no significant covariance among the terms of **x**. As an example, if mean rates of ecosystem productivity could be functionally predicted with a single, mean value of soil moisture (or other abiotic, biotic, and social drivers), Equation (13.1) would be correct and no information on the joint distribution function of moisture and other parameters would be necessary. Many environmental models implicitly assume Equation (13.1) and assign mean or modal values of **x** over extensive areas (e.g., spatially lumped watershed models, land-atmosphere schemes within global or mesoscale atmospheric model grid cells).

If Equation (13.1) is not correct, we can next explore whether a simple approach to integrating $p(\mathbf{x})$ over the distribution of **x** yields the expected or emergent ecosystem behavior

$$E[p(x)] = \int p(x)\, f(x)\, \partial x. \tag{13.2}$$

Here, Equation (13.2) represents a multiple integration over the joint distribution function of x. An implicit assumption is that $p(x)$ is a spatially independent process such that spatial patterns (e.g., neighborhood effects, connectivity) are not important. It is only under very limited circumstances that the process function, $p$, the joint frequency distribution, $f$, and the integration are fully known and can be carried out analytically. Therefore, the integration is typically carried out as a discrete summation over empirical or modeled frequency distributions of patch conditions.

As discussed in the hybrid approach, above, Equation (13.2) can be partitioned in space into landscape subregions with simpler or known distributions of $x$. These subregions could include hillslopes with more uniform exposure and microclimate, specific land forms, land uses, or urban zoning or management units. In this case, the development of land classification schemes in which the classes have known or easily parameterized distributional forms for characterizing the variance of key controlling variables and factors would be an efficient, hybrid approach. This approach is used in the Regional Hydroecological Simulation System (RHESSys) family of models (Band et al. 1991, 1993, 2000), making use of hillslope, microclimate zones, and land-use elements for simulation of carbon, water, and nutrient cycling. Although the explicit attention here is to the patch structure of the landscape, the control of fluxes within the landscape is of ultimate interest (Cadenasso et al. 2003a).

An alternative of this approach is the "mosaic" strategy (e.g., Avissar 1991) in which the surface is partitioned into a set of spatially exhaustive and mutually exclusive patches that can be regular (e.g., rectangular or triangular grids) or irregular polygons. Specific parameters can be assigned to each patch, and the function evaluation can either be carried out over all patches or the patches can be classified into a smaller number of types with the integration carried out over the set of classes and mapped back to the patch network. If the number of patch types or the range of conditions is too large, the number of distinct patch states that must be integrated or summed may approach the number of patches in an area, negating the advantage of the frequency distribution approach.

Note that a number of environmental models use the "mosaic" approach and simulate ecosystem state and flux processes over a two-dimensional extent. However, the assumption of spatial independence renders this approach insensitive to pattern, and the patches could be reshuffled with no influence on aggregate ecosystem behavior. If neighborhood or global pattern or connectivity is integral to the function of the ecosystem, the patch network needs to be represented as a flux network, explicitly computing exchange of mass, energy, or information, or with parameters representing the effects of position and pattern. An example of the latter approach is given by the wetness index proposed by Beven and Kirkby (1979), which describes a relative topographic position pertinent to soil moisture patterns. In this case, the wetness index gives a relative "wetness" or soil water deficit, that can be used to scale catchment mean water deficits to any topographic position. The approach attempts to approximate the effects of soil water lateral redistribution without actually computing patch to patch flux.

## Urban Patch Dynamics

As discussed above, development of a patch dynamics "profile" or typology of urban spaces would provide an efficient means of characterizing urban ecosystem heterogeneity without the need to measure or map all biophysical and social attributes of each patch. The large body of research on ecological patch dynamics over the past two decades cannot be discussed in detail here, but insights from the literature are summarized in these key points (Pickett and White 1985; Shugart and Seagle 1985; Levin 1989; Kolasa and Rollo 1991; Tilman 1994; Wiens 1995; Wu and Loucks 1995; Pickett and Rogers 1997; Shachak et al. 1999; Band et al. 2000; Cadenasso et al. 2003b):

1. Biogeophysical habitats are spatially heterogeneous. Patches of contrasting composition, structure, and function can be recognized.
2. Patches are created by a variety of persistent environmental templates, by transient organism actions, or by ephemeral disturbances and sudden shifts in system drivers including human activities and constructions.
3. Patches change due to the movement of organisms and materials into and out of them, as well as succession within them.
4. A mosaic of patches exhibits complex dynamics due to within-patch dynamics, different rates and starting conditions in different patches, and fluxes between patches. This recognizes that patches alter through time, including by birth, death, and catastrophic or gradual transformation into different patch types.
5. Boundaries not only delimit patches but also may control important lateral flows between adjacent patches.

The basic concepts of patch dynamics apply to both urban and non-urban systems (Grimm et al. 2000). Urban patch delimitations or models are more inclusive than those of classical ecology, which do not account for people and their effects (Pickett et al. 1997). Of course, urban patches must still account for heterogeneity in soil, hydrology, and vegetation as the major ecological and geophysical components. The only difference with *urban* patch dynamics is the range of factors that must be drawn on to construct urban patch models, including institutional, economic, and social drivers (Machlis et al. 1997).

Adding institutional components to urban patches adds an understanding of the way human individuals are organized into larger aggregations. Institutions range from nuclear families or households, through aggregations of people that share leisure activities, to community groups and neighborhood associations, religious congregations, businesses, to government agencies at various levels (Ostrom 1998). This listing is only exemplary, and many more kinds of institutions could be named. The point is that there are many ways people are organized to achieve different functions. These organizations can differ in distribution and impact across space, leading to institutional patchiness (Gottdiener and Hutchison 2000). Economic drivers of system structure and function can also result in urban patchiness (Krugman 1996). Patches may be affected by such specific economic factors as income, capital accumulation in different sectors, or public and private investment (Harvey 2000). Social drivers include capital that people build up through interactions, the degree of neighborhood cohesion, or ethnic, racial, and other factors by which group identity is forged. Changes in these factors over time within urban systems promotes the dynamic evolution of the patch network (Grove and Burch 1997).

All patches may be defined on the basis of substrates and flowpaths that absorb *versus* disperse a particular set of resources (Pickett et al. 2000). For example, in forested ecosystems, patches may be defined on the basis of their hydrological function. The substrates within different patches can differ in infiltration capacity, groundwater transmissivity, and capacity to return water to the atmosphere via evapotranspiration. These functions can be controlled by soil properties, the proportion and nature of canopy and litter cover, slope, aspect and hillslope position, and so forth (Band et al. 1993). Within urban ecosystems, similar controls on soil-landform catenae interact with additional controls at the household, neighborhood, and municipal levels and include the flux of information and capital (Olsen 1982). In the mixed land use patches of a metropolis, these functions are affected by pavement, building footprints, roofing materials, the presence of storm drains and curb channels, and other such built components. Figure 13.1

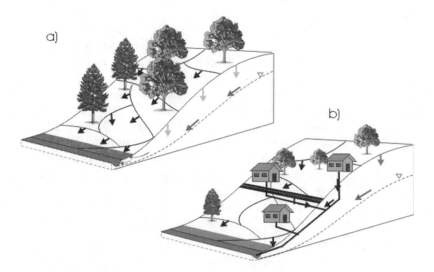

Hillslope level representation of hydrologic flowpaths for
a) a forested catchment with topographically driven drainage pattern and
b) an urbanizing catchment, where roads and storm sewers redirect flow routing,
increase quickflow, reduce groundwater recharge, bypass riparian zones

FIGURE 13.1. Two hillslope drainage sequences in undeveloped and suburban ecosystem settings. In the undeveloped scenario, repeated patterns of soil, vegetation, water, and nutrient cycling co-evolve along topographic gradients, providing a structure that can be used to characterize sub-hillslope patch structure. The implementation of drainage infrastructure, land cover, roads, and individual patch fertilization and irrigation strongly alters the hillslope scale sequence of water and nutrient sources and sinks.

shows two hillslope drainage patterns, with and without the additional over-lay of human development. Within the mixed urban context, these types of elementary hillslopes, or sub-hillslope components such as the ridge, mids-lope, slope toe/riparian zone, can constitute the landscape partitioning required in the hybrid approach to represent ecosystem heterogeneity. Patch structures within the hillslopes may be explicitly delineated or statis-tically summarized and need to contain sufficient information on both nat-ural and built components, as well as human individual and institutional activity.

Regulatory, cultural, and social norms, as well as embedded economic sys-tems contribute to the structure and characteristics (size, shape, connectivity) of the patch networks (Grove 1997; Machlis et al. 1997). The resulting patch networks in turn influence the distribution of inputs and cycling of mass, energy, capital, and information. Human management of the landscape homogenizes the landscape at certain levels (Forman 1995) while significantly increasing the heterogeneity of the landscape at other levels (Clay 1973). For example, the sinuosity of edges typically decreases in managed as compared to wild landscapes, whereas in urban systems, a new, high intensity of fine-scale differentiation in hard surfaces, building volumes, and vegetation lay-ering is typical. Alternatively, in some cases the urbanization process may reduce the significance of a set of ecosystem factors, such as when areas are extensively paved. These shifts in the length scales of ecosystem variance reflect the manners in which the social identity and social hierarchies that characterize different people are distributed over space. The differences in identity and social ranking produce specific patterns of services and uses of the landscape. For example, within a neighborhood there is often a homo-geneity of land cover patterns at the neighborhood scale, reflecting social and economic norms, whereas at the parcel or streetscape level there are mixes of pervious and impervious surfaces such as rooftops, roads, walkways, irrigated and fertilized gardens and lawns, and less managed vegetation including residual woods and street trees. Compared to less developed land-scapes, there is an increased level of heterogeneity introduced at the scale of meters, with a potential drop in the heterogeneity at scales one to two orders of magnitude larger, depending on the spatial extent and similarity of devel-opments and sociodemographic groups (Grove and Burch 2002).

## Criteria for Urban Patches

The influence of individuals and institutions in modifying land cover, vege-tation, soils, and drainage systems can be represented by classical methods of describing patch form and composition. Most of these landscape modifi-cations can be mapped or remotely sensed and incorporated as structural elements of the landscape. However, the direct influence of individuals and institutions on ecosystem energy and mass balances requires consideration of individual and institutional behavior. The need to include economic, social, and institutional drivers in mass and energy budget models of urban patches requires a novel strategy for integration. The first step in this strategy is to identify a shared conceptual spatial model. For example, a model might focus on an area that includes structures that reflect both the influence of social, vegetation, and micrometeorological processes. Therefore, patch delineation in the very mixed urban landcover may focus on boundaries of human activity as embedded within functional ecosystem units (e.g., catchments). Human activity can be indexed as consumer choices, environmental decisions, and environmental recreation (Grove et al. in press)

Neighborhoods or subdivisions are widely recognized patch types in urban systems and may efficiently describe spatial variations in both land cover and sociodemographic groups. They are used to help define census data collection, organize the efforts of community groups, and focus commercial activity. In Baltimore, there are 276 officially recognized neighborhoods distinguished on the basis of housing stock (e.g., connected or detached), size of building footprint relative to open space, density of buildings, connections to the street and transportation grid, presence and nature of vegetation, building and lot vacancy, and density and nature of institutional properties (e.g., schools and hospitals). Of course, social differences exist among neighborhoods as well, and these differences include demography, ethnicity, access to public and private capital, educational resources, and various kinds of social capital, among others.

Delineating patch types that adequately describe the heterogeneity of the system must encompass characteristics of the ecological, social, and physical realms (Grove and Burch 1997). An example of one such delineation is given in Figure 13.2, showing a patch network characterized by natural and human components of the system. Identifying criteria include building type and density, amount and arrangement of pavement representing private *versus* shared access to buildings, and the proportion, heterogeneity, and layering of vegetation in the area. A parcel database, census information, and household survey can add socioeconomic information to this patch network.

Although landscape elements of these neighborhoods can be explicitly mapped at high resolution, human characteristics are uniformly available only at the census block or larger level unless surveys of household actions and values are used. Law et al. (2004) collected household level information on lawn fertilizer, irrigation, and soil chemistry in a set of subdivisions in Baltimore County, including those shown in Figure 13.2, in order to characterize the statistics of fertilizer application rates and their relation to soil, parcel, and subdivision characteristics. Analysis revealed nonlinear dependencies of fertilizer application rates on parcel size, home value and age, and soil characteristics when aggregated to the subdivision level (Law et al. 2004). This information has then been used to parameterize individual nutrient additions to lawn areas in a set of suburban catchments in the Baltimore Ecosystem Study for use in hydroecological models.

(a)

(b)

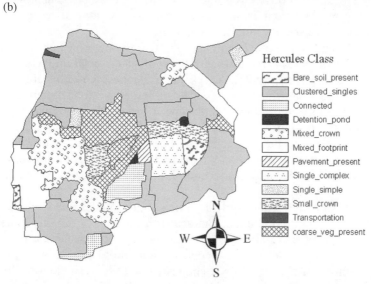

Hercules Class

- Bare_soil_present
- Clustered_singles
- Connected
- Detention_pond
- Mixed_crown
- Mixed_footprint
- Pavement_present
- Single_complex
- Single_simple
- Small_crown
- Transportation
- coarse_veg_present

FIGURE 13.2. A suburban catchment at the head of the Gwynns Falls watershed, Baltimore County, Maryland. (a) High-resolution aerial photograph shows a mix of landscapes and developments in this site, including housing stock spanning 150 years in age, school grounds, athletic fields, commercial development, and aggrading forest of different composition and size classes. (b) A functional classification of the area captures information on housing density, development characteristics, and vegetation structure.

Following this example, individual activity at the patch level, such as the direct additions or abstractions of material, can best be characterized by a stochastic process conditional on a set of biogeophysical and social factors and applicable to land elements within an area (neighborhoods, census blocks). This can be expressed as

$$F(\geq H \mid \mathbf{BG}, \mathbf{SE}, \mathbf{R}) \tag{13.3}$$

where the probability function for $H$, a direct loading or abstraction, rate, (e.g., irrigation, fertilization) is conditioned by arrays of biogeophysical ($\mathbf{BG}$), socioeconomic ($\mathbf{SE}$), and regulatory ($\mathbf{R}$) attributes at multiple scales. These terms can include information pertinent to the household, neighborhood, and municipal levels, as well as landscape position. As above, we assume that there is an irreducible level of spatial variance (or spatial "unknowability") that exists at the patch, parcel, or individual levels. The form of $F$ is conditional on the specific landscape element (e.g., lawn, garden, septic spreading field, woodlot), as well as the spatially varying $\mathbf{BG}$, $\mathbf{SE}$, and $\mathbf{R}$ terms. Therefore, the patch delineation may not attempt to map each landscape element but instead focus on larger multielement areas within which the distribution functions describing individual and institutional activity is stationary (for similar landscape elements). This may occur at the census block level or more likely the neighborhood or subdivision level within which demographic and landscape architectural characteristics may be more uniform, again in keeping with the hybrid approach to representation of landscape heterogeneity.

## Scaling Urban Ecosystem Patch Dynamics

For an approach such as that given by Equation (13.3) to be applicable to a larger region, spatial information needs to be developed that can be used to estimate expected individual and institutional activity. One potential approach to the estimation of spatial information on individual and institutional activity is through the construction of statistical links between available demographic information from census and commercial consumption information. Grove and Burch (2002) have discussed the use of specific lifestyle clusters—PRIZM (Potential Rating Index for Zip Markets)—produced by Claritis, Inc. The clusters are based on tastes, attitudes, and consumer profiles at the household level and use information from census, household survey, public opinion polls, and consumer purchasing data. Although this demographic clustering was devised originally to characterize consumer preferences and behavior for marketing purposes, information on such activities as garden and lawn maintenance expenditures can potentially be added to PRIZM classes at the census block level. This information would need to be calibrated with the type of household surveys reported by Law et al. (2004) but could potentially be used in conjunction with parcel information to generate stochastic realizations of such activities as lawn fertilization and irrigation rates at the patch or parcel level across a metropolis.

## Sources and Impacts of Heterogeneity in Urban Ecosystems

Given the framework to represent the characteristics of ecosystem hetero-geneity discussed above, we review a set of important effects of landscape patch structure on local to aggregate ecosystem behavior. The sources and impacts we discuss are not an exhaustive set but illustrate the multiscale interactions between patch heterogeneity and local to landscape scale processes. We illustrate sources of heterogeneity that occur at scales below the parcel, at the parcel, at the neighborhood scale, and above.

### Heterogeneity in Nutrient Sources

Consider the generalized quantities for nitrogen mass loading within subur-ban ecosystems in Baltimore, in this case low density (2-acre zoning) septic and well-water serviced areas (Figure 13.3). Aggregated to the level of a full landscape or catchment, lawn fertilization provides the greatest N load, applied to up to ~50% of the landscape. Septic loading aggregated to the full catchment is significantly lower, roughly equivalent to atmospheric dep-osition. However, it is concentrated in 1–2% of the area (spreading fields). Within this area, loading rates are an order of magnitude higher than N

## Nested (sub)urban flux fields

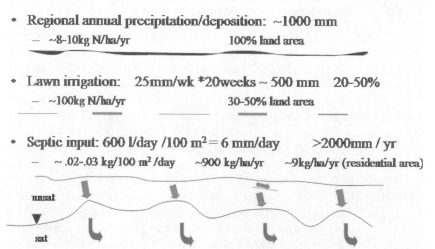

- **Regional annual precipitation/deposition:** ~1000 mm
  - ~8-10kg N/ha/yr                    100% land area

- **Lawn irrigation:**   25mm/wk *20weeks ~ 500 mm   20-50%
  - ~100kg N/ha/yr                     30-50% land area

- **Septic input:** 600 l/day /100 m² = 6 mm/day        >2000mm / yr
  - ~ .02-.03 kg/100 m² /day   ~900 kg/ha/yr   ~9kg/ha/yr (residential area)

unsat

sat

FIGURE 13.3. Loading rates of different sources of water and nitrogen from subur-ban landscapes as a function of the unit areas: regional or watershed, lawn, and spreading fields. Values are typical for low-density suburban landscapes in Baltimore County. The values show that lawn fertilization is the largest load per unit watershed area but that localized septic input is an order of magnitude higher in "hot spots."

loading per unit catchment area. In addition, compared to precipitation and expected lawn irrigation, septic effluent volumes within the small areas of the septic spreading fields are very high per unit area, sufficient to produce recharging soil profiles most of the time. Therefore, adding all sources of N loading and expressing this load as a mean rate per watershed area does not capture spatial distribution or covariance terms. Although a small part of the overall catchment water and nitrogen budget, the process produces a hot-spot effect and may dominate N delivery to the streams. Stream sampling in the Baltimore Ecosystem Study has shown the highest nitrate concentrations occur in the suburban fringe in septic-serviced, low-density development, significantly in excess of streams in central cities or sanitary sewer–serviced suburbs with similar lawn fertilization rates. Although the precise source of the nitrate has not been fully established, Groffman et al. (2002) have found suburban lawn soils to be very retentive of nitrogen, with very low leaching rates measured below the rooting zone.

## Urban Hydrology and Land/Atmosphere Interactions

The impacts of heterogeneity on land-atmosphere water and energy exchange are a function of scale. Three spatial scales, micro-, local-, and meso-, are commonly recognized in urban areas (Figure 13.4). Following the discussion above, current urbanization tends to increase microscale heterogeneity. Two of the most important impacts of urbanization on the hydrology and land-atmosphere exchange of urban ecosystems are (1) the introduction of impervious patches that do not absorb or store water and have very different thermal properties than soil or vegetated patches, and (2) the large-scale import of water for both indoor use and irrigation of specific patch types (e.g., lawns). In addition to the outdoor water use, there is a substantial inadvertent subsidy of water to the urban hydrologic cycle due to supply and sanitary water leaks, with supply leaks potentially accounting for up to 20–30% of the total water volume diverted from the original water supply source in older North American cities. Both the leaks, which often occur beneath the rooting zone, and targeted irrigation concentrate water in specific locations, adding to system patchiness and augmenting specific flowpaths at the microscale level.

Analogous to plant canopies, the roof level through the ground level is termed the *urban canopy layer* (UCL). At the microscale, the elements that make up the UCL such as buildings, trees, roads, lawns, and so forth may create their own microclimate. At this microscale, the close proximity of contrasting surface materials with different radiative, thermal, and moisture properties and surface concentrations of moisture or temperature create gradients that enhance vertical and horizontal fluxes between the materials. Contrasts in surface properties are also enhanced by potential runoff–run-on processes from impervious to pervious surfaces. Urban and suburban design includes curb and storm sewer drainage, which concentrates flow into an extended engineered drainage net and can create a more xeric environment

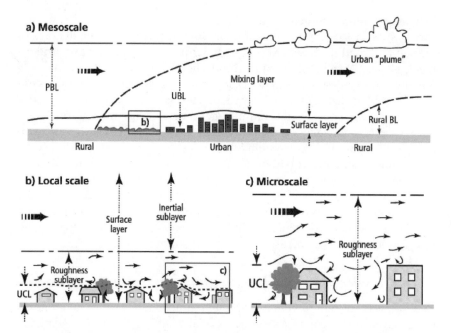

FIGURE 13.4. Scales used to distinguish atmospheric processes in urban areas. Idealized vertical structure of the urban atmosphere shown for the three primary scales used to distinguish atmospheric processes in urban areas: (a) an urban region at the scale of the whole city (mesoscale); (b) a land-use zone (local scale); and (c) a street canyon (microscale). BL refers to urban boundary layer (P, planetary; U, urban; R, rural) and UCL refers to urban canopy layer (see text for further explanation). [Figure modified after Oke (1997), reprinted with kind permission of Springer Science + Business Media, Inc. from Piringer et al. (2002), Figure 1, p. 3].

by interrupting and redirecting downslope flow directly into streams. The runoff–run-on processes would subsidize moisture and latent heat flux from the pervious materials. This effect has been recognized for some time and has been incorporated into standard urban runoff models as the percent directly connected impervious area, which distinguishes impervious area that may contribute to pervious surfaces.

Figure 13.5 shows percent impervious surface in the Glyndon catchment (Figure 13.2) as estimated from Thematic Mapper imagery and obtained from the Mid Atlantic Regional Earth Science Applications Center at the University of Maryland. Older commercial and residential neighborhoods to the north and west of the large vegetated region in the middle of the catchment predate stormwater infrastructure and contribute runoff to the pervious area. Surface soil moisture content sampled through an annual cycle was consistently higher than in a control catchment without impervious run-on (Tenenbaum et al. in press). Run-on infiltration can be observed during storm events from specific portions of the surrounding area as

FIGURE 13.5. Percent impervious for the Glyndon catchment area (shown in Figures 13.2 and 13.3) at the head of the Gwynns Falls watershed. Impervious area was estimated by the University of Maryland RESAC as the % per pixel (black is 0%) using ETM imagery. The central, undeveloped area receives significant run-on infiltration from a set of surrounding communities that do not have curb and storm sewer drainage.

diffuse flow, although we cannot yet estimate how much of the enhanced soil moisture is due to run-on. Alternatively, areas with stormwater infrastructure and sanitary systems that absorb groundwater and stormflow appear to have lower than expected soil moisture due to the enhanced concentration and drainage of surface and groundwater.

The temporal variability of the focal, unbuilt patches discussed above often increases as a result of development, although the magnitude differs depending on the connectivity to adjacent patch types and on the drainage infrastructure. Likewise, the medium-scale spatial heterogeneity of these landscapes is typically increased by development. However, the implementation of water resources infrastructure may tend to decrease heterogeneity in soil water contents at specific space and time scales by the provision of drainage or irrigation. Whether a patch of interest becomes developed and serviced by infrastructure or not or developed early or late in the urbanization process partially depends on its inherent capacity to absorb or shed water (Boone 2003). In other words, heterogeneity may have different expressions in time and in space. We have primarily focused on the spatial component of heterogeneity while recognizing its temporal counterpoint.

Moist vegetation surrounded by built materials, such as asphalt, results in microscale differential heating and circulations. Advective (horizontal) heat fluxes from drier/warmer built surfaces can lead to an enhanced evaporative (this includes both evaporation and transpiration) or latent heat flux from the moister vegetated surface. The enhanced evaporation rates may then lead to additional microscale spatial heterogeneity as the vegetated area is no longer uniformly wet. Dry surfaces may develop around the edge of the vegetation, although some edges of pervious surfaces may also be wetter due to runoff from adjacent impervious surfaces.

The height profile of elements on the landscape influence land-atmosphere interactions. The apparent chaotic motion of the atmosphere at the microscale becomes more organized at the local scale ($10^2$–$10^4$ m). A residential neighborhood, for example, comprising several similar street canyons with similar combinations of surface materials and morphology, plus intervening buildings, gardens, and so forth, creates a local-scale climate that extends horizontally, forming atmospheric heterogeneity at the subdivision or neighborhood scale. In general, the microclimate patch contrast is linked to the distinctiveness and contrast of surface materials at these scales, as well as the degree of atmospheric mixing as determined by the height of the UCL. Subdivisions recently converted from agricultural fields or forests often have very little mature vegetation and contain a fairly uniform housing stock. In such settings, the houses are the major "roughness elements." Older neighborhoods, in contrast, may consist of rows of houses, with mature gardens predominantly at the rear with trees along the streets. In such cases, the tallest roughness elements are the trees, and the "surface" as viewed from above is predominately trees in rows.

Just as at the microscale, edge effects between neighborhoods and advection are important but difficult to study and resolve. Neighborhoods may or may not have a distinct boundary. Contrasts may be evident between residential and commercial or industrial areas or more diffuse as residential neighborhoods grade into lower and lower density developments. The greater the number of edges, the greater the enhancement of advective transport of energy and mass due to contrasts in surface properties. As one moves away from the edges into the center of the area, the fluxes return to that expected for that area (Figure 13.6). The key in any study of surface-atmosphere exchanges is to consider the characteristics of the roughness elements (e.g., land cover, building structure) rather than land use, although land use will influence the amount and type of anthropogenic heat that is generated both internally and externally. Once again, neighborhoods of similar housing types would be useful spatial partitions for statistically describing roughness fields.

At the mesoscale, the local-scale effects of neighborhoods within a city combine, and the integrated presence of the city influences surface-atmosphere exchanges. The nature of urban effects and the strength of

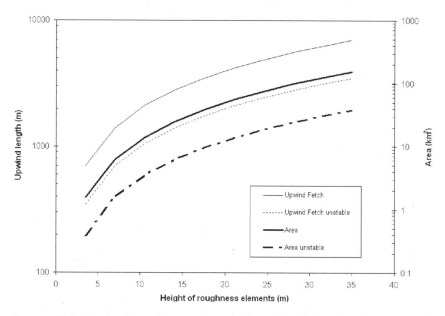

FIGURE 13.6. Approximate distances (upwind length) or dimensions (upwind area) a site must be from the edge of a patch in order to measure a flux characteristic of the patch type (i.e., independent of upwind advected influences). Results are shown for urban areas with different building and vegetation heights (roughness elements) and different conditions of atmospheric stability: all conditions averaged (solid lines) and unstable conditions (dashed lines). The more stable the atmospheric conditions and the taller the roughness elements, the larger the patch needs to be to obtain representative flux measurements of that patch.

advection between the city and its surrounding are functions in part of the larger topographic/regional setting of the city, notably whether the surroundings are water (ocean, lake) or vegetation (e.g., agriculture, forest, desert flora) and the extent of human influence on the surrounding regional/rural landscape. In agricultural or desert regions, for example, there may be more vegetation in the urban area than the rural surroundings and/or irrigation by urban residents may mean that cities are oases of wetter surfaces (e.g., Brazel et al. 2000; Martin and Stabler 2002). In such circumstances, urban areas may have enhanced latent heat fluxes and suppressed sensible heat fluxes compared to their rural surroundings. These patterns, caused by direct human subsidy to the ecosystem in specific patches, can be altered by regulatory instruments influencing individual behavior. Grimmond and Oke (2000) showed that when Vancouver, British Columbia, experienced a drought, an irrigation ban was put into effect, resulting in an increase in the Bowen ratio (ratio of sensible to latent heat flux) for residential neighborhoods compared to other years when extensive garden irrigation was in effect.

# Conclusions

A set of major points have emerged from this discussion of heterogeneity in urban ecosystems:

- The distribution and scale of surface properties are important controls of urban function: Urban heterogeneity exists at all scales, but suburban landscapes may shift variance toward shorter length scales, which can have significant impacts on water and nutrient budgets by creating strong heterogeneity and concentration in surface material and loading rates, as well as significant local gradients in moisture and energy states and flux. This process creates "hotspots" in moisture and nutrient loading.
- Specific patch patterns and connectivity influences ecosystem scale water and nutrient retention: Natural landscapes evolve spatial gradational patterns in which downslope patches can retain resources, such as water and nutrients, draining from upslope patches. The modification of patch flux connectivity by the engineered drainage system in suburban and city landscapes and specific pattern of pervious and impervious surfaces can reduce overall ecosystem retention by bypassing riparian zones and other "absorbing" patches (Groffman et al. 2003). Alternatively, patterns of impervious and pervious patches promoting run-on infiltration may mitigate some of the effects of impervious development. Spatially inexplicit (statistical) descriptions of patch distributions are typically insufficient to capture these effects.
- Urban/suburban patch boundaries can serve to absorb, concentrate, or diffuse flow: Urbanization can produce important *concentration* and drainage effects of water and its constituents from the terrestrial ecosystem or redistribution and subsidy of pervious patches by impervious patches, depending on the nature of urban infrastructure at the micro- and mesoscale of parcels to neighborhoods. These small-scale "streetscape" boundaries can be absorbing or transmitting and have important ecosystem-level effects on moisture conditions and terrestrial/aquatic coupling.
- Direct individual and institutional activity need to be quantified and integrated as part of the urban ecosystem: Human responses to variable environments are important system feedbacks. Conceptual frameworks for representing sources and impacts of ecosystem heterogeneity in urban environments need to develop methods to link the disparate scales and patterns of available social science and ecosystem observations. Urban ecosystems should be treated like any other ecosystem but with greater attention paid to the active role of human individuals and institutions. An approach to characterizing individual and household activity may be drawn from combinations of household survey sampling and parcel databases, potentially extended using commercial consumption information that can be spatially distributed at the census block or neighborhood scale.

These points summarize a spatially explicit view of urban ecosystems that accounts for their physical, biotic, and human components and the connections between them. The manifest infrastructural, social, and biological heterogeneities of metropolitan systems present a compelling opportunity to promote the understanding of the role of heterogeneity in ecosystems.

*Acknowledgements.* This research is supported by the Baltimore Ecosystem Study Project, National Science Foundation Long-Term Ecological Research Program, grant number DEB 9714835, and by the EPA-NSF joint program in water and watersheds, project number GAD R825792. We thank the USDA Forest Service Northeastern Research station for site management, and in kind services to the BES.

## References

Avissar, R. 1991. A statistical-dynamical approach to parameterize subgrid-scale land surface heterogeneity in climate models. Surv. Geophys. 12: 155–178.

Band, L.E., Peterson, D.L., Running, S.W., Coughlan, J.C., Lammers, R., Dungan, J., and Nemani, R.R. 1991. Ecosystem processes at the watershed level: basis for distributed simulation. Ecol. Modeling 56: 171–196.

Band, L.E., Patterson, P., Nemani, R., and Running, S.W. 1993. Forest ecosystem processes at the watershed scale: incorporating hillslope hydrology. Forest Meterol. 63: 93–126.

Band, L.E., Tague, C.L., Brun, S.E., Tenenbaum, D.E., and Fernandes, R.A. 2000. Modeling watersheds as spatial object hierarchies: structure and dynamics. Trans. Geographic Information Systems 4: 181–196.

Berry, B.J.L. 1990. Urbanization. In The earth as transformed by human action: global and regional changes in the biosphere over the past 300 years, ed. B.L. Turner, pp. 130–120. New York: Cambridge University Press.

Beven and Kirkby 1979. A physically based, variable contributing area model of basin hydrology, Hydrol. Sci. Bull, 24(1), 43–69.

Black, P.E. 1991. Watershed hydrology. Prentice Hall, Englewood Cliffs.

Boone, C.G. 2003. Obstacles to infrastructure provision: the struggle to build comprehensive sewer works in Baltimore. Historical Geography 31: 151–168.

Brazel, A., Selover, N., Vose, R., and Heisler, G. 2000. The tale of two cities—Baltimore and Phoenix urban LTER sites. Climate Res. 15: 123–135.

Bresler, E., and Dagan, G. 1988. Variability of yield of an irrigated crop and its causes, 1. Statement of the problem and methodology. Water Resources Res. 24, 381–388.

Brush, G.S. 2003. Case 2: The Chesapeake Bay Estuarine System. In The changing environment, ed. N. Roberts, pp. 397–416. Cambridge, MA: Blackwell Publishers.

Cadenasso, M.L., Pickett, S.T.A., Weathers, K.C., Bell, S.S., Benning, T.L., Carreiro, M.M., and Dawson, T.E. 2003a. An interdisciplinary and synthetic approach to ecological boundaries. BioScience 53: 717–722.

Cadenasso, M.L., Pickett, S.T.A., Weathers, K.C., and Jones, C.G. 2003b. A framework for a theory of ecological boundaries. BioScience 53: 750–758.

Clay G. 1973. Close up: how to read the American city. New York: Praeger.

Effland, W.R., and Pouyat, R.V. 1997. The genesis, classification, and mapping of soils in urban areas. Urban Ecosystems 1: 217–228.

Fahrig, L. 1992. Relative importance of spatial and temporal scales in a patchy enviroment. Theor. Popul. Biol. 41: 300–314.

Fisher, S.G. 1993. Pattern, process, and scale in freshwater systems: some unifying thoughts. In Aquatic ecology: scale, pattern, and process, ed. P.S. Giller, pp. 575–597. Oxford: Blackwell Scientific Publishers.

Forman, R.T.T. 1995. Some general principles of landscape and regional ecology. Landscape Ecol. 10: 133–142.

Garreau, J. 1991. Edge city: life on the new frontier. Doubleday, New York.

Gottdiener, M., and Hutchison, R. 2000. The new urban sociology, 2nd ed. New York: McGraw Hill.

Grimm, N.B., Grove, J.M., Pickett, S.T.A., and Redman, C.L. 2000. Integrated approaches to long-term studies of urban ecological systems. BioScience 50: 571–584.

Grimmond, C.S.B., and Oke, T.R. 2000. Variability of evapotranspiration rates in urban areas. In Biometeorology and Urban Climatology at the Turn of the Millenium, eds. R.J. de Dear, J.D. Kalma, T.R. Oke, and A. Auliciems, pp. 475–480. Geneva: World Meteorological Organization.

Grimmond, C.S.B., and Oke, T.R. 2002: Turbulent heat fluxes in urban areas: Observations and local-scale urban meteorological parameterization scheme (LUMPS). J. Appl. Meteorol. 41, 792–810.

Grimmond, C.S.B., and Oke, T.R. 1999. Rates of evaporation in urban areas: Impacts of urban growth on surface and ground waters. International Association of Hydrological Sciences Publication 259: 235–243.

Groffman, P.S., Pouyat, R.V., and Williams, C.O. 2002. Soil nitrogen cycling in urban forests and grasslands. Paper given at the Spring Meeting of the American Geophysical Union, Washington, DC, May 2002.

Groffman, P.M., Bain, D.J., Band, L.E., Brush, G.S., Grove, J.M., Pouyat, R.V., Yesilonis, I.C., and Zipperer, W.C. 2003. Down by the riverside: urban riparian ecology. Frontiers Ecol. Environ. 1: 315–321.

Grove, J.M. 1997. New tools for exploring theory and methods in human ecosystem and landscape analyses: computer modeling, remote sensing and geographic information systems. In Integrating social sciences and ecosystem management, ed. H. K. Cordell, Champaign, IL: Sagamore.

Grove, J.M., and Burch, Jr. W.R. 1997. A social ecology approach and application of urban ecosystem and landscape analyses: a case study of Baltimore, Maryland. Urban Ecosystems 1: 259–275.

Grove, J.M., and Burch, Jr. W.R. 2002. A social patch approach to urban patch dynamics. Proceedings of the Ninth International Symposium on Society and Resource Management (ISSRM) June 2–5, 2002, Bloomington, Indiana. Bloomington, IN: Indiana University.

Grove, J.M., Troy, A.R., O'Neil-Dunne, J.P.M., Burch, Jr., W.R., Cadenasso, M.L., and Pickett, S.T.A. in press. Characterization of households and its implications for the vegetation of urban ecosystems. Ecosystems, in press.

Harvey, D. 2000. Spaces of hope. Berkeley: University of California Press.

Kolasa, J., and Pickett, S.T.A. (eds.). 1991. Ecological heterogeneity. New York: Springer-Verlag.

Kolasa, J., and Rollo, C.D. 1991. Introduction: the heterogeneity of heterogeneity: a glossary. In Ecological heterogeneity, ed. J. Kolasa, pp. 1–23. New York: Springer-Verlag.

Krugman, P. 1996. The self organizing economy. Blackwell, Cambridge.

Law, N.L., Band, L.E., and Grove, J.M. 2004. Nutrient inputs to urban watersheds from fertilizer usage. J. Environ. Planning Manage. V. 47, p. 737–755.

Levin, S.A. 1989. Challenges in the development of a theory of community and ecosystem structure and function. In Perspectives in ecological theory, ed. J. Roughgarden, pp. 242–255. Princeton, NJ: Princeton University Press.

Machlis, G.E., Force, J.E., and Burch, W.R. 1997. The human ecosystem .1. The human ecosystem as an organizing concept in ecosystem management. Society Natural Resources 10: 347–367.

Martin, C.A., and Stabler, L.B. 2002. Plant gas exchange and water status in urban desert landscapes. J. Arid Environ. 51: 235–254.

Oke, T.R. 1997. Urban environments. In The surface climates of Canada, eds. W.G. Bailey, T.R. Oke, and W.R. Rouse, Montreal: McGill-Queens University Press, 303–327.

Olson, S.H. 1982. Urban metabolism and morphogenesis. Urban Geography 3: 87–109.

Ostrom, E. 1998. A behavioral approach to the rational choice theory of collective action presidental address. Am. Political Sci. Rev. 92: 1–22.

Pickett, S.T.A., Burch, Jr. W., Dalton, S., Foresman, T.W., and Rowntree, R. 1997. A conceptual framework for the study of human ecosystems in urban areas. Urban Ecosystems 1: 185–199.

Pickett, S.T.A., and Cadenasso, M.L. 2002. Ecosystem as a multidimensional concept: meaning, model and metaphor. Ecosystems 5: 1–10.

Pickett, S.T.A., Cadenasso, M.L., and Grove, J.M. 2004. Resilient cities: meaning, models, and metaphor for integrating the ecological, socio-economic, and planning realms. Landscape and Urban Planning 69(4): 369–384.

Pickett, S.T.A., Cadenasso, M.L., and Jones, C.G., 2000. Generation of heterogeneity by organisms: creation, maintenance, and transformation. In Ecological consequences of habitat heterogeneity, ed. M. Hutchings, pp. 33–52. New York: Blackwell.

Pickett, S.T.A., and Rogers, K.H. 1997. Patch dynamics: the transformation of landscape structure and function. In Wildlife and landscape ecology: effects of pattern and scale, ed. J.A. Bissonette, pp. 101–127. New York: Springer-Verlag.

Pickett, S.T.A., and White, P.S. (eds.). 1985. The ecology of natural disturbance and patch dynamics. Orlando, FL: Academic Press.

Pouyat, R.V., Yesilonis, I., and Russell-Anelli, J. submitted. Soil chemical and physical properties in an urban landscape. Soil Science Society America Journal.

Piringer, M., Grimmond, C.S.B., Joffre, S.M., Mestayer, P., Middleton, D.R., Rotach, M.W., Baklanov, A., De Ridder, K., Ferreira, J., Guilloteau, E., Karppinen, A., Martilli, A., Masson, V., and Tombrou, M. 2002. Investigating the surface energy balance in urban areas – recent advances and future needs. Water Air Soil Pollution Focus 2,(5–6), 1–16.

Rastetter, E.B., King, A.W., Cosby, B.J., Hornberger, G.M., O'Neill, R.V., and Hobbie, J.E. 1992a. Aggregating fine-scale ecological knowledge to model coarser-scale attributes of ecosystems. Ecol. Applications 2: 55–70.

Shachak, M., Pickett, S.T.A., Boeken, B., and Zaady, E. 1999. Managing patchiness, ecological flows, productivity, and diversity in drylands. In Arid lands management: toward ecological sustainability, ed. T.W. Hoekstra, pp. 254–263. Urbana: University of Illinois Press.

Shugart, H.H., and Seagle, S.W. 1985. Modeling forest landscapes and the role of disturbance in ecosystems and communities. In The ecology of natural disturbance and patch dynamics, ed. S.T.A. Pickett, pp. 353–368. Orlando, London: Academic Press.

Strayer, D.L., Ewing, H.A., and Bigelow, S. 2003. What kind of spatial and temporal details are required in models of heterogeneous systems? Oikos 102, 654–662.

Tenenbaum, D., Band, L.E., Kenworthy, S., and Tague, C.L. in press. Analysis of soil moisture patterns in forested and suburban catchments using high-resolution photogrammetric and LIDAR digital elevation datasets. Hydrol. Processes (in press).

Tilman, D. 1994. Competition and biodiversity in spatially structured habitats. Ecology 75: 2–16.

Wiens, J.A. 1995. Landscape mosaics and ecological theory. In Mosaic landscapes and ecological processes, ed. L. Hansson, pp. 1–26. New York: Chapman and Hall.

Wiens, J.A. 2000. Ecological heterogeneity: In an ontogeny of concepts and approaches, eds. M.J. Hutchings, E.A. John, and A.J.A. Stewart, pp. 9–31. Malden, MA: Blackwell.

Wu, J., and Loucks, O.L. 1995. From balance of nature to hierarchical patch dynamics: a paradigm shift in ecology. Q. Rev. Biol. 70: 439–466.

Zipperer, W.C., Foresman, T.W., Sisinni, S.M., and Pouyat, R.V. 1997. Urban tree cover: an ecological perspective. Urban Ecosystems 1: 229–247.

# 14
# Origins, Patterns, and Importance of Heterogeneity in Riparian Systems

Robert J. Naiman, J. Scott Bechtold, Deanne C. Drake, Joshua
J. Latterell, Thomas C. O'Keefe, and Estelle V. Balian

## Abstract

Riparian systems epitomize heterogeneity. As transitional semiterrestrial areas influenced by water, they usually extend from the edges of water bodies to the edges of upland terraces. Riparian systems often exhibit strong biophysical gradients, which control energy and elemental fluxes, and are highly variable in time and space. These attributes contribute to substantial biodiversity, elevated biomass and productivity, and an array of habitats and refugia. Focusing on riparian systems of medium-sized floodplain rivers, we describe heterogeneity at multiple space and time scales, illustrate interactions among scales, and propose a conceptual model integrating major system components. We show how climatic and geologic processes shape an array of physical templates, describe how disturbances redistribute materials, and illustrate how soils and subsurface processes form and are sustained. Collectively, these processes strongly influence plant productivity and fluxes of channel-shaping large woody debris (LWD). Ultimately, riparian ecosystem function integrates climate (past and present), geologic materials and processes, soil development and attendant microbial transformations, subsurface characteristics, plant productivity, animal activities, and LWD—and the active, continuous and variable feedbacks between the individual components.

## Introduction

Riparian communities respond continuously in time and space to a complex array of hydrologic (e.g., water regimes, hydraulic shear stress, sediment deposition, erosion, deposition of large woody debris) and biotic (e.g., animal activities, plant production, microbially mediated nutrient cycling) influences. The resulting mosaic of riparian subcommunities are composed of species with contrasting life history strategies that moderate downstream fluxes of water, materials and energy —and fundamentally influence nutrient and organic matter dynamics. Collectively, the biophysical processes support numerous types of aquatic, semiaquatic, and terrestrial food webs, resulting in high biodiversity. Indeed, riparian systems appear to be sites of focused storage and dissipation of materials and energy within the larger, regional landscape. They are likely "hotspots" for regional heterogeneity owing to the inherently dynamic and nonlinear processes linking the flux and retention of water and materials to interactive landscapeforming processes (Benda et al. 1998; Naiman et al. 1998).

Riparian systems provide an unusually rich array of lessons on the origins, patterns, and ecological importance of biophysical heterogeneity. Heterogeneity is manifested in a diverse array of landscape elements and processes operating on several spatial and temporal scales. These include longitudinal, lateral, and vertical gradients in geomorphic features (e.g., gravel bars, terraces, islands), surface and subsurface flows of water and nutrients, and disturbance regimes (e.g., floods, drought, fire, wind). Fluvial actions (e.g., erosion, transport, deposition) are the dominant agents of riparian change and constitute one suite of the natural disturbance processes primarily responsible for sustaining the high level of heterogeneity (Poff et al. 1997; Ward et al. 2002). It is the hydrologic connectivity—the flux of matter, energy, and biota via water—in combination with animal activities (Naiman and Rogers 1997), microbial processes, and vegetation dynamics that largely sustain riparian heterogeneity. Although individual features such as a specific vegetative patch type may exhibit dynamic transitions fueled by interactions between fluvial dynamics and plant succession, their relative abundance within a catchment tends to remain in quasiequilibrium over decades to centuries. In general, riparian systems are highly heterogeneous, as well as central nodes for biodiversity, for energy and elemental fluxes, and for elevated biomass and productivity. Throughout a catchment, riparian systems exhibit strong biophysical gradients, high variability in time and space, and provide a diversity of habitats and refugia (Naiman and Décamps 1997; Naiman et al. 2005).

This chapter describes riparian heterogeneity at multiple scales of space and time, illustrates interactions among the scales, and offers a conceptual model integrating the major ecosystem components. We accomplish our objectives by showing how climatic and geologic processes shape physical templates, and by illustrating how soils and subsurface processes form and are sustained on the major physical templates. We then discuss how the latter processes influence biodiversity and plant productivity and, ultimately, the generation of channel-shaping large woody debris (LWD) and the disturbances driving fluxes of LWD from the forest to streams. Integration at the ecosystem scale is accomplished through a conceptual model relating climate, geology, soils, subsurface characteristics, plant productivity, and LWD with the active, continuous, and variable feedbacks between the individual components. Examples herein draw heavily on lessons we have learned from mid-sized alluvial rivers draining the rainforests of the North American

Pacific Coastal Ecoregion. Riparian heterogeneity in other regions may differ from our examples, at least in form and function. Likewise, examples presented here may not accurately represent the nature of riparian areas transformed by human activities. Nevertheless, we believe that fundamental principles governing the vitality and heterogeneity of riparian systems are similar.

## Setting the Stage: Geologic and Climatic Processes

Riparian heterogeneity is a product of history. Geology, topography, climate, vegetation, and animals interact to create and maintain physical and biological heterogeneity over the full spectrum of time and space. Geological and biological legacies of long-term climate cycles persist at many scales and have lagged effects. For example, lithotopography (geological parent material and landforms) is a legacy of the distant past, while disturbances such as floods and fires produce effects that have immediate impacts but also may be expressed for centuries. The legacies of geology, climate, and biogeography shape riparian heterogeneity—including soil processes, subsurface flows, forest biodiversity and productivity, and the dynamics of LWD.

### Geologic History

Coarse-scale heterogeneity is produced in riparian systems by parent material and landforms. Lithotopography places basic constraints on riparian assemblages and their heterogeneity: elevation, exposure, slope, groundwater dynamics, and parent material fundamentally shape system processes. The geology of tectonically active regions is especially complex, resulting in patchy distributions of parent material that strongly influence development of soils and biota (Figure 14.1A)—and the resulting topography interacts strongly with climate to influence weather patterns (Figure 14.1B). Ultimately, geologically driven heterogeneity is expressed in the character of stream corridors and their biota, as disturbance regimes change with channel geomorphology and climate (Vannote et al. 1980; Benda et al. 1998). These patterns are fully reflected in the biophysical heterogeneity of riparian zones (Figure 14.2) and are easily illustrated in large river basins such as the Amazon where the geology of mountainous headwaters exerts strong controls on downstream environments (e.g., McClain and Naiman 2005).

Although many geological processes operate over long time periods and large scales (e.g., formation of floodplains), important examples of decadal-scale geomorphic processes include sediment accumulations and avulsions, landslides, vertical channel adjustments, and debris flows (Montgomery and Buffington 1998). Medium to large river corridors are generally long-lived geologic features that exist in a quasiequilibrium where the lateral channel movements within floodplains maintain successional and geofluvial

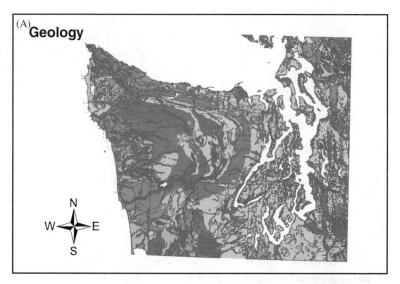

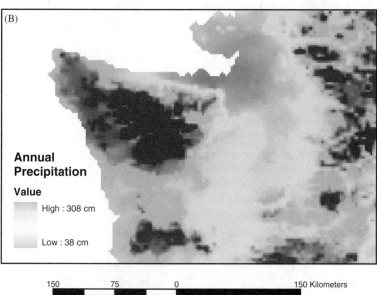

FIGURE 14.1. (A) The underlying geology of Washington's Olympic Peninsula illustrates the complexity of parent materials over macro and meso spatial scales. More than 600 geological formations are present (only a few major formations are represented here; color image available at (http://www.fish.washington.edu/people/naiman/cv/reprints/naiman_2004_cc/02_geology_precip_map.pdf). Adapted from Washington State Geospatial Archive (http://wagda.lib.washington.edu). (B) Total annual precipitation on the Olympic Peninsula is strongly influenced by topography, varying from only 38 cm/yr in the rainshadow of the northeastern peninsula to 308 cm/yr on the Pacific Coast (© 2000–2003 The Climate Source, http://www.climatesource.com).

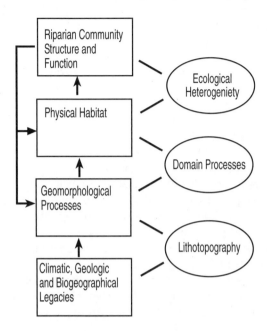

FIGURE 14.2. Schematic representation of relationships between drivers of ecological heterogeneity in riparian forests (from Naiman et al. 2000 Copyright, American Institute of Biological Sciences).

heterogeneity. However, at the catchment scale, riparian community structure and dynamics reflect the nature of catchment-scale variation in discrete disturbance regimes, or *process domains* (Montgomery and Buffington 1998).

## Climate

It is well-known that large-scale climate fluctuations during the Quaternary drove the development of modern geomorphic features and plant assemblages (Davis 1981). In North America, at least 15 glaciations carved high-latitude landscapes, lowered sea levels, and dramatically altered the distributions of plants and animals. Conspicuous glacial features of modern riparian systems include terminal and lateral moraines, glacially carved river valleys filled with outwash, lake systems, and distinctive fluvial topographies caused by catastrophic glacial floods (e.g., the Red River Valley of Minnesota and the Channeled Scablands of eastern Washington). Lower latitude landscapes were also shaped by the cooler and wetter climate of Pleistocene glacial periods (e.g., the basin-and-range province of southwestern North America).

However, climate cycles also operate at relatively high frequencies. Examples include the El Niño Southern Oscillation (ENSO) with a 5–7 year period (Philander 1990), the Pacific Decadal Oscillation (PDO) which operates on an ~60-year period (Mantua et al. 1997), and the North Atlantic

Oscillation (NAO) at shorter than decadal timescales (Ottersen et al. 2001). These climate cycles influence precipitation patterns, temperature, stream flow, and storm frequency over relatively short scales, which in turn influence the frequency and scale of mass wasting, windthrow, and flooding—which are important drivers of heterogeneity in riparian forests.

Hydrologic regimes vary considerably even with relatively subtle climatic changes; for example, in the Pacific Northwest, average winter El Niño surface temperature are 0.4 °C to 0.7 °C higher than La Niña temperatures. Warmer winter temperatures result in lower snowpack and earlier hydrographic peak flows in El Niño years (University of Washington, Joint Institute for Study of the Atmosphere and Ocean, Climate Data; http://tao.atmos.washington.edu/PNWimpacts). Collectively, the physical drivers, along with temperature-regulated biological processes (e.g., decay rates and plant growth) shape the ecological properties of riparian systems, clearly demonstrating the strong influences of climate and climate change on heterogeneity.

Climate gradients and fire regimes are additional sources of riparian heterogeneity. Climate-driven heterogeneity resulting from sharp precipitation gradients are seen where dominant weather patterns intercept elevated landforms. For example, precipitation on Washington's Olympic Peninsula (Figure 14.1B) varies from 308 cm/yr of largely orographic precipitation on the west side of the Olympic Mountains to only ~38 cm/yr in the rainshadow of the Olympics—a linear distance of 130 km. Fire regimes are correlated with climate-related heterogeneity in riparian systems (Agee 1993). But fire, which can dictate nutrient cycling and forest succession dynamics, also affects erosion and sedimentation processes (Wondzell and King 2003). Wildfires burn even in very wet coastal temperate rainforests, albeit not at a high frequency. In British Columbia's Clayoquot River catchment (~554 cm/yr of precipitation) nearly all exposed, south-facing slopes within 1 km of the river burned in the past 800 years with only ~20% of all sites remaining unburned over the 6000-year history (Gavin et al. 2003). At small scales, fire can result in decreased habitat heterogeneity (e.g., as seen in substrate embeddedness and near-bed velocities; Minshall et al. 1989, 1997), whereas at larger scales, fire is an important natural source of heterogeneity playing a crucial role in creating and maintaining aquatic diversity (Bisson et al. 2003).

## Biotic Responses to Geologic and Climatic History

Perceptions of heterogeneity are frequently based on vegetation, which is easily observed, sessile, and has an air of permanence. In reality, plant distributions are plastic and respond to geomorphic substrate (e.g., Gregory et al. 1991), changes in climate (Davis 1986), pressure from herbivores (e.g., Naiman and Rogers 1997), distributions of pollinators (e.g., Cox and Elmqvist 2000), and competition from other plants (e.g., DiTomasco 1998). Early successional riparian plants, in particular, are generally adapted to flooding, which facilitates reestablishment after disturbance (Naiman et al.

2005); succession after flooding illustrates a familiar manifestation of vegetative heterogeneity. As the riparian assemblage develops, it affects soil formation and fertility and helps control erosion. This is well illustrated in a recently deglaciated river valley in Glacier Bay, Alaska, where it took only ~100 years for woody vegetation to stabilize stream banks and provide points for LWD accumulation. The increased LWD retention and bank stabilization led to pool formation and improvement of fish habitat within 150 years (Sidle and Milner 1989).

Life history traits such as long-term seed dormancy, N-fixation in some early successional species, physiological adaptations to inundation, and water-borne dispersal of propagules allow riparian plants to thrive in heterogeneous and frequently disturbed environments (Naiman et al. 2005). Many riparian plants are specifically adapted to cope with flooding, sediment deposition, physical abrasion, and stem breakage (Blom et al. 1990; Mitsch and Gosselink 1993; Naiman et al. 1998). By the nature of their heterogeneity, riparian systems may support a greater diversity of organisms and life-history strategies than surrounding upland forests.

In summary, patterns of geology and climate interact with biological components of riparian systems to produce characteristic patterns and feedbacks in soils, hyporheic zones, plant communities, and woody debris processes. We now turn to a discussion of riparian system patterns and processes and their relationships to heterogeneity, drawing primarily on themes and examples from our research on forested small- to mid-sized rivers while fully recognizing that every river has a unique combination of processes, organisms, and conditions.

## Heterogeneity in Floodplain Soils

Salient characteristics of riparian floodplains include spatial and temporal heterogeneity resulting from fluvial redistribution of sediments, organic matter, and other materials, as well as temporal heterogeneity in the form of cumulative soil alterations by vegetation. As the primary reservoir of nutrients and carbon, and as a growth medium for plants, soil heterogeneity is reflected in patterns of production, community composition and terrestrial-aquatic transfers of carbon and nutrients.

### Complexity in Sediment Distribution

Differences in the mobility of eroded materials lead to depositional patterns observable at multiple scales. Within an overall longitudinal pattern of decreasing particle size with distance downstream, floodplain soils form where hydrologic energy is dispersed in lateral directions, and landforms reflect the energies of stream flows at the time of deposition. Coarse sediments form levees, abandoned and secondary channels, and a variety of bar forms in areas of intense stream flows (Leopold et al. 1964), especially when coincident with roughness elements, such as riparian vegetation or LWD. Elevation of floodplain surfaces occurs in areas subjected to less intense flows. At macro- and mesoscales, depositional patterns are controlled by channel morphology, while at the microscale resistance to flow created by emergent vegetation and LWD can be especially effective at trapping fine sediments (Walling et al. 1996). This sorting also affects the distribution of organic matter and various chemical compounds that become associated with mineral sediments. Both iron and phosphorus are transported in microaggregates of organic matter, silt and clay particles, and tend to concentrate in overbank deposits in alluvial soils (Walling et al. 2000; Rhoton et al. 2002).

The size distributions of sediments in fluvial landforms influence soil nutrient and moisture dynamics in two important ways. First, the complex surfaces of fine sediments provide large amounts of surface area for adsorption of organic and inorganic materials, including organic matter and bioavailable nutrients. Adsorption to mineral surfaces and incorporation within stable aggregates reduces leaching losses (Sollins et al. 1996) and inhibits decomposition (Christensen 1996), contributing to short and long-term organic matter (OM) retention (Raich and Schlesinger 1992; Trumbore 1993). Strong correlations between OM and soil clay and/or silt concentration (Burke et al. 1989; Schimel et al. 1994; Epstein et al. 1997; Hook and Burke 2000) have been measured in a variety of well drained soils. The contribution of this to floodplain heterogeneity is well illustrated by the distribution of silt, clay, and OM in Washington's Queets River soils (Figure 14.3). Soil OM is strongly related to silt and clay concentration in both young and old soils despite large changes in plant production and community composition, suggesting that fluvial deposition of fine particles plays a primary role in the rapid development of OM-rich, productive soils that are characteristic of floodplains.

Soil particle size also affects nutrient and moisture dynamics by determining the size of pore spaces between sediment particles. This influences the movement of liquids and gases—most importantly water and oxygen—through the soil. Coarse soils tend to be droughty and are especially prone to leaching of nitrate (Vitousek et al. 1982) and dissolved organic carbon (Nelson et al. 1993). Anoxic conditions in poorly drained, fine-textured soils reduce decomposition of organic matter, facilitate anaerobic microbial processes, and can lead to accumulation of toxic chemicals (Mitsch and Gosselink 1993). On the Garonne River (France) floodplain, denitrification increases linearly with soil silt and clay content in fine-textured (>65% silt and clay) soils but is not measurable in medium and coarse soils (Pinay et al. 2000). Denitrification may occur more broadly in coarse-textured soils in "microsites," especially where buried wood supports high rates of microbial activity (Jacinthe et al. 1998). A better understanding of the distribution of buried wood in floodplains may be useful in predicting where denitrification is likely to occur.

Spatial variation in soil texture has important implications for growth rates and species compositions of floodplain forests. Fine-textured soils tend to

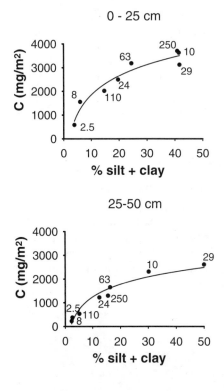

Figure 14.3. Correlation of soil C with silt and clay concentration in Queets River floodplain soils (C = 1163.7 * ln(% silt + clay) − 831; $r^2$ = 0.94, p < 0.01). Silt and clay range from 4% to 42% of total soil volume and are well correlated with soil OM throughout the 250-year chronosequence despite large changes in plant production and community composition, suggesting that particle size rather than plant production is the primary factor controlling soil OM retention. Numbers indicate soil patch ages and are not correlated with C or % silt and clay (S. Bechtold and R.J. Naiman, unpublished data).

have higher rates of net N mineralization and primary productivity, which may be reflected in faster within-species growth rates (Reich et al. 1997; Prescott et al. 2000) or species replacements to faster growing species (Pastor et al. 1984). Although the complexity of contributing factors makes it difficult to distinguish causal relations, correlations suggest that primary production is fundamentally influenced by interactions with sedimentary templates (Reich et al. 1997; Prescott et al. 2000); buffering of nutrient and water exchange conferred by fine sediments and associated OM may lead to more efficient cycling through soil-plant pathways and reduced leaching losses.

Landform heterogeneity subsumes many potential influences on soil biogeochemistry besides sediment size. Because floodplain topography is shaped by interactions between water flows, sediments, vegetation, and LWD across a low-gradient environment, fine-scale topographic variation often results in large contrasts in soil properties and access to water. Interaction with ground/hyporheic water is frequently an important influence on soil dynamics. Water table depth in combination with soil/subsoil texture can influence soil moisture and organic matter regimes. Saturated soils in swales and other wetland areas not subjected to scour can become sinks for river-borne OM (Johnston et al. 2001) and phosphorus (Stoeckel and Miller-Goodman 2001) and are often hotspots for denitrification (Groffman and Tiedje 1989; Farrell et al. 1996). In arid environments, evaporation from the surface of bare sediments in areas of shallow ground water may initiate a capillary pump leading to the formation of a salt crust at the sediment surface, inhibiting plant colonization of sediments (Van Cleve et al. 1993). Even where soil moisture is not influenced by ground/hyporheic water, enhanced plant production resulting from root exploitation of subsurface water can enhance soil organic matter inputs (Décamps 1996). The thickness of a soil cap overlying coarse bed sediments can also influence plant production through modification of moisture and nutrient regimes (Binkley et al. 1995).

## Stages of Floodplain Evolution

Temporal heterogeneity is illustrated by manifold changes in physical, chemical, and biological properties accompanying riparian system development (Figure 14.4). Evolution of the floodplain landscape begins with the formation of depositional landforms, such as bars and terraces. Terraces form as floodplain elevation increases through aggradation or channel incision, isolating sediments on terraces from surface flows. Terrace formation can occur

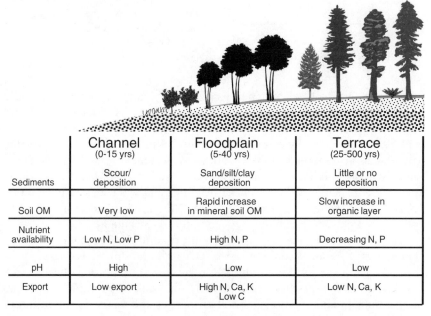

| | Channel (0-15 yrs) | Floodplain (5-40 yrs) | Terrace (25-500 yrs) |
|---|---|---|---|
| Sediments | Scour/ deposition | Sand/silt/clay deposition | Little or no deposition |
| Soil OM | Very low | Rapid increase in mineral soil OM | Slow increase in organic layer |
| Nutrient availability | Low N, Low P | High N, P | Decreasing N, P |
| pH | High | Low | Low |
| Export | Low export | High N, Ca, K Low C | Low N, Ca, K |

FIGURE 14.4. Idealized floodplain evolution of mid-sized rivers in the Pacific Northwest. Deposition of progressively finer sand, silt, and clay over coarse bed sediments leads to rapid formation of floodplain soils. P-rich silt and clay deposits, N-fixation by red alder, and rapid mineral soil OM accumulation contribute to soil fertility. However, high leaching losses of N and base cations and reduced availability of P moderates nutrient dynamics in older conifer soils on terraces. Organic layer OM accumulates in older conifer patches.

within a few decades in small- to medium-sized rivers with a large sediment supply (Schumm 1977; Richards et al. 1993) and may be especially rapid in forested catchments where LWD contributes to localized aggradation/ degradation of the river channel (Montgomery et al. 1996). Isolation of terraces from flooding forms a crucial juncture in the evolution of soils. Although soils continue to undergo gradual pedological alteration relative to the rapid changes that result from fluvial deposition, the distribution of sediments in terraces can be considered a fixed physical template relative to its subsequent influence on soil dynamics. The legacy of channel movement across its floodplain is manifested in a complex mosaic of sedimentary landforms of varying ages. This mosaic includes both highly dynamic components continuously altered by interaction with surface waters and components that have become isolated from surface waters but bear the legacies of past disturbances.

Following deposition, soils undergo alteration via interactions with riparian vegetation. On-site production of soil OM inputs increases rapidly in seral forests, usually reaching a maximum when the canopy achieves full leaf area. N-fixing plants are common colonizers of early successional ecosystems (Walker 1993). In the Pacific Northwest, red alder (*Alnus rubra*) has a profound effect on soils during the first 60–80 years of floodplain development (Bormann et al. 1994). In addition to providing a rich source of N to terrestrial vegetation, large amounts of nitrate are often leached to streams (Bechtold et al. 2003). Soil acidification results from formation of organic acids during decomposition, and is especially strong in conifer forests or where there is a vigorous N-fixer (Johnson 1992). In addition to weathering of primary minerals, this results in displacement of cations from soil exchange sites (Foster et al. 1989; Homann et al. 1994). Although large amounts of P are frequently deposited on floodplains, little is known of its availability for uptake. N-fixation is frequently P-limited (Crews 1993; Vitousek 1999). Although high P availability in new sediments may initially enable high rates of N-fixation, subsequent complexation of P with hydrous Fe and Al oxides as pH decreases could have important influences on both P and N availability.

The end point for floodplain soils is disintegration by erosion. Relative to most upland soils, riparian floodplain soils have limited—and variable—life spans. Estimated 200–2000 year turnover rates of Washington's western Olympic Peninsula river floodplains (O'Connor et al. 2003) maintain soils in early states of pedogenic development. Channel migration is an important control on soil development even in very large rivers. For example, >26% of the lowland Amazon forest is maintained in early successional stages by river migration (Salo et al. 1986).

## Interactions Between Patches

Exchange of nutrients across patch boundaries is often an important influence on the biogeochemistry of adjacent areas. Although some exchange may occur between adjoining soil patches, by far the most significant fluxes are from soils to subsurface (hyporheic/ground) waters and between soils and surface waters. Overall, leaching of lithogenic nutrients tends to reduce the productive capacity of soils, and leaching of atmospherically fixed C and N leads to increases in aquatic productivity. As noted above, base cation leaching is driven by soil acidification and is highly responsive to vegetation type. The mobility of many chemical species is influenced by redox reactions. Fluxes of Fe, Mg, S, P, and other elements tend to occur where fine soils, high OM, and/or shallow water tables facilitate the formation of reducing conditions.

Relative to subsurface water, soils are usually rich in C and N. There are large differences in the mobility of different chemical species in soils affecting their transfer to aquatic systems. The physical, chemical, and biological factors controlling these exchanges often vary dramatically between adjacent soil patches. Nitrate will freely move through soils, as well as subsurface and surface waters, where it exists in excess of biotic demand. For example, in Oregon forests, nitrate export is more strongly related to the proportion of entire watershed with red alder cover than to the amount of riparian alder (Compton et al. 2003). Carbon is of particular interest as its availability usually limits groundwater/hyporheic microbial activity and is, by comparison, very abundant in soils (Findlay and Sinsabaugh 2003). Soils tend to be efficient at retaining dissolved organic matter. Sorption dynamics and hydrology play important roles in controlling dissolved organic carbon (DOC) export (Neff and Asner 2001), with coarse soils high in organic content more likely to leach DOC into subsurface hyporheic waters. Direct lateral transfer of DOC from soils immediately bordering streams frequently occurs during elevated stream flows (Boyer et al. 2000).

Many biogeochemical transformations occur primarily along well defined boundaries. The vertical boundary between soils, subsoil sediments, and ground/hyporheic water is of particular interest. Convergence of deeper nitrate-bearing flow paths with a C source, which frequently occurs as groundwater upwells into soils near stream edges, may lead to greatly increased denitrification (Hedin et al. 1998). Where the water table is deeper and does not directly interact with soils, soil leaching leaves an imprint but does not determine hyporheic/ground water productivity.

## Hidden Heterogeneity: Hyporheic Processes

The hyporheic zone encompasses saturated sediment below and adjacent to the river channel and represents an interface where surface water mixes with groundwater (Stanford and Ward 1993; Edwards 1998). The hyporheic zone is thus a three-dimensional subsurface component of the riparian landscape, connected to the stream channel (Figure 14.5) and is important because of its large interstitial volume and surface area. Inputs of carbon

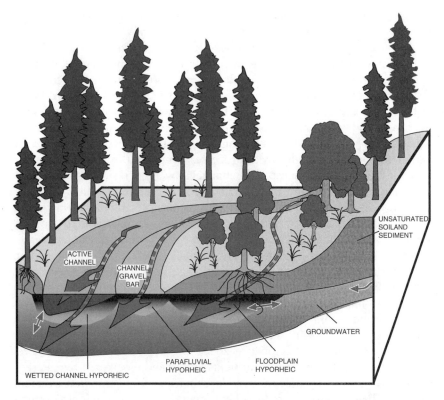

FIGURE 14.5. The hyporheic zone forms the subsurface interface between the river channel and the riparian forest and, in many cases, it modulates the fluxes of water, nutrients, and energy between the river and the riparian soils and vegetation (from Naiman et al. 2000. Copyright, American Institute of Biological Sciences).

and nutrients from upriver, groundwater, and from the overlying riparian soils, along with variable concentrations of oxygen and microbial diversity in the saturated sediments, drive the important microbial and physical processes responsible for nutrient transformations.

This subsurface component of the riparian landscape displays considerable spatial and temporal heterogeneity at catchment, individual reach, and sediment interface levels. Heterogeneity in the hyporheic zone may be manifested in extent and biophysical characteristics, as dictated by agents like parent geology, variation in stream flow, groundwater recharge, and channel changes. Interactions between floods, sediment, LWD, and historical legacies serve as controllers to create heterogeneous patterns of subsurface flow.

To understand what controls spatial heterogeneity in the physical and functional attributes of hyporheic zones, the river's geomorphology must be examined through an historic lens where parent geology and past glaciations play important roles. The amount and rate of exchange between the surface water and groundwater is dependent on local geology, as this establishes the degree of hydraulic conductivity and the water residence time (Valett et al. 1996). In alluvial rivers such as Montana's Flathead River this translates to $\sim 3 \times 10^8$ m$^3$ of hyporheic habitat compared to $\sim 1 \times 10^5$ m$^3$ of channel habitat, with the hyporheic area demonstrating significant heterogeneity relative to the river channel as measured by biophysical characteristics (Stanford and Ward 1988).

The shape and extent of the hyporheic zone exhibits strong heterogeneity over space and time in response to channel movement and alluvial deposition, which control the formation, evolution, and blockage of hyporheic flowpaths. Sediment sorting through fluvial processes stimulated by LWD and other roughness elements and subsequent channel movement results in paleo-channels (i.e., abandoned channels) that become preferential flowpaths for hyporheic flow (Stanford and Ward 1988). As new preferential flowpaths are formed, previous areas of hyporheic activity become increasingly isolated from the active channel. These processes have important functional significance for stream biota—when the proportion of surface water decreases in the hyporheic zone, there is a concurrent change in the interstitial invertebrate community (Marmonier and Chatelliers 1991). Seasonal variation in surface flow is also important in driving the temporal and spatial heterogeneity in the shape and extent of the hyporheic zone. With increasing discharge in winter, the primary flow of water shifts from focused flowpaths to sheet flows throughout the terrace (Clinton et al. 2002). This triggers spatial and temporal heterogeneity in the delivery of nutrients and C, and the oxygen regime (i.e., redox environment) by affecting residence time and interaction with sediment surfaces. These changes in redox play a major role in determining nutrient transformations and subsequently affect microbial production and associated nutrient processing.

Spatial and temporal variability in biogeochemical and physical processes within the hyporheic zone creates heterogeneity in nutrient concentrations, which has important implications for system productivity. Nitrate leaching from overlying soils, delivery to the hyporheic zone, and subsequent emergence of nitrate-rich water from focused subsurface flow paths results in patchy "hotspots" of aquatic primary production (see Fevold 1998; Dent and Grimm 1999). Heterogeneity in nutrient sources, microbial assemblages, and inherent physical characteristics may cause the hyporheic zone to act as either a source or sink for nutrients. The process of denitrification is particularly important because it results in a loss of available N and illustrates the interaction among nutrients, microbes, and physical characteristics of the hyporheic zone. Specifically, hot spots of denitrification arise where overlying soils or an influx of organic-rich stream water provide ample carbon and nitrogen while microbial respiration depletes oxygen resulting in anoxic conditions (Figure 14.6). Rates of biogeochemical transformations are greatest in the boundary layers, which are found at the interfaces between upland areas, riparian zones, and the active channel. These form steep gradients between nutrients (particularly N and P), C, and oxygen (Hedin et al. 1998),

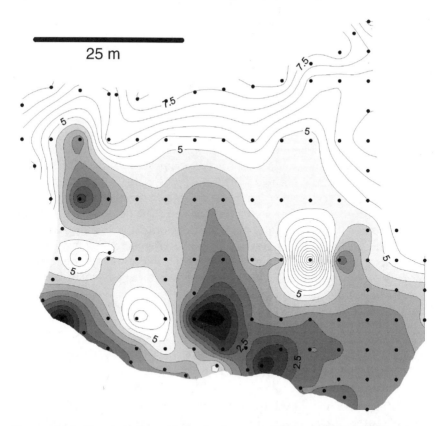

FIGURE 14.6. Concentrations of dissolved oxygen (DO) are highly heterogeneous within hyporheic zones. This figure illustrates DO (mg $O_2$/L) variability in two dimensions with darker shading illustrating areas of low oxygen, although DO also varies with depth (a third dimension). Biochemical processes such as denitrification exhibit high spatial variability relative to gradients in DO.

and these interfaces thus serve as functionally important locations for nutrient processing. Strong chemical gradients generally occur at the heads, intersections, and tails of flowpaths but may be heterogeneously distributed across the riparian landscape. Flowpath heads occur where surface water enters the hyporheic zone while the tails of flow paths are found where hyporheic water emerges in the main channel and throughout side channels. Heterogeneity in the rate of biogeochemical transformations that occur along these chemical or flow gradients is controlled by variation in the particle size of the substrate, which plays an important role in determining hydraulic conductivity. Gradients tend to be strong in areas with fine substrate and low hydraulic conductivity as these sites are characterized by slower moving water, which preserves anoxic conditions. In contrast, water chemistry in hyporheic zones is most similar to surface water in reaches that are relatively unstable (i.e., high levels of bed movement). These areas are dominated by larger substrate with greater hydraulic conductivity (Fowler and Death 2001).

Variability in C flux, driven by variation in seasons, discharge, and overlying vegetation (Clinton et al. 2002), creates strong heterogeneity in the hyporheic zone and has important consequences for microbial communities. Dissolved organic carbon (DOC) concentrations are controlled by the levels of surface water input, leaching from the overlying forest soil patches, metabolic uptake, and adsorption to sediments. Leaching of C from soils is seasonal and highly dependent on soil characteristics. The discharge regime helps determine both the direction of dominant flow paths and residence time of water in the hyporheic zone. Attributes of the overlying riparian forest are largely responsible for the quantity and quality of C and associated nutrients that leach through overlying soils. The patchwork of vegetation and soils is thus a controller of microbial production. Spatial variability in hydraulic conductivity creates additional heterogeneity in flow rates that lead to diverse redox conditions. Buried wood is likely an additional source of C in these environments (Nanson et al. 1995), although its relative importance remains difficult to quantify.

## Heterogeneity in the Diversity and Productivity of Riparian Vegetation

Physical heterogeneity in riparian systems sustains high levels of biodiversity and creates spatial heterogeneity in productivity of riparian vegetation. Various landforms, from micro- to macroscales, provide physical templates for vegetation with differing species-specific life history characteristics (Walker et al. 1986; Pollock et al. 1998). It is the interrelation of disturbance regimes, soil characteristics, biological processes, and life-histories variation that is responsible for the high degree of heterogeneity observed in riparian vegetation at the reach and catchment scale (Figure 14.7).

Heterogeneity in microclimate, particularly temperature and humidity (Chen et al. 1999), soil characteristics (see earlier), and microtopography in riparian landscapes has important habitat implications for a wide range of taxa and drives variability in rates of microbially mediated processes. For example, in riparian wetlands, sites with intermediate flood frequencies and high spatial variation in flood frequency are species-rich in plant life, whereas sites that are frequently, rarely, or permanently flooded are species-poor (Pollock et al. 1998). These data suggest that small-scale spatial variation in physical processes, which is characteristic of riparian zones, can dramatically alter the ecological consequences of disturbances. In southeast Alaska, 78% of the variability in plant species richness (vascular plants and mosses) in riparian forests can be explained by the interaction between flood level and microtopography, with high species richness occurring at sites subject to

# 1960

# 1993

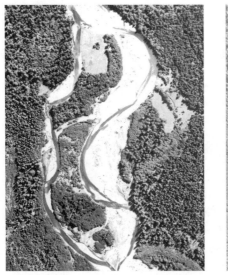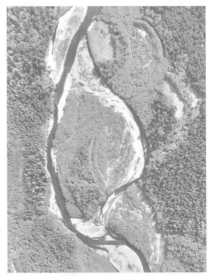

FIGURE 14.7. Dramatic shifts in the Queets River over the past 30 years illustrate the ability of the river to drive spatial and temporal heterogeneity in its riparian system.

intermediate flood duration and having the most diverse microtopography. This variability in local environmental conditions also influences development of different vegetative communities on same-age surfaces in floodplains that comprise patchworks of species composition and productivity rates (Balian 2001; Bartz and Naiman 2005).

In addition to forming patchworks of landforms, the natural flow regime also creates templates with a variable range of biophysical characteristics that are important factors for vegetative production and composition—which in turn favor high levels of biodiversity. Heterogeneity in flood frequency and magnitude creates gradients of disturbance and soil moisture that favor variety in the spatial arrangement of riparian species. In periodically flooded riparian systems, for example, the physiological responses of some plants allow them to maintain sufficient aeration when roots are flooded while other species survive by adjusting their timing of reproduction (e.g., Blom et al. 1990). These hydrologic controllers create a concentrated and highly diverse assemblage of plant species, including trees, shrubs, forbs, grasses, and epiphytes. In some cases up to 90% of the plant species within a catchment are represented in riparian areas (Naiman et al. 2005). Riparian plant species typically exhibit a spectrum of life-history strategies that are either tolerant of flooding and associated sedimentation, only found on surfaces no longer receiving overbank flows, or those that are adapted to some intermediate condition (Nanson and Beach 1977). The flow regime thus plays an important role in driving the structure of riparian vegetation patches, illustrated by declines in diversity where flow regimes are transformed by dams (Nilsson et al. 2002).

Limited data exist on tree production in riparian areas. Forest productivity in upland forests and the characteristics of individual species have been studied extensively. However, understanding riparian forest productivity and assemblages of riparian species is important in quantifying organic matter production and the rate at which riparian trees attain sufficient sizes to initiate the formation of stable LWD jams when they fall into rivers. The generation of LWD triggers feedbacks that create high levels of landscape heterogeneity (described below). Our data from Washington's Queets River indicate that productivity on floodplain terraces is within the range found in upland forests (Balian 2001), but considerable spatial and temporal variability, often more than an order of magnitude, exists within patches of approximately the same age at adjacent sites.

Successional processes associated with disturbance events and subsequent recolonization are responsible for much of the heterogeneity observed in riparian tree production, with annual rates closely related to dominant tree species and their associated understory communities. On the Queets River, the first stage is often a fast growing community of willow (*Salix* spp.) that reaches maximum productivity (4.3 Mg/ha/year) at approximately 10 years of age (Balian 2001). Willow is soon replaced by red alder (*Alnus rubra*) that reaches maximum productivity (7.9 Mg/ha/year) ~40 years after stand initiation. Finally, Sitka spruce (*Picea sitchensis*) out-competes red alder with a continuous increase in production (~13.9 Mg/ha/year) between 40 and 150 years. The first 40–60 years in the development of the riparian forest are thus the most dynamic. This is significant because fluvial disturbance resets stand development at scattered locations throughout the floodplain. These heterogeneous patterns of disturbance create high spatial variability in production relative to upland stands though the range of observed values may be similar. As a result of the episodic nature of disturbance, riparian vegetation patches can be found at different states of succession at the reach scale (Figure 14.8). Even with patches at similar states of succession, productivity can vary by an order of magnitude due to local soil characteristics, driven by deposition and the local accumulation of organic material. The lateral channel movement eventually captures riparian trees, where they continue to play important roles in shaping the dynamics of the system.

## Large Woody Debris and Riparian Heterogeneity

The ruins of riparian forests form the prominent LWD jams that shape the next generation of riparian forest. Depending on the scale of examination, LWD jams may simultaneously create, exhibit, control, or respond to ecological heterogeneity. Within forested floodplain rivers, at mesoscales of space (e.g., stream reaches) and time (e.g., decades to centuries), stable LWD jams

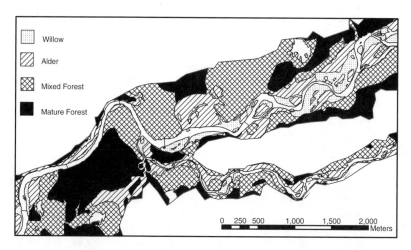

FIGURE 14.8. The migrating river channel is responsible for the establishment of a heterogeneous assemblage of vegetative patches within the channel migration zone. The patches shown represent young willow, mixed mid-age alder and young spruce, and old-growth spruce-dominated forests.

play key ecological roles in creating heterogeneity. At these scales, differentiation in form and dispersion of LWD jams strongly influences the availability of resources (e.g., habitat quantity and quality) for aquatic organisms and riparian vegetation, and subsequently riparian system function.

In gently sloping floodplain rivers, most LWD originates from lateral channel movement through forested terraces, and stream power is often sufficient to redistribute the pieces into jams during high flows (Murphy and Koski 1989). LWD jams in floodplain rivers typically form along channel margins or mid-channel and may be composed of LWD from the adjacent riparian forest, upstream forests, or a combination of the two (Abbe and Montgomery 1996, 2003). Jams vary widely in size, stability, arrangement, and functional significance. The dynamic behavior of forested floodplain landscapes is stimulated, in part, by a web of small-scale feedbacks set in motion by the localized scour and deposition of sediments that result from alteration of flow hydraulics by stable LWD jams. By simultaneously creating localized areas of stability and instability, LWD jams are the infrastructure for a complex web of positive and negative feedbacks that mold the riparian landscape into a patchwork of LWD-rich landforms, forest patches, and channel features (Figure 14.9). This biologically generated heterogeneity builds upon the heterogeneity imposed by physical processes at larger scales.

## Influence of LWD on Fluvial Landforms

Interactions between LWD jams and the stream channel (e.g., the physical substrate) stimulate structural heterogeneity in fluvial landforms. Acting

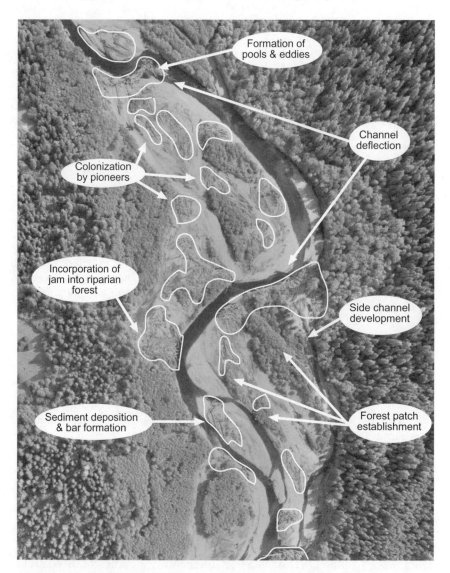

FIGURE 14.9. LWD jams (outlined) stimulate heterogeneity in structural and functional attributes of the flood plain of the Queets River, Washington. Only exposed jams are indicated, though LWD is also abundant within riparian forest patches.

alone, or synergistically, LWD jams cultivate new landforms, reinforce existing landforms, and transform or reconfigure existing landforms within stream reaches. In general, where LWD increases flow velocity and turbulence, channel stability is reduced. By redirecting flow and elevating shear stress, LWD jams create pools and side channels and promote bank erosion, channel avulsions, and channel switching (Nakamura and Swanson

1994; Abbe and Montgomery 1996). Where jams reduce flow velocity and turbulence, channel stability and resource availability for pioneering vegetation is enhanced. LWD jams may form backwaters and eddies and encourage deposition of sediments and organic matter in areas of enhanced stability. Over time, or in areas with high sediment supply, alluvial bars form downstream from stable LWD jams and may enlarge and coalesce into islands or extend from floodplain margins as terraces (Fetherston et al. 1995; Gurnell et al. 2001; Abbe and Montgomery 2003). In addition to triggering the formation of new fluvial landforms, LWD jams may armor landform margins against erosion, enabling the long-term persistence of some landforms within an otherwise volatile landscape (Abbe and Montgomery 1996). Floodplain terraces may arise from aggrading LWD-reinforced landforms adjacent to areas where the active channel is downcutting. As the relative elevation of the stream channel is lowered, patterns of sedimentation and disturbance on the floodplain terraces become less volatile and intense.

## Biophysical Controls on Interactions Between LWD and Stream Channels

The outcome of interactions between LWD jams and stream channels (jam-channel interactions) is determined by a complex web of linkages between LWD, flow hydraulics, channel morphology, sediment supply, landforms, and riparian vegetation. The flow regime, the physical attributes of the valley, and the composition of riparian forest, though interwoven, each have strong and recognizable influences on the intensity and dispersion of interactions between LWD jams and stream channels.

Spatial and temporal variation in the incidence and intensity of jam-channel interactions result when natural fluctuations in flood magnitude, frequency, and duration are superimposed upon variable patterns of LWD jam distribution. For example, in a river where the seasonal hydrograph is driven by snowmelt, jams may form and dissipate relatively frequently in areas inundated on an annual basis—these jams may be engaged in sustained, relatively intense interactions with the stream channel throughout periods of snowmelt runoff. In contrast, jams higher on the floodplain may form infrequently and only interact with the stream channel during large, episodic intense floods (e.g., rain-on-snow events).

Channel gradient, confinement, and width control the intensity of jam-channel interactions and strongly influence jam dispersion. The intensity of jam-channel interactions increases with jam size and stability and depends on the orientation of the jam relative to the channel axis (Abbe and Montgomery 2003). LWD jams tend to be larger and more isolated as channel size increases and confinement is alleviated (Swanson et al. 1982) and as the capacity for fluvial transport of LWD is enhanced (Lienkaemper and Swanson 1986). The capacity for LWD jams to transform reach-scale channel bed morphology is also constrained by valley confinement (Montgomery and Buffington 1998). LWD jams strongly influence channel morphology at small scales within large, unconfined alluvial reaches. In addition, LWD jams may transform simple stream reaches underlain by coarse sediments or bedrock into complex reaches, rich with patchy sediment accumulations and pools (Montgomery et al. 1995, 1996).

The size, species composition, and density of floodplain forest communities control the intensity of interactions between LWD jams and the stream channel, as well as the dispersion of jams. These factors contribute to heterogeneity in the residence time of LWD, which limits the longevity of the ecological effects of LWD. The functional significance of LWD may rapidly diminish or extend for centuries, according to the rate at which LWD is depleted from the river corridor through decay, fragmentation, abrasion, and transport (see Hyatt and Naiman 2001). The longevity and stability of LWD in the stream channel is enhanced in trees of large dimensions and high decay resistance. For example, coniferous riparian forests typically produce larger LWD with greater longevity, whereas hardwoods produce smaller debris more susceptible to flushing, fragmentation, and decay (Harmon et al. 1986; Bilby and Wasserman 1989). Jams are likely to form adjacent to mature stands of trees that contribute LWD sufficiently large to initiate jams. Even though the amount of LWD in stream channels typically increases with riparian tree density (Bilby and Wasserman 1989), the capacity for LWD jams to transform reach-scale channel bed morphology is also constrained by riparian forests (Montgomery and Buffington 1998). Extensive riparian vegetation reduces the susceptibility of stream reaches to transformation by LWD jams, as vegetation enhances bank stability, particularly in unconfined reaches (Smith 1976).

## Interplay Between LWD-Driven Heterogeneity and System Function

Spatial and temporal variation in LWD jams and related landforms result in a rich mosaic of aquatic habitats, forested landforms, and microclimates. LWD jams create heterogeneity in the low- to moderate-gradient streams by enhancing the variety and abundance of pools, riffles, and side-channels inhabited by aquatic organisms. Likewise, LWD jams enhance reach-scale heterogeneity in the variety, abundance, and spatial configuration of forest patches in various stages of initiation, establishment, growth, and destruction (Fetherston et al. 1995; Abbe and Montgomery 1996; Gurnell et al. 2001).

Interplay between LWD jams, organic matter, and stream substrates enhance the productivity of many stream organisms by creating resource-rich aquatic habitats. LWD encourages retention of particulate organic material, where it can be processed and used by aquatic invertebrates (Gregory et al. 1987; Bilby and Bisson 1998). Likewise, LWD can slow the downstream transport of spawning substrate, which can benefit fish production (House and Boehne 1989). Some fish benefit from the energetically profitable habitat (see Fausch 1984) within backwater pools, side channels, and eddies flanking marginal LWD jams (Moore and Gregory 1988). Pools created by LWD may contribute to fish productivity by providing refuge during climatic extremes, cover from predators, and encouraging habitat portioning among sympatric species (McMahon and Hartman 1989; Reeves et al. 1997). Fish and aquatic invertebrates often decline after LWD removal (Dolloff 1986; Elliott 1986; Fausch and Northcote 1992) and increase in response to LWD additions (House and Boehne 1989; Wallace et al. 1995; Cederholm et al. 1997). These contrasting responses underscore the functional importance of LWD, which extends beyond the margins of riparian forests to streams.

Patterns of LWD-related floodplain forest development follow predictable trajectories that differ according to the type of jam that initiated landform development (Fetherston et al. 1995; Abbe and Montgomery 1996, 2003) (Figure 14.10). Moisture-rich, sandy alluvial deposits in the lee of LWD jams may be colonized after floods by pioneering riparian vegetation such as red alder and willow. When LWD jams establish mid-channel, the establishment and growth of pioneering vegetation enhances local hydraulic roughness, encouraging additional sediment deposition. Persistent bars accumulate sediments with successive floods—burying most of the

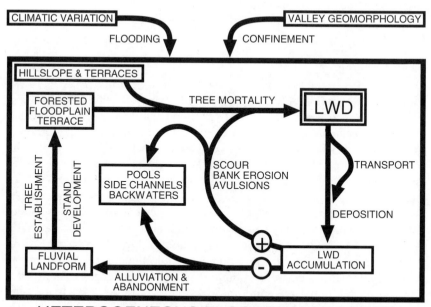

HETEROGENEOUS FLOODPLAIN FOREST

FIGURE 14.10. Linkages between LWD and heterogeneity in patterns and processes within a forested floodplain valley. Positive feedbacks stimulate instability, whereas negative feedbacks enhance stability.

original wood jam—and emerge as forested islands (Naiman et al. 2000). LWD jams may also form wedge-shaped bars along the floodplain margin (i.e., flow-deflection jams; Abbe and Montgomery 2003). These jams may form as bank erosion accelerates around large trees that have toppled into the channel—surrounding trees topple like dominoes as the river cuts deeper into the forest, gradually forming a jam that deflects flow from the previous channel axis. Patches of riparian forest establish on the resulting sediment accumulations, typically with the oldest trees located at the distal portion of the bar where the jam was initiated. Alternatively, swaths of even-aged floodplain forest may form as riparian vegetation establishes in abandoned channels created by LWD-related channel avulsions or channel switching. Eventually, exposed jams soften with decay and conifers (*Picea, Tsuga*) germinate the largest remaining of LWD, which provide moist microsites and refugia from competition and disturbance (McKee et al. 1982). Jams incorporated into riparian forest patches may strengthen the landform against further erosion, providing refuge for patches of mature forest within a highly dynamic corridor (Abbe and Montgomery 1996; Naiman et al. 2002). Ultimately, conifers overtop the alder and reach sufficient size to initiate new LWD jams upon their death and delivery to the channel. Many landforms are eventually eroded at various stages of development by channel movement, adding the living trees and accumulated LWD to the channel and starting the process of floodplain forest development anew.

## Conclusions

Riparian forests are highly complex systems exhibiting substantial heterogeneity—and ecosystem linkages—over broad spatial and temporal scales. Geology and climate interact to set the physical template, and that template shapes subsequent riparian structure and processes, from soils to LWD. The concepts presented here represent a rudimentary beginning. The collective understanding of riparian corridors as dynamic biophysical systems increases substantially every few years—and there remain immediate uncertainties as well as great challenges to be met. The chief uncertainties relate to determining what aspects of heterogeneity are most ecologically meaningful as well as predicting how system heterogeneity will respond to changes in system components due to the nonlinearity and apparent stochasticity of complex interactions between components. Major challenges relate to setting meaningful spatial and temporal scales on heterogeneity and to focusing on multiple factors as drivers of heterogeneity. Identifying meaningful scales is important because intellectual paradigms or management guidelines are often based on acceptable minimums, which can lead to system simplification if operative scales are not clearly identified. Finally, heterogeneity and the responses of the riparian system arise from many factors, some of which may dominate at particular scales of space and time—and it is this perspective that needs quantification for achieving a predicable understanding.

*Acknowledgments.* We are grateful to the Andrew W. Mellon Foundation, Weyerhaeuser Company, Pacific Northwest Research Station of the U.S. Forest Service, National Science Foundation, and the Olympic National Park for support and cooperation. H. Décamps, M. McClain, J.F. Franklin, and two anonymous reviewers provided many useful comments on an earlier version.

## References

Abbe, T.B., and Montgomery, D.R. 1996. Large woody debris jams, channel hydraulics, and habitat formation in large rivers. Regulated Rivers Res. Manage. 12: 201–221.

Abbe, T.B., and Montgomery, D.R. 2003. Patterns and processes of wood debris accumulation in the Queets River basin, Washington. Geomorphology 51: 81–107.

Agee, J.K. 1993. Fire ecology of Pacific Northwest forests. Washington DC: Island Press.

Balian, E. 2001. Growth and production of dominant riparian tree species along the Queets River, Washington. U.S. Thesis, University of Washington, Seattle, WA.

Bartz, K.K., and Naiman, R.J. 2005. Impacts of salmon-borne nutrients on riparian soils and vegetation in southwest Alaska. Ecosystems (in press).

Bechtold, J.S., Edwards, R.T., and Naiman, R.J. 2003. Biotic versus hydrologic control over seasonal nitrate leaching in a floodplain forest. Biogeochemistry 63: 53–72.

Bilby, R.E., and Bisson, P.A. 1998. Function and distribution of large woody debris. In River ecology and management: lessons from the Pacific Coastal Ecoregion, eds. R.J. Naiman and R.E. Bilby, pp. 324–346. New York: Springer-Verlag.

Binkley, D., Suarez, F., Rhoades, C., Stottlemeyer, R., and Valentine, D.W. 1995. Parent material depth controls ecosystem composition and function on a riverside terrace in northwestern Alaska. Ecoscience 2: 377–381.

Benda, L.E., Miller, D.J., Dunne, T., Reeves, G.H., and Agee, J.K. 1998. Dynamic landscape systems. In River ecology and management: lessons from the Pacific Coastal Ecoregion, eds. R.J. Naiman, and R.E. Bilby, pp. 261–288. New York: Springer-Verlag.

Bilby, R.E., and Wasserman, L.J. 1989. Forests practices and riparian management in Washington State: Data based regulation development. In Practical approaches to riparian resource management, eds. R.E. Gresswell, B.A. Barton, and J.L. Kershner, pp. 87–94. Billings, Montana: United States Bureau of Land Management.

Bisson, P.A., Rieman, B.E., Luce, C., Hessburg, P.F., Lee, D.C., Kershner, J.L., Reeves, G.H., and Gresswell, R. 2003. Fire and aquatic ecosystems of the western USA: Current knowledge and key questions. Forest Ecol. Manage. 178: 213–229.

Blom, C.W.P.M., Bogemann, G.M., Laan, P., van der Sman, A.J.M., van de Steeg, H.M., and Voesenek, L.A.C.J. 1990. Adaptations to flooding in plants from river areas. Aquatic Botany 38: 29–47.

Bormann, B.T., Cromack, K., and Russell III, W.O. 1994. Influences of red alder on soils and long-term ecosystem productivity. In The biology and management of red alder, ed. R.F. Tarrant, pp. 47–56. Corvallis, OR: Oregon State University Press.

Boyer, E.W., Hornberger, G.M., Bencala, K.E., and McKnight, D.M. 2000. Effects of asynchronous snowmelt on flushing of dissolved organic carbon: a mixing model approach. Hydrol. Processes 14: 3291–3308.

Burke, I.C., Yonker, C.M., Parton, W.J., Cole, C.V., Flach, K., and Schimel, D.S. 1989. Texture, climate and cultivation effects on soil organic matter content in U.S. grassland soils. Soil Sci. Soc. Am. J. 53: 800–805.

Cederholm, C.J., Bilby, R.E., Bisson, P.A., Bumstead, T.W., Fransen, B.R., Scarlett, W.J., and Ward, J.W. 1997. Response of juvenile coho salmon and steelhead to placement of large woody debris in a coastal Washington stream. North Am. J. Fisheries Manage. 17: 947–963.

Chen, J., Saunders, S., Crow, T., Naiman, R.J., Brosofske, K., Mroz, G., Brookshire, B., and Franklin, J.F. 1999. Microclimate in forest ecosystem and landscape ecology. BioScience 49: 288–297.

Christensen, B.T. 1996. Carbon in primary and secondary organomineral complexes. In Structure and organic matter storage in agricultural soils, eds. M. Carter and B. Stewart, pp. 97–165. Boca Raton, FL: CRC Press.

Clinton, S.M., Edwards, R.T., and Naiman, R.J. 2002. Forest-river interactions: Influence on hyporheic dissolved organic carbon concentrations in a floodplain terrace. J. Am. Water Resources Assoc. 38: 619–631.

Compton, J.E., Church, M.R., Larned, S.T., and Hogsett, W.E. 2003. Nitrogen export from forested watersheds in the Oregon Coast Range: the role of $N_2$-fixing alder. Ecosystems 6: 773–785.

Cox, P.A., and Elmqvist, T. 2000. Pollinator extinction in the Pacific Islands. Conservation Biol. 14: 1237–1240.

Crews, T.E. 1993. Phosphorus regulation of nitrogen fixation in a traditional Mexican agroecosystem. Biogeochemistry 21: 141–166.

Davis, M.B. 1981. Quaternary history and the stability of forest communities. In Forest succession concepts and application, eds. D.C. West, H.H. Shugart, and D.B. Botkin, pp. 132–153. New York: Springer-Verlag.

Davis, M.B. 1986. Climatic stability, time lags and community disequilibrium. In Community ecology, eds. J.M. Diamond and T. Case, pp. 269–284. New York: Harper and Row.

Décamps, H. 1996. The renewal of floodplain forests along rivers: a landscape perspective. Verhandlungen der Internationalen Vereinigung für Theoretische und Angewandte Limnologie 26: 35–59.

Dent, L., and Grimm, N.B. 1999. Spatial heterogeneity of stream water nutrient concentrations over successional time. Ecology 80: 2283–2298.

Di Tomasco, J.M. 1998. Impact, biology, and ecology of saltcedar (Tamarix spp.) in the southwestern United States. Weed Technol. 12: 326–336.

Dolloff, C.A. 1986. Effects of stream cleaning on juvenile coho salmon and Dolly Varden in southeast Alaska. Trans. Am. Fisheries Soc. 115: 743–755.

Edwards, R.T. 1998. The hyporheic zone. In River ecology and management: lessons from the Pacific Coastal Ecoregion, eds. R.J. Naiman and R.E. Bilby, pp. 399–429. New York: Springer-Verlag.

Elliott, S.T. 1986. Reduction of a Dolly Varden population and macrobenthos after removal of logging debris. Trans. Am. Fisheries Soc. 115: 392–400.

Epstein, H.E., Lauenroth, W.K., and Burke, I.C. 1997. Effects of temperature and soil texture on ANPP in the U.S. Great Plains. Ecology 78: 2628–2631.

Fausch, K.D. 1984. Profitable stream positions for salmonids: Relating specific growth rate to net energy gain. Can. J. Zool. 62: 441–451.

Fausch, K.D., and Northcote, T.G. 1992. Large woody debris and salmonid habitat in a small coastal British Columbia stream. Can. J. Fisheries Aquatic Sci. 49: 682–693.

Farrell, R.E., Sandercock, P.J., Pennock, D.J., and van Kessei, C. 1996. Landscape-scale variations in leached nitrate: relationship to denitrification and natural nitrogen-15 abundance. Soil Sci. Soc. Am. J. 60: 1410–1415.

Fetherston, K.L., Naiman, R.J., and Bilby, R.E. 1995. Large woody debris, physical process, and riparian forest development in montane river networks of the Pacific Northwest. Geomorphol. 13: 133–144.

Fevold, K. 1998. Sub-surface controls on the distribution of benthic algae in floodplain backchannel habitats of the Queets River. M.S. Thesis, University of Washington, Seattle, WA.

Findlay, S.E.G., and Sinsabaugh, R.L. editors. 2003. Aquatic Ecosystems: Interactivity of Dissolved Organic Matter. Elsevier, San Diego.

Foster, N.W., Nicolson, J.A., and Hazlett, P.W. 1989. Temporal variation in nitrate and nutrient cations in draining waters from a deciduous forest. J. Environ. Quality 18: 238–244.

Fowler, R.T., and Death, R.G. 2001. The effect of environmental stability on hyporheic community structure. Hydrobiologia 445: 85–95.

Gavin, D.G., Brubaker, L.B., and Lertzman, K.P. 2003. Holocene fire history of a coastal temperate rainforest based on soil charcoal radiocarbon dates. Ecology 84: 186–201.

Gregory, S.V., Lamberti, G.A., Erman, D.C., Koski, K.V., Murphy, M.L., and Sedell, J.R. 1987. Influence of forest practices on aquatic production. In Streamside management: forestry and fisheries interactions. Contribution No. 57, eds. E.O. Salo and T.W. Cundy, pp. 233–255. Seattle, WA: University of Washington, Institute of Forest Resources.

Gregory, S.V., Swanson, F.J., and McKee, W.A. 1991. An ecosystem perspective of riparian zones. BioScience 40: 540–551.

Groffman, P.M., and Tiedje, J.M. 1989. Denitrification in north temperate forest soils: relationships between denitrification and environmental factors at the landscape scale. Soil Biol. Biochem. 21: 621–626.

Gurnell, A., Petts, G.E., Hannah, D.M., Smith, B.P.G., Edwards, P.J., Kollman, J., Ward, J.V., and Tockner, K. 2001. Riparian vegetation and island formation along the gravel-bed Fiume Tagliamento, Italy. Earth Surface Processes Landforms 26: 31–62.

Harmon, M.E., Franklin, J.F., Swanson, F.J., Sollins, P., Gregory, S.V., Lattin, J.D., Anderson, N.H., Cline, S.P., Aumen, N.G., and Sedell, J.R., et al. 1986. Ecology of coarse woody debris in temperate ecosystems. Adv. Ecol. Res. 15: 133–302.

Hedin, L.O., Fischer, J.C.v., Ostrom, N.E., Kennedy, B.P., Brown, M.G., and Robertson, G.P. 1998. Thermodynamic constraints on nitrogen transformations and other biogeochemical processes at soil-stream interfaces. Ecology 79: 684–703.

Homann, P.S., Cole, D.W., and Van Miegroet, H. 1994. Relationships between cation and nitrate concentrations in soil solutions from mature and harvested red alder stands. Can. J. Forest Res. 24: 1646–1652.

Hook, P.B., and Burke, I.C. 2000. Biogeochemistry in a shortgrass landscape: control by topography, soil texture, and microclimate. Ecology 81: 2686–2703.

House, R.A. and Boehne, P.L. 1986. Effects of instream structure on salmonid habitat and populations in Tobe Creek, Oregon. North Am. J. Fisheries Manage. 6: 38–46.

Hyatt, T.L., and Naiman, R.J. 2001. The residence time of large woody debris in the Queets River, Washington, USA. Ecol. Applications 11: 191–202.

Jacinthe, P.A., Groffman, P.M., Gold, A.J., and Mosier, A. 1998. Patchiness in microbial nitrogen transformations in groundwater in a riparian forest. J. Environ. Quality 27: 156–164.

Johnson, D.W. 1992. Base cation distribution and cycling. In Atmospheric deposition and forest nutrient cycling, eds. D.W. Johnson and S.E. Lindberg, pp. 275–333. New York: Springer-Verlag.

Johnston, C.A., Bridgham, S.D., and Schubauer-Berigan, J.P. 2001. Nutrient dynamics in relation to geomorphology of riverine wetlands. Soil Sci. Soc. Am. J. 65: 557–577.

Leopold, L.B., Wolman, M.G., and Miller, J.P. 1964. Fluvial processes in geomorphology. W.H. Freeman, San Francisco.

Lienkaemper, G.W., and Swanson, F.J. 1986. Dynamics of large woody debris in streams in old-growth Douglas-fir forests. Cana. J. Forest Resources 17: 150–156.

Mantua, N.J., Hare, S.R., Zhang, Y., Wallace, J.M., & Francis, R. 1997. A Pacific interdecadal climate oscillation with impacts on salmon production. Bull. Am. Meteorol. Soc. 78: 1069–1079.

Marmonier, P., and Chatelliers, M.C.D. 1991. Effects of spates on interstitial assemblages of the Rhone River—importance of spatial heterogeneity. Hydrobiologia 210: 243–251.

McClain, M.E., and Naiman, R.J. 2005. Andean influences on the environment of the Amazon River. BioScience. (submitted)

McKee, A., Laroi, G., and Franklin, J.F. 1982. Structure, composition, and reproductive behavior of terrace foress, South Fork Hoh River, Olympic National Park. In Ecological research in National Parks of the Pacific Northwest. eds. E.E. Starkey, J.F. Franklin, and W.J. Matthews, pp. 22–29. Corvallis, OR: National Park Cooperative Studies Unit.

McMahon, T.E., and Hartman, G.F. 1989. Influence of cover complexity and current velocity on winter habitat use by juvenile coho salmon (*Oncorhynchus kisutch*). Can. J. Fisheries Aquatic Sci. 46: 1551–1557.

Minshall, G.W., Brock, J.T., and Varley, J.D. 1989. Wildfires and Yellowstone's stream ecosystems. BioScience 39: 707–715.

Minshall, G.W., Robinson, C.T., and Lawrence, D.E. 1997. Post fire responses of lotic ecosystems in Yellowstone National Park, USA. Can. J. Fisheries Aquatic Sci. 54: 2509–2525.

Mitsch, W.J., and Gosselink, J.G. 1993. Wetlands, Second edition. Van Nostrand, Reinhold, New York.

Montgomery, D.R., Buffington, J.M., Smith, R.D., Schmidt, K.M. and Pess, G.R. 1995. Pool spacing in forest channels. Water Resources Res. 31: 1097–1105.

Montgomery, D.R., Abbe, T.B., Buffington, J.M., Peterson, N.P., Schmidt, K.M., and Stock, J.D. 1996. Distribution of bedrock and alluvial channels in forested mountain drainage basins. Nature 381: 578–589.

Montgomery, D.R., and Buffington, J.M. 1998. Channel processes, classification, and response. In River ecology and management: lessons from the Pacific Coastal Ecoregion, eds. R.J. Naiman and R.E. Bilby, pp. 13–42. New York: Springer-Verlag New York, Inc.

Moore, K.M.S., and Gregory, S.V. 1988. Response of young-of-the-year cutthroat trout to manipulation of habitat structure in a small stream. Trans. Am. Fisheries Soc. 117: 162–170.

Murphy, M.L., and Koski, K.V. 1989. Input and depletion of woody debris in Alaska streams and implications for streamside management. North Am. J. Fisheries Manage. 9: 427–436.

Naiman, R.J., Balian, E.V., Bartz, K.K., Bilby, R.E., and Latterell, J.J. 2002. Dead wood dynamics in stream ecosystems. In Symposium on the Ecology and Management of Dead Wood in Western Forests, eds. P.J. Shea, W.F. Laudenslayer, B. Valentine, C.P. Weatherspoon, and T.E. Lisle, pp. 23–48. USDA Forest Service, Pacific Southwest Research Station, Albany, CA.

Naiman, R.J., Bilby, R.E., and Bisson, P.A. 2000. Riparian ecology and management in the Pacific coastal rain forest. BioScience 50: 996–1011.

Naiman, R.J., and Décamps, H. 1997. The ecology of interfaces—riparian zones. Annu. Rev. Ecol. Systematics 28: 621–658.

Naiman, R.J., Décamps, H., and McClain, M.E. 2005. Riparia: Ecology, Conservation and Management of Streamside Communities. San Diego: Elsevier Academic Press.

Naiman, R.J., Fetherston, K.L., McKay, S., and Chen, J. 1998. Riparian forests. In River ecology and management: lessons from the Pacific Coastal Ecoregion, eds. R.J. Naiman and R.E. Bilby, pp. 289–323. New York: Springer-Verlag.

Naiman, R.J., and Rogers, K.H. 1997. Large animals and the maintenance of system-level characteristics in river corridors. BioScience 47: 521–529.

Nakamura, F., and Swanson, F.J. 1994. Distribution of coarse woody debris in a mountain stream, western Cascade Range, Oregon. Can. J. Forest Resources 24: 2395–2403.

Nanson, G.C., Barbetti, M., and Taylor, G. 1995. River stabilization due to changing climate and vegetation during the Quaternary in western Tasmania, Australia. Geomorphology 12: 145–158.

Nanson, G.C., and Beach. 1977. Forest succession on a meandering-river floodplain, northeast British Columbia, Canada. J. Biogeogr. 4: 229–251.

Neff, J.C., and Asner, G.P. 2001. Dissolved organic carbon in terrestrial ecosystems: synthesis and a model. Ecosystems 4: 29–48.

Nelson, P.N., Baldock, J.A., and Oades, J.M. 1993. Concentration and composition of dissolved organic carbon in streams in relation to catchment soil properties. Biogeochemistry 19: 27–50.

Nilsson, C., and Svedmark, M. 2002. Basic principles and ecological consequences of changing water regimes: riparian plant communities. Environ. Manage. 30: 468–480.

O'Connor, J.E., Jones, M.A., and Haluska, T.L. 2003. Flood plain and channel dynamics of the Quinault and Queets rivers, Washington, USA. Geomorphology 51: 31–59.

Ottersen, G., Planque, B., Belgrano, A., Post, E., Reid, P.C., and Stenseth, N.C. 2001. Ecological effects of the North Atlantic Oscillation. Oecologia. 128: 1–14.

Pastor, J., Aber, J.D., McClaugherty, C.A., and Melillo, J.M. 1984. Aboveground production and N and P cycling along a nitrogen mineralization gradient on Blackhawk Island, Wisconsin. Ecology 65: 256–268.

Philander, S.G. 1990. El Nino, La Nina, and the Southern Oscillation. New York: Academic Press.

Pinay, G., Black, V.J., Planty Tabacchi, A.M., Gumiero, B., and Décamps, H. 2000. Geomorphic control of denitrification in large river floodplain soils. Biogeochemistry 50: 163–182.

Poff, N.L., Allan, J.D., Bain, M.B., Karr, J.R., Prestegaard, K.L., Richter, B.D., Sparks, R.E., and Stromberg, J.C. 1997. The natural flow regime. BioScience 47: 769–784.

Pollock, M.M., Naiman, R.J., and Hanley, T.A. 1998. Plant species richness in riparian wetlands—A test of biodiversity theory. Ecology 79: 94–105.

Prescott, C.E., Chappell, H.N., and Vesterdal, L. 2000. Nitrogen turnover in forest floors of coastal Douglas-fir at sites of differing in soil nitrogen capital. Ecology 81: 1878–1886.

Raich, J.W., and Schlesinger, W.H. 1992. The global carbon dioxide flux in soil respiration and its relationship to vegetation and climate. Tellus 44B: 81–99.

Reeves, G.H., Hall, J.D., and Gregory, S.V. 1997. The impact of land-management activities on coastal cutthroat trout and their freshwater habitats. In Sea-run Cutthroat Trout: Biology, Management, and Future Conservation. eds. P.A. Bisson and R.E. Gresswell, pp. 138–144. Corvallis, OR: Oregon Chapter, American Fisheries Society.

Reich, P.B., Grigal, D.F., Aber, J.D., and Gower, S.T. 1997. Nitrogen mineralization and productivity in 50 hardwood and conifer stands on diverse soils. Ecology 78: 335–347.

Rhoton, F.E., Bigham, J.M., and Lindbo, D.L. 2002. Properties of iron oxides in streams draining the loess uplands of Mississippi. Appl. Geochem. 17: 409–419.

Richards, K., Chandra, S., and Friend, P. 1993. Avulsive channel systems: characteristics and examples. eds. J. Best and C. Bristow, pp. 195–203. In Braided rivers, London: Geographical Society Special Publication No. 75.

Salo, J., Kalliola, R., Hakkinen, I., Makinen, Y., Niemela, P., Puhakka, M., and Coley, P.D. 1986. River dynamics and the diversity of Amazon lowland forest. Nature 322: 254–259.

Schimel, D.S., Braswell, B.H., Holland, E.A., McKeown, R., Ojima, D.S., Painter, T.H., Parton, W.J., and Townsend, A.R. 1994. Climatic, edaphic and biotic controls over storage and turnover of carbon in soils. Global Biogeochem. Cycles 3: 279–293.

Schumm, S. A. 1977. The Fluvial System. New York: John Wiley & Sons.

Smith, D.G. 1976. Effect of vegetation on lateral migration of anastamosed channels of a glacier meltwater river. Geological Soc. Am. Bull. 87: 857–860.

Sollins, P., Homann, P., and Caldwell, B.A. 1996. Stabilization and destabilization of soil organic matter: mechanisms and controls. Geoderma 74: 65–105.

Stanford, J.A., and Ward, J.V. 1993. An ecosystem perspective of alluvial rivers: Connectivity and the hyporheic corridor. J. North Am. Benthological Soc. 12: 48–60.

Stoeckel, D.M., and Miller-Goodman, M.S. 2001. Seasonal nutrient dynamics of forested floodplain soil influenced by microtopography and depth. Soil Sci. Soc. Am. J. 65: 922–931.

Swanson, F.J., Gregory, S.V., Sedell, J.R., and Campbell, A.G. 1982. Land-water interactions: The riparian zone. In Analysis of coniferous forest ecosystems in the western United States, ed. R.L. Edmonds, pp. 267–291. Stroudsburg, PA: Hutchinson Ross.

Trumbore, S.E. 1993. Comparison of carbon dynamics in tropical and temperate soils using radiocarbon measurements. Global Biogeochem. Cycles 7: 275–290.

Valett, H.M., Morrice, J.A., Dahm, C.N., Campana, M.E. 1996. Parent lithology, surface-groundwater exchange, and nitrate retention in headwater streams. Limnol. Oceanogr. 41: 333–345.

Van Cleve, K., Viereck, L.A., Marion, G.M., Yarie, J. and Dyrness, C.T. 1993. Role of salt-affected soils in primary succession on the Tanana River floodplain, interior Alaska. Can. J. Forest Res. 23: 877–1018.

Vannote, R.L., Minshall, G.W., Cummins, K.W., Sedell, J.R., and Cushing, C.E. 1980. The river continuum concept. Can. J. Fisheries Aquatic Sci. 37: 130–137.

Vitousek, P.M. 1999. Nutrient limitation to nitrogen fixation in young volcanic soils. Ecosystems 2: 505–510.

Vitousek, P.M., Gosz, J.R., Grier, C.C., Melillo, J.M., and Reiners, W.A. 1982. A comparative analysis of potential nitrification and nitrate mobility in forest ecosystems. Ecol. Monogr. 52: 155–177.

Walker, L.R. 1993. Nitrogen fixers and species replacements in primary succession. In Primary succession on land, eds. J. Miles and D.W.H. Walton, pp. 249–272. Blackwell Scientific, Oxford.

Walker, L.R., Silver, W.L., Willig, M.R., and Zimmerman, J.K, editors. 1996. Long-term responses of Caribbean ecosystems to disturbance. Biotropica 28: 414–613.

Walker, L.R., Zasada, J.C., and Chapin III, F.S. 1986. The role of life history processes in primary succession on an Alaskan floodplain. Ecology 67: 1243–1253.

Wallace, J.B., Webster, J.R., and Meyer, J.L. 1995. Influence of log additions on physical and biotic characteristics of a mountain stream. Can. J. Fisheries Aquatic Sci. 52: 2120–2137.

Walling, D.E., He, Q., and Blake, G.R. 2000. River flood plains as phosphorus sinks. In The role of erosion and sediment transport in nutrient and contaminant transfer, ed. M. Stone, pp. 211–218. Wallingford, UK: IAHS Publication No. 263.

Walling, D.E., He, Q., and Nicholas, A.P. 1996. Floodplains as suspended sediment sinks. In Floodplain processes. ed. M.G. Anderson, pp. 399–440. Chichester, UK: Wiley.

Ward, J.V., Tockner, K., Arscott, D.B., and Claret, C. 2002. Riverine landscape diversity. Freshwater Biol. 47: 517–539.

Wondzell, S.M., and King, J.G. 2003. Post-fire erosional processes in the Pacific Northwest and Rocky Mountain region. Forest Ecol. Manage. 178: 75–87.

# 15
# Flowpaths as Integrators of Heterogeneity in Streams and Landscapes

STUART G. FISHER and JILL R. WELTER

## Abstract

Streams are heterogeneous in both space and time. Hydrologic flowpaths along which biogeochemical processing occurs integrate different patches of the stream. Disturbance events (flood and drying) change these patches, alter connectivity, and reinforce spatial heterogeneity. Heterogeneity within patches (surface stream, hyporheic zone, sand bars, and riparian zone) is generated by the interaction of nitrogen (the limiting nutrient) in transport and organisms such as algae and bacteria. These organisms store nitrogen as they grow, alter N forms and concentrations in transport, and in some cases (e.g., denitrification) export it to the atmosphere. Changes in nitrogen in transport can be large, as are community responses to nitrogen availability, thus reinforcing spatial heterogeneity in successional time. Flowpaths connect patches as well and generate changes in recipient patches as a function of nitrogen delivery rate. This is especially evident at patch boundaries. In streams, flow is markedly linear and inexorably downstream in orientation; however, landscapes are drained by coalescing, dendritic networks that intimately connect stream channels with terrestrial flowpaths over and beneath soils. We propose that a unified theory of landscapes will require a focus on spatial linkage, a consideration of both spatial and temporal heterogeneity, and a blurring of distinctions between terrestrial and aquatic elements.

## Introduction

The concept of heterogeneity has been used variously in stream ecology to describe habitat variability (e.g., sediments) and effects on invertebrate communities (Palmer et al. 1997) or more broadly as patch structure and dynamics at multiple scales (Pringle et al. 1988). Poff et al. (1989) considered heterogeneity of forcing variables such as flood and drought in shaping stream function, again with an emphasis on invertebrates. Dent and Grimm

(1999) considered spatial heterogeneity of nutrient concentration using spatial autocorrelation analysis in a desert stream and applied this approach at three scales to deduce scale-specific causation of resultant patterns (Dent et al. 2001). Results of this approach lend insight into stream structure and function and permit an objective determination of operant hierarchical scales. Fractal analysis has been used to determine patterns of algal distribution in streams (Sinsabaugh et al. 1991) and to infer causes of spatial heterogeneity of invertebrate communities resulting from biotic interactions (Cooper et al. 1997).

Many stream ecologists have acknowledged that streams are spatially variable and have considered how these subsystems interact. Stanford and Ward (1993) have shown how the stream channel interacts with flood plains and how this variability and connectivity are central to stream function and biodiversity. Poole (2002) used a hierarchical approach adapted from Frissell et al. (1986) to examine longitudinal changes in solutes and community organization in streams and to thereby define an integrative approach to fluvial landscape ecology. Fisher et al. (1998a) developed a model of lateral interaction of stream elements in disturbance time to show how subsystem interactions shape whole system function, in that case, in terms of nutrient retention and spiraling.

Although these efforts represent substantial progress in understanding streams as spatially complex ecosystems, the field is still struggling with the challenge of linking heterogeneity with whole ecosystem functioning (Palmer and Poff 1997), determining how and when heterogeneity, in all its manifestations, matters.


## Objectives

The purpose of this paper is to examine the consequences of heterogeneity for ecosystem function using streams as an example; in particular, results of our work in Sycamore Creek in Arizona. We will attempt to develop a concept of patch integration to determine when heterogeneity generates higher order properties by virtue of patch interaction. Several terms are essential to this discussion. First, *structure* refers to the configuration of the ecosystem in space. *Patch structure* refers to a situation in which variance changes abruptly at boundaries that enclose patches that are themselves relatively homogeneous. Gradients may occur within patches or may characterize entire ecosystems wherein boundaries do not exist (although they may be arbitrarily imposed). Patch *integration* refers to an interaction among patches and may take several forms (hydrology, organismal movements, wind action) and involve several distinct *currencies* such as nitrogen, caribou, bird song, pheromones, and visual images (Reiners et al. this volume). We think of *integrator* as the mode of connection among patches and currency as the entity moved by the integrator. More broadly, an integrator can be viewed as a set of rules or an algorithm for summing patch influence on the whole. In this paper, we will emphasize integration via hydrologic *flowpaths*.

Heterogeneity exists when the ecosystem is not uniform and patches are present. This is virtually always the case. However, if the whole-system consequence of this heterogeneity is merely additive, the result is arguably less interesting than if whole-system consequences "emerge" as more than the sum of parts and are not deducible from patch structure alone. By integration we mean lateral interaction among patches sensu Turner and Chapin (this volume). Integration occurs when patches interact in a nonadditive way, resulting in nonlinear interaction among patches. The resulting nonlinear function may be contingent upon patch configuration and arrangement as well as specific routing of the integrator among various structural patches. Integration algorithms may change over time at different temporal scales and may themselves vary with time. Furthermore, integration varies as a function of currency, thus hydrologic integration for nitrogen may have a different influence on ecosystem function than for phosphorus. Migration (an integrating mechanism) of wildebeest will affect ecosystem function differently than will swarms of locusts. Ecosystem function is the holistic property that integrated patches of heterogeneous systems influence. In our stream research, material retention is an ecosystem function (emergent property) of special interest, but other ecosystem properties such as primary production or biodiversity or carbon processing efficiency can be influenced by patch structure and integration as well.

In the sections below, we will describe how hydrologic integration acts through the currency of nitrogen to integrate patches in Sycamore Creek, a well studied stream of the Sonoran desert of central Arizona (Fisher et al. 1982). The ultimate issue is nitrogen retention in arid landscapes. We know that only a small fraction of nitrogen entering desert landscapes is hydrologically exported (Grimm and Fisher 1992). In this paper, we define any process preventing hydrologic export to be retention. This includes both storage (e.g., as soil organic nitrogen) and loss to the atmosphere (e.g., ammonia volatilization and denitrification). We do not know where in this heterogeneous landscape the nitrogen is lost or retained, nor do we know the relative importance of various processes operating to retain or export it. Our goal below is to illustrate several issues that arise from attempts to apply heterogeneity-integrating ideas to streams at the level of surface stream, hyporheic zone, sand bars, riparian zones, drainage networks, and catchments (Figure 15.1). In each of these, nitrogen is transported in various chemical forms along hydrologically defined flowpaths.

Multiple disturbances, most notably drying and flooding, influence Sycamore Creek. Successional changes between disturbance events are pronounced. Thus patch structure, flow (integrator force and pattern), and nitrogen concentration (currency magnitude) change rapidly. Sycamore Creek has been described in detail elsewhere (Fisher et al. 1998b).

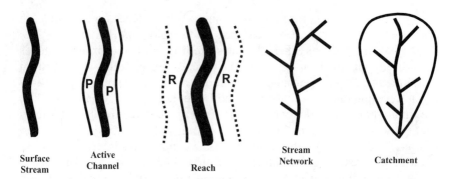

FIGURE 15.1. Depiction of aquatic ecosystem components as a function of increasing spatial extent: surface stream, defined as the wetted perimeter; active channel, comprised of the surface stream and parafluvial zone (P) or sand and gravel bars; stream reach, comprised of the active channel and adjacent riparian zone (R); stream network; and catchment, which includes all nested stream segments in a given area, as well as the land area they drain.

## Surface Stream

At the level of the surface stream channel, defined as the wetted perimeter, water flows on the sediment surface and connects patches represented by different substrates (cobbles and sand, for example) and superimposed benthic communities of algae and cyanobacteria mixed with organic detritus and assorted invertebrates. Hydrologic flowpaths connect these patches, and nitrogen in transport is removed or augmented by organismal uptake, assimilation, growth, excretion, or decomposition.

Flash floods obliterate and then restore patch structure. In postflood successional time, benthic algae recolonize sediments at a rate determined by the availability of inorganic nitrogen (largely nitrate), the limiting element in this system. As growth requires nitrogen, concentration declines in a downstream direction (Grimm 1987). Eventually, nitrogen is so low that N-fixing cyanobacteria gain a competitive advantage and replace green algae, gradually dominating the stream bottom in a downstream to upstream direction (Figure 15.2; Grimm 1994). In this case, patchiness in terms of algal coverage develops and changes over time as a function of the flowpath integrator. Changes in the form of the currency ($NO_3^-$ to atmospheric $N_2$) shifts community composition to cyanobacteria. In this manner, ecosystem function (N retention) simultaneously causes and responds to patchiness (heterogeneity). Interestingly, a positive rate of nitrogen accretion continues after hydrologically supplied N is depleted. Diffusion of atmospheric $N_2$ supplants hydrology as the integrator controlling nitrogen uptake later in successional time. This dynamic at the scale of 100 m and 100 days is both a cause and consequence of heterogeneity and involves a shift in integrators in time (hydrologic to atmospheric), both operating on the same currency (N).

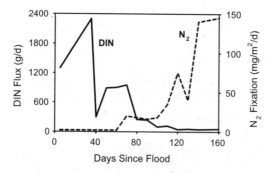

FIGURE 15.2. Temporal shifts in stream water DIN flux (g/d) and $N_2$ fixation $(mg/m^{-2}d^{-1})$ by cyanobacteria in the surface stream as a function of days post flood (Grimm 1994). Stream water DIN concentration is typically high immediately after floods; however, as algae begin to recolonize and take up nitrogen, DIN concentration declines, often to undetectable levels. When DIN concentration is low, cyanobacteria, which can fix atmospheric N, have a competitive advantage, and cyanobacterial biomass begins to increase over other green alga species. This leads to a shift in the integrator linking patches from hydrologic to atmospheric and a shift in producer community composition from green algae to cyanobacteria. In stream segments, nitrogen is depleted first in downstream reaches, thus space can be substituted for time on the X-axis (Grimm 1994 and Dent et al. 2001).

During periods of drought, surface flow may be lost as the stream dries (Stanley et al. 1997). Although the larger desert landscape may continue to be connected hydrologically, flow occurs deeper and more slowly in fluvial sediments. Mortality is high among stream organisms. At the scale of the catchment, surface drying represents a structural patch dropping out of the integration. In this case, the physical structure of the landscape does not change; rather, the integrator shifts horizontally and vertically over time as the surface stream shrinks and then dries completely. Any model of heterogeneity and ecosystem function must be able to deal with patches that come and go (algae) but also with patches that remain but lose connections with others (the surface stream as a whole during drought). At a variety of scales, the relationship between heterogeneity and ecosystem function will change in time as well as space. Heterogeneity can have a strong temporal component.

## Hyporheic Zone

Even when drying eliminates the surface stream, hyporheic flow continues. Water always moves beneath stream sediments whether surface flow is present or not. Vertical up- and downwelling zones exist, due to geomorphology, in particular the run-riffle sequence (Dent et al. 1999). Upwelling

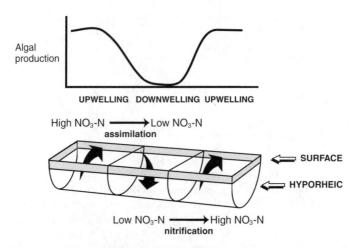

FIGURE 15.3. Patterns in algal production and nitrate concentrations in hyporheic upwelling and downwelling zones. When stream water DIN concentration is low, patterns of algal production in the surface stream are tightly linked to zones of hyporheic upwelling and downwelling. Water downwelling into the hyporheic zone from the surface stream is low in DIN and thus algal production is low in these areas. However, downwelling water is often high in DON (dissolved organic nitrogen), and mineralization and subsequent nitrification of organic nitrogen in the hyporheic zone increases nitrate concentrations in subsurface water. Where this high nitrate water upwells into the surface stream, algal production is high (Valett et al. 1994). Patterns of upwelling and downwelling are hydrologically driven and may lead to positive (net uptake and storage of N) retention or negative (N-fixation and increased N export), depending on their configuration. (Valett et al. 1994).

zones terminate hyporheic flowpaths of usually tens of meters and, because nitrification is high in stream sediments, contribute water high enough in nitrate to stimulate algal growth (and nutrient retention) on stream bottom sediments there (Figure 15.3; Valett et al. 1994). Downwelling zones receive surface water low in nitrate, and algal growth is much lower there and often is dominated by cyanobacteria that fix nitrogen rather than sequester inorganic nitrogen in transport. This pattern of up and down welling is hydrologically driven and, depending on its configuration, may result in nutrient retention by algae and a decrease in transported nitrogen or in nitrogen fixation and atmospheric linkage.

Surface stream–hyporheic interaction is an example of integration of heterogeneous patches by hydrology and a quantitative change in currency (N concentration), resulting in uptake or augmentation. In this example, activity is localized at the interface between subsystems. Vertical connectivity both generates heterogeneity (in algae and in nitrogen cycling) and is accentuated by it, thus is a positive feedback. We argue that simply adding hyporheic and surface rates to determine system function would miss this important interfacial property. Instead, knowledge of their connection is needed to fully understand the fluvial system as a whole.

## Sand Bars and the Parafluvial Zone

Main channel sand bars also exchange water and nutrients with the surface stream but they do so laterally rather than vertically, as is the case with the hyporheic zone. Transformations of nutrients are similar, and outwelling edges of sandbars support dense algal communities, which may retain up to 80% of outwelling nitrogen (Henry and Fisher 2003). Nitrogen fixers dominate inwelling and nonwelling zones, at least during low flow when dissolved inorganic nitrogen in stream water is below limiting levels. This generates a spatial pattern—"hot" spots of nitrogen retention and "cold" spots of nitrogen fixation (Fisher et al. 1998b). In postflood successional time, the streambed is a mosaic of N-fixing and N-retaining photosynthesizers. Relative abundance of these patches will determine whole system N retention of the active channel subsystem (surface stream plus hyporheic zone plus sand bars). Again, ecosystem function is determined by flow-path dynamics.

As it turns out, nitrification across sand bars is nonlinear (Figure 15.4a) presumably because dissolved organic nitrogen derived from the surface stream is depleted by microbes, whereas mineralization of phosphorous (from apatite minerals in sand) is linear (Figure 15.4b; Holmes et al. 1994; Holmes 1995). As a result, long flowpaths through sand bars decrease N:P ratios and have the capacity to shift potential nutrient limitation from N to P. We have not yet seen N:P drop below Redfield ratios wherein phosphorus limits productivity, but were this to happen, algal growth and nitrogen removal could be controlled by phosphorus concentration, not nitrogen.

Distribution of sand in bars may affect the outcome (consequence) of heterogeneity because of the change in nutrient concentration along the flowpath. Many small bars will increase N:P while the same amount of sand in one large bar will decrease N:P in comparison (Figure 15.5). This is an example of the same integrator (water) working on multiple currencies (N and P) simultaneously but in different ways owing to their chemical properties. The question is, how does their interaction influence an ecosystem property (e.g., N retention), and the answer is through control via stoichiometry involving a shift in the limiting nutrient.

## Riparian Zone

Riparian zones are important in that they represent an interface between upland areas and streams and may serve as a filter (via uptake or transformation). Riparian zones thereby influence the rate of input of nutrients into

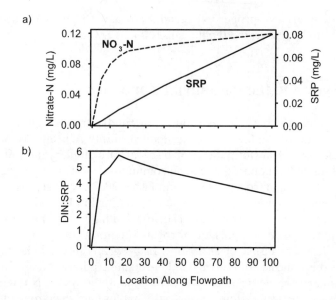

FIGURE 15.4. Changes in sand bar subsurface water (a) nitrate and soluble reactive phosphorus concentrations and (b) DIN:SRP ratio, as a function of location along the flowpath. Adapted from Holmes 1995. (a) Increases in SRP are linear along the flowpath while increases in nitrate are nonlinear, reaching a plateau. (b) As a result, long flowpaths through sand bars decrease N:P ratios and have the capacity to shift potential nutrient limitation from N to P.

the stream channel just as sand bar edges and hyporheic-surface interfaces influence fluxes across ecosystem components. In many areas, riparian surfaces intercept water and solutes as they move into the stream for the first time and thus represent a lateral filter. In arid streams such as Sycamore Creek, water enters stream channels first via tributaries and then moves into

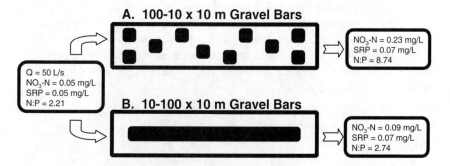

FIGURE 15.5. Consequences of sand bar configuration on reach-scale nitrogen retention. Adapted from Holmes 1995. Many small bars (A) will increase N:P while the same amount of sand in one large bar (B) will decrease N:P in comparison.

riparian zones (and back) from the stream side. While water and its load move back and forth with flowpaths determined by geomorphology, desert streams are net hydrologic "losing reaches," and much stream flow is lost by transpiration of riparian trees (Culler et al. 1982). Biologically active nutrients are stored long term in riparian trees. Nitrogen may be lost via denitrification at these interfaces (Schade et al. 2001). Because net flow occurs out of the stream most of the time, the riparian zone is a sink for nutrients at base flow. During floods, however, exchange can be large as water inundates riparian terraces, mobilizes nutrients accumulated there by soil processes such as nitrification, and transports them back to the stream channel (Marti et al. 2000; Schade et al. 2002; Heffernan and Sponseller 2004). This exchange is punctuated in arid lands but is important in all streams and has been called ROSS (region of seasonal saturation) by Baker et al. (2000).

We see then that riparian zones represent another patch contributing to heterogeneity and integrated by hydrology in a manner dictated by the interaction of geomorphology and hydrology. Depending on the regime of exchange during base flow or flood stage, nutrients vary, stoichiometric changes can occur, and nitrogen may be retained in biomass or lost to the atmosphere. The magnitude of landscape level nutrient retention is thus a function of spatial and temporal patterns of flowpaths and associated currencies.

As stated earlier, water enters larger desert streams not underground across the riparian zone at base flow but down tributary channels during storms. These tributary channels form networks wherein flowpaths in the form of surface flow are highly organized as a convergent, branched network. Depending on climate, this network may itself be highly intermittent, as is integration of heterogeneity at this scale.

## Network Structure

Up to this point we have discussed streams as if they were linear systems with longitudinally and laterally (and in some cases vertically) dispersed subsystems connected by flowing water and its load. Heterogeneity exists within each subsystem and in the larger stream of which they are a part. This linear view of streams has been productive in helping us understand upstream-downstream linkages, lateral connections, and size-related changes in stream segment function (Vannote et al. 1980). Only recently have stream ecologists begun to treat streams as branched structures (Osborne and Wiley 1992; Fisher 1997; Nakamura et al. 2000; Power and Dietrich 2002), a view prevalent among geomorphologists for more than a half century (Horton 1945; Strahler 1952).

Stream flow in channels coalesces in a convergent network, the structure of which can vary considerably depending on geomorphology, hydrology, slope, and catchment age. The stream network perfuses the terrestrial watershed and integrates ecosystem properties from headwaters to the sea, should the catchment be large enough. Clearly at the landscape scale, the stream network, through transport and processing, reflects whole ecosystem function, especially in the case of material retention, as small watershed budgets have shown (Likens et al. 1970). Depending on climate and geomorphology, low-order stream channels may be dry most of the time. This is especially true in arid land streams such as Sycamore Creek where small streams may transport water for only a few hours a year.

Storm size, intensity, and duration influence the extent of flow in intermittent networks, and the majority of events generate runoff that is "absorbed" by this component of the landscape—only large, rare events generate flow that extends into large perennial streams. In addition, storm events may generate flow in some segments of the network, whereas others remain dry and hydrologically unconnected. In a sense, the network is variously integrated from storm to storm by a set of meteorologic and hydrologic variables that were largely irrelevant to integration at the level of stream segments (discussed earlier). Hydrologic models exist of stream network operation in terms of water flows (Tague and Band 2001; D'Odorico and Rigon 2003). We are suggesting that these transport functions be combined with order-specific processing rates to generate a holistic picture of material retention at the network level.

At Sycamore Creek, spatial and temporal patterns of surface runoff for a summer storm in the low-order network are illustrated in Figure 15.6a. Hydrographs show substantial change in the runoff signal from order to order and its complete loss (presumably by absorption in sediments) in some cases (fifth and seventh order). DOC, $NH_4^+$, and $NO_3^-$ concentrations are quite high in transport (Figure 15.6b), and when flow stops, these materials stop as well and produce a legacy of materials that may jump-start biological processes with the advent of water associated with the next storm.

Potential denitrification in channel sediments reflects this legacy. Rates vary with order and sediment depth and indicate maximum activity that might occur in networks after storms (Welter 2004). The network-specific rates of denitrification will be a complex function of order-specific rates; mobilization and deposition of raw materials fueling denitrification; the geometry of the network, which will determine how and when water and materials are routed; and drying rate, which will limit the duration of biological activity. We developed a hot-spot index that takes into account the potential for gaseous loss of N and the time that each site is wet or active (Figure 15.7). According to this scheme, potential for N loss is highest in deep sediments of intermediate orders. Surface activity is depressed by comparison, probably due to more rapid drying and lower potential for denitrification, although transport-related legacies may also play a role. As we continue to move down the network into progressively larger channels, we will eventually reach perennial streams. Although the intermittent upland networks experience more discrete flow

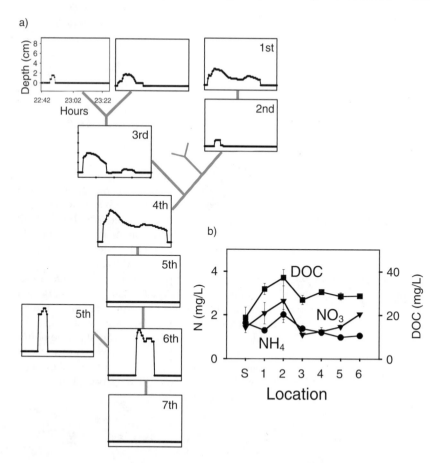

FIGURE 15.6. Temporal patterns in (a) surface runoff (depth in cm) and (b) runoff chemistry ($NO_3$-N, $NH_4$-N, and DOC mg/L) for a single 1-cm storm event in July 2000. (a) Runoff signal changes with location and from order to order in the network. X- and Y-axis scales are identical for all depth panels, indicating that runoff water is "absorbed" in some locations (fifth- and seventh-order channels). (b) Inorganic N and DOC concentrations (mg/L) also change with location (S indicates sheetflow collected directly from terrestrial hill slopes, and 1–7 represent different stream orders). Concentrations are quite high and represent a significant source of carbon and nitrogen for microbes in channel sediments. Thus, hot spots of microbial metabolism are likely to occur where flow is "absorbed" in the network. These materials may also produce a legacy of available resources that may jump-start biological processes with the advent of water associated with the next storm.

events, they dry quickly. Perennial stream-riparian systems remain wet or active most of the time. Further research is needed to determine how these different network positions compare in terms of their contribution to net N retention seasonally, annually, or on longer temporal scales.

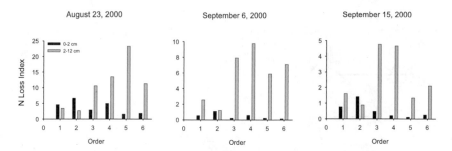

FIGURE 15.7. N loss hot-spot index values as a function of channel order and depth based on areal rates of potential denitrification and percent soil moisture after a single storm in August 2000. Potential denitrification rates were measured in the lab using a method similar to the assay of denitrification enzyme activity (DEA; Smith and Tiedje 1979). Samples were collected separately for soil moisture over time following the storm. Hot-spot index values were calculated by multiplying potential denitrification rates by percent soil moisture on each of three dates after the storm and used as an index of potential loss at each site. Upon wetting, index values closely resemble patterns in potential rates; however, *in situ* rates would likely vary depending on storm-specific delivery of DOC and $NO_3^-$ to different locations in the network. As sediments begin to dry, index values shift across orders and depths, with higher rates maintained where sediments remain wet for the longest period of time. Index values change dramatically over time (note shift in Y-axis scale) as a result of drying. Hot spots for denitrification in intermittent networks turn and off in the landscape in response to moisture and may shift spatially from storm to storm based on patterns in legacies of DOC and $NO_3^-$ availability.

## Some Overarching Issues

By discussing how the heterogeneous stream-riparian landscape is connected (integrated) by flowpath and how fluxes of the currency (nitrogen) changes as a function of connectivity in space and time, we can begin to understand how heterogeneity can influence ecosystem functioning (nutrient retention) in a manner not evident by simply adding up patch-specific processes. These interactions are of course complicated and ever changing, but several conceptual issues have emerged from our examination of the several subsystems of which streams are composed. We will summarize these general issues briefly below in hopes that they are general enough to apply widely to a range of landscapes, integrators, and currencies.

First, heterogeneity may apply to integrators as well as structural patches. In our studies of desert streams, we see that many patches are involved in net function, yet hydrologic connectivity also varies greatly, and in time of drought may be absent. Flood and drought can change the nature of the relationship among patches as much or more as changing the array of patches themselves. The nature of integration changes seasonally to be sure but may also respond to longer term schedules such as climate change or geologic cycles.

Second, in the fluvial system water is a primary integrator and is undoubtedly the major force connecting patches of the landscape. Other integrators may operate simultaneously in parallel, interacting networks. For example, in some fluvial systems, fish and invertebrate movements connect patches. Salmon migrations are famous for moving nutrients (Bilby et al. 1996; Helfield and Naiman 2001), but insect emergences may represent substantial terrestrial subsidies; for example, with the riparian zone via bird or spider predation (Sanzone et al. 2003). It would be interesting to compare the relative effects of multiple integrators such as hydrology and animal movements (and their interactions) in other landscapes that experience substantial migrations.

Material movement by spatial fluxes of animals and water can take many forms—as many as there are elements. Because the vector (water) moves many things, an opportunity exists to compare patch integration in the context of different currencies (chemical elements, diseases, or propagules, for example).

With chemical elements moving across the landscape in a single integrator, water, a lucrative opportunity for application of stoichiometric concepts and models (Sterner and Elser 2002) exists at the landscape level. Our example of N:P changes across sand bars is a simple one, and more work using multiple elements is needed. It is likely that landscape integrator interactions adjust nutrient ratios in such a way that shifting control will occur. Rather than thinking of control by a single key element, a better question is when, where, and under what conditions are elements X, Y, and Z key? The answer to this question will not only vary in space and time, but shifts in key elements will *determine* patterns in space and time.

There are a host of ecosystem functions that can be examined as well and no reason to think these will respond in parallel even to a single integrator and a single currency. In Sycamore Creek, for example, movement of nitrogen through sand bars by water results in an increase in nitrogen uptake without a concomitant increase in productivity (a second ecosystem function). The reason for this is cyanobacteria were able to grow just fine at inwelling edges using atmospheric nitrogen and thereby not retaining N in transport (Henry and Fisher 2003).

Connectivity among the patches that confer heterogeneity at any hierarchical level is itself heterogeneous in time and, as a result, movement of water and materials through the stream network is halting and saltatorial. The intermittent, uneven movement generates a spatial pattern that reflects this transport history and is therefore a legacy of events past. Legacies may provide insight into past episodic transport dynamics, but more importantly, they influence future ecosystem functioning when flow, and biological activity, resume. Because of the temporal separation of deposition and restored activity, functional lags are characteristics of this system.

Finally, our consideration of the interaction of flowpath and spatial heterogeneity suggest to us that patch shape and configuration may be crucial descriptors of heterogeneity when the integrator impinges on patches in a spatially oriented way. Not only would geometry of single patches relative to flow direction be relevant but also the sequence or order of patches linked by flow. Landscape ecology has provided a rich toolbox and lexicon for dealing with patch shape. With the addition of the concept of integration, can a science of shape be far behind?

## Flow-Integrated Landscapes

Although networks are the true shape of streams, (as depicted in Figure 15.1), they are not planar, but three-dimensional (Figure 15.8). Taking this view, we can see that these are flowpath-integrated landscapes—including both the terrestrial and aquatic components of the watershed. From ridge tops to valley bottoms and within the stream network, all landscape elements are integrated by hydrology via flowpaths. Thus, in many ways, the separation between terrestrial and aquatic landscape elements is artificial.

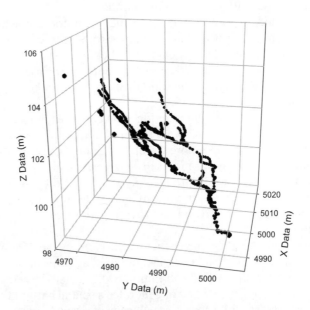

FIGURE 15.8. Three-dimensional image of network structure. Although stream ecologists focus on hydrologic integration of aquatic patches in the landscape, hydrology connects all landscape patches (both terrestrial and aquatic) via flowpaths. These elements lie along a terrestrial-aquatic continuum; varying in the directionality of flow, vertical *versus* horizontal. Resulting hydrologic flows likely result in nonlinear patch interactions in all catchments and motivate study of flow-integrated landscapes and a blurring of distinctions between terrestrial and aquatic elements.

These elements lie along a terrestrial-aquatic continuum; varying in the directionality of flow (vertical *vs.* horizontal) and time wet (and therefore biologically active). But all of these elements are linked via hydrology. This is an example of a common perspective borrowed from one field (stream ecology) and used to "capture" conceptually a larger whole. The influence of terrestrial-aquatic linkages on material transport and retention has been considered from both terrestrial (Peterjohn and Correll 1984; Giblin et al. 1991) and aquatic (Hynes 1975; McDowell and Likens 1988; Boyer et al. 2000) points of view.

Although terrestrial ecologists have primarily focused on vertical fluxes (e.g., percolation, soil development) and stream ecologists have historically emphasized horizontal fluxes, merging of the two approaches may be fruitful. To understand factors that influence material transport and retention in flow-integrated landscapes would require integration of vertical and horizontal flows, calculation of resulting vectors (vertical *vs.* horizontal), determination of residence times and process rates, dissection of flowpaths to determine control points, and assembly via modeling to determine higher level emergent effects of network structure and segment/node configuration. To do this we need to borrow from soil science, hydrology, biogeochemistry, fluvial dynamics, and geomorphology. We think that separation of aquatic and terrestrial ecology is counterproductive in this context and for these reasons.

*Acknowledgments.* We thank the organizers of the conference for the opportunity to present these ideas verbally and in this volume. Sycamore Creek research has been funded over the years by several grants from the National Science Foundation, most recently DEB 0075650. Several of our colleagues did the basic research on which these ideas are based and worked with us in developing our thinking about them, most notably Nancy Grimm, Max Holmes, Jim Heffernan, Julia Henry, Jay Jones, Eugenia Marti, John Schade, Ryan Sponseller, Emily Stanley, and Maurice Valett. We thank two anonymous reviewers and M.G. Turner for valuable suggestions that improved the manuscript.

# References

Baker, M.A., Valett, H.M., and Dahm, C.N. 2000. Organic carbon supply and metabolism in a shallow groundwater ecosystem. Ecology 81: 3133–3148.

Bilby, R.E., Fransen, B.R., and Bisson, P.A. 1996. Incorporation of nitrogen and carbon from spawning coho salmon into the trophic system of small streams: evidence from stable isotopes. Can. J. Fisheries Aquatic Sci. 53: 164–173.

Boyer, E.W., Hornberger, G.M., Bencala, K.E., and McKnight, D.M. 2000. Effects of asynchronous snowmelt on flushing of dissolved organic carbon: a mixing model approach. Hydrol. Processes 14: 3291–3308.

Cooper, S.D., Barmuta, L., Sarnelle, O., Kratz, K., and Diehl, S. 1997. Quantifying spatial heterogeneity in streams. J. North Am. Benthol. Soc. 16: 174–188.

Culler, R.C. et al. 1982. Evapotranspiration before and after clearing phreatophytes, Gila River flood plain, Graham County, Arizona. USGS Professional Paper 655-P. 67 pp.

Dent, C.L., and Grimm, N.B. 1999. Spatial heterogeneity of stream water nutrient concentrations over successional time. Ecology 80: 2283–2298.

Dent. C.L., Grimm, N.B., and Fisher, S.G. 2001. Multiscale effects of surface-subsurface exchange on stream water nutrient concentrations. J. North Am. Benthol. Soc. 20: 162–181.

D'Odorico, P., Rigon, R. 2003. Hillslope and channel contributions to the hydrologic response. Water Resources Res. 39(5): 1–9.

Fisher, S.G. 1997. Creativity, idea generation and the functional morphology of streams. J. North Am. Benthol. Soc. 16: 305–318.

Fisher, S.G., Gray, L.J., Grimm, N.B., and Busch, D.E. 1982. Temporal succession in a desert stream following flash flooding. Ecol. Monogr. 52: 93–110.

Fisher, S.G., Grimm, N.B., Marti, E., Holmes, R.M., and Jones, J.B. Jr. 1998a. Material spiraling in stream corridors: a telescoping ecosystem model. Ecosystems 1: 19–34.

Fisher, S.G., Grimm, N.B., Marti, E., and Gomez, R. 1998b. Hierarchy, spatial configuration, and nutrient cycling in a desert stream. Aust. J. Ecol. 23: 41–52.

Frissell, C.A., Liss, W.J., Warren, C.E., and Hurley, M.D. 1986. A hierarchical framework for stream habitat classification: viewing streams in a watershed context. Environ. Manage. 10: 199–124.

Giblin, A.E., Nadelhoffer, K.J., Shaver, G.R., Laundre, J.A., and McKerrow, A.J. 1991. Biogeochemical diversity along a riverside toposequence in arctic Alaska. Ecol. Monogr. 61: 415–435.

Grimm, N.B. 1987. Nitrogen dynamics during succession in a desert stream. Ecology 68: 1157–1170.

Grimm, N.B. 1994. Disturbance, succession, and ecosystem processes in streams: a case study from the desert. In Aquatic ecology: scale, pattern and process, eds. P.S. Giller, A.G. Hildrew, and D.G. Raffaeli, pp. 93–112. Joint Symposium of the British Ecological Society and the American Society of Limnology and Oceanography. Oxford: Blackwell Scientific.

Grimm, N.B., and Fisher, S.G. 1992. Responses of arid-land streams to changing climate. In Climate change and freshwater ecosystems, eds. P. Firth, and S.G. Fisher, pp. 211–233. New York: Springer Verlag.

Heffernan, J.B., and Sponseller, R.A. 2004. Nutrient re-mobilization and processing in Sonoran Desert riparian soils following artificial re-wetting. Biogeochemistry 70(1): 117–134.

Helfield, J.M., and Naiman, R.J. 2001. Effects of salmon-derived nitrogen on riparian forest growth and implications for stream productivity. Ecology 82: 2403–2409.

Henry, J.C., and Fisher, S.G. 2003. Spatial segregation of periphyton communities in a desert stream: causes and consequences for nitrogen cycling. J. North Am. Benthol. Soc. 22: 511–527.

Holmes, R.M. 1995. Parafluvial nutrient dynamics in a desert stream ecosystem. Ph.D. Dissertation, Arizona State University. 270 pp.

Holmes, R.M., Fisher, S.G., and Grimm, N.B. 1994. Parafluvial nitrogen dynamics in a desert stream ecosystem. J. North Am. Benthol. Soc. 13: 468–478.

Horton, R.E. 1945. Erosional development of streams and their drainage basins: hydrophysical approach to quantitative morphology. Geol. Soc. Am. Bull. 56: 281–300.

Hynes, H.B. 1975. The stream and its valley. Internat. Vereiningung fur Theoretische und Angewandte Limnologie, Verhandlungen. 19: 1–15.

Likens, G.E., Bormann, F.H., Johnson, N.M., Fisher, D.W., and Pierce, R.S. 1970. Effects of forest cutting and herbicide treatment on nutrient budgets in Hubbard Brook Watershed-ecosystem. Ecol. Monogr. 40: 23–47.

Marti, E., Fisher, S.G., Schade, J.D., and Grimm, N.B. 2000. Flood frequency, arid land streams, and their riparian zones. In Streams and groundwaters, eds. J.B. Jones, and P.J. Mulholland, pp. 111-136. San Diego: Academic Press.

McDowell, W.H., and Likens, G.E. 1988. Origin, composition and flux of dissolved organic carbon in the Hubbard Brook Valley. Ecol. Monogr. 58: 177–195.

Nakamura, F., Swanson, F.J., and Wondzell, S.M. 2000. Disturbance regimes of stream and riparian systems—a disturbance-cascade perspective. Hydrol. Processes 14: 2849–2860.

Osborne, L.L., and Wiley, M.J. 1992. Influence of tributary spatial position on the structure of warmwater fish communities. Can. J. Fisheries Aquatic Sci. 49: 671–681.

Palmer, M.A., and Poff, N.L. 1997. The influence of environmental heterogeneity on patterns and processes in streams. J. North Am. Benthol. Soc. 16: 169–173.

Palmer, M.A., Hakenkamp, C.C., and Nelson-Baker, K. 1997. Ecological heterogeneity in streams: why variance matters. J. North Am. Benthol. Soc. 16: 189–202.

Peterjohn, W.T., and Correll, D.L. 1984. Nutrient dynamics in an agricultural watershed–observations on the role of a riparian forest. Ecology 65: 1466–1475.

Poff, N.L., and Ward, J.V. 1989. Implications of stream flow variability and predictability for lotic community structure: a regional analysis of stream flow patterns. Can. J. Fisheries Aquatic Sci. 46: 1805–1818.

Poole, G.C. 2002. Fluvial landscape ecology: addressing uniqueness within the river discontinuum. Freshwater Biol. 47: 641–660.

Power, M.E., and Dietrich, W.E. 2002. Food webs in river networks. Ecol. Res. 17: 451–471.

Pringle, C.M., Naiman, R.J., Bretschko, G., Karr, J.R., Oswood, M.W., Webster, J.R., Welcomme, R.L., and Winterbourn, M.J. 1988. Patch dynamics in lotic systems: the stream as a mosaic. J. North Am. Benthol. Soc. 7: 503–524.

Sanzone, D., Meyer, J.L., Marti, E., Gardiner, E.P., Tank, J.L., and Grimm, N.B. 2003. Carbon and nitrogen transfer from a desert stream to riparian predators. Oecologia 134: 238–250.

Schade, J.D., Fisher, S.G., Grimm, N.B., and Seddon, J.A. 2001. The influence of a riparian shrub on nitrogen cycling in a Sonoran Desert stream. Ecology 82: 3363–3376.

Schade, J.D., Marti, E., Welter, J.R., Fisher, S.G., and Grimm, N.B. 2002. Sources of nitrogen to the riparian zone of a desert stream: implications for riparian vegetation and nitrogen retention. Ecosystems 5: 68–79.

Smith, M.S., and Tiedje, J.M. 1979. Phases of denitrification following oxygen depletion in soil. Soil Biol. Biochem. 11: 262–267.

Sinsabaugh, R.L., Weiland, T., and Linkins, A.E. 1991. Epilithon patch structure in a boreal river. J. North Am. Benthol Soc. 10: 419–429.

Stanford, J.A., and Ward, J.V. 1993. An ecosystem perspective of alluvial rivers: connectivity and the hyporheic zone. J. North Am. Benthol. Soc. 12: 48–60.

Stanley, E.H., Fisher, S.G., and Grimm, N. B. 1997. Ecosystem expansion and contraction: a desert stream perspective. BioScience 47: 427–435.

Sterner, R.W., and Elser, J.J. 2002. Ecological stoichiometry: The biology of elements from molecule to biosphere. Princeton, NJ: Princeton University Press.

Strahler, A.N. 1952. Hyposometric (area-altitude) analysis of erosional topography. Geol. Soc. Am. Bull. 63: 1117–1142.

Tague C.L., and Band, L.E. 2001. Evaluating explicit and implicit routing for watershed hydro-ecological models of forest hydrology at the small catchment scale. Hydrol. Processes 15: 1415–1439.

Turner, M.G., and Chapin III, F.S. 2004. Causes and consequences of spatial heterogeneity in ecosystem function. In Ecosystem function in heterogeneous landscapes, eds. pp. 1–4.

Valett, R.M., Fisher, S.G., Grimm, N.B., and Camille, P. 1994. Vertical hydrologic exchange and ecological stability of a desert stream ecosystem. Ecology 75: 548–560.

Vannote, R.L., Minshall, G.W., Cummins, K.W., Sedell, J.R., and Cushing, C.E. 1980. The river continuum concept. Can. J. Fisheries Aquatic Sci. 37: 130–137.

Welter, J.R. 2004. Nitrogen transport and processing in the intermittent drainage network: linking terrestrial and aquatic ecosystems. Ph.D. Dissertation, Arizona State University.

# 16
# Causes and Consequences of Spatial Heterogeneity in Lakes

Timothy K. Kratz, Sally MacIntyre, and Katherine E. Webster

## Abstract

Lakes, far from being the homogeneous environments we might expect, offer a rich and dynamic heterogeneity at multiple spatial and temporal scales that we are just beginning to understand. At the within-lake scale, a complex set of phenomena such as internal waves and stream intrusions leads to both horizontal and vertical heterogeneity. Developing an understanding of whether and how this heterogeneity affects ecosystem processes is in its early stages, but nutrient movement both horizontally and vertically may be more structured than previously conceptualized and will depend on interactions among nutrient loading, stratification, surface meteorology, and basin morphometry. Within a landscape, lakes often differ from each other both in their average characteristics and in their among-year dynamics. Much of this heterogeneity has been linked to how water flows across the landscape. In landscapes dominated by groundwater flow, there is often more heterogeneity in lake characteristics and response to climatic events than in landscapes where exposed bedrock leads to rapid horizontal transport of water. Humans can affect heterogeneity across lakes by causing changes in land use and cover and within lakes by simplifying the physical structure of the littoral zone.

## Introduction

Lakes exhibit spatial heterogeneity at many different spatial scales. From the parallel Langmuir streaks (Langmuir 1938) commonly seen by airline passengers, to among-lake differences in chemical and biological properties (Juday and Birge 1933), to regional differences in origin and setting (Soranno et al. 1999), lakes are not uniform across space (Richerson et al. 1978). Understanding the causes of spatial heterogeneity within and among lakes has been a long-standing goal of limnologists.

For the casual observer, lakes appear as discrete units delineated by their shoreline and defined by surface phenomena. This perspective implies a disconnection from external forces and internal processes. Instead, we know that lake ecosystems are shaped by abiotic and biotic forces resulting in substantial heterogeneity both within and among lakes. At broad spatial scales, geomorphic setting constrains the expression of lake features (Magnuson and Kratz 2000; Riera et al. 2000). The geology and landforms of a region dictate hydrologic flowpaths and the biogeochemical transformations that occur as water flows from the terrestrial to aquatic system (Winter 2001). Landforms characterized by steep slopes, high elevation, or hydrologic isolation can set barriers to dispersal of organisms. Climate influences seasonality of hydrologic flows, affecting the delivery of water and solutes to lakes, and sets temperature regimes that physiologically constrain species distributions. Taken together, these factors set limits on the physical properties, biogeochemistry, and biotic assemblages of lakes within a region.

Within a region, the seeming uniformity of lakes observed from the air is belied by significant among-lake variation. Lakes are hydrologically connected to their catchments, and thus their chemistry reflects inputs of nutrients and other solutes in runoff. Adjoining wetlands supply humic material, influencing lake-dissolved organic carbon and water color (Gergel et al. 1999) and, subsequently, the attenuation of light energy through the water column (Snucins and Gunn 2000). Lake size has a fundamental influence on a range of ecosystem properties of lakes including the relationship between lake depth and nutrient cycling and between lake area and productivity (Fee et al. 1992), thermal regimes, and species richness (Magnuson et al. 1998).

Within individual lakes, substantial heterogeneity exists in both near-shore (littoral) and deeper open-water (pelagic) zones. Since the late 1800s, limnologists have recognized the vertical thermal stratification characterizing many north temperate lakes during the summer (Kalff 2002). Warming of surface waters sets up density gradients that eventually separate the warm upper layer from denser and colder bottom waters. This seasonal stratification cycle generates considerable spatial and temporal heterogeneity in the open water zone of lakes in temperature, dissolved oxygen and other gases, nutrient cycling, and distributions of fish, invertebrates, and algae. The littoral zone exhibits patchy structural complexity as slope and wave action affect sediment composition, macrophyte community composition, and woody structure (Kalff 2002). Changing physics and chemistry induce a dynamic template for biological interactions such as competition and predation, as well as cycling of energy and nutrients throughout food webs. All these factors shape biological communities and ecosystem processes.

Here we examine heterogeneity in lakes, its causes, and its effects on ecosystem processes at two spatial scales that have been a focus of intense research over the past decade. First, we focus on within-lake heterogeneity related to physical processes operating at fine spatial and temporal scales. We address variation in hydrodynamics within lakes related to basin size and bathymetry. For instance, the rates of heating and cooling at a lake's surface vary with proximity to boundaries. These differences drive circulation both at the surface and at depth. Inflows from streams and groundwater occur as intrusions. The extent of the intrusions and the concentrations of dissolved organic matter and nutrients within them vary with intensity of rainfall in the watershed. Similarly, wind forcing generates internal waves. The amplitude and stability of these waves varies with bottom slope and topographic features. Where wave amplitudes are larger, or where their shape varies due to topography, instabilities can develop and lead to turbulence and intrusion formation. Consequently, vertical fluxes of nutrients and rates of associated biogeochemical processes vary with the temporal variability of wind forcing and with bottom bathymetry. This type of within-lake spatial heterogeneity is dynamic, and its ecological implications depend on how long it persists.

Second, we use a landscape perspective to interpret the spatial structure of lake properties within a region. Essentially, we explore how and why neighboring lakes differ and whether a spatially explicit template that is repeatable across regions underlies these differences. In particular, we address how broad-scale processes such as water flow across a landscape can lead to systematic spatial patterns in lake characteristics and ecosystem dynamics. We draw on examples from lake districts from Wisconsin, Alaska, Alberta, Ontario, and other regions to examine generalities and location-specific differences in the relationship between spatial heterogeneity and lacustrine ecosystem structure and dynamics. Finally, we consider briefly the role human activities play in altering patterns of heterogeneity in lakes and their watersheds.

In considering both scales, we consider the following two questions. When does spatial heterogeneity matter for ecosystem function in lakes and when does it not matter? What causes spatial heterogeneity in lakes and what is the interplay between heterogeneity and ecosystem function?

## Within-Lake Heterogeneity

Thermal structure creates vertical heterogeneity in lakes. This structure, due to density stratification, is determined by meteorological forcing at the air-water interface, basin size and geometry, and the concentrations of colored solutes and particulates that determine the absorption of irradiance. When heat losses exceed heat gains (as in winter or in shallow systems), the water column mixes. When heat gains exceed heat losses and the lake is sufficiently deep, the water column is divided into an epilimnion, or upper mixed layer, a strongly stratified metalimnion, and a weakly stratified hypolimnion. The strength and persistence of stratification depends on lake size and latitude (Lewis 1983). The seasonal changes in stratification influence species succession of phytoplankton and zooplankton and the primary productivity of tropical, temperate, and arctic lakes (Goldman and Horne 1994; Wetzel 2001; Kalff 2002). Although our understanding of the implications of changing seasonal thermal structure has developed over the century, our knowledge of lake physics has increased dramatically over the past two decades along with the advent of rapid profiling instrumentation and increased use of remote sensing. The new insights and instruments increase our capacity to look at within-lake structure. The challenge ahead is to link our new insights on physical processes to species composition and rates of biogeochemical cycling. In the following, we discuss the new understandings and their ecological implications.

## Upper Mixed Layer

Despite its name, the upper mixed layer (or epilimnion) is not always mixing (Falkowski 1983; Imberger 1985a). While the surface layer, the region of the upper mixed layer directly affected by wind forcing and surface heating (Imberger 1985a), is actively mixing, the waters at the base of the mixed layer may not be. The demarcation between zones where mixing does and does not occur can be discriminated by temperature differences as small as a few hundredths of a degree. Even before high-resolution physical measurements were available, Lewis (1973, 1978) observed persistent structure within a weakly stratified upper mixed layer and illustrated its implications for phytoplankton ecology. The persistent stratification allows phytoplankton to be organized in layers (Talling 1981) in what was previously considered a turbulent environment.

The structure of the upper mixed layer, the depth of vertical mixing within it, and the overall extent of stratification within a lake are influenced by a series of inherently spatial factors. In a lake with complex morphometry, differences in sheltering from wind and/or sunlight will lead to horizontal differences in temperature, the depth of the surface mixing layer, the thickness of the thermocline, and the strength of the stratification (MacIntyre et al. 2002). Hence, the light climate of phytoplankton will vary both vertically and horizontally with potential implications for species composition. Even within classes of autotrophs, some species prefer a more constant light field, and others are adapted for ones with fluctuations (Litchman 1998, 2000). It is also common to find populations of phytoplankton deeper in the water column at the base of the metalimnion. The abundance of phytoplankton within these layers may depend on the degree of sheltering, with higher abundance in more sheltered areas where mixing occurs less frequently.

Spatial variations in the rate of heating, cooling, and the depth of wind mixing cause horizontal density differences that generate horizontal overflows and gravity currents at depth (Talling 1963; Imberger 1985b; Imberger and Parker 1985; Monismith et al. 1990; Coates and Ferris 1994; Sturman et al. 1996; Wells and Sherman 2001; MacIntyre et al. 2002). Although these flows could lead to homogenization within the lake basin, in fact they may lead to more persistent vertical stratification and can initiate development of layered communities of organisms. For instance, overflows and underflows occur during wind forcing (Parker and Imberger 1986). Overflows, generally of warmer water from a site with less wind exposure, may lead to subtle temperature differences within the mixed layer that restrict the depth of surface mixing (MacIntyre 1998; Sander et al. 2000). If sub-basins of a lake are not only distinguished by thermal characteristics but also by differences in nutrient loading or species composition, the intrusions from one basin to another may promote formation of small scale layering of organisms.

## Spatial Heterogeneity Due to Inflows

Knauer et al. (2000) identified three factors that will determine whether horizontal patchiness due to an inflow will occur in the mixed layer. First, the volume of the inflow must be low enough that the entire mixed layer is not inundated. Second, the horizontal mixing rate must not be rapid enough for complete dispersion. And third, timescales of chemical or biological transformations must be rapid relative to the physical processes that would tend to erase their signature.

The density of inflowing water regulates the depth at which an inflow will penetrate into a lake. When lakes are weakly stratified in the upper mixed layer, or intrusions occur in the metalimnion, fine-scale layering of solutes, bacteria, and phytoplankton may persist if the turbulence is not sufficient to disperse them. Profiling with high-resolution sensors facilitates discrimination of such fine-scale layering. Although layered communities in the metalimnia and hypolimnia of small, sheltered lakes are well known, we have only just begun to see this structure in larger water bodies (Alldredge et al. 2002; Rines et al. 2002; Lovejoy et al. 2002; McManus et al. 2003).

A high-resolution profile in 1.5 km² Toolik Lake, Alaska, provides an example of fine-scale heterogeneity caused by stream inflow (Figure 16.1). Due to several days of rain in the Toolik catchment, stream inflow to the lake had increased to one of the highest amounts recorded for the site. The lowered conductivity in the metalimnion indicates the intrusion from the stream inflow. Fluorescence illustrates the layering of phytoplankton into one layer formed from the original chlorophyll maximum and another formed as a result of entrainment by vertical mixing (MacIntyre, Sickmon, Goldthwaite, and Kling, unpublished data). Loading of inorganic and organic nutrients depended on time since the discharge began and, due to the temporal variation in stream temperature, led to layering of these nutrients vertically in Toolik Lake. Whether such events lead to layers of different phytoplankton communities, each determined by competition for these different resources, depends on the persistence of the stratification.

Many studies illustrate vertical and horizontal heterogeneity caused by inflows in lakes as small as tens of hectares and as large as Lake Superior (Imberger 1985b; Vincent et al. 1991; Nepf and Oldham 1997; Spigel and

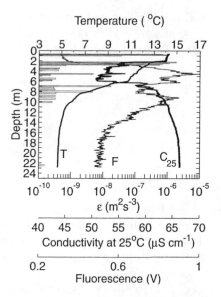

FIGURE 16.1. Temperature-gradient microstructure profile from Toolik Lake, AK, 1300 h 19 July 1999, taken 37 hours after high discharge from a rainstorm. Profile shows temperature (T), conductivity ($C_{25}$), relative fluorescence (F), and turbulence as quantified by the rate of dissipation of turbulent kinetic energy ($\varepsilon$, gray histograms). An intrusion of lower conductivity water was flowing in the metalimnion. Algal biomass was elevated in two layers: one associated with the intrusion, the other a deep chlorophyll maximum. Chlorophyll $a$ concentrations were ~2 $\mu$g $L^{-1}$ in each. Although dissipation was high at its upper boundary, the turbulence was low within the intrusion, $\varepsilon < 3 \times 10^{-8}$ $m^2$ $s^{-3}$, likely due to the strong stratification. The initial surface overflow from the storm caused an increase in primary productivity; the subsequent intrusions introduced terrestrial organic matter and nutrients at different depths in the metalimnion (MacIntyre, Sikmon, Goldthwaite, and Kling, unpublished data).

Priscu 1998; Simek 2001). When inflows have nutrient concentrations or plankton assemblages distinct from the ambient water into which they flow, hot spots are created vertically, horizontally, and temporally in which species interact. Patches with elevated concentrations of food may persist (Lasker 1978; Wroblewski and Richmond 1987; Hembre and Megard 2003) with implications for resultant heterogeneity in competition and growth rates.

## Internal Waves, Turbulence, and Intrusions

One of the most important developments in physical limnology in the past 10 years has been the linkage of the internal wave field to turbulence production. Internal waves are supported where the water column is stratified; hence, this finding is important for fluxes of solutes and particles through the metalimnion. Internal wave amplitudes are largest near sloping boundaries and topographic features, and because increased instabilities are associated with increased wave amplitudes, turbulence is enhanced by up to four orders of magnitude near topographic features (Goudsmit et al 1997; Saggio and Imberger 1998; MacIntyre et al. 1999).

Internal waves occur in lakes of all sizes, although amplitudes are larger and turbulence production is greater in larger lakes (Figure 16.2).

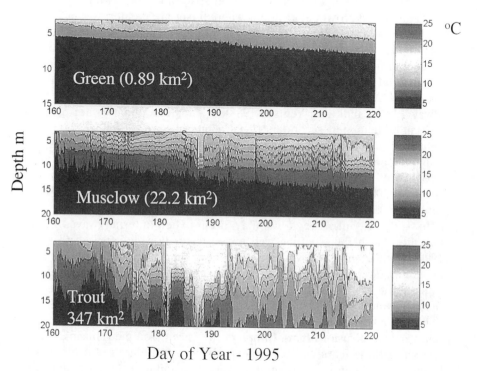

FIGURE 16.2. Time series of isotherms in three lakes of the Northern Ontario Lake Size Series, a set of lakes in the same geological setting exposed to similar meteorology but ranging in size from Green Lake (0.89 km²) to Trout Lake (347 km²) (Fee et al. 1992). Internal wave amplitude increases with lake size. The upper mixed layer is warmer and the thermocline more highly stratified in the smaller lakes (S. MacIntyre, J.R. Romero, and E.J. Fee, unpublished data).

In moderately sized ($150$ km$^2$) Mono Lake, when internal wave amplitudes increased after wind forcing of $10$ m s$^{-1}$, $62\%$ of the metalimnion was turbulent at sites where bottom slopes exceeded $0.02$. In contrast, at sites where bottom slopes were less than $0.001$, less than $6\%$ of the water column was turbulent (S. MacIntyre, J. Clark, and R. Jellison, unpublished data). During calmer periods, even less of the water column was turbulent. Based on the spread of a conservative tracer ($SF_6$), approximately an order of magnitude more material was transported through the thermocline during a 2-day storm event than during 6 calm days. These findings indicate that nutrient and particle fluxes will vary over space and time with the possibility of enhanced growth at locations where fluxes are greater.

Due to the increased internal wave amplitudes and mixing in the metalimnion at lake margins, the boundaries of lakes are likely to be hot spots of biogeochemical activity. Rates of primary production are likely to be higher due to greater nutrient fluxes (MacIntyre et al. 1999; MacIntyre and Jellison 2001) and overall higher irradiance due to internal wave movements (Holloway 1984; Lande and Yentsch 1988). If anoxic boundaries are present, rates of methanogenesis or denitrification are likely to be higher due to the enhanced mixing of reactants. Turbulence in the central portions of lakes only appears to lead to significant turbulent transport immediately after strong wind forcing (MacIntyre and Jellison 2001; Etemad-Shahidi and Imberger 2001; Saggio and Imberger 2001). Consequently, reactions in these central portions will proceed at a slower pace.

In addition, intrusive flows may occur when internal waves break nearshore with the resulting well mixed water flowing offshore (Thorpe 1998; McPhee-Shaw and Kunze 2002). Due to the reduced vertical mixing in metalimnetic waters offshore, these intrusions, similar to the ones induced by stream flows, also have the potential to be hot spots for biogeochemical reactions and to develop a distinct species assemblage.

Within lakes, spatial heterogeneity exists on the scale of gyres (km) to thin vertical intrusions (cm). Coupling the insights on mechanisms of formation of these features with the experimental paradigms of ecology will lead to a better understanding of the factors leading to species patchiness, biodiversity, and spatial variations in biogeochemical cycling within lakes. Furthermore, because lake size is not random across the landscape, that is, lakes higher in the flow system tend in general to be smaller than lakes lower in the flow system (see next section), lakes with thin epilimnia and highly stratified metalimnia may dominate high in the flow system. In these lakes, intrusive flows may be a crucial determinant of within-lake variability. In contrast, internal waves are likely to have larger amplitudes and have the potential to lead to greater mixing near boundaries in lakes lower in the landscape. Horizontal flows, long believed to be an agent that would reduce heterogeneity, are frequently organized into coherent features by the interactions of lake morphometry and wind (Melack and Gastil 2001; Rueda et al. 2003; Stocker and Imberger 2003; Dodson 2005, Plate 41). With our ability to discriminate physical features over a broad range of sizes, we are poised to determine their consequences for spatial heterogeneity of biogeochemical processes and species composition.

## Landscape-Scale Heterogeneity

Over the past 50 years, it has become obvious that lakes are strongly influenced by the characteristics of their watersheds. Numerous studies, for example, have shown that nutrient (e.g., Soranno et al. 1996) and dissolved organic carbon (e.g., Dillon and Molot 1997) loading to lakes depends on the size, land-use, geology, and hydrology of their watersheds. This research has been crucial in developing a better understanding of important lake issues such as cultural eutrophication.

More recently, there has been an increased interest in understanding how and why neighboring lakes differ in their characteristics and dynamics (e.g., Kratz et al. 1997; Hershey et al. 1999). Lakes are often prominent and abundant features of formerly glaciated landscapes worldwide. Locally, neighboring lakes share the same climate, geologic setting, age, process of origin, and watershed characteristics. Yet, these lakes often differ markedly in physical, chemical, and biological attributes and in how these attributes change over time. In this section, we discuss how water flow across a landscape can cause heterogeneity among lakes in their physical, chemical, and biological characteristics and how this heterogeneity can affect lake dynamics.

Any process that has a heterogeneous distribution across the landscape can potentially lead to differences among neighboring lakes. Many recent studies have concluded that water movement across the landscape is an important contributor to spatial heterogeneity among lakes (Kratz et al. 1997; Soranno et al. 1999; Kling et al. 2000; Webster et al. 2000; Winter 2001). Although precipitation may fall relatively uniformly at local scales, lateral movement of water either through surface flow or groundwater can cause spatial patterning across lakes in a landscape. In a study of northern Wisconsin lakes situated in a groundwater-dominated hydrologic setting, Riera et al. (2000) correlated a number of lake characteristics, including area, specific conductance, pH, water clarity, and fish species richness with lake order, a measure of the relative position of a lake in the flow system (Figure 16.3). Lakes high in the flow system tend to be smaller, more clear, less used by humans, have lower ionic strength and acid neutralizing capacity, and have fewer fish species than lakes lower in the flow system (Riera et al. 2000). In a series of northern Alaska lakes connected by streams, Kling et al. (2000) found increasing patterns of conductivity, acid neutralizing capacity, calcium, and magnesium from high to low in the lake chain. Similarly, in a study of lake chains from areas throughout North America, Soranno et al. (1999) found that conductivity generally, but not always, was greater in lakes lower in the lake chain (Figure 16.4).

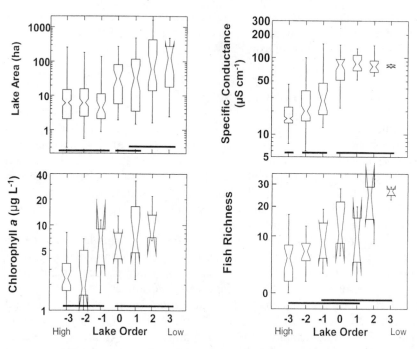

FIGURE 16.3. The relationship between position of lakes in the hydrologic flow system and four limnological variables. Lake order is a measure of hydrologic position with negative orders indicating lakes high in the flow system and positive numbers indicating lakes lower in the flow system (see Riera et al. 2000). "High" and "Low" refer to position in the flow system. Horizontal bars indicate lake orders that are not statistically different according to multiple means tests. From Magnuson et al. (2005). Used by permission of Oxford University Press.

One likely mechanism to explain some of these systematic differences in lake characteristics with position in the flow system involves chemical evolution of water as it flows through the landscape. For example, in northern Wisconsin, where groundwater is an important part of the hydrologic system, silicate hydrolysis leads to increases in acid neutralizing capacity, silica, and other chemical constituents in water as it flows through the noncalcareous sandy tills and outwash (Kenoyer et al. 1992a, 1992b). Similarly, Kling et al. (2000) suggest that material processing that occurs in the surface waters of lakes and streams as water moves from headwaters to downstream lakes in northern Alaska can lead to systematic spatial differences in the water chemistry of lakes. Some of the biological differences observed among lakes as a function of position in the flow system, such as snail abundance and community structure (Lewis and Magnuson 2000), owe indirectly to these chemical differences among lakes (Kratz et al. 1997; Riera et al 2000). In addition, in systems where seasonal or annual migration between streams and lakes is an important part of the life cycle of organisms, factors controlling this

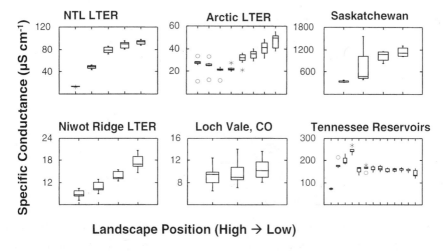

FIGURE 16.4. The relationship between position of a lake in a lake chain and specific conductance for six lake chains in North America. From Magnuson et al. (2005). Used by permission from Oxford University Press.

migration, such as the size and gradient of streams, or controlling the viability of the organisms, such as the size or depth of the lakes, can lead to systematic spatial differences in biological community structure across a landscape (Hershey et al. 1999).

Movement of water across a landscape also has the potential to create spatial patterns in how lakes respond to short- and long-term changes in external drivers such as atmospheric deposition of solutes, land use, or weather. Lakes within a lake district are exposed to a complex set of external drivers that differ in return frequency, magnitude, duration, and spatial scale. Much of the interannual variation we observe in lakes is related to weather; year-to-year shifts in temperature and precipitation generate substantial background "noise" in limnological variables. Interacting with this background variation are dynamics induced by press-type external drivers such as acid deposition and by pulsed events such as drought.

Does the spatial heterogeneity of lake features produced by water flow across the landscape create spatial patterns in how lakes respond to these press or pulse events? The response of any one lake to interannual climatic variability, to press drivers such as atmospheric deposition, or to pulsed events such as drought can be thought of as determined by a hierarchy of filters that progressively constrain the range of possible outcomes. These filters are essentially properties that can attenuate or amplify responses to the external driver or generate time lags in response (Magnuson et al. 2005; Webster et al. 2000). At the highest level of the hierarchy are regional filters like geology, hydrologic setting, and climatology (Figure 16.5). At the lowest level are internal lake properties like morphometry and food web structure that ultimately shape the final expression of a given lake's response to the

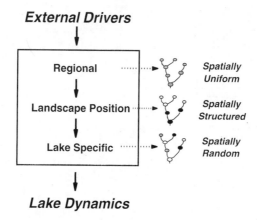

FIGURE 16.5. Diagram of hierarchy of controls on lake dynamics and expected spatial patterns in lake dynamics generated at each level of control (modified from Webster et al. 2000). Lines between lakes (shown by dots) indicate connections along the flowpath; dots with similar shading have similar dynamics.

external signal. At an intermediate level are filters set by a lake's position in hydrologic flow system, constrained by regional controls and subject to the variation generated by internal lake processes.

Using this concept of hierarchical filters, we expect that the spatial configuration of lake responses to external drivers will differ depending on the level of control (Figure 16.5) (Webster et al. 2000). If the dynamics of a given variable are controlled primarily by factors acting at the regional scale, we would expect dynamics to be uniform across the landscape—all lakes would respond in the same way. If control were by internal, lake-specific factors, we would expect dynamics across lakes to be spatially random. If, however, position in the flow system was important, we would expect a response pattern that was spatially structured in that lakes at similar positions in the landscape would show similar dynamics.

Acid deposition, a well studied regional-scale external driver, provides a useful illustration of the application of this hierarchical approach. Lakes whose dynamics are influenced by acid deposition are primarily those located in geologic regions where sources of buffering capacity to lakes are low (Kalff 2002)—this regional filter constrains the expression of the acidification response. However, not all lakes within a geologically sensitive region are affected by acid inputs. For the most part, lakes that respond to changes in acid deposition are either headwater lakes or precipitation-dominated seepage lakes (which lack surface water inlets or outlets) (Eilers et al. 1983). Spatial heterogeneity imposed by position within the flow system here reflects the relative dominance of precipitation compared to other water sources such as groundwater that are enriched with solutes that buffer lakes from incoming acids. But even at this level of refinement, long-term trends in water chemistry variables of interest like pH, acid neutralizing capacity (ANC), and sulfate can differ across lakes in a region and at similar positions in the landscape (Stoddard et al. 1999). This suggests that lake-specific factors also modulate the strength of response to changing deposition rates. The response of an individual lake to long-term acid deposition is constrained hierarchically by the lake's geologic setting, position in the flow system, and finally a set of factors specific to the lake itself.

Other examples of how the spatial heterogeneity of water flowpaths across the landscape can affect lake dynamics come from studies of lake responses to drought and experimental logging. During a sustained drought in midwestern North America in the late 1980s, Webster et al. (1996) observed that northern Wisconsin lakes low in the flow system accumulated calcium and magnesium whereas lakes high in the flow system either showed no change or a decrease. Because weathering reactions in the groundwater system are the primary source of these ions, this spatial pattern suggests that the way drought influences lake-groundwater connections depends on the location of the lake in the flow system. A consequence of prolonged drought would be the disconnection of lakes high in the flow system from groundwater inputs, their major source of buffering capacity, making these lakes even more sensitive to acidification by acid deposition (Webster et al. 1990, 1996). Similar spatial patterning in dynamics was observed in a groundwater-dominated hydrologic setting in northern Alberta by Devito et al. (2000). Recharge lakes, located higher in the flow system, showed the largest increase in total phosphorus (P) after experimental logging in their catchments. In this P-enriched boreal forest, surface and near-surface flowpaths are significant sources of P. In lakes located lower in the flow system, increased P-export after logging activities was modulated by discharge of groundwater from regional and local flowpaths which were unaffected by logging activities. Such spatial heterogeneity is lacking in the Experimental Lakes Area (ELA) in northwestern Ontario (Webster et al. 2000). The ELA lakes are dominated by surface water flow system in a region with exposed bedrock and relatively homogeneous geology that may limit spatially explicit patterns.

Collectively, these results suggest that geological setting is an important determinant of spatial heterogeneity in many physical, chemical, and biological attributes among lakes within a lake district. In lake districts where chemical evolution of water along horizontal flowpaths is significant, spatial heterogeneity among lakes is likely to occur. These conditions are met in lake districts with deep porous soils or glacial materials such as those where groundwater flow is important. In contrast, in lake districts set in shallow soils or with exposed bedrock, where chemical evolution of water is not strong because of relatively rapid horizontal movement of water, lakes are more likely to be homogenous in their average conditions and their dynamics. We would expect that, all else being equal, lake districts with a more heterogeneous assemblage of lakes would have overall a greater species richness and a broader array of dominant biogeochemical processes than lake districts that are more homogeneous.

## Human-Induced Changes in Heterogeneity

The reciprocal interactions between people and lakes can influence heterogeneity and ecosystem processes both among and within lakes. In an analysis of land-use change trajectories in southeastern Michigan, Walsh et al. (2003) found that lakes are attractors of residential development. Areas within up to 800 m of a lake had higher proportions of residential land-use than that in the entire 1720 km² study area. Interestingly, the disproportionately high residential development around lakes has increased from 1938 to 1995, suggesting that lakes have become a more powerful influence on land-use patterns over the past 60 years. Similarly, Schnaiberg et al. (2002) report that since the 1960s, more than half of all new homes were built on lakeshores in a recreational lake district in northern Wisconsin.

Although these differential patterns of residential development may act to increase heterogeneity in land use at the scale of hundreds of km², the effect on individual lakes may be one of homogenization. For example, complexity of physical structure of the littoral zone appears to be inversely related to residential development of lakes. In northern Wisconsin, lakes with high amounts of residential development have up to 10-fold less coarse woody habitat than undeveloped lakes (Christensen et al. 1996). In a study of northern Minnesota lakes, macrophyte density was reduced in lakes that had high levels of residential development (Radomski and Goeman 2001). This simplification of the physical structure of the littoral zone has been linked with reduced growth rates of bluegills and largemouth bass, two common fish species in these lakes (Schindler et al. 2000). As lakes become developed, their littoral zones may become more similar as humans modify the inshore areas of lakes. Such habitat modifications combined with species introductions by humans contribute to the homogenization of aquatic communities (Rahel 2002). The effect of human activities on the heterogeneity of lakes may differ according to the scale considered and certainly warrants attention in the future.

## Conclusions

Lakes exhibit spatial heterogeneity at multiple spatial scales. At within-lake scales, heterogeneity can be subtle, fine scale, and highly dynamic. Stream intrusions, differential heating and cooling of inshore and offshore surface waters, and internal waves breaking against the lake boundary can all lead to spatial and temporal heterogeneity in physical, chemical, and biological characteristics. Across a landscape, lakes can differ systematically as a function of their position in the hydrologic flow system. These among-lake differences occur in average conditions as well as in among-year dynamics. We currently know much more about the forces shaping heterogeneity in physical and chemical aspects of lakes than we do about biological and ecosystem-level consequences. Such analyses are further complicated by human activities that potentially alter patterns of spatial and temporal heterogeneity driven by physical and geomorphic forces. Ongoing research in these areas will continue to expand our understanding of the interplay between spatial heterogeneity and lake ecosystems.

## References

Alldredge, A.L., Cowles, T.J., MacIntyre, S., Rines, J.E.B., Donaghay, P. L., Greenlaw, C.F., Holliday, D.V., Dekshenieks, M.M., Sullivan, J.M., and Zaneveld. J.R.V. 2002. Occurrence and mechanisms of formation of a dramatic thin layer of marine snow in a shallow Pacific fjord. Marine Ecol. Progress Ser. 233: 1–12.

Christensen, D.L., Herwig, B.R., Schindler, D.E., and Carpenter, S.R. 1996. Impacts of lakeshore residential development on coarse woody debris in north temperate lakes. Ecol. Appl. 6: 1143–1149.

Coates, M.J., and Ferris, J., 1994. The radiatively driven natural convection beneath a floating plant layer. Limnol. Oceanogr. 39: 1186–1194.

Devito, K.J., Creed, I.F., Rothwell, R.L., and Prepas, E.E. 2000. Landscape controls on phosphorus loading to boreal lakes: implications for the potential of forest harvesting. Can. J. Fisheries Aquatic Sci. 57: 1977–1984.

Dillon, P.J., and Millot, L.A., 1997. Dissolved organic and inorganic carbon mass balances in central Ontario lakes. Biogeochemistry 36: 29–42.

Dodson, S.I. 2005. Introduction to limnology. New York: McGraw-Hill. 400 pp.

Eilers, J.M., Glass, G.E., Webster, K.E., and Rogalla J.A. 1983. Hydrologic control of lake susceptibility to acidification. Can. J. Fisheries Aquatic Sci. 40: 1896–1904.

Etemad-Shahidi, A., and Imberger, J. 2001. Anatomy of turbulence in thermally stratified lakes. Limnol. Oceanogr. 46: 1158–1170.

Falkowski, P.G. 1983. Light-shade adaptation and vertical mixing of marine phytoplankton: a comparative field study. J. Marine Res. 41: 215–237.

Fee, E.J., Shearer, J.A., DeBruyn, E.R., and Schindler, E.U. 1992. Effects of lake size on phytoplankton photosynthesis. Can. J. Fisheries and Aquatic Sci. 49: 2445–2459.

Gergel, S.E., Turner, M.G., Kratz, T.K. 1999. Dissolved organic carbon as an indicator of the scale of watershed influence on lakes and rivers. Ecol. Applications. 9: 1377–1390.

Goldman, C.R., and Horne, A.J. 1994. Limnology, 2nd edition. New York: McGraw-Hill. 576 pp.

Goudsmit, G.-H., Peeters, F., Gloor, M., and Wuest, A. 1997. Boundary versus internal diapycnal mixing in stratified natural waters. J. Geophys. Res. 102: 27903–27914.

Hembre, L.K., and Megard, R.O. 2003. Seasonal and diel patchiness of a Daphnia population: an acoustic analysis. Limnol. Oceanogr. 48: 2221–2233.

Hershey, A. E., Gettel, G. M., McDonald, M. E., Miller, M. C., Mooers, H., O'Brien, W.J., Pastor, J., Richards, C., and Schuldt, J. A. 1999. A geomorphic-trophic model for landscape control of arctic lake food webs. BioScience 49: 887–897.

Holloway, G. 1984. Effects of velocity fluctuations on vertical distributions of phyto-plankton. J. Marine Res. 42: 559–571.

Imberger, J. 1985a. The diurnal mixed layer. Limnol. Oceanogr. 30: 737–770.

Imberger, J. 1985b. Thermal characteristics of standing waters: an illustration of dynamic processes. In Perspectives in Southern Hemisphere Limnology, eds. B.R. Davies and R.D. Walmsley, pp. 7–29. Boston: W. Junk Publishers.

Imberger, J., and Parker, G. 1985. Mixed layer dynamics in a lake exposed to a spatially variable wind field. Limnol. Oceanogr. 39: 473–488.

Juday, C., and Birge, E.A. 1933. The transparency, the color and the specific conductance of the lake waters of northeastern Wisconsin. Trans. Wisconsin Acad. Sci. Arts Lett. 28: 205–259.

Kalff, J. 2002. Limnology: inland water ecosystems. Englewood Cliffs, NJ: Prentice Hall. 592 pp.

Kenoyer, G.J., and Bowser, C.J. 1992a. Groundwater chemical evolution in a sandy silicate aquifer in northern Wisconsin: 1. Patterns and rates of change. Water Resources Res. 28: 579–89.

Kenoyer, G.J., and Bowser, C.J. 1992b. Groundwater chemical evolution in a sandy silicate aquifer in northern Wisconsin: 2. Reaction modeling. Water Resources Res. 28: 591–600.

Kling, G.W., Kipphut, G.W., Miller, Michael M., and John. O'Brien, W. 2000. Integration of lakes and streams in a landscape perspective: the importance of material processing on spatial and temporal coherence. Freshwater Biol. 43: 477–497.

Knauer, K., Nepf, H.M., and Hemond, H.F. 2000. The production of chemical heterogeneity in Upper Mystic Lake. Limnol. Oceanogr. 45: 1647–1654.

Kratz, Timothy K., Webster, Katherine E., Bowser, Carl J., Magnuson, John J., and Benson, Barbara J. 1997. The influence of landscape position on northern Wisconsin lakes. Freshwater Biol. 37: 209–217.

Lande, R., and Yentsch, C.S. 1988. Internal waves, primary production, and the compensation depth of marine phytoplankton. J. Plankton Res. 10: 565–571.

Langmuir, I., 1938. Surface motion of water induced by wind. Science 87: 119–123.

Lasker, R. 1978. The relation between oceanographic conditions and larval anchovy food in the California Current: identification of factors contributing to recruitment failure. Rapp. P.-V. Reun. Cons. Int. Explor. Mer. 173: 212–230.

Lewis, W.M., Jr. 1973. The thermal regime of Lake Lanao (Philippines) and its theoretical implications for tropical lakes. Limnol. Oceanogr. 18: 200–217.

Lewis, W.M., Jr. 1978. Spatial distribution of the phytoplankton in a tropical lake (Lake Lanao, Philippines). Int. Rev. Ges. Hydrobiol. 63: 619–635.

Lewis, W.M. 1983. A revised classification of lakes based on mixing. Can. J. Fisheries Aquatic Sci. 40: 1779–1787.

Lewis, D.B., and Magnuson, J.J. 2000. Landscape spatial patterns in freshwater snail assemblages across Northern Highland catchments. Freshwater Biol. 43: 409–420.

Litchman, E. 1998. Population and community responses of phytoplankton to fluctuating light. Oecologia 117: 247–257.

Litchman, E. 2000. Growth rates of phytoplankton under fluctuating light. Freshwater Biol. 44: 223–235.

Lovejoy, Carmack, C.E.C., Legendre, L., and Price, N.M. 2002. Water column interleaving: A new physical mechanism determining protist communities and bacterial states. Limnol. Oceanogr. 47: 1819–1831.

MacIntyre, S. 1998. Turbulent mixing and resource supply to phytoplankton. In Physical Processes in Lakes and Oceans, ed. J. Imberger, pp. 539–567. Coastal and Estuarine Studies. Washington, DC: American Geophysical Union.

MacIntyre, S., and Jellison, R. 2001. Nutrient fluxes from upwelling and enhanced turbulence at the top of the pycnocline in Mono Lake, CA. Hydrobiologia 466: 13–29.

MacIntyre, S., Flynn, K.M., Jellison, R., and Romero, J.R. 1999. Boundary mixing and nutrient flux in Mono Lake, CA. Limnol. Oceanogr. 44: 512–529.

MacIntyre, S., Romero, J.R., and Kling, G.W. 2002. Spatial-Temporal Variability in Mixed Layer Deepening and Lateral Advection in an Embayment of Lake Victoria, East Africa. Limnol. Oceanogr. 47: 656–671.

Magnuson, J.J., and Kratz, T.K. 2000. Lakes in the landscape: approaches to regional limnology. Internationale Vereinigung Fur Theoretische Und Angewandte Limnologie 27: 74–87.

Magnuson, J.J., Tonn, W.M., Banerjee, A., Toivonen, J., Sanchez, O., and Rask, M. 1998. Isolation vs. extinction in the assembly of fishes in small northern lakes. Ecology 79: 2941–2956.

Magnuson, J.J., Kratz, T.K., Benson, B.J., and Webster, K.E. 2005. Coherent dynamics among lakes. In Long-term dynamics of lakes in the landscape, eds. J.J. Magnuson, T.K. Kratz, and B.J. Benson. New York: Oxford University Press.

McManus, M.A., Alldredge, A.L., Barnard, A., Boss, E., Case, J., Cowles, T.J., Donaghay, P.L., Eisner, L.B., Gifford, D.J., and Greenlaw, C.F., et al. 2003. Changes in Characteristics, Distribution and Persistence of Thin Layers Over a 48 Hour Period. Marine Ecol. Progress Ser. 261: 1–19.

McPhee-Shaw, E.E., and Kunze, E. 2002. Boundary layer intrusions from a sloping bottom: A mechanism for generating intermediate nepheloid layers. J. Geophys. Res. 107: 3050.

Melack, J.M., and Gastil, M. 2001. Airborne remote sensing of chlorophyll distributions in Mono Lake, California. Hydrobiologia 466: 31–38.

Monismith, S.G., Imberger, J., and Morrissey, T. 1990. Horizontal convection in the sidearm of a small reservoir. Limnol. Oceanogr. 35: 1676–1702.

Nepf, H.M., and Oldham, C.E. 1997. Exchange dynamics of a shallow contaminated wetland. Aquatic Sci. 59: 193–213.

Parker, G.J., and Imberger, J. 1986. Differential mixed-layer deepening in lakes and reservoirs. In Limnology in Australia, eds. P. De Deckker and W.D. Williams, pp. 63–92. Boston: W. Junk Publishers.

Radomski, P., and Goeman, T. J. 2001. Consequences of human lakeshore development on emergent and floating-leaf vegetation abundance. North Am. J. Fisheries Manage. 21: 46–61.

Rahel, F. J. 2002. Homogenization of freshwater faunas. Annu. Rev. Ecol. Systematics 33: 291–315.

Richerson, P.J., Powell, T.M., Leigh-Abbott, M.R., and Coil, J.A. 1978. Spatial heterogeneity in closed basins. Spatial pattern in plankton communities, ed. J.H. Steele, pp. 239–276. New York: Plenum Press.

Riera, J.L., Magnuson, J.J., Kratz, T.K., and Webster, K.E. 2000. A geomorphic template for the analysis of lake districts applied to the Northern Highland Lake District, Wisconsin, USA. Freshwater Biol. 43: 301–318.

Rines, J.E.B., Donaghay, P.L., Dekshenieks, M.M., Sullivan, J.M., Twardowski, M.S. 2002. Thin layers and camouflage: hidden Pseudo-nitzschai (Bacillariophyceae) populations in a fjord in the San Juan Islands, Washington, USA. Marine Ecol. Progress Ser. 225: 123–137.

Rueda., F.J., Schladow, S.G., Monismith, S.G., and Stacey, M.T. 2003. Dynamics of a large polymictic lake. I: Field observations. ASCE J. Hydrologic Eng. 129: 82–91.

Saggio, A., and Imberger, J. 1998. Internal wave weather in a stratified lake. Limnol. Oceanogr. 1780–1795.

Saggio, A., and Imberger, J. 2001. Mixing and turbulent fluxes in the metallimnion of a stratified lake. Limnol. Oceanogr. 46: 392–409.

Sander, J., Simon, A., Jonas, T., and Wuest, A. 2000. Surface turbulence in natural waters: A comparison of large eddy simulations with microstructure observations. J. Geophys. Res. 105: 1195–1207.

Schindler, Daniel E., Geib, Sean I., and Williams, Monica R. 2000. Patterns of fish growth along a residential development gradient in north temperate lakes. Ecosystems 3: 229–237.

Schnaiberg, J., Riera, J., Turner, M.G., and Voss. P.R. 2002. Explaining human settlement patterns in a recreational lake district: Vilas County Wisconsin, USA. Environ. Manage. 30: 24:34.

Simek, K., Armengol, J., Comerma, M., Garcia, J.C., Kojecka, P., Redoma, J., and Hejzlar, J. 2001. Changes in the epilimnetic bacterial community composition, production, and protest-induced mortality along the longitudinal axis of a highly eutrophic reservoir. Microbial Ecol. 42: 359–371.

Snucins, E., and Gunn, J. 2000. Interannual variation in the thermal structure of clear and colored lakes. Limnol. Oceanogr. 45: 1639–1646.

Soranno, P.A., Hubler, S.L., and Carpenter, S.R. 1996. Phosphorous loads to surface waters: a simple model to account for spatial pattern of land-use. Ecolo. Applications 6: 865–878.

Soranno, P.A., Webster, K.E., Riera, J.L., Kratz, T.K., Baron, J.S., Bukaveckas, P.A., Kling, G.W., White, D.S., Caine, N., Lathrop, R.C., and Leavitt, P.R. 1999. Spatial variation among lakes within landscapes: ecological organization along lake chains. Ecosystems 2: 395–410.

Spigel, R.H., and Priscu, J.C. 1998. Physical limnology of the McMurdo Dry Valley Lakes. In Ecosystem dynamics in a polar desert: The McMurdo Dry Valleys, Antarctica, ed. J.C. Priscu, pp. 153–187. AGU Antarctic Research Series vol. 72. Washington, DC: American Geophysical Union.

Stocker, R., and Imberger, J. 2003. Horizontal transport and dispersion in the surface layer of a medium-sized lake. Limnol. Oceanogr. 48: 971–982.

Stoddard, J.L., Jeffries, D.S., Lukewille, A., Clair, T.A., Dillon, P.J., Driscoll, C.T., Forsius, M., Johannessen, M., Kahl, J.S., and Kellogg, J.H., et al. 1999. Regional trends in aquatic recovery from acidification in North America and Europe 1980–95. Nature. 401: 575–578.

Sturman, J.J., Ivey, G.N., and Taylor, J.R. 1996. Convection in a long box driven by heating and cooling on the horizontal boundaries. J. Fluid Mechanics 30: 61–87

Talling, J.F. 1963. Origin of stratification in an African rift lake. Limnol. Oceanogr. 8: 68–78.

Talling, J.F. 1981. The development of attenuance depth-profiling to follow the changing distribution of phyptoplankton and other particulate material in a productive English lake. Archiv fur Hydrobiologie 93: 1–20.

Thorpe, S.A. 1998. Some dynamical effects of the sloping sides of lakes. In Physical processes in lakes and oceans, ed. J. Imberger, pp. 441–460. Coastal and Estuarine Studies. Washington, DC: American Geophysical Union.

Vincent, W.F., Gibbs, M.M., and Spigel, R.H. 1991. Eutrophication processes regulated by a plunging river inflow. Hydrobiologia 226: 51–63.

Walsh, S.E., Soranno, P.A., and Rutledge, D.T. 2003. Lakes, wetlands, and streams as predictors of land use/cover distribution. Environ. Manage. 31: 198–214.

Webster, K.E., Newell, A.D., Baker, L.A., and Brezonik, P.L. 1990. Climatically induced rapid acidification of a softwater seepage lake. Nature 6291: 374–376.

Webster, K.E., Kratz, T.K., Bowser, C.J., Magnuson, J.J., and Rose, W.J. 1996. The influence of landscape position on lake chemical responses to drought in northern Wisconsin, USA. Limnol. Oceanogr. 41: 977–984.

Webster, K.E., Soranno, P.A., Baines, S.B., Kratz, T.K., Bowser, C.J., Dillon, P.J., Campbell P., Fee, E.J., and Hecky, R.E. 2000. Structuring features of lake districts: geomorphic and landscape controls on lake chemical responses to drought. Freshwater Biol. 43: 499–516.

Wells, M.T., and Sherman, B. 2001. Stratification produced by surface cooling in lakes with significant shallow regions. Limnol. Oceanogr. 46: 1747–1759.

Wetzel, R.G. 2001. Limnology: lake and river ecosystems, 3rd edition. New York: Academic Press. 1006 pp.

Winter, T.C. 2001. The concept of hydrologic landscapes. J. Am. Water Resources Assoc. 37: 335–49.

Wroblewski, J.S., and Richman, J.G. 1987. The non-linear response of plankton to wind mixing events—implications for survival of larval northern anchovy. J. Plankton Res. 9: 103–123.

# Section IV

## Application of Frameworks and Concepts

# Editors' Introduction to Section IV: Application of Frameworks and Concepts

The previous sections have provided a rich repertoire of conceptual approaches to, perspectives on, and examples of ecosystem functioning in heterogeneous landscapes. Nevertheless, these chapters have largely ignored or only tangentially addressed the implications for natural resource management. In the following section, three chapters tackle this difficult issue head on in three different arenas: fire management, water management, and conservation planning.

Managers, faced with external constraints and limited resources, are often forced to treat the world as more homogeneous than it is. Clearly, incorporating important aspects of functional heterogeneity has the potential to enhance ecosystem and landscape management. But how much heterogeneity of what type do they need to incorporate? What is the best, most parsimonious way to incorporate such heterogeneity into management plans?

As one might expect, there is no simple answer to this question. However, all of the authors argue that there is a real need for research on functional heterogeneity that is of direct use to managers; a real need to get our current understanding incorporated into management; and a real need for frameworks integrating functional heterogeneity into management planning.

Bill Romme (Chapter 17) shows us that wildland fire regimes have diverse, scale-dependent causes, from regional control by climate and vegetation characteristics to local control by fuel mass and structure periodically overridden by the influence of extreme weather. Further, almost all fires burn heterogeneously, creating mosaics of fire severity with spatially variable consequences for plant community structure, soil characteristics, energy flow, and biogeochemistry. In contrast, current fire management practices often treat fire as relatively homogenous, with fuel load control as the primary management tool. He argues that although we know much about the role of heterogeneity in the causes of fire, there is a pressing need to incorporate this understanding into wildland fire policy and management. He also argues that we urgently need more research on spatial patterns in fire history, fire effects, and organismal and ecosystem responses to fire spatial variability, as a basis for the enhanced wildland fire management in the future.

Alan Steinmann and Rodney Denning (Chapter 18) also point out that spatial heterogeneity in ecosystem structure and function is rarely considered in water management, despite the fact that its incorporation provides a new way to view freshwater resources with potentially useful management strategies. They begin by using the general concept of landscape connectivity as the overarching feature of aquatic systems, examining upstream-downstream, hydrogeomorphic, floodplain-river, hillslope-river, surface water-groundwater, and within-ecosystem linkages. They then develop a conceptual framework relating these linkages to the diversity of ecosystem services provided by water. The framework is used to explore how the importance of compositional and configurational heterogeneity within these linkage types varies depending on the particular ecosystem good or service selected. The result is a series of postulates as to what type of heterogeneity will influence what type of ecosystem good or service—a potential road map for managers. They then show how this framework can be applied using the Greater Everglades Ecosystem in south Florida. Here, spatial heterogeneity and the configuration of this highly interconnected and heavily managed system affects numerous ecosystem services—water supply, water quality, navigation, and Everglades restoration—and they explore the socioeconomic and environmental consequences of this connectivity under different flow conditions.

Hugh Possingham and colleagues (Chapter 19) ask to what extent an understanding of landscape spatial heterogeneity can inform conservation decisions, using reserve design and population viability analysis as their foci. With reserve design they show that the fundamental conservation planning principles of comprehensiveness and representativeness very much depend on compositional landscape heterogeneity. However, the planning principle of adequacy, which relies on understanding configurational heterogeneity, is not well incorporated into planning and deserves increased attention. They suggest and illustrate one way of enhancing historic pattern-based approaches to configurational heterogeneity by incorporating spatially explicit ecosystem processes, but admit much more work needs to be done. Within the context of population viability analysis (PVA) they call for the development of theory and decision support tools that integrate population viability with spatially explicit ecological processes. Although PVA invariably includes spatial population processes, it has largely focused on landscape configurational heterogeneity. They astutely point out that this focus might only be justified when the scale of planning coincides with either the scale of habitat heterogeneity or the scale at which small populations are self-sustaining. They conclude that integrating PVA into conservation planning with both compositional and configurational heterogeneity are important future challenges.

# 17
# The Importance of Multiscale Spatial Heterogeneity in Wildland Fire Management and Research

WILLIAM H. ROMME

## Abstract

The occurrence and effects of fire vary greatly over multiple spatial and temporal scales. At a regional scale, variation in synoptic climate and associated vegetation characteristics results in diverse fire regimes, ranging from systems having frequent, low-severity fires (e.g., pine forests of the southwestern and southeastern United States) to systems characterized by infrequent but stand-replacing fires (e.g., subalpine and boreal forests of North America). At a finer scale, spatial variability in fuel mass and structure may influence fire ignition and severity under a middle range of weather conditions, but effects of fuels may be overwhelmed by effects of extreme weather—either extremely wet (no fire) or extremely dry and windy (large, severe fires). Almost all fire events exhibit a heterogeneous pattern of burning and create a mosaic of fire severity within the burned area, resulting in spatially variable changes in plant community structure, soil characteristics, and ecosystem processes of energy and biogeochemistry. We have a pressing need to better incorporate our understanding of spatial heterogeneity into wildland fire policy and management and to address urgent research questions about spatial patterns in fire history, fire effects, and responses of organisms and ecosystems to the spatial variability of fire.

## Introduction

The devastating fire season of 2000 awakened the American people to the need for better understanding and management of wildland fires. More than 120,000 fires burned over 8.4 million acres and destroyed over 860 structures, while firefighting efforts cost approximately $1.3 billion (Machlis et al. 2002; iii). The U.S. Departments of Interior and Agriculture responded by developing the National Fire Plan, and the U.S. Congress implemented the plan with an appropriation of approximately $2.8 billion in 2001 (Machlis et al. 2002; 26). Implementation was barely underway when the devastating 2002

353

and 2003 wildfire seasons occurred, resulting in calls for even more aggressive action to reduce fire hazards, with a particular emphasis on fuels reduction. Although fire hazard is indeed acute in many areas, the widely touted Healthy Forests Restoration Act of 2003 appears seriously oversimplified from an ecological perspective and may in fact result in little protection from damaging fire but serious damage to the land in many places. Most troubling about the plan is its failure to explicitly acknowledge ecological heterogeneity: it appears to assume tacitly that (i) the fire hazard and its root causes are essentially the same in all forests, so the same basic approach to fire mitigation can be applied almost everywhere; and (ii) fire behavior and effects are controlled primarily by fuel conditions in all types of forests, hence, reduction of fuel mass by any means will reduce damaging fire behavior.

Our current understanding of fire ecology in western U.S. forests is sufficient to begin developing more effective fire management programs that are tailored to unique ecological conditions. A crucial task is simply to incorporate this knowledge into specific policy actions and to integrate the science with the social and economic concerns unique to each community facing a threat of wildfire damage. However, there are many aspects of fire ecology that we do not yet understand adequately, and so we need to identify and prioritize the key scientific questions that bear on major issues of fire policy and management (Veblen 2003).

In this chapter, I examine one component of this developing research framework, viz., the importance of spatial heterogeneity in fire occurrence and fire effects. All fire events, except perhaps the tiniest ones, exhibit a heterogeneous pattern of burning, in response to variation in, and interactions among, ambient weather, fuels, and topography at multiple spatial and temporal scales. For example, relative humidity and moisture content of fine fuels vary over the course of a day (generally lower humidity during the high temperatures of mid-day, then higher humidity when temperatures drop at night) and from day to day as regional air masses bring in wetter or drier air. Thus, fire behavior and fire effects at any point on the ground are influenced in part by the hour and day at which the fire occurs. Moreover, the fuel matrix varies in composition, mass, and arrangement; for example, between younger and older stands, and from moist to dry micro-site conditions. The upshot is that nearly all fires create a heterogeneous mosaic of fire severity; that is, patches of greater and lesser plant mortality, organic matter consumption, and effects on the post-fire microclimate and dynamic processes of energy and matter.

I approach the issue of spatial heterogeneity and fire in three steps. First, I illustrate how gradients in biotic and abiotic conditions influence fire regimes. A fire regime is a summary of central tendencies and variation in the major parameters of fire occurrence, behavior, and effects, including frequency, extent, seasonality, behavior, and effects on soils and biota (Agee 1998; Brown 2000; Heyerdahl et al. 2001; Morgan et al. 2001). Second, I illustrate how the inherent variability of fire itself creates spatial heterogeneity in the responses of organisms and ecosystems. Finally, I consider the implications of heterogeneity in fire regimes and fire effects for fire management policy and identify some of the most urgent research needs. My examples come primarily from western North America, but the principles probably apply to all regions where fire is an important ecological process.

Patterns and major mechanisms underlying spatial heterogeneity in fire regimes are strongly influenced by the scale of analysis (e.g., Heyerdahl et al. 2001). Lertzman and Fall (1998) discuss the hierarchical nature of interactions among various scales of forest patterns and processes, including "top-down" controls exerted by more coarse-scale processes (e.g., regional climate) and "bottom-up" controls exerted by more fine-scale processes (e.g., local topographic influences on fine fuel moisture). Similarly, Baker (2003) contrasts a "broad-scale" view of climate, fuels, and fire with a "contingent" view that emphasizes local variability and history. I first examine the broadest scale: regional or geographic patterns of fire occurrence and fire effects. I then consider the intermediate scale of landscapes or individual mountain ranges and finally deal with the finest scale variation within and between individual forest stands.

## Spatial Heterogeneity in Forest Fire Regimes

### Regional-Scale Variability

The ecological role of fire differs dramatically in different regions of North America, primarily in response to regional variation in synoptic climate (Agee 1998; Schmidt et al. 2002). The very moist deciduous forests of the northeast rarely burn (e.g., Clark and Royall 1996), whereas the dry pine forests of the southeast and southwest are unsustainable without periodic fire (e.g., Myers 2000; Covington and Moore 1994). Ecosystems with a long history of fire are populated with organisms having diverse adaptations and other mechanisms for tolerating fire (e.g., Whelan 1995), whereas species that have rarely experienced fire during their evolutionary history tend to be extremely vulnerable to damage from fire (e.g., tropical rain forests: Kinnaird and O'Brien 1998, but see Uhl 1998).

Even within a single forest type, we see substantial regional variation in fire regimes. For example, ponderosa pine forests (*Pinus ponderosa*) are distributed throughout the Rocky Mountain region from Mexico to Canada. Prior to disruption of western fire regimes by Euro-American settlers in the late 1800s, ponderosa pine forests in northern Arizona were characterized by very frequent fires, with mean fire intervals of 10 years or less (Moore et al. 1999). Mean fire intervals become longer with increasing latitude, reaching 10–30 years in ponderosa pine forests of the Black Hills in western South Dakota (Brown and Sieg 1996, 1999; Brown and Shepperd 2001). Fire severity also varied along the same latitudinal gradient: fires were predominantly low-severity (little or no canopy mortality) in Arizona but were of mixed severity (patches of high-severity, stand-replacing fire intermingled with low-severity fire) in at least some ponderosa pine forests of the Colorado Front Range (Brown et al. 1999; Veblen et al. 2000; Ehle and Baker 2003) and Black Hills (Shinneman and Baker 1997). The primary mechanism underlying this latitudinal variation in fire regimes of ponderosa pine forests probably is a top-down control related to synoptic climatic variation (Heyerdahl et al. (2001). Although total annual precipitation is similar in ponderosa pine forests of the three areas just described, the seasonality is strikingly different. Arizona typically experiences a pronounced dry season in early summer, which coincides with high temperatures, abundant lightning—and abundant fire activity (Friederici 2003). In contrast, the Black Hills have a wet spring and summer, and most fires occur in late summer or fall (Brown 1996, 1999). Lightning frequency also decreases from south to north in the Rocky Mountain region (Baker 2003).

## Landscape-Scale Variability in Forest Fire Regimes

Within a single mountain range, fire regimes typically vary with elevation, topography, and vegetation. In the San Juan Mountains of southwestern Colorado, low-elevation ponderosa pine forests burned every 10–30 years prior to Euro-American settlement in the 1870s (Grissino-Mayer et al. 2004). In contrast, fire intervals in high-elevation spruce-fir forests (*Picea engelmannii* and *Abies lasiocarpa*) were measured in centuries (Romme et al. 2000). Fires in ponderosa pine forests were predominantly understory burns that killed few of the large canopy trees and maintained an open forest structure, whereas the high-elevation fires were usually stand-replacing.

Two principal mechanisms are responsible for the profoundly different historical fire regimes in low *versus* high elevation forests of the San Juan Mountains. The first is climate. Mean annual precipitation is ca. 50–60 cm in the ponderosa pine forest zone, compared with 75–100 cm in the spruce-fir forest zone (Romme et al. 1992). The snowpack usually melts in April or early May in the ponderosa pine zone, and fuels are dry enough to support fires when high temperatures and lightning arrive in June. In contrast, snowpacks typically persist in the spruce-fir zone well into May and June and saturate the fuels as they melt. By the time the high-elevation fuels have dried sufficiently to carry fire, the summer monsoon has usually arrived (typically in early to mid July), again wetting fuels and prohibiting extensive fire. Rain usually is frequent and often heavy throughout the summer and fall in the high country, until the first autumn snows again preclude extensive fire activity. Indeed, extensive high-elevation fires only occur in years of below-average snowpack or delayed arrival of the summer monsoon.

The second mechanism influencing fire regimes in low *versus* high elevation forests involves the physical characteristics of the dominant species (Stephens 2001; Baker 2003). Mature ponderosa pine trees have thick, insulating bark and shed their lower branches, producing a gap between the fuels on the ground and in the canopy. The long needles create a loose, well aerated fuel bed when they fall to the ground. The combination of climate and fuels characteristics in ponderosa pine forests is conducive to frequent, low-intensity surface fires, and the mature trees are very tolerant of this kind of fire behavior. In contrast, spruce and fir have short needles that form a dense, poorly aerated fuel bed when they fall. The mature trees also tend to retain their lower branches, resulting in continuous fuels between the ground and the canopy. In the rare years when climatic conditions are dry enough to permit fire, the fuel structure of spruce-fir forests tends to support high-intensity crown fires rather than the low-intensity surface fires of the ponderosa pine zone.

Topography also may influence fire frequency and severity (Baker and Kipfmueller 2001). Northerly and easterly aspects may support less frequent but more severe fire regimes than southerly and westerly aspects: the drier aspects support more frequent fires because of more rapid drying of fuels, but fires tend to be less severe because of lower site productivity and less time for fuel accumulation between fire events (Heyerdahl et al. 2001; Rollins et al. 2002; Fule et al. 2003). Moist valley bottoms may burn less frequently than adjacent drier slopes (Romme and Knight 1981). Areas of gentle topography may burn more frequently because fires spread into the area from outside, whereas areas of more rugged topography may burn less often because cliffs, valleys, and other natural barriers inhibit fire spread and create more patchy burning patterns (e.g., Floyd et al. 2000; Heyerdahl et al. 2001; Grissino-Mayer et al. 2004).

## Stand-Level Variability in Forest Fire Regimes

Whereas broad-scale heterogeneity in climate and vegetation characteristics almost always exerts a powerful control over fire regimes, finer scale spatial heterogeneity in fuel mass and arrangement has an important influence on fire ignition and behavior under some weather conditions—but not others. Renkin and Despain (1992) analyzed >200 fires that were allowed to burn without interference in subalpine forests of Yellowstone National Park from 1972 to 1988. The best predictor of fire ignition and extent was the moisture content of large dead woody fuels (>7.5 cm in diameter). Fires generally remained <1 ha in size whenever the fuel moisture exceeded about 13%. Only when fuel moisture was <13% did fires burn >1 ha (although even then many ignitions extinguished naturally while still very small). Under these drier conditions, spatial variability in vegetation and fuel conditions significantly influenced fire spread and extent. Late-successional forests, with heavy ground and ladder fuels, were most likely to burn (Renkin and Despain 1992).

In the summer of 1988, however, fuel moistures in Yellowstone were extremely low (<10%; Renkin and Despain 1992) as a result of low snowfall in the previous winter followed by limited summer rain, and the dry conditions were accompanied by frequent episodes of strong dry winds. In contrast to fires observed in previous dry years, the 1988 fires exhibited little response to local differences in forest structure, fuel conditions, or fuel moisture, especially during late summer when the greatest amount of area burned. Rather, fire patterns on the landscape were shaped primarily by wind direction, apparently because the vegetation was equally flammable almost everywhere (Turner and Romme 1994). The 2002 Hayman fire in Colorado behaved similarly: on one day of extreme fire weather conditions, 25,000 ha burned at high severity with almost no influence of topography or local fuel structure; but fire behavior and severity were much more responsive to local variability on subsequent days of more moderate fire weather (Graham 2003).

Thus, in many vegetation types, weather (both current and antecedent) exerts the overriding control over fire behavior at the two extremes of moisture conditions: under wet conditions (when fuel moisture is too high for ignition) and under extremely dry conditions (when essentially all plant material can burn). Only between these weather extremes does spatial variability in vegetation structure and fuel characteristics have an important influence on fire ignition and spread. This pattern appears to characterize many subalpine and boreal forests, piñon-juniper woodlands, chaparral, and probably other major vegetation types [Bessie and Johnson 1995; Moritz 1997; Floyd et al. 2000; Keeley and Fotheringham 2001 (but see Minnich et al. 2001); Schoennagel et al. 2004].

## Spatial Heterogeneity in Fire Effects

Spatial variation in fire frequency and in heat release and duration during a single fire event can lead to significant heterogeneity in plant injury, species adaptations, and post-fire community and ecosystem recovery. For example, historical fires (pre-1900) in a ponderosa pine-dominated landscape in central Colorado typically burned with mixed severity, resulting in patches of complete canopy mortality interspersed with patches of only partial mortality (Brown et al. 1999). In the patches of complete mortality (from <1 ha to ca. 100 ha in size), a dense stand of roughly even-aged young trees usually developed soon after the fire, but where the fire was especially severe, little or no tree regeneration might occur for up to several decades. Where canopy mortality was only partial or negligible, the stand maintained an all-aged, all-sized structure (Kaufmann et al. 2000, 2003). One of the consequences of twentieth-century fire exclusion in this area has been homogenization of the landscape, as all of the formerly distinct patches have developed a similar dense stand structure.

During the first 3 years after the 1988 Yellowstone fires, total biotic cover was greatest in areas of lowest fire severity (understory burning with little canopy mortality) and least in areas of crown fire and complete consumption of the soil litter layer. Biotic cover also was lower in large patches of crown fire than in small patches (Turner et al. 1997, 1999, 2003). The principal mechanism responsible for these patterns was related to the magnitude and duration of heat release and plant mortality during the fire. Most herbaceous species had rhizomes or deep roots that survived even though the fire consumed the aboveground portions of the plants (e.g., *Epilobium angustifolium*, *Lupinus argenteus*, *Arnica cordifolia*). New shoots sprouting from these surviving belowground structures contributed most of the aboveground plant cover during the first 2–3 years after the fire. Where soil heating was greatest, mortality of belowground plant structures also was greatest, and post-fire cover was therefore lowest.

Post-fire densities of lodgepole pine (*Pinus contorta* var. *latifolia*), the dominant canopy species, also varied with local fire severity and size of burned patch following the 1988 Yellowstone fires (Turner et al. 1997). Highest densities were seen in large burned patches and in areas where surface fire scorched but did not consume needles and small twigs. Moderate fire severity evidently stimulated serotinous cones to release their seed and also created a suitable seed bed for seedling survival, whereas high-severity crown fires killed much of the canopy seed bank, and low-severity fires did not remove enough litter and herbaceous plant cover to create a suitable seed bed for pine seedlings.

An even more important predictor of post-fire pine seedling density was the proportion of serotinous trees in the canopy at the time of the fire, which varied at multiple scales across the Yellowstone landscape from 0 to >80% (Tinker et al. 1994). These spatial patterns in serotiny appear to reflect spatial and temporal heterogeneity in historical fire frequency and severity, which in turn has influenced long-term selective pressures on local lodgepole pine populations. For example, a high proportion of serotinous trees is commonly seen in stands at lower elevations, where fire historically recurred within the life span of the trees that established immediately after the previous fire event (median fire interval <200 years). In contrast, low serotiny characterizes most higher elevation stands, where historical fire intervals typically exceeded the life spans of individual trees, and where tree recruitment into canopy gaps is an important population process during long (300+ years) periods without fire (Schoenaggel et al. 2003). Thus, striking spatial patterns in post-fire community composition and structure were produced by spatially variable fire severity in 1988, interacting with spatial patterns in the local abundance of serotinous trees, which in turn were largely a product of past heterogeneity in fire frequency and severity.

Fire also produces rapid changes in microenvironment, including altered insolation, albedo, air temperature, and relative humidity. These changes result in secondary effects on soil temperature, moisture, water absorption, biogeochemical processes, and microbial activity (e.g., Neary et al. 1999). The magnitude of these changes varies greatly at both broad and very fine scales. For example, the heat energy released by a fire volatilizes organic compounds in litter and soil organic matter. These compounds then condense on soil particles to form a waxy coating that repels water. The combination of reduced litter and plant cover on the soil surface, plus a water-repellent soil layer, creates the potential for significant post-fire erosion and sedimentation. However, the strength of the water-repellent effect varies spatially in response to variation in heat release and substrate (DeBano 2000; Huffman et al. 2001; Pierson et al. 2001). Moreover, a major erosion event usually requires a heavy, localized rainfall event within the first 2 years of a fire, after which time soils and plant cover generally recover (Meyer et al. 2001; Moody and Martin 2001). Thus, a stochastic, top-down climatic process (heavy local rain storm) may have more impact on post-fire soil erosion than the heterogeneous direct effects of the fire per se—just as hot, dry, windy weather conditions (another top-down climatic effect) can overwhelm the effects of local fuel heterogeneity on fire severity.

## Implications of Spatial and Temporal Heterogeneity for Wildland Fire Management

It is hardly surprising that regions with vastly different climate and vegetation have very different fire regimes. What is noteworthy, however, is that fire regimes within the "same" vegetation type (e.g., "ponderosa pine forest") can vary so substantially along regional-scale climatic gradients. It follows that even though historical fire regimes and ecological restoration strategies have been very well documented and developed for one region, for example, ponderosa pine forests of northern Arizona (Moore et al. 1999; Friederici 2003), the concepts developed in this region cannot be exported uncritically to other regions where ponderosa pine is also a dominant species without local research into historical fire regimes and controls on local fire regimes (e.g, in northern Colorado). Extensive, uncritical extrapolation from a few well studied sites may result in unfortunate outcomes for managers and the public alike. For example, the fine-grained northern Arizona restoration prescription applied in northern Colorado will not re-create the coarser grained pre-1900 patch mosaic that characterized the latter area and may not even afford much protection from damaging fire behavior under the extreme weather conditions that periodically recur in this region, as in the 2002 Hayman fire (Graham 2003).

In part because of the spatial heterogeneity in historical fire regimes, the magnitude and impact of twentieth-century fire exclusion varies greatly with elevation and topographic conditions throughout the West. Understanding these patterns is essential for prioritizing management activities, especially those intended to restore or maintain natural ecological structure and function (e.g., in national parks and wilderness areas). The greatest impacts of twentieth-century fire exclusion generally have occurred in lower elevation ponderosa pine forests, where the former fire regime of frequent, low-intensity fires ended abruptly around 1850–1880 (Allen et al. 2002). In contrast, high-elevation spruce-fir and lodgepole pine forests historically burned at centuries-long intervals—intervals that commonly exceeded the current fire-free period (Veblen 2000). Many such forests today are indeed very dense, contain heavy fuel loads, and can support intense fire behavior under dry weather conditions, but these are all characteristics of the natural fire regimes in these systems—they are *not* primarily artifacts of twentieth-century fire exclusion. Nor will thinning and low-intensity prescribed burning "restore" a high-elevation forest; on the contrary, such treatment will almost surely move the system out of its historical range of variability (Romme et al. 2004).

Consensus is growing among fire managers and researchers that we need to tolerate and reintroduce—at a landscape scale—the full range of normal fire behavior in ecosystems where fire was an important historical process (e.g., Allen et al. 2002). Our current policies of complete fire exclusion are very effective under weather and fuels conditions conducive to low and moderate severity fires—but fire exclusion *cannot* be achieved under extreme weather and fuels conditions, such as we saw in the large western fires of 2000, 2002, and 2003. The result of this disparity in effectiveness of fire control is that we now have extensive fires *only* under the most extreme conditions. Consequently, the beneficial effects of low and moderate severity fires are excluded, or restricted to very small areas, but the damaging effects of large, high-severity fires still occur (Finney and Cohen 2003).

## Urgent Research Needs

In many places we now have a good base of information from which to implement broad-scale restoration of fire and fire-related processes [e.g., northern Arizona ponderosa pine forest (Friederici 2003)]. In other geographic areas, however, the information base is inadequate to support confident actions (Veblen 2003). Moreover, important questions remain about heterogeneous fire effects in all ecosystems. I suggest two general subject areas in which research on the spatial heterogeneity of fire occurrence and fire effects is most urgently needed:

(1) We need a better understanding of historical fire regimes and the magnitude and causes of twentieth-century change in several extensive but poorly studied vegetation types. The goal is not necessarily to reestablish disturbance processes *exactly* as they were in the historical period, but to identify environmental and evolutionary constraints on the kinds of ecological structures and processes likely to be sustainable in a particular setting (Swetnam et al. 1999; Morgan et al. 2001; Allen et al. 2002). One such poorly understood type is the "mixed conifer" forest (e.g., Agee 1998) found at middle elevations in many western mountain ranges. Another major vegetation type that may be fundamentally misunderstood in many places is the piñon-juniper woodland (*Pinus edulis, P. monophylla, Juniperus osteosperma, J. occidentalis*), which covers millions of square kilometers in the foothills

and intermountain basins across the West (Schmidt et al. 2002; Romme et al. 2003; Baker and Shinneman 2004). A related need is a better understanding of the relative importance of climatic variability *versus* fuel conditions in controlling fire frequency and severity along gradients of local climate and vegetation structure (e.g., Bessie and Johnson 1995; Schoennagel et al. 2004). Aggressive fire mitigation and "restoration" carried out in the absence of adequate local information about historical fire regimes and the major controls on those fire regimes may in fact cause more serious ecological degradation than waiting for adequate local information.

(2) We need a better understanding of how spatial variability in fire frequency and severity affects individual organisms and ecological processes. Although a number of fascinating case studies of fire-dependant species exist (e.g., Preston and Baldwin 1999), surprisingly few studies have focused on how spatial or temporal *heterogeneity* in fire frequency and effects may structure local populations, communities, or ecosystems. Without this kind of knowledge, our efforts to restore fire as a natural ecological process (or to remove fire from the system) may have unintended and unfortunate consequences for biodiversity and ecosystem function. Previous work suggests, for example, that the serotinous Australian shrub *Banksia hookeriana* has evolved to tolerate not just fire but a remarkably narrow range of fire intervals (Enright et al. 1996), whereas jack pine (*Pinus banksiana*) and lodgepole pine in North America can tolerate great variability in fire interval and fire severity (Muir and Lotan 1985; Gauthier et al. 1996; Schoennagel et al. 2003). Similarly, ant communities in Australia, and soil invertebrate communities in Sweden, exhibited different responses to different combinations of fire frequency and severity (Andersen 1991; Wikars and Schimmel 2001), and bird species exhibited individualistic preferences for more severely or less severely burned portions of coniferous forests in the Rocky Mountains (N. Kotliar, personal communication). Finally, it is clear that variation in fire intensity and duration of soil heating leads to variable changes in soil microbial composition, nutrient availability, and water repellency (Neary et al. 1999; Pierson et al. 2001), but many of the details are poorly worked out (e.g., Smithwick et al. 2005). It is beyond the scope of this paper to comprehensively review the pertinent literature on this topic, but such a review is urgently needed, both to guide future research and to inform managers and policymakers of potential opportunities and pitfalls in returning fire to local landscapes.

*Acknowledgments.* My thinking about spatial heterogeneity in fire regimes and fire effects has been influenced greatly by my 25 years of studying fire in Yellowstone National Park. I wish to acknowledge all that I have learned from my Yellowstone colleagues, especially Monica Turner, Dan Tinker, Don Despain, Bob Gardner, and Dennis Knight. I also thank my colleagues in southwestern Colorado, especially Albert Spencer, Lisa Floyd-Hanna, David Hanna, Bill Baker, and Jeff Redders. Finally, my recent experience as a team leader reviewing the Hayman fire in northern Colorado has given me many new insights, for which I especially thank Merrill Kaufmann, Tom Veblen, Mark Finney, Pete Robichaud, Claudia Regan, Peter Brown, Lee MacDonald, and Russ Graham. My research on causes and consequences of spatial heterogeneity in wildland fire has been supported by grants from the National Science Foundation, U.S. Forest Service, National Park Service, U.S. Department of Agriculture, Mellon Foundation, and University of Wyoming–National Park Service Research Station.

## References

Agee, J.K. 1998. The landscape ecology of western forest fire regimes. Northwest Sci. 72(special issue): 24–34.

Allen, C.D., Savage, M., Falk, D.A., Suckling, K.F., Swetnam, T.W., Schulke, T., Stacey, P.B., Morgan, P., Hoffman, M., and Klingel, J.T. 2002. Ecological restoration of southwestern ponderosa pine ecosystems: a broad perspective. Ecol. Applications 12: 1418–1433.

Andersen, A.N. 1991. Responses of ground-foraging ant communities to three experimental fire regimes in a savanna forest of tropical Australia. Biotropica 23: 575–585.

Baker, W.L. 2003. Fires and climate in forested landscapes of the U.S. Rocky Mountains. In Fire and climatic change in temperate ecosystems of the western Americas, eds. T.T. Veblen, W.L. Baker, G. Montenero, and T.W. Swetnam, pp. 120–157. New York: Springer.

Baker, W.L., and Kipfmueller, K.F. 2001. Spatial ecology of pre-Euro-American fires in a southern Rocky Mountain subalpine forest landscape. Professional Geographer 53: 248–262.

Baker, W.L., and Shinneman, D.J. 2004. Fire and restoration of piñon -juniper woodlands in the western United States: a review. Foest Ecol. Manage. 189: 1–21.

Bessie, W.C., and Johnson, E.A. 1995. The relative importance of fuels and weather on fire behavior in subalpine forests. Ecology 76: 747–762.

Brown, J.K. 2000. Introduction and fire regimes. In Wildland fire in ecosystems: effects on flora, eds. J.K. Brown, and J.K. Smith, pp. 1–7. USDA Forest Service General Technical Report RMRS-GTR-42 volume 2.

Brown, P.M., and Seig, C.H. 1996. Fire history in interior ponderosa pine communities of the Black Hills, South Dakota, USA. Int. J. Wildland Fire 6: 97–105.

Brown, P.M., and Seig, C.H. 1999. Historical variability in fire at the ponderosa pine–northern Great Plains ecotone, southeastern Black Hills, South Dakota. Ecoscience 6: 539–547.

Brown, P.M., and Shepperd, W.D. 2001. Fire history and fire climatology along a $5^0$ gradient in latitude in Colorado and Wyoming. Paleobotanist 50: 133–140.

Brown, P.M., Kaufmann, M.R., and Shepperd, W.D. 1999. Long-term, landscape patterns of past fire events in a montane ponderosa pine forest of central Colorado. Landscape Ecol. 14: 513–532.

Clark, J.S., and Royall, P.D. 1996. Local and regional sediment charcoal evidence for fire regimes in presettlement north-eastern North America. J. Ecol. 84: 365–382.

Covington, W.W., and Moore, M.M. 1994. Southwestern ponderosa forest structure: changes since Euro-American settlement. J. Forestry 92: 39–47.

DeBano, L.F. 2000. The role of fire and soil heating on water repellency in wildland environments: a review. J. Hydrol. 231–232: 195–206.

Ehle, D.S., and Baker, W.L. 2003. Disturbance and stand dynamics in ponderosa pine forests in Rocky Mountain National Park, USA. Ecol. Monogr. 73: 543–566.

Enright, N.J., Lamont, B.B., Marsula, R. 1996. Canopy seed bank dynamics and optimum fire regime for the highly serotinous shrub, *Banksia hookeriana*. J. Ecol. 84: 9–17.

Finney, M.A., and Cohen, J.D. 2003. Expectation and evaluation of fuel management objectives. In Proceedings of the conference on Fire, Fuel Treatments, and Ecological Restoration: Proper Place, Appropriate Time, eds. P.N. Omi, and L.A. Joyce (technical editors), pp. 353–366. USDA Forest Service Proceedings RMRS-P-29.

Floyd, M.L., Romme, W.H., and Hanna, D. 2000. Fire history and vegetation pattern in Mesa Verde National Park. Ecol. Applications 10: 1666–1680.

Friederici, P., ed. 2003. Ecological restoration of southwestern ponderosa pine forests. Washington, DC: Island Press.

Fule, P.Z., Crouse, J.E., Heinlein, T.A., Moore, M.M., Covington, W.W., and Verkamp, G. 2003. Mixed-severity fire regime in a high-elevation forest: Grand Canyon, Arizona. Landscape Ecol. 18(5): 465–485.

Gauthier, S., Bergeron, Y., Simon, J.P. 1996. Effects of fire regime on the serotiny level of jack pine. J. Ecol. 84: 539–548.

Graham, R.T., technical ed. 2003. Hayman fire case study. USDA Forest Service General Technical Report RMRS-GTR-114. Available at www/fs.fed.us/rm/ hayman_fire.

Grissino-Mayer, H.D., Romme, W.H., Floyd, M.L., and Hanna, D.D. 2004. Climatic and human influences on fire regimes of the Southern San Juan Mountains, Colorado. Ecology 85(6): 1708–1724.

Heyerdahl, E.K., Brubaker, L.B., and Agee, J.K. 2001. Spatial controls of historical fire regimes: a multiscale example from the Interior West, USA. Ecology 82: 660–678.

Huffman, E.L., MacDonald, L.H., and Stednick, J.D. 2001. Strength and persistence of fire-induced soil hydrophobicity under ponderosa and lodgepole pine, Colorado Front Range. Hydrol. Processes 15: 2877–2892.

Kaufmann, M.R., Regan, C.M., and Brown, P.M. 2000. Heterogeneity in ponderosa pine / Douglas-fir forests: age and size structure in unlogged and logged landscapes of central Colorado. Can. J. Forest Res. 30: 98–117.

Kaufmann, M.R., Huckaby, L.S., Fornwalt, P.J., Stoker, J.M., and Romme, W.H. 2003. Using tree recruitment patterns and fire history to guide restotation of an unlogged ponderosa pine / Douglas-fir landscape in the southern Rocky Mountains after a century of fire suppression. Forestry 76: 231–241.

Keeley, J.E., and Fotheringham, C.J. 2001. Historic fire regime in southern California shrublands. Conservation Biol. 15: 1536–1548.

Kinnaird, M.F., and O'Brien, T.G. 1998. Ecological effects of wildfire on lowland rainforest in Sumatra. Conservation Biol. 12: 954–956.

Lertzman, K., and Fall, J. 1998. From forest stands to landscapes: spatial scales and the roles of disturbances. In Ecological scale: theory and applications, eds. D.L. Peterson, and V.T. Parker, pp. 339–367. New York: Columbia University Press.

Machlis, G.E., Kaplan, A.B., Tuler, S.P., Bagby, K.A., and McKendry, J.E. 2002. Burning questions: a social science research plan for federal wildland fire management. Contribution number 943, Idaho Forest, Wildlife and Range Experiment Station, University of Idaho, Moscow.

Meyer, G.A., Pierce, J.L., Wood, S.H., and Jull, A.J.T. 2001. Fire, storms, and erosional events in the Idaho batholith. Hydrol. Processes 15: 3025–3038.

Minnich, R.A. 2001. An integrated model of two fire regimes. Conservation Biol. 15: 1549–1553.

Moody, J.A., and Martin, D.A. 2001. Initial hydrologic and geomorphic response following a wildfire in the Colorado Front Range. Earth Surface Processes Landforms 26: 1049–1070.

Moore, M.M., Covington, W.W., and Fule, P.Z. 1999. Reference conditions and ecological restoration: a southwestern ponderosa pine perspective. Ecol. Applications 9: 1266–1277.

Morgan, P., Hardy, C.C., Swetnam, T.W., Rollins, M.G., and Long, D.G. 2001. Mapping fire regimes across time and space: understanding coarse and fine-scale patterns. Int. J. Wildland Fire 10: 329–342.

Moritz, M.A. 1997. Analyzing extreme disturbance events: fire in Los Padres National Forest. Ecol. Applications 7: 1252–1262.

Muir, P.S., and Lotan, J.E. 1985. Disturbance history and serotiny of Pinus contorta in western Montana. Ecol. 66: 1658–1668.

Myers, R.L. 2000. Fire in tropical and subtropical ecosystems. In Wildland fire in ecosystems: effects on flora. eds. J.K. Brown, and J.K. Smith, pp. 161–173. USDA Forest Service General Technical Report RMRS-GTR-42 volume 2.

Neary, D.G., Klopatek, C.C., DeBano, L.F., and Ffolliott, P.F. 1999. Fire effects on belowground sustainability: a review and synthesis. Forest Ecol. Manage. 122: 51–71.

Pierson, F.B., Robichaud, P.R., and Spaeth, K.E. 2001. Spatial and temporal effects of wildfire on the hydrology of a steep rangeland watershed. Hydrol. Processes 15: 2905–2916.

Preston, C.A., and Baldwin, I.T. 1999. Positive and negative signals regulate germination in the post-fire annual, *Nicotiana attenuata*. Ecol. 80: 481–494.

Renkin, R.A., and Despain, D.G. 1992. Fuel moisture, forest type, and lightning-caused fire in Yellowstone National Park. Can. J. Forest Res. 22: 37–45.

Rollins, M.G., Morgan, P., and Swetnam, T. 2002. Landscape-scale controls over 20th century fire occurrence in two large Rocky Mountain (USA) wilderness areas. Landscape Ecol. 17: 539–557.

Romme, W.H., and Knight, D.H. 1981. Fire frequency and subalpine forest succession along a topographic gradient in Wyoming. Ecol. 62: 319–326.

Romme, W.H., Jamieson, D.W., Redders, J.S., Bigsby, G., Lindsey, J.P., Kendall, D., Cowen, R., Kreykes, T., Spencer, A.W., and Ortega, J.C. 1992. Old-growth forests of the San Juan National Forest in southwestern Colorado. In Old-growth forests in the southwest and Rocky Mountain regions, pp. 154–165. USDA Forest Service General Technical Report RM-213.

Romme, W.H., Floyd, L., Hanna, D., and Redders, J.S. 2000. Using natural disturbance regimes to mitigate forest fragmentation in the central Rocky Mountains, pp. 377–400. In Forest fragmentation in the southern Rocky Mountains, eds. R.L. Knight, F.W. Smith, S.W. Buskirk, W.H. Romme, and W.L. Baker. Boulder, Colorado: University Press of Colorado.

Romme, W.H., Floyd-Hanna, M.L., and Hanna, D.D. 2003 Ancient piñon-juniper forests of Mesa Verde and the West: A cautionary note for forest restoration programs. In Proceedings of the conference on Fire, Fuel Treatments, and Ecological Restoration: Proper Place, Appropriate Time. eds. P.N. Omi, and L.A. Joyce (technical editors). USDA Forest Service Proceedings RMRS-P-29.

Romme, W.H., Turner, M.G., Tinker, D.B., and Knight, D.H. 2004 Emulating natural forest disturbance in the wildland-urban interface of the Greater Yellowstone Ecosystem of the United States pp. 243–250, In Emulating natural forest landscape disturbances: concepts and applications, eds. A.H. Perera, L.J. Buse, and M.G. Weber. New York: Columbia University Press.

Sampson, R.N., Atkinson, R.D., and Lewis, J.W. eds. 2000. Mapping wildfire hazards and risks. New York: Food Products Press, Haworth Press.

Schmidt, K.M., Menakis, J.P., Hardy, C.C., Hann, W.J., Bunnell, D.L. 2002. Development of coarse-scale spatial data for wildland fire and fuel management. USDA General Technical Report RMRS–87.

Schoennagel, T., Turner, M.G., and Romme, W.H. 2003. The influence of fire interval and serotiny on postfire lodgepole pine density in Yellowstone National Park. Ecol. 84: 2967–2978.

Schoennagel, T., Veblen, T.T., and Romme, W.H. 2004. The interaction of fire, fuels, and climate across Rocky Mountain forests. BioScience 54(7): 661–676.

Shinneman, D.J., and Baker, W.L. 1997. Nonequilibrium dynamics between catastrophic disturbances and old-growth forests in ponderosa pine landscapes of the Black Hills. Conservation Biol. 11: 1276–1288.

Smithwick, E.A.H., Turner, M.G., Mack, M.C., and Chapin III, F.S. 2005. Post-fire soil N-cycling in northern conifer forests affected by severe, stand-replacing wildfires. Ecosystems.

Stephens, S.L. 2001. Fire history differences in adjacent Jeffrey pine and upper montane forests in the eastern Sierra Nevada. Int. J. Wildland Fire 10: 161–167.

Swetnam, T.W., Allen, C.D., and Betancourt, J.L. 1999. Applied historical ecology: using the past to manage for the future. Ecol. Applications 9: 1189–1206.

Tinker, D.B., Romme, W.H., Hargrove, W.W., Gardner, R.H., and Turner, M.G. 1994. Landscape-scale heterogeneity in lodgepole pine serotiny. Can. J. Forest Res. 24: 897–903.

Turner, M.G., and Romme, W.H. 1994. Landscape dynamics in crown fire ecosystems. Landscape Ecol. 9: 59–77.

Turner, M.G., Romme, W.H., Gardner, R.H., and Hargrove, W.W. 1997. Effects of fire size and pattern on early succession in Yellowstone National Park. Ecol. Monogr. 67: 411–433.

Turner, M.G., Romme, W.H., and Gardner, R.H. 1999. Prefire heterogeneity, fire severity, and early postfire plant reestablishment in subalpine forests of yellowstone National Park, Wyoming. International Journal of Wildland Fire 9(1): 21–36.

Turner, M.G., Romme, W.H., and Tinker, D.B. 2003. Surprises and lessons from the 1988 Yellowstone fires. Frontiers Ecol. Environ. 1: 351–358.

Uhl, C. 1998. Perspectives on wildfire in the humid tropics. Conservation Biol. 12: 942–943.

Veblen, T.T. 2000. Disturbance patterns in southern Rocky Mountain forests. In Forest fragmentation in the southern Rocky Mountains, eds. R.L. Knight, F.W. Smith, S.W. Buskirk, W.H. Romme, and W.L. Baker, pp. 31–54. Boulder, Colorado: University Press of Colorado.

Veblen, T.T., Kitzberger, T., and Donnegan, J. 2000. Climatic and human influences on fire regimes in ponderosa pine forests in the Colorado Front Range. Ecol. Applications 10: 1178–1195.

Veblen, T.T. 2003. Key issues in fire regime research for fuels management and ecological restoration. In Fire, fuel treatments, and ecological restoration, technical eds. P.N. Omi, and L.A. Joyce, pp. 259–275. Conference Proceedings; 2002, 16–18 April, Fort Collins, CO. USDA Forest Service Proceedings RMRS-P-29: Available at http://www.fs.fed.us/rm/pubs/rmrs_p029.pdf.

Whelan, R.J. 1995. The ecology of fire. Cambridge, UK: Cambridge University Press.

Wikars, L., and Schimmel, J. 2001. Immediate effects of fire-severity on soil invertebrates in cut and uncut pine forests. Forest Ecol. Manage. 141: 189–200.

# 18
# The Role of Spatial Heterogeneity in the Management of Freshwater Resources

ALAN D. STEINMAN and RODNEY DENNING

## Abstract

Spatial heterogeneity of ecosystem structure and function is rarely taken into consideration in the management of our planet's freshwater resources. Incorporation of spatial heterogeneity provides a new way to view freshwater resources and leads to potentially useful management strategies. In this chapter, we address the relationship between the management of freshwater resources and spatial heterogeneity by introducing landscape concepts as they apply to water management, developing a conceptual framework, describing how this relationship applies to ecosystem services provided by fresh water, and using a case study that explains the potential relevance of spatial heterogeneity to water management.

Landscape connectivity can be viewed as the overarching feature linking aquatic systems, with six types of landscape connections: two types of longitudinal (upstream-downstream and hydrogeomorphic), lateral (floodplain-river), lateral-vertical (hillslope-river), vertical (surface water-groundwater), and within-ecosystem linkages. Hypothesized relationships between landscape linkages and ecosystem services provided by fresh water are explored that address the role of spatial heterogeneity.

The south Florida ecosystem is used as a case study to show how spatial heterogeneity in this system affects a variety of ecosystem services. Spatial heterogeneity of nutrient sources will influence the location of restoration efforts in the Everglades. In addition, ecosystem services of water supply, water quality, and navigation are influenced by the configuration of the highly interconnected and heavily managed Greater Everglades Ecosystem in south Florida. The socioeconomic and environmental consequences of this connectivity are explored under high and low flow conditions.

## Introduction

There is increasing recognition that fresh water is a crucial and imperiled resource (Naiman et al. 1995; Jackson et al. 2001; Baron et al. 2002). Assessments on the state of global water conditions have alerted the public that freshwater resources are seriously threatened and that we need to start thinking about water quantity and quality in new ways (Gleick 1998, 2000, 2002). One innovative approach is the improved integration of the ecological, engineering, social, and economic sectors dealing with water resources, which is resulting in a more holistic approach to water management (Naiman et al. 1998; Falkenmark 1999; Johnson et al. 2001; Baron et al. 2002; Steinman et al. 2002). However, further advances are needed in understanding how best to manage the planet's limited supply of fresh water.

Spatial heterogeneity of water resources is an often overlooked, but potentially important, factor in water management. For example, water bodies are often located at the lowest topographic point in the landscape, allowing them to serve as integrators of landscape processes (Naiman et al. 2002). This spatial placement has been exploited by ecologists, who have used aquatic biota as indicators of ecosystem change (Plafkin et al. 1989; Karr and Chu 1999). This chapter will focus on the concept of spatial heterogeneity and how it can be applied to water management, including a conceptual framework that assesses the potential role of spatial heterogeneity in water management, how the concept of spatial heterogeneity applies to ecosystem services provided by fresh water, and a brief case study that explains the potential relevance of spatial heterogeneity to water management.

## Spatial Heterogeneity in Water Management: Providing a Landscape Context

Aquatic ecosystems are characterized by connectivity. Hydrologic connectivity is the water-mediated linkage of matter, energy, or organisms within or between elements of the hydrologic cycle (Pringle 2001). Connectivity can be viewed as an overarching landscape feature of aquatic ecosystems, which can be decomposed into other landscape connections (see below). Both the types of connected water bodies involved in the connections (composition) and the spatial arrangement (configuration) of the linkages help characterize watersheds and set the boundaries for the management of water resources. Table 18.1 identifies six distinct landscape connections with potential bearing on water management, classified based on geographic scale. For the purposes of this chapter, geographic scale is classified as either broad ($>1$ km$^2$) or local ($<1$ km$^2$), but in reality these scales form a continuum. Thus, some landscape connections apply at both broad and local scales (Table 18.1).

TABLE 18.1. Examples of landscape features, the key components of spatial hetero-geneity associated with each feature, and the relevant spatial array of features in the landscape.

| Landscape feature Type of connectivity | Scale | Example | Key components of heterogeneity |
|---|---|---|---|
| Longitudinal (upstream-downstream) | Broad ($>1$ km$^2$) | River Continuum Concept | Configuration/composition |
| | Broad | Cultural eutrophication | Configuration/composition |
| | Broad | Invasive species | Configuration/composition |
| Longitudinal (hydrogeomorphic) | Broad | North Temperate Lakes Region | Configuration/composition |
| Lateral (floodplain-river) | Broad/local | Kissimmee River | Configuration |
| Lateral-vertical (hillslope-river) | Broad/local | Logging | Configuration |
| Vertical (surface water-groundwater) | Local ($<1$ km$^2$) | Wells | Configuration |
| | Local | Septic systems | Configuration |
| Internal (within ecosystem) | Variable | Lake sediments | Configuration/composition |

## Broad-Scale Landscape Features ($>1$ km$^2$)

Broad-scale landscape features focus on connections across multiple aquatic ecosystems and usually extend across regional geographic bound-aries. Examples include large drainage basins (encompassing first- to sixth-order streams; cf. Ward 1997) and lake chains in a region (Soranno et al. 1999).

### Longitudinal (Upstream-Downstream Linkage)

Longitudinal linkages (Figure 18.1A) refer to the influence of upstream processes on downstream structure and function. Both the configuration and composition of this linkage are crucial to issues of water management (Table 18.1). Configurations of potential importance include (1) the shape of the stream network (e.g., small *vs.* large number of headwater streams influencing the degree of land-water interaction); (2) disruption of normal flow patterns by reservoirs (cf. Ward and Stanford 1983); and (3) disruption of normal flow patterns due to riverine wetlands, upwelling zones, and downwelling zones. These last two examples also have a compositional com-ponent, as the type and features of the discontinuities in the longitudinal gradient (e.g., dam size or type; wetland volume or vegetation structure) can all strongly influence ecosystem function. The River Continuum Concept (Vannote et al. 1980) is an excellent example of heterogeneity defined largely by a continuous gradient (Table 18.1). However, at a finer scale, it is evident that both the configuration and composition of discrete elements,

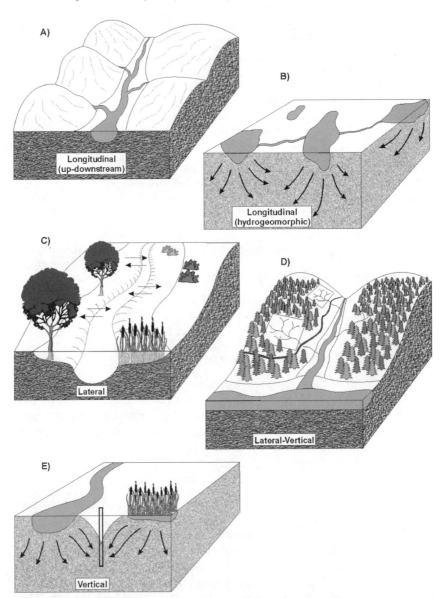

FIGURE 18.1. Representation of different types of landscape features that determine spatial heterogeneity in aquatic ecosystems. (A) Longitudinal (upstream-downstream) linkages; (B) longitudinal (hydrogeomorphic) linkages; (C) lateral (floodplain-river) linkages; (D) lateral-vertical (hillslope-river) linkages; (E) vertical (surface water-groundwater) linkages. See text for more detail.

such as dams, channel units, or upwelling/downwelling zones, also influence system dynamics. For example, nitrogen cycling in desert streams is influenced by periphyton, whose spatial distribution is determined by the direction of hydrologic exchange with the subsurface flow (Henry and Fisher, 2003; Fisher and Welter this volume).

Longitudinal linkages also have implications for water quantity and quality (Table 18.1). The configuration of water extractions can profoundly influence aquatic habitat, as has been shown for the Colorado River, where upstream water diversions and extractions have reduced river flows into the Colorado River delta by nearly 75% during the twentieth century (Pitt et al. 2000). The configuration of pollutant loading in a watershed such as the location of sources (upstream, downstream), as well as compositional factors such as the type and concentration of pollutants, determine the location of potential ecological impairments (Steinman et al. 2003a) and biological hot spots. In turn, these locations will influence the placement, cost, and efficacy of remediation sites (e.g., constructed wetlands) in the landscape.

Longitudinal linkages also can influence dispersal patterns of species. Hydrologic connectivity strongly influences the spread of invasive species (Vanderploeg et al. 2002), one of the greatest threats to biodiversity and ecosystem integrity (Walker and Steffen 1997). In aquatic ecosystems, the Great Lakes have served as the poster child for invasive species (Ricciardi and Rasmussen 1998; Ricciardi and MacIsaac 2000); impacts include habitat loss, food chain disruption, and alterations to native fisheries.

Both the spatial configuration of the Great Lakes and the compositional heterogeneity of nuisance species are relevant to ecosystem function (Table 18.1). Despite the generally upstream-downstream longitudinal configuration of the Great Lakes (e.g., Lake Superior→ Lake Erie), the spatial trajectory of dispersal by invaders has not always followed this longitudinal gradient. Discontinuous dispersal patterns (large movements not explained by natural spread of the organism; Vanderploeg et al. 2002) have been identified for the zebra mussel and the zooplankter *Cercopagis* (Vanderploeg et al. 2002). Once established, these invasive species show a combination of continuous and discontinuous dispersal; species with planktonic stages will be moved by water currents and will be more likely to follow a continuous distribution, whereas species that are capable of surviving intralake ship transport (Grigorovich et al. 2003) are more likely to follow a discontinuous distribution. This compositional heterogeneity in species, each with potentially different life histories and distribution patterns, results in tremendous challenges to water management.

## Longitudinal (Hydrogeomorphic Linkage)

A second type of longitudinal linkage involves surface water bodies that are hydrologically connected via surface or subsurface water movement. An example of this type of longitudinal linkage is the North Temperate Lakes

(Magnuson and Frost 1982; Figure 18.1B), where lakes share a common groundwater flow system. Spatial location within the landscape (i.e., configuration), such as position within the chain of connected lakes, as well as the amount of groundwater entering a lake, is directly related to the lake's position within the landscape (Webster et al. 1996). In addition, the type of water body (lake, wetland, aquifer; i.e., composition) will influence local environmental conditions, which in turn determines habitat quality (Table 18.1).

## Broad/Local Landscape Features

These landscape features can operate at both the larger ($>1$ km$^2$) and smaller ($<1$ km$^2$) scales and involve connections across different ecosystem types, such as between a river and its floodplain.

### Lateral (Floodplain-River Linkage)

Linkages between rivers and their floodplains (Figure 18.1C) are vital for maintaining a healthy lotic ecosystem. The configuration of these connections will influence the degree to which energy associated with floodwaters is dissipated, high-quality habitat is maintained, and materials are exchanged across the ecotone (Junk et al. 1989; Bayley 1995; Benke 2001) (Table 18.1). In large river systems where levee construction prevents rivers from overflowing their banks, the disconnection with the floodplain has resulted in the loss of biodiversity and ecological functions (Galat et al. 1998; Toth et al. 1998). This hydrologic linkage can occur at both broad (floods) and local (levee breaches, naturally low topography) scales.

### Lateral-Vertical (Hillslope-River Linkage)

Human activities on hillslopes, such as logging or road construction, can profoundly affect water bodies below. The spatial configuration of these activities (Figure 18.1D) along the hillslope gradient will influence nutrient and sediment fluxes throughout the catchment (Likens et al. 1970). This, in turn, can affect habitat quality and stream productivity (Gregory et al. 1987). The configuration of road networks in montane watersheds (i.e., hillslope location) can influence floods and debris flows (Table 1; Jones et al. 2000).

## Local Landscape Features

These landscape features focus on localized ($<1$ km$^2$) connections between individual aquatic ecosystems, denoting their relationship to the land or surface/subsurface flow, such as surface water-groundwater linkages.

### Vertical (Surface Water-Groundwater Linkage)

The linkage between surface water and groundwater (Figure 18.1E) is receiving increasing attention (cf. Winter et al. 1998; Steinman et al. 2003b), largely because of loosely regulated withdrawals, or problems associated with contamination in shallow aquifers connected directly to surface waters. Surface water-groundwater interactions have usually been ignored with respect to water management concerns, possibly because of the practical difficulties in managing this complex issue. However, increasing urban sprawl, with concomitant increases in septic systems for waste disposal and well-drilling for potable water, indicates that this linkage is very relevant to water supply and waste processing (Table 18.1). The configuration of these linkages is particularly relevant to water management issues. For example, septic systems placed too close to private wells have obvious human health implications, whereas septic systems placed too close to pristine watersheds may result in contamination. Wells located in aquifers that are hydraulically connected to surface water may result in stream flow depletion where withdrawal rates exceed recharge rates.

## Internal Landscape Feature

This landscape feature focuses on within-ecosystem spatial heterogeneity, such as the position of different sediment types in a lake. The scale will vary depending on the size of the ecosystem.

Internal Linkage

Spatial heterogeneity within an ecosystem can have important functional consequences. For example, the sediment composition of Lake Okeechobee, one of the largest lakes in the United States, is not uniform (Reddy et al. 1995). The lake's large surface area and shallow mean depth result in wind-driven dynamics that strongly influence sediment resuspension. Because each of the lake's five major sediment types has very different P stores, uptake rates, exchange rates, and assimilative capacities (Reddy et al. 1995; Steinman et al. 1999), the compositional heterogeneity of these sediments affects the chemistry and biology of the lake. For example, mud sediments are easily resuspended and in that region, abiotic turbidity controls light attenuation (Phlips et al. 1997). Configurational heterogeneity also is important. Mud sediments located near the center of Lake Okeechobee have expanded in area throughout the 1990s. Resuspended mud at the center of the lake has relatively little ecological impact. However, advection of mud sediments toward the shoreline negatively impacts the ecologically productive and economically important nearshore regions—increased phosphorus levels and suspended solids reduce light transmissivity through the water column, thereby threatening the submersed aquatic vegetation and associated fisheries in the nearshore regions of the lake (Havens and James 1999).

## Conceptual Framework

Inclusion of spatial heterogeneity creates a new way to conceptualize the management of freshwater resources. The complexity of managing water makes it unlikely that any one conceptual approach will be universally applicable. We attempt to bypass this situation by identifying a framework that uses ecosystem services and functions as the focal point. In this approach, we describe how an ecosystem service provided by fresh water is linked to elements of spatial heterogeneity through management goals, operational scales, and landscape features (Figure 18.2). This hierarchical approach is admittedly a simplification of what might occur in the real world, where there will usually be multiple ecosystem services, management goals, and scales at work. However, there is inherent value in providing a conceptual framework within which relevant management questions can be framed and discussed.

The first step in the conceptual model is to define the ecosystem service or function of interest (Figure 18.2). This provides a general framework for the other hierarchically arranged elements. The next steps are to identify the management goal relevant to the ecosystem service of interest and to define the appropriate management strategy and operating scale for the management goal in question. The operational scale will normally be a natural outcome of the management goal and helps define the relevant landscape features (Table 18.1). Finally, based on the landscape features, one can

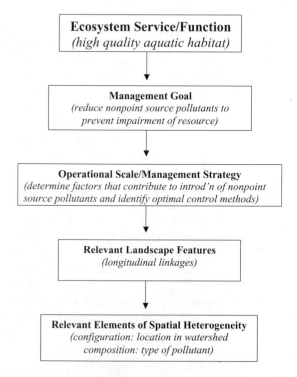

FIGURE 18.2. Conceptual model showing the relationships between the management of fresh water resources and spatial heterogeneity in the landscape. An example of this model (habitat function) is shown in italics.

determine the influence of spatial heterogeneity in the model, and if important, the relative significance of compositional or configurational heterogeneity (Figure 18.2).

An example of how the conceptual framework can be applied involves the ecosystem function of preserving or restoring high quality aquatic habitat (Figure 18.2). The management goal is to reduce nonpoint-source pollution to the point where the beneficial use of the resource is not impaired. As a management strategy, it is crucial that we determine the factors that contribute to the introduction of these pollutants, identify optimal control methods, and implement them to the best of our ability. The landscape feature that is most crucial in this example is longitudinal linkage, as the upstream pollutant may have profound influences on downstream habitat. However, if the source of the nonpoint pollutant is from impervious surfaces, hillslope-river linkages may be important, and if the pollutant is contaminating groundwater, then the surface water-groundwater linkage also comes into play.

The relevant components of spatial heterogeneity in this example include the configuration and composition of the nonpoint sources relevant for source control strategies. For example, the configurational heterogeneity of source locations in the watershed will influence whether the impacts will be isolated or compounded as they move downstream. Land use/land cover distributions help in developing control strategies. Impacted areas, such as those with high percentages of agricultural and urban/developed land use, may require a gradient of buffer strips to control input to water bodies. However, in those areas where nutrient loads exceed the assimilative capacity of these management practices, constructed wetlands may be needed to treat nutrients from the entire basin. Compositional heterogeneity is also important; management of toxic chemicals may be very different from nutrient pollution. In this case, sediment heterogeneity is important as concentrations will be greater in silt and organic sediments compared to sand.

The concept of spatial heterogeneity will not be relevant to all water management problems, either because the system is relatively homogeneous or other factors override the influence of heterogeneity. However, in those situations where the configuration or composition of landscape components influence water quality, quantity, or timing, this conceptual model should be of value.

## Applications of Spatial Heterogeneity to Freshwater Ecosystem Services and Functions

Fresh water provides numerous ecosystem services and functions, including water for drinking, industry, agriculture, the production of fish and waterfowl, navigation, recreation, waste processing, hydroelectric power, and habitat (Daily 1997). However, little attention has been paid to how the

TABLE 18.2. Hypothesized relevance of key landscape attributes on select ecosystem services and functions provided by fresh water.

| Ecosystem service | Landscape feature | | | | |
| --- | --- | --- | --- | --- | --- |
| | Longitudinal (upstream-downstream) | Longitudinal (hydrogeomorphic) | Lateral (floodplain-river) | Lateral-vertical (hillslope-river) | Vertical (surface water-groundwater) |
| Potable water | Moderate | Low | Low | Low | High |
| Water supply for agriculture, industry, and municipalities | High | Low | Low | Moderate | High |
| Navigation | Low | Low | Low | Low | Low |
| Recreation | Moderate | Moderate-high | Low | Low | Low |
| Flood control | Low | Low | High | Low | Moderate |
| Groundwater recharge | Moderate | Moderate | Moderate | Low-moderate | High |
| Waste processing | Moderate | Moderate | Low | Low | High |
| Hydroelectric power | High | Low | Low | Low | Low |

spatial heterogeneity of the landscape, as described earlier, relates to the services and functions provided by water. The relevance of spatial heterogeneity will likely vary among the ecosystem services and functions provided by fresh water. However, we are aware of no work that has been done evaluating these relationships. In Table 18.2, we propose a hypothetical relevance for each of the landscape features (as defined earlier) relative to various ecosystem services and functions provided by fresh water; note that the relevance is not consistent among landscape features.

## Longitudinal

The configuration of longitudinal linkages is viewed as highly relevant for water supply and hydroelectric power (Table 18.2). The areal extent and orientation (e.g., north- *vs.* south-facing) of the watershed collecting precipitation, the climatic conditions influencing precipitation and evapotranspiration, and where in the watershed human withdrawals occur all influence the quantity and timing of water supply.

Humans have been harnessing hydropower for centuries, but until recently the alterations to longitudinal flow were relatively minor. Construction of large, capital-intensive water development projects in the twentieth century has resulted in profound changes to the natural flow regime of rivers. The number and location of dams within a watershed have important implications for ecosystem function. In the Colorado River basin, upstream dams trap so much sediment that tidal action at the river delta actually removes more sediment than the river deposits (Kowaleski et al. 2000; Cohen 2002). In addition, more water is legally apportioned from the Colorado River than actually flows in most years, a problem that derives from basing estimates of annual flow on unusually wet years (Cohen 2002). Thus, upstream habitats are likely to remain wetter longer than downstream habitats.

## Longitudinal (Hydrogeomorphic)

Spatial heterogeneity associated with hydrogeomorphic linkages is hypothesized to be highly relevant to recreation (Table 18.2). The location of lakes within a landscape, including both their hydrologic position within the local to regional flow system and their placement relative to neighboring lakes, has been shown to influence a number of limnological characteristics (Kratz et al. 1997; Kratz et al. this volume). Soranno et al. (1999) examined spatial variation among a series of lake chains throughout North America and found that in general, the lakes located further down the lake chain contained more nutrients and chlorophyll. This is consistent with the observation of Riera et al. (2000) that lakes located toward the bottom of a chain tended to have a higher density of human settlement due to their enhanced recreational opportunities (e.g., more accessible, larger, better fishing) compared to lakes high in the landscape.

## Lateral (Floodplain-River)

Flood control is the ecosystem service hypothesized to be influenced most by floodplain-river linkages (Table 18.2). The Kissimmee River basin, located in south-central Florida, was a meandering river with a 1.5- to 3-km-wide floodplain that in its natural conditions was intimately connected to the river channel (Koebel 1995). However, an inundated floodplain was not compatible with the increased development and agricultural activity in the basin after World War II. Between 1962 and 1971, the Kissimmee River was channelized by the U.S. Army Corps of Engineers and transformed into a series of five impoundments. The channelization was a success in terms of flood control, but it was almost immediately recognized as an ecological disaster as it resulted in the loss of 12,000 to 14,000 ha of floodplain wetlands that were drained, covered with spoil, or converted into canal (Koebel 1995; Toth et al. 1998). As a consequence, a river restoration project was initiated.

Two particular aspects of spatial heterogeneity were considered in the restoration plan. First, longitudinal linkages were deemed crucial to restoration success. It was essential that the headwaters of the Kissimmee River have a more natural hydroperiod. In turn, this would result in seasonal inflows to the Kissimmee that were more characteristic of the prechannelized state. Increased stage height in the upper chain of lakes, variable discharges based on season and water level, greater discharge capacity to the river, and land acquisition of more than 6,500 ha were components crucial for ensuring that restoration to the river channel was not compromised by insufficient consideration of upper basin needs. Second, floodplain-river linkages were assessed. Prior to full-scale restoration, several projects were conducted to determine the feasibility of backfilling the channel and whether inundation of the floodplain would occur with the frequency and spatial extent that were planned (Loftin et al. 1990).

## Lateral-Vertical (Hillslope-River)

We hypothesize that the ecosystem service most affected by the spatial arrangement of hillslope activities is water supply quantity and timing (Table 18.2). Road construction is of particular concern, as road-stream crossings can alter natural flow regimes (Bilby et al. 1989; Montgomery 1994), block the movement of fish and aquatic mammals (Furniss et al. 1991; Warren and Pardew 1998), and, via alteration of runoff patterns, increase nonpoint-source pollution (Trombulak and Frissell 2000).

Roads are pervasive. In the coterminous United States, there are approximately 63 million km of public roads and 5.3 million km of streams and rivers (Riitters and Wickham 2003). Jones et al. (2000) noted that roads located near the ridge of a hillslope would have much less direct interaction with streams than roads located in middle and lower hillslope positions. As road construction has been associated with a higher frequency of landslides, especially in steep forest landscapes (Swanson and Dyrness 1975), the spatial distribution of roads can profoundly affect their impact on the water resource.

## Vertical (Surface Water-Groundwater Linkage)

The configuration of surface water-groundwater linkages is hypothesized to be especially relevant to water supply, groundwater recharge, and waste processing. At a coarse scale, excessive extraction of groundwater is known to have a number of habitat-related impacts to hydraulically connected surface waters, including declines in base flow of rivers, reductions in the spatial extent of stream habitat, increases in summer stream temperatures, and impairments of water quality (Jones and Mulholland 2000). However, at a finer scale, the spatial position of wells can have localized impacts, such as when they intercept contaminated aquifers (from septic or industrial waste).

This section has illustrated that the relevance of a landscape feature—whether it be spatial heterogeneity (configurational or compositional heterogeneity), type of landscape linkage (longitudinal, horizontal, lateral, vertical), or scale—depends on the ecosystem service or function being examined. This conclusion has several management implications. First, the "one size fits all" approach to management of water resources with respect to spatial heterogeneity will result in failure. Second, the approach laid out above, which involves understanding and appreciating how different forms of heterogeneity and types of linkages influence separate ecosystems services and functions, will allow greater flexibility in decision making by water resource managers.

## Case Study: The South Florida Ecosystem

The south Florida ecosystem illustrates how different water management scenarios emerge when the system is viewed in terms of spatial heterogeneity, landscape linkage, and ecosystem service and function. Lake Okeechobee and its watershed in south-central Florida exceed 11,400 km$^2$ and serve as the headwaters for the Florida Everglades (Figure 18.3). This system provides water supply for agriculture, municipalities, industry, and the environment; navigation; recreation; flood control; wellfield recharge; and habitat for fish and wildlife (Steinman et al. 2002). The influence of spatial heterogeneity on these ecosystem services is illustrated for habitat quality, water supply, navigation, recreation, and flood control.

Increased agricultural land use and improved drainage resulted in changes to system structure and function. Wetland cover declined, and with less nutrient assimilation within the watershed, phosphorus loads increased to downstream receiving waters resulting in degraded biotic communities (Flaig and Havens 1995; Steinman and Rosen 2000; Steinman et al. 2001, 2003a). These

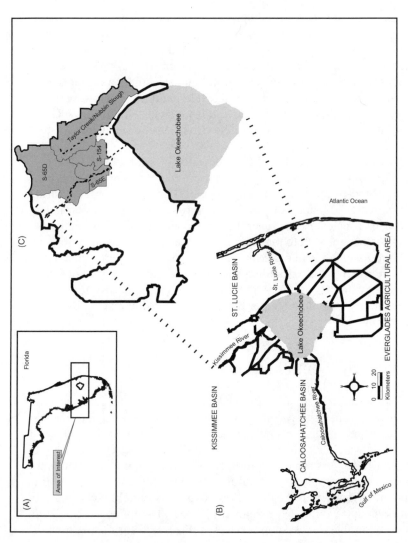

FIGURE 18.3. Spatial arrangement of watersheds (i.e., components) in south Florida. (A) State of Florida with area of interest delimited by the rectangle. (B) Enlargement of the area of interest, showing the Kissimmee basin to the north, Caloosahatchee basin to the west, and St. Lucie basin to the east, of Lake Okeechobee. Dark lines represented rivers and canals. (C) Enlargement of 3(B) showing detail of the Lake Okeechobee watershed (black outline), including the four priority sub-basins (based on phosphorus contributions).

increased loads worked their way to Lake Okeechobee and eventually into the Everglades. Four basins (Taylor Creek/Nubbin Slough, S-154, S-65D, and S-65E; Figure 18.3), which total approximately 1200 km$^2$, contribute the highest phosphorus concentrations and loads to the lake (35% of total). Therefore, the majority of proposed restoration activities, as currently contained in the Comprehensive Everglades Restoration Plan (CERP), are concentrated in these priority basins.

The specific locations for these facilities within each basin have not been defined, even though the spatial distribution of these project components in the landscape is crucial, as their location can profoundly influence their overall effectiveness. Zedler (2003) noted the relative efficacy of wetlands on flood abatement, water quality, and biodiversity would vary depending on their location within a watershed. For example, restored wetlands along riparian floodplains may be most effective for flood control, but if the major concern is improving water quality, restored wetlands immediately downstream of tributaries with high nutrient loads may make more sense.

Different interests compete for the highly managed water resources in south Florida. Compartmentalizing the region into different spatial components helps illustrate how location influences the ecosystem services provided by fresh water under different conditions. The region can be coarsely subdivided into the Kissimmee basin north of the lake, the lake itself, the Caloosahatchee basin west of the lake, the St. Lucie basin east of the lake, the Everglades Agricultural Area (EAA) immediately south of the lake, and the remnant Everglades farther to the south (Figure 18.4). At a very coarse scale, these regions of south Florida can be viewed as patches, and hence, the configuration of these patches strongly influences the degree to which ecosystems services can be provided.

Under high flow conditions (Figure 18.4a), the increased nutrient runoff from the watershed results in impaired water quality, while higher water levels lead to reduced light reaching the lake bottom and result in the loss of submerged aquatic vegetation (SAV). Under these same conditions, the lake serves its intended purpose of providing flood control for the region, and boat navigation is maintained. Downstream, habitat in both estuaries and the Everglades is impaired because of salinity imbalances associated with discharges from the lake (Kraemer et al. 1999) and phosphorus loads to the Everglades (McCormick and Scinto 2001; Miao and DeBusk 2001). However, this condition does provide sufficient water for irrigation, sprinkling, aquifer recharge, and human consumption.

Under low flow conditions (Figure 18.4b), fewer nutrients enter the lake thereby improving water quality, and the lower water levels result in more light reaching the lake bottom (Havens et al. 2002). However, navigation may be threatened if lake levels become too low. Habitat is improved in the estuaries and Everglades due to reduced inflows from the lake, but only up to a certain point. When water levels become too low, the estuaries can become too saline, the Everglades too dry, and water supply cannot be fully

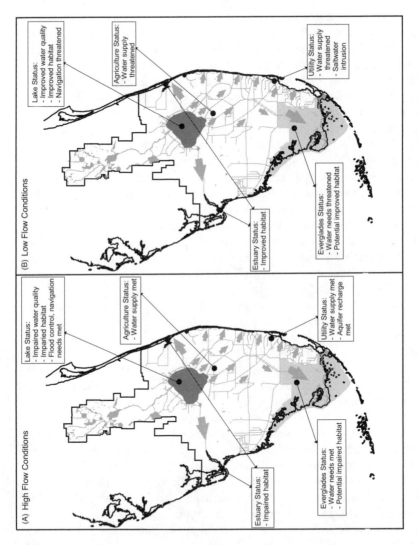

FIGURE 18.4.  Flow chart illustrating impacts to different ecosystem services within each south Florida watershed (component). (A) System under high water conditions in Lake Okeechobee. (B) System under low water conditions in Lake Okeechobee.

met under very low water conditions, resulting in water use restrictions (cf. Steinman et al. 2002).

Clearly, the configuration of these compartments is crucial to water management given their highly managed and artificially connected nature. This has been formally recognized in south Florida by the adoption of a new regulation schedule for Lake Okeechobee, which was specifically designed to optimize environmental benefits at minimal or no impact to competing lake purposes. New adaptive protocols for the operation of the regulation schedule attempt to balance the competing demands on the lake while ensuring that water releases from the lake do not impair downstream ecosystems. Both the spatial location and the type of ecosystem (i.e., the compositional heterogeneity) are crucial because the perceived value of ecosystem services varies within the region and among users in each region. For example, if climate models project an extended period of high water, surface waters in the upper regions of the ecosystem could be drawn down, allowing more storage during the runoff events and reducing the environmental threats to the estuaries because less water is being released from the lake.

## Conclusions

Effective and efficient management of our freshwater resources is one of the most important challenges facing this planet. By taking compositional and configurational heterogeneity into account, we provide a new approach for viewing the management of freshwater resources. Future needs regarding this framework include applying it conceptually to different water management scenarios to assess its viability and applicability, modifying the conceptual framework as necessary, and determining if it is amenable to computational modeling, such as a decision-support model. The modeling component may involve developing (1) a user-friendly, spatially explicit interface, (2) weighting functions (driven by management needs) to address what aspects of heterogeneity are most important in that system, (3) dynamic modeling to focus on key processes, and (4) user feedback to address limitations. Although we believe there is heuristic value in this conceptual framework, its ultimately utility—as with any conceptual or computational model—will depend on whether it works in the real world.

*Acknowledgments.* The authors are grateful to Gary Lovett, Pat Soranno, two anonymous reviewers, and especially Clive Jones for their thoughtful and constructive ideas and comments, Lori Nemeth and Mark Brady for their graphical assistance, and Xuefeng Chu for his input. The senior author appreciates his past interactions with Karl Havens, Nick Aumen, Lou Toth, Peter Doering, Bob Chamberlain, and Tom Fontaine, all of whom influenced his thinking about water management issues in Florida.

## References

Baron, J., Poff, N.L., Angermeier, P.L., Dahm, C.N., Gleick, P.H., Hairston, N.G., Jr., Jackson, R.B., Johnston, C.A., Richter, B.G., and Steinman, A.D. 2002. Meeting ecological and societal needs for freshwater. Ecol. Applications 12: 1447–1460.

Bayley, P.B. 1995. Understanding large river-floodplain ecosystems. BioScience 45: 153–158.

Benke, A.C. 2001. Importance of flood regime to invertebrate habitat in an unregulated river-floodplain ecosystem. J. North Am. Benthol. Soc. 20: 225–240.

Bilby, R.E., Sullivan, K., and Duncan, S.H. 1989. The generation and fate of road-surface sediment in forested watersheds in southwestern Washington. Forest Sci. 35: 453–468.

Cohen, M. 2002. Managing across boundaries: the case of the Colorado River delta. In The world's water 2002–2003. The biennial report on freshwater resources, ed. P. Gleick, pp. 133–148. Washington, DC: Island Press.

Daily, G.C., ed. 1997. Nature's services: societal dependence on natural ecosystems. Washington, DC: Island Press.

Falkenmark, M. 1999. Forward to the future: a conceptual framework for water dependence. Ambio 28: 356–361.

Flaig, E.G., and Havens, K.E. 1995. Historical trends in the Lake Okeechobee ecosystem I. Land use and nutrient loading. Archiv für Hydrobiologie Monographische Beiträge 107: 1–24.

Furniss, M.J., Roeloffs, T.D., and Yee, C.S. 1991. Road construction and maintenance. In Infuences of forest and rangeland management on salmonid fishes and their habitats. Special Publication 19, ed. W.R. Meehan, pp. 297–323. Bethesda, MD: American Fisheries Society.

Galat, D.L., Fredrickson, L.H., Humberg, D.D., Bataille, H.J., Brodie, J.R., Dohrenwend, J., Gelwicks, G.T., Havel, J.E., Helmers, D.L., Hooker, J.B., et al., 1998. Flooding to restore connectivity of regulated, large-river wetlands. BioScience 48: 721–733.

Gleick, P.H. 1998. Water in crisis: paths to sustainable water use. Ecol. Applications 8: 571–579.

Gleick, P.H. 2000. The World's Water 2000–2001: The biennial report on freshwater resources. Washington, DC: Island Press.

Gleick, P.H. 2002. Environment and security water conflict chronology 2002. In The World's Water 2002–2003. The biennial report on freshwater resources, ed. P.H. Gleick, pp. 194–206. Washington, DC: Island Press.

Gregory, S.V., Lamberti, G.A., Erman, D.C., Koski, K.V., Murphy, M.L., and Sedell, J.R. 1987. Influence of forest practices on aquatic production. In Streamside management: forestry and fishery interactions, eds. E.O. Salo, E.O., and T.W. Cundy, pp. 233–255. Seattle, WA: Institute of Forest Resources, Contribution 57, University of Washington.

Grigorovich, I.A., Colautti, R.I., Mills, E.L., Holeck, K., Ballert, A.G., and MacIsaac, H.J. 2003. Ballast-mediated animal introductions in the Laurentian Great Lakes: retrospective and prospective analyses. Can. J. Fisheries Aquatic Sci. 60: 740–756.

Havens, K.E., and James. R.T. 1999. Localized changes in transparency linked to mud sediment expansion in Lake Okeechobee, Florida: ecological and management implications. Lake Reservoir Manage. 15: 54–69.

Havens, K.E., Harwell, M.C., Brady, M.A., Sharfstein, B., East, T.L., Rodusky, A.J., Anson, D., and Maki, R.P. 2002. Large-scale mapping and predictive modeling of submerged aquatic vegetation in a shallow eutrophic lake. Sci. World 2: 949–965.

Henry, J.C., and Fisher, S.G. 2003. Spatial segregation of periphyton communities in a desert stream: causes and consequences for N cycling. J. North Am. Benthol. Soc. 22: 511–527.

Jackson, R.B., Carpenter, S.R., Dahm, C.N., McKnight, D.M., Naiman, R.J., Postel, S.L., and Running, S.W. 2001. Water in a changing world. Ecol. Applications 11: 1027–1045.

Johnson, N., Revenga, C., and Echevaerria, J. 2001. Managing water for people and nature. Science 292: 1071–1072.

Jones, J.A., Swanson, F.J., Wemple, B.C., and Snyder, K.U. 2000. Effects of roads on hydrology, geomorphology, and disturbance patches in stream networks. Conservation Biol. 14: 76–85.

Jones, J.B., and Mulholland, P.J. editors. 2000. Streams and ground waters. San Diego, CA: Academic Press.

Junk, W.J., Bayley, P.B., and Sparks, R.E. 1989. The flood-pulse concept in river-floodplain systems. In Proceedings of the International Large River Symposium, ed. D.P. Dodge, pp. 10–27. Canadian Special Publication of Fisheries and Aquatic Sciences 106. Ottawa, Ontario, Canada.

Karr, J.R., and Chu, E.W. 1999. Restoring life in running waters. Better biological monitoring. Washington, DC: Island Press.

Koebel, J.W. Jr. 1995. An historical perspective on the Kissimmee River restoration project. Restoration Ecology 3: 149–159.

Kowaleski, M., Guillermo, E.A.S., Flessa, K.W., and Goodfriend, G.A. 2000. Dead delta's former productivity: two trillion shells at the mouth of the Colorado River. Geology 28: 1059–1062.

Kraemer, G.P., Chamberlain, R.H., Doering, P.H., Steinman, A.D., and Hanisak, M.D. 1999. Physiological responses of *Vallisneria americana* transplants along a salinity gradient in the Caloosahatchee Estuary (SW Florida). Estuaries 22: 138–148.

Kratz, T.K., Webster, K.E., Bowser, C.J., Magnuson, J.J., and Benson, B.J. 1997. The influence of landscape position on lakes in northern Wisconsin. Freshwater Biol. 37: 209–217.

Likens, G.E., Bormann, F.H., Johnson, N.M., Fisher, D.W., and Pierce, R.S. 1970. Effects of forest cutting and herbicide treatment on nutrient budgets in the Hubbard Brook Catchment ecosystem. Ecol. Monogr. 40: 23–47.

Loftin, M.K., Obeysekera, J., Neidrauer, C., and Sculley, S. 1990. Hydraulic performance of the phase I demonstration project. In Proceedings of the Kissimmee River Restoration Symposium, eds. M.K. Loftin, L.A. Toth, and J. Obeysekera, pp. 197–209. West Palm Beach, FL: South Florida Water Management District.

Magnuson, J.J., and Frost, T.M. 1982. Trout Lake Station: a center for north temperate lake studies. Bull. Ecol. Soc. Am. 63: 223–225.

McCormick, P.V., and Scinto, L.J. 2001. Influence of phosphorus loading on wetlands periphyton assemblages: a case study from the Everglades. In Phosphorus Biogeochemistry of Subtropical Ecosystems: Florida as a case example, eds. K.R. Reddy, G.A. O'Connor, and C.L. Schelske, pp. 301–320. New York: CRC/Lewis Publishers.

Miao, S.L., and W.F. DeBusk. 2001. Effects of phosphorus enrichment on structure and function of sawgrass and cattail communities in the Everglades. In Phosphorus Biogeochemistry of Subtropical Ecosystems: Florida as a case example, eds. K.R. Reddy, G.A. O'Connor, and C.L. Schelske, pp. 275–300. New York: CRC/Lewis Publishers.

Montgomery, D.R. 1994. Road surface drainage, channel initation, and slope instability. Water Resources Res. 30: 1925–1932.

Naiman, R.J., Magnuson, J.J., McKnight, D.M., and Stanford, J.A., eds. 1995. The
freshwater imperative: A research agenda. Washington, DC: Island Press.

Naiman, R.J., Magnuson, J.J., and Firth, P.L. 1998. Integrating cultural, economic, and
environmental requirements for fresh water. Ecol. Applications 8: 569–630.

Naiman, R.J., Bunn, S.E., Nilsson, C., Petts, G.E., Pinay, G., and Thompson, L.C. 2002.
Legimitizing fluvial ecosystems as users of water: an overview. Environ. Manage.
30: 455–467.

Phlips, E.J., Cichra, M., Havens, K.E., Hanlon, C., Badylak, S., Rueter, B., Randall, M.,
and Hansen, P. 1997. Relationships between phytoplankton dynamics and the availability of light and nutrients in a shallow subtropical lake. J. Plankton Res. 19: 319–342.

Pitt, J., Luecke, D.G., Cohen, M.J., Glenn, E.P., and Valdés-Casillas, C. 2000. Two
countries, one river: managing for nature in the Colorado River delta. Natural
Resources J. 40: 819–864.

Plafkin, J.L., Barbour, M.T., Porter, K.D., Gross, S.K., and Hughes, R.M. 1989. Rapid
bioassessment protocols for use in streams and rivers: benthic macroinvertebrates
and fish. EPA/440/4-89-001. Assessment and Water Protection Divison, U.S.
Washington, DC: Environmental Protection Agency.

Pringle, C.M. 2001. Hydrologic connectivity and the management of biological
reserves: a global perspective. Ecol. Applications 11: 981–998.

Reddy, K.R., Sheng, Y.P., and Jones, B.L. 1995. Lake Okeechobee Phosphorus
Dynamics Study. Volume I. Summary. Contract C91-2554. South Florida Water
Management District, West Palm Beach, FL.

Ricciardi, A., and MacIsaac, H.J. 2000. Recent mass invasion of the North American
Great Lakes by Ponto-Caspian species. Trends Ecol. Evolution 16: 62–65.

Ricciardi, A., and Rasmussen, J.B. 1998. Predicting the identity and impact of future
biological invaders: a priority for aquatic resource management. Can. J. Fisheries
Aquatic Sci. 55: 1759–1765.

Riera, J.L., Magnuson, J.J., Kratz, T.K., and Webster, K.E. 2000. A geomorphic template for the analysis of lake districts applied to the Northern Highland Lake District, Wisconsin, U.S.A. Freshwater Biol. 43: 201–318.

Riitters, K.H., and Wickham, J.D. 2003. How far to the nearest road? Frontiers Ecol.
Environ. 1: 125–129.

Soranno, P.A., Webster, K.E., Riera, J.L., Kratz, T.K., Baron, J.S., Bukaveckas, P.A.,
Kling, G.W., White, D.S., Caine, N., Lathrop, R.C., and Leavitt, P.R. 1999. Spatial
variation among lakes within landscapes: ecological organization along lake
chains. Ecosystems 2: 395–410.

Steinman, A.D., and Rosen, B.H. 2000. Lotic-lentic linkages associated with Lake
Okeechobee, Florida. J. North Am. Benthol. Soc. 19: 733–741.

Steinman, A.D., Havens, K.E., Aumen, N.G., James, R.T., Jin, K.-R., Zhang, J., and
Rosen, B. 1999. Phosphorus in Lake Okeechobee: sources, sinks, and strategies. In
Phosphorus Biogeochemistry of Subtropical Ecosystems: Florida as a case exam-
ple, eds. K.R. Reddy, G.A. O'Connor, and C.L. Schelske, pp. 527–544. New York:
CRC/Lewis Publishers.

Steinman, A.D., Havens, K.E., Carrick, H.J., and VanZee, R. 2001. The past, present,
and future hydrology and ecology of Lake Okeechobee and its watersheds. In The
Everglades, Florida Bay, and Coral Reefs of the Florida Keys. An ecosystem handbook, eds. J. Porter and K. Porter, pp. 19–37. Boca Raton, FL: CRC Press.

Steinman, A.D., Havens, K., and Hornung, L. 2002. The managed recession of Lake Okeechobee, Florida: integrating science and natural resource management. Conservation Ecology 6(2): 17. Available at http://www.consecol.org/vol6/iss2/art17.

Steinman, A.D., Conklin, J., Bohlen, P.J., and Uzarski, D.G. 2003a. Influence of cattle grazing and pasture land use on macroinvertebrate communities in freshwater wetlands. Wetlands 23: 877–889.

Steinman, A.D., Luttenton, M., and Havens, K.E. 2003b. Sustainability of surface and subsurface water resources: case studies from Florida and Michigan, U.S.A. Water Resources Update 126: 54–59.

Swanson, F.J., and Dyrness, C.T. 1975. Impact of clear-cutting and road construction on soil erosion by landslides in the western Cascade Range, Oregon. Geology 3: 393–396.

Toth, L.A., Melvin, S.L., Arrington, D.A., and Chamberlain, J. 1998. Hydrologic manipulation of the channelized Kissimmee River. BioScience 48: 757–764.

Trombulak, S.C., and Frissell, C.A. 2000. Review of ecological effects of roads on terrestrial and aquatic communities. Conservation Biol. 14: 18–30.

Vanderploeg, H.A., Nalepa, T.F., Jude, D.J., Mills, E.L., Holeck, K.T., Liebig, J.R., Grigorovich, I.A., and Ojaveer, H. 2002. Dispersal and emerging ecological impacts of Ponto-Caspian species in the Laurentian Great Lakes. Can. J. Fisheries Aquatic Sci. 59: 1209–1228.

Vannote, R.L., Minshall, G.W., Cummins, K.W., Sedell, J.R., and Cushing, C.E. 1980. The river continuum concept. Can. J. Fisheries Aquatic Sci. 37: 130–137.

Walker, B., and Steffen, W. 1997. An overview of the implications of global change for natural and managed terrestrial ecosystems. Conservation Ecol. 1(2): 2. Available at http://www.consecol.org/vol1/iss2/art2.

Ward, J.V. 1997. An expansive perspective of riverine landscapes: pattern and process across scales. River Ecosystems 6: 52–60.

Ward, J.V., and Stanford, J.A. 1983. The serial discontinuity concept of lotic ecosystems. In Dynamics of Lotic Ecosystems, eds. T.D. Fontaine and S.M. Bartell, pp. 347–356. Ann Arbor, MI: Ann Arbor Science Publishers.

Warren, M.L. Jr., and Pardew, M.G. 1998. Road stream crossings as barriers to small-stream fish movement. Trans. Am. Fisheries Soc. 127: 637–644.

Webster, K.E., Kratz, T.K., Bowser, C.J., Magnuson, J.J., and Rose, W.J. 1996. The influence of landscape position on lake chemical response to drought in northern Wisconsin, USA. Limnol. Oceanogr. 41: 977–984.

Winter, T.C., Harvey, J.W., Franke, O.L., and Alley, W.M. 1998. Ground water and surface water, a single resource. USGS Circular 1139. Denver, CO.

Zedler, J.B. 2003. Wetlands at your service: reducing impacts of agriculture at the watershed scale. Frontiers Ecol. Environ. 1: 65–72.

# 19

# The Roles of Spatial Heterogeneity and Ecological Processes in Conservation Planning

HUGH P. POSSINGHAM, JANET FRANKLIN, KERRIE WILSON, and TRACEY J. REGAN

## Abstract

In this chapter we ask the question: To what extent does an understanding of landscape spatial heterogeneity inform conservation decisions? We answer this question in the context of two central decision-making fields within conservation biology: systematic conservation planning and population viability analysis. The conservation planning principles of comprehensiveness and representativeness are fundamentally reliant on data and concepts of compositional landscape heterogeneity. The principle of adequacy is not accommodated in conservation planning very well and it relies on an understanding of the configurational heterogeneity of the landscape. A major challenge for conservation planning scientists is to develop theory and decision support tools that incorporate ideas of population viability and spatially explicit ecological processes. Population viability analysis invariably includes spatial population processes, and as a field has largely focused on the importance of the configurational heterogeneity of landscapes. We argue that this focus might only be justified when the scale of planning coincides with either the scale of habitat heterogeneity or the scale at which small populations operate. Integrating population viability analysis into conservation planning, and showing a balanced interest in compositional and configurational heterogeneity, are important future challenges.

## Introduction

Ecological heterogeneity comes in many forms ranging from the biophysical to the ecological. Substrates like soil type are highly variable but relatively static on an ecological time frame. Other aspects of heterogeneity, for example species distributions and ecological processes, can exhibit greater temporal variation. There are two components of heterogeneity: composition and configuration. Compositional heterogeneity refers to the number of different elements in the landscape, and configurational heterogeneity refers to the spatial arrangement of these elements. The elements can be discrete (patches) or continuous (gradients). We discuss two areas of application: systematic reserve system design and population management using population viability models.

In the field of reserve system design, the overall objective is to create a system of protected areas that conserves as much of a region's biodiversity as possible in the long term (McNeely 1994). To do this we first need to sample as much of the biodiversity as possible. Hence, an understanding of compositional spatial heterogeneity is absolutely central to reserve system design. In contrast, the role of configurational spatial heterogeneity is discussed, but poorly dealt with, in the systematic conservation planning literature.

The only way we know how to determine the adequacy of a reserve system is to assess the viability of key species. Population viability analysis (PVA) is a tool for choosing between different management options for threatened species. Traditionally, PVA has dealt with compositional heterogeneity by assuming there are only two habitat types: suitable and unsuitable. This is clearly inadequate as habitat quality will, in general, vary continuously (Franklin this volume). Configurational heterogeneity is believed to be important to the viability of populations, but the evidence is equivocal (Fahrig this volume). Ultimately good conservation planning will involve a marriage of reserve system design principles and population viability principles.

For both reserve system design and population management, we postulate that spatial heterogeneity is relatively unimportant to conservation decision-making when the spatial scale of management (the spatial extent of typical planning actions or reserves) is significantly different to the spatial scale of the underlying heterogeneity or the population processes of the species of concern. We suggest that spatial heterogeneity is most important when its scale of variation is roughly the same as the scale of management *and* the scale of population and other ecosystem processes.

In this paper, we will (1) describe the general reserve system design problem, (2) look at how heterogeneity at different scales has or could influence reserve system design, (3) consider the role of spatial processes in reserve system design, (4) examine how spatial heterogeneity at different scales influences conservation plans derived from population models, and (5) present an initial general framework for how we might deal with heterogeneity considerations in conservation planning.

## The General Reserve System Design Problem

In its broadest sense, conservation planning is about allocating parts of a landscape to a management regime. For example, in forestry we could allocate any 50-ha compartment to one of the following: no harvesting, no harvesting and predator control, selective harvesting, clear-fell at 30-year rotation for woodchips, clear-fell at 70-year rotation for construction timber, conversion to native plantation, conversion to exotic plantation, or conversion to infrastructure (buildings, mills, houses). In the narrow sense, only the first two treatments would be a necessary, but not sufficient, to allocate a compartment to a reserve system. Here we will consider the more restricted question of reserve system design where parcels of land are selected for the reserve system.

The overall objective of reserve system design is to create a system of protected areas that conserves as much of a region's biodiversity as possible in the long term. To do this, we first need to sample as much of the biodiversity as possible. A simple solution is to select every parcel for the reserve system. Clearly, this is socially and economically infeasible in most regions so we add an additional objective, that of efficiency. Our economically prudent objective is to conserve as much biodiversity as possible in the long-term as efficiently as possible.

Three further notions are important here: comprehensiveness, representativeness, and adequacy (Margules and Pressey 2000). A comprehensive reserve system is one that captures every known element of biodiversity. Given our interest in efficiency, we can only hope to sample each element and we should do that in a representative fashion (i.e., the set of samples that capture each element of biodiversity should be "typical" or representative). The adequacy of a reserve system refers to how well it meets the management goal of preserving biodiversity in the long term.

These concepts define the classical reserve system design problem—referred to as "gap analysis" in the United States (Scott et al. 1993) and a "CAR reserve system" (comprehensive, adequate, and representative) in Australia (JANIS 1997). In this classic form, conservation planners take whatever data they have on any aspect of biodiversity (e.g., species distributions, habitat types, land systems) and seek to represent a fixed proportion of the original extent of each of these features in a reserve system (Margules and Pressey 2000). In this sense, conservation planning is highly reliant on patterns of heterogeneity, especially those that appear to be invariant in the short term (e.g., vegetation types, altitude, soil type, etc). Although spatial heterogeneity in species distributions and habitat types is commonly measured in landscape ecology (O'Neill et al. 1988; Haines-Young and Chopping 1996; Gustafson 1998; McAlpine et al. 2002; McGarigal, 2002), within-feature variation, which should be dealt with under the conservation planning principle of represenativeness, is rarely explicitly considered in reserve system design. So that we can explore the role of spatial heterogeneity at different spatial scales in reserve system design, let us formulate the problem mathematically and consider a specific example.

## The Basic Reserve Design Problem

Let the total number of sites that could be in the reserve system be $m$ and the number of features (e.g., species, vegetation types, etc.) be represented by $n$. The information about whether or not a feature is found in a site is contained in a site-by-feature ($m \times n$) matrix $\mathbf{A}$ whose elements $a_{ij}$ are

$$a_{ij} = \begin{cases} 1 & \text{if feature } j \text{ occurs in site } i \\ 0 & \text{otherwise} \end{cases}$$

$$\text{for } i = 1, \ldots, m \text{ and } j = 1, \ldots, n.$$

Next, define a control variable (the part of the system that we control), that determines whether or not a site is included in the reserve system, as the vector $\mathbf{X}$ with dimension $m$ and elements $x_i$, given by

$$x_i = \begin{cases} 1 & \text{if site } i \text{ is included in the reserve} \\ 0 & \text{otherwise} \end{cases}$$

$$\text{for } i = 1, \ldots, m.$$

With these definitions, the basic minimum representation problem is

$$\text{Minimize } \sum_{i=1}^{m} x_i \quad \{\text{minimize the number of sites in the reserve system}\}$$

$$\text{subject to } \sum_{i=1}^{m} a_{ij} x_i \geq 1, \text{ for } j = 1, \ldots, n$$

$$\{\text{subject to each feature being represented at least once}\}$$

where $a_{ij}, x_i \in \{0,1\}$.

This is the integer linear programming formulation of the set-covering problem (Possingham et al. 2000). In many cases, the feature by site data is not simply zeros and ones, but could represent the number of occurrences of the feature in the site, the estimated population size of a species, or the area of a feature like habitat type. In this case, the targets for each feature are likely to be different from one. However, the basic structure of the problem remains unaltered. The key issue is that this basic approach does not explicitly deal with representativeness because we do not know if the sites captured to meet a target for a feature are representative, or typical, of that feature. To explore these ideas, let us consider a particular example.

Consider a planning area with 16 sites and 5 different features (Figure 19.1a; Table 19.1). Figure 19.1 shows the spatial location of the five features, two of which are species, represented by point data, and three of which are habitats that are mutually exclusive and cover the entire planning area. Each of the 16 sites may, or may not, be selected for the final reserve system. Table 19.1 captures all the information in the map except it ignores spatial relationships between both the sites and the habitats. Our task is to comprehensively represent each of the five features in the reserve system as efficiently as possible.

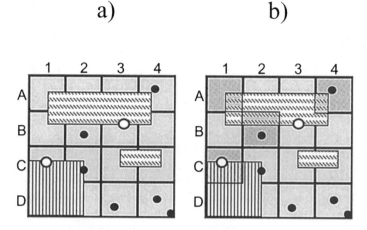

FIGURE 19.1. (a) A hypothetical planning area with 16 sites; the grid cells A1–D4. The polygons of different sizes and textures represent three different habitat types, and the black and white dots represent populations of two species. (b) The most efficient solution that meets a 25% target coverage of each feature is highlighted by the darkest squares (A1, A4, B2, C1). See Table 19.1 for a representation of the map as a data matrix.

TABLE 19.1. The amount of each feature in each planning unit, and overall conservation targets, for the planning landscape described in Figure 19.1.

| | Features | | | | |
| | Habitats | | | Species | |
| Site code | Dash | Stripe | Dots | White | Black |
|---|---|---|---|---|---|
| A1 | 20 | 0 | 80 | 0 | 0 |
| A2 | 50 | 0 | 50 | 0 | 0 |
| A3 | 50 | 0 | 50 | 0 | 0 |
| A4 | 15 | 0 | 85 | 0 | 1 |
| B1 | 15 | 0 | 85 | 0 | 0 |
| B2 | 40 | 0 | 60 | 0 | 1 |
| B3 | 40 | 0 | 60 | 1 | 0 |
| B4 | 10 | 0 | 90 | 0 | 0 |
| C1 | 0 | 60 | 40 | 1 | 0 |
| C2 | 0 | 30 | 70 | 0 | 1 |
| C3 | 25 | 0 | 75 | 0 | 0 |
| C4 | 35 | 0 | 65 | 0 | 0 |
| D1 | 0 | 100 | 0 | 0 | 0 |
| D2 | 0 | 50 | 50 | 0 | 0 |
| D3 | 0 | 0 | 100 | 0 | 1 |
| D4 | 0 | 0 | 100 | 0 | 2 |
| Total | 300 | 240 | 1060 | 2 | 6 |
| 25% Target | 75 | 60 | 265 | 1 | 2 |

Much has been written about this sort of problem. However, most authors have focused on issues such as the efficiency and speed of algorithms to solve the problem (Pressey et al 1997), whether certain types of feature can act as surrogates for other types of feature (e.g., if we conserve a sample of all habitat types will that guarantee conservation of all species; Ferrier and Watson 1996; Andelman and Fagan 2000; Ferrier 2002), and whether the reserve system is adequate in the long term (Cabeza and Moilanen 2001). We will consider a different issue and use this example to explore the role of spatial heterogeneity in reserve system design.

Because our discussion of the role of spatial heterogeneity will be couched in the context of efficient solutions to the reserve design problem, we will first need to determine what those efficient solutions are, given the data at hand. If our goal is to conserve at least one example of each feature, the classical minimum set problem, then there are several equally efficient two-site solutions to the problem displayed in Figure 19.1. For example, solution sets with two sites that meet the five targets for single representation include {B2, C1} and {C2, B3}. If we want to conserve at least 25% of the original extent of each feature, then there is only one most efficient reserve system comprising four sites {A1, A4, B2, C1} (Figure 19.1b). We will use this small sample problem to explore the role of spatial heterogeneity and the conservation of ecological processes in reserve system design.

# Reserve System Design and Static Spatial Heterogeneity

## Scale of Habitat Mapping

Habitat mapping is scale dependent (Davis et al. 1991; Franklin and Woodcock 1997). Exactly how many types of habitat (ecoregion, land system, vegetation) an ecologist chooses to define depends on the spatial scale at which they are working and the intended application of the data. If heterogeneity is mapped at a very coarse scale, then what appears to be a single habitat may indeed be several different habitats. For example, the striped habitat in Figure 19.1 is conserved by selecting site C1, but is this a representative sample of the striped habitat? If we look more closely, the striped habitat may comprise several types of habitat or it may contain a feature, like small rocky outcrops, that is not mapped. If those types and/or features are well mixed throughout the striped habitat (i.e. they display fine scale patchiness), then we may not miss our 25% target for each type of feature by much (Figure 19.2a). If however they display coarse-grained patchiness, then site C1 will fail to represent the variability in the striped habitat by a long margin (Figure 19.2b). This results in the selection of an unrepresentative sample of habitat and is a consequence of unmapped compositional heterogeneity.

For conservation planning, we can minimize this problem by mapping different features, such as drainage lines or rocky outcrops, at the finest scale

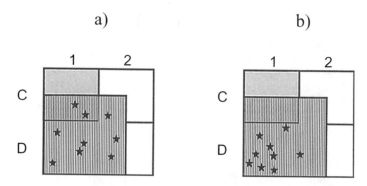

FIGURE 19.2. Enlarged section of Figure 19.1 (C1–C2–D1–D2) illustrating that (a) fine-scale or (b) coarse-scale patchiness in the location of rocky outcrops (stars) in the striped habitat affects our ability to represent this feature in a reserve planning exercise, because the compositional heterogeneity at that fine a scale is unmapped.

required. Where this is not possible, then data on biophysical features (e.g., altitude, aspect, soil type) should be combined with habitat maps to create more fine-scale heterogeneity in order for the heterogeneity to be sampled in the reserve system. This partitioning of habitats into smaller classes is facilitated by the ability of our software tools, such as geographical information system (GIS) software tools, to support different geographical data models (*sensu* Goodchild 1994), and by the use of hierarchical systems of habitat classification (Bailey et al. 1994; Küchler and Zonneveld 1988). However, it can also present a challenge because classification systems that are nested categorically do not necessary correspond to mapped entities that are nested spatially (reviewed in Franklin and Woodcock 1997). Further, despite the proliferation of spatially referenced environmental datasets (Estes and Mooneyhan 1994; Franklin 2001), spatially explicit information on fine-scale habitat features for large regions can still be difficult (and/or expensive) to develop (e.g., Elith and Burgman 2002). From this example and discussion we can conclude that spatial compositional heterogeneity at all scales is important for conservation planning, and the issue of conserving representative samples of biodiversity features remains challenging.

## The Scale of Conservation Planning Units

Although maps of biodiversity features at appropriate scales may not be as ubiquitously available as we may wish, a further issue is that the scale and positioning of the planning units (sites) used in reserve design is often quite unrelated to the underlying spatial heterogeneity. The planning unit layer may be imposed by external socioeconomic considerations—such as property boundaries—or it may be simply a "convenient" tessellation of the planning region. Limitations on algorithms to find good solutions to conservation

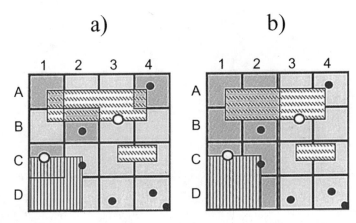

FIGURE 19.3. Optimal solutions to the reserve design problem where the planning units are defined at two different scales. (a) The same optimal reserve system for conserving 25% of every feature as in Figure 19.1. (b) The optimal reserve system for the same problem but where planning units are four times bigger, an increase in linear scaling of a factor of two. The result is reduced efficiency. See the legend of Figure 19.1 for feature descriptions.

planning problems means that tessellations with more than 100,000 planning units can be computationally difficult (although continual advances in speed and storage space of desktop computers mean this approximate upper bound is continually relaxed). This computational consideration often fixes the scale of the planning units, which in turn impacts the efficiency of solutions. For example, if the size of planning units was four times bigger in our example (Figures 19.1b and 19.3a), the most efficient solution would involve selecting 50% of the planning region, rather than 25% (Figure 19.3b).

Although an increase in the size of planning units will always make the reserve system less efficient, a coarser planning unit scale has two benefits (Pressey and Logan 1998): the reserve system is more cohesive, and the problem of representing hidden habitat and feature heterogeneity is reduced. To overcome the fragmented nature of reserve systems designed using fine-scale planning units, and thus address landscape configuration explicitly, several researchers have devised ways of incorporating spatial cohesion into the reserve design problem and have developed algorithms that implement these solutions (Bedward et al. 1992; Possingham et al. 2000).

## Spatial Design in Conservation Planning

A major problem with the basic reserve design problem is that it relies only on the data in a matrix (Table 19.1) that ignores spatial relationships between planning units and hence it ignores configurational spatial heterogeneity. One solution to this problem is to include information about spatial relationships by using the boundaries that sites share with each other. The boundary length of a reserve system is an important part of the cost of managing that system, so it makes sense to make the boundary length as small as possible. Reducing boundary length reduces the impact of edge effects and reduces fragmentation, both of which are core conservation planning principles (Noss and Csuti 1994; Fagan et al. 1999). To include boundary length in the problem we can amend the objective in the basic reserve design problem to be a weighted sum of the number of sites selected and the boundary length of the entire reserve system

$$\text{Minimize } \sum_{i=1}^{m} x_i + (\text{BLM} \times \text{the boundary length of the system})$$

(which means minimize the number of sites in the reserve system and add a constant, BLM, multiplied by the boundary length of the system; Possingham et al. 2000).

If the parameter BLM (acronym for boundary length multiplier) is large, then the emphasis will be on reducing the boundary length of the reserve system. If the BLM is relatively small, then the reserve systems will have a small area with compactness given secondary consideration. For example, in our sample problem the optimal solution had four sites with a boundary length of 16 units (Figures 19.1b and 19.4a). Two solutions with one more site (less efficient in terms of the area) but with a boundary length of only 10 (more efficient from the perspective of boundary length) can be found (Figures 19.4b and 19.4c). The best solution will depend on the relative costs of boundary length and area. In practice, some method of making the reserve system spatially cohesive at a relevant management scale is essential.

Adding a consideration of reserve system boundary length in the reserve design problem is only one of many ways that space can play a role in conservation planning. Allowing for connectivity and minimizing different measures of isolation can also be included. For example, Siitonen et al. (2002) considers three measures of connectivity: the total continuous area

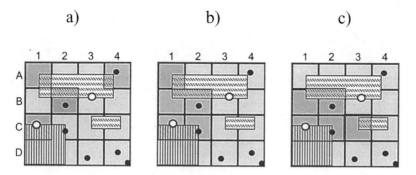

FIGURE 19.4. Three solutions to the reserve design problem where the planning objective differs. In solution (a), the focus is on minimizing area, whereas (b) and (c) are two different solutions for minimizing boundary length. See the legend of Figure 19.1 for feature descriptions.

of a single reserve, the level of isolation of that reserve from other reserves, and a measure of connectivity that assesses the size of a cluster of reserves considered to be sufficiently close to be deemed "connected." One problem with these measures and the boundary length approach to achieving compactness and connectivity is that none are explicit about why we value compactness and/or connectivity. Ideally, the objective of the conservation planning problem would be to optimize a direct functional relationship between species viability and landscape spatial pattern. This is an active area of research that should allow more explicit configurational objectives to be incorporated into conservation planning (Frank and Wissel 2002).

## Reserve System Design and Spatial Processes

Although the previous section provides insight into how issues of spatial heterogeneity can play a role in reserve system design, the focus was on pattern rather than process. There is an increasing desire to incorporate spatial processes explicitly into conservation planning, but there are few examples of this being achieved. Here we look at three situations where the boundary length problem formulation described above can be modified to take spatial processes in to account.

   Consider the situation where we have data on the movement of organisms or propagules of an organism in space. Connectivity is a crucial issue in conservation biology and planning (e.g., Beier 1993; Schadt et al. 2002) and often the connectivity observed is directional. Assume that the sample planning region (Figure 19.4) is a marine planning region and prevailing currents are from left to right (west to east) (Figure 19.5a). Under this circumstance, we can see that the more compact reserve systems in Figures 19.4b and 19.4c are both preferable to the fragmented system. We also know

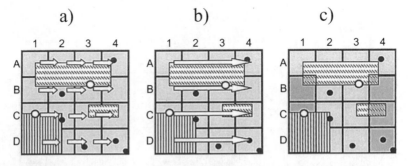

FIGURE 19.5. Two different sorts of connectivity between cells. In (a), connectivity is west-east to adjacent cells (arrows)—the solution displayed in Figure 19.4c is a good solution to this problem. In (b), propagules flow mostly to cells three sites to the east (arrows), and solution (c) is a good solution. See the legend of Figure 19.1 for feature descriptions.

the reserve system solution in Figure 19.4c is better than that in Figure 19.4b because it has three internal boundaries that favor east-west connectivity rather than just two. If we allow the cost of free east-west boundaries that reduce connectivity to be more expensive than free north-south boundaries, then the solution in Figure 19.4c will automatically be favored over the solution in Figure 19.4b.

Having relaxed our thinking that the boundary length between two sites is the physical length of the boundary between two sites, we can now include a boundary length between any two sites in the system. For example, if we know that propagules tend to disperse three sites to the east, then we can make the "boundary length" between column 1 sites and column 4 sites large and all other boundary lengths small (Figure 19.5b). In this case, the best reserve solutions should preferentially contain sites from columns 1 and 4 if they are in the same row (Figure 19.5c).

By modifying our interpretation of what a boundary is, we can see that connectivity issues can be formulated and solved in the context of the reserve design problem. This can be achieved by describing spatial relationships between cells using "boundary lengths" that measure how favorable connections are. In essence, we have placed a cost on each site being included in the reserve system, which is reduced each time a connected site is also included in the reserve system. There are many other ways in which this could be mathematically formulated; for example, we could list pairs of sites that are favorably connected for every species and have a target for the number of favorable connections included in the reserve system for each species. We can also incorporate propagation of unfavorable spatial processes; for example, fire may preferentially move north-south across the region. If we were to spread the risk of a reserve system being affected by a single fire—risk spreading—we could put a low cost on north-south boundaries and hence favor reserve systems that are aligned east-west. The process of including spatially explicit ecological processes in reserve system design is in its infancy. Reiners (this volume) provides a detailed summary of the kinds of spatial processes that we could incorporate into conservation planning.

The full consideration of spatially explicit ecosystem processes in reserve system design will necessitate that objectives are set for the processes themselves. For example, if we know something about how land management influences the flow of nutrients through a system, or a population across a landscape, then such flows should be included in the problem formulation as part of the objective or as constraints. Although this has not occurred, we believe that such explicit incorporation of ecosystem processes into the conservation planning problem is the best way forward.

One of the most crucial processes to consider in conservation planning is spatial population processes; for example, the regular flow of migratory individuals or favored flows of natal dispersers. These need to be included in the design of reserve systems, as the issue of the long-term viability of species (or other features) within reserve systems that are commonly ignored. Interest in designing adequate reserve systems is increasing (Cowling et al. 1999; Cabeza and Moilanen 2001; Araújo and Williams 2000). Although a thorough discussion of the role of population modeling in conservation planning is beyond the scope of this chapter, we briefly raise some interesting issues with respect to the issue of compositional and configurational spatial heterogeneity.

## Conservation Planning Based on Population Models

A properly posed reserve system design problem could include either an objective of minimum species loss or a constraint for every species to meet a specified level of viability. If we have an estimate of the number of individuals of each species in each site and we are willing to accept a target number of individuals as a surrogate for viability, then the reserve design problem remains unchanged (Pressey et al. 2003). However, a simple target number of individuals ignores the issues of fragmentation and connectivity discussed above, and we have no simple formula that adequately relates the configurational spatial heterogeneity of a reserve system to the viability of a species. In short, we have no adequate way of optimally designing reserve systems that incorporate the configurational aspects of population viability.

Population viability analysis (PVA) has been used to devise conservation plans for single species and to assess reserve systems and different sorts of landscape management (e.g., Lindenmayer and Possingham 1996). Typically, these models treat configurational heterogeneity by exploring the consequences of different patterns of just two habitats: suitable habitat and matrix habitat (Franklin this volume). The role of compositional heterogeneity with a landscape that is more than binary is less often explored, although interest in this topic is growing (Fahrig this volume).

Reviewing conservation planning using population models is beyond the scope of this paper. However, from a pragmatic perspective one question we can ask is: How often is configurational heterogeneity important to viability? In an extensive review, Fahrig (2003) suggests that the spatial pattern of habitat may rarely be important, and even where spatial pattern is a significant factor it is not always true that more fragmented systems decrease viability. So an interesting and important question is: When can we ignore spatial heterogeneity in reserve system design and still incorporate notions of viability?

In an extensive research program, Fox et al. (2004) developed population viability analysis models for 11 forest-dependent rare, threatened, or sensitive species in a forestry planning region in northeast Tasmania. The species vary from invertebrates to wide-ranging large birds, and from epiphytes to mammals. The project was designed to help Forestry Tasmania assess the viability of different species, given a range of region-wide forestry management scenarios including native forest harvesting and conversion of native forest to pine or eucalypt plantations. The scale of overall management is a region of about 200 km by 100 km, and the scale of operational management is a logging coupe (compartment)—about 50 ha. Intriguingly, they found that the configuration of the management was only important in one of the 11 species, and for most species simply determining the total amount of suitable habitat was sufficient to determine the viability of the species. It is worth briefly exploring why configuration was rarely important.

For wide-ranging mobile species like the Tasmanian wedge-tailed eagle, *Aquila audax,* and yellow-tailed black-cockatoo, *Calyptorhynchus funereus*— which both can cross the management region in a few hours—management at the 50-ha logging unit scale is too fine to be of concern. Similarly, these species range so broadly that compositional heterogeneity is not important above and beyond the patch-matrix (uncleared-cleared forest) dichotomy. At the other end of the spectrum, with invertebrates such as beetles, each 50-ha logging coupe (area that is logged in a single event) is large enough to maintain such a large population that it is only activities at a small scale that matter. A study by Smith et al. (2000) investigated the importance of landscape configuration in a heavily forested area on the persistence of a rare carnivorous snail *Tasmaphena lamproides.* They applied a spatially explicit population viability analysis model for the species (Regan et al. 2001) to several spatial and temporal landscape configurations. They also found that at the management scale of 50–100 ha, the total amount of available habitat was sufficient to determine the species' viability and the spatial configuration was not important. These species see heterogeneity at the scale of meters. Subtle differences in forest type or other small-scale compositional heterogeneity can be quite critical, but the precise spatial relationships are unimportant because such large populations occur in small areas.

The only species for which configuration appeared to matter was the spotted-tailed quoll, which uses landscapes at the scale of a few hundred meters. This coincides with the scale of management and much of the compositional heterogeneity. Similar observations were made in a study by Smith (2000) who investigated, using population viability models, the impact of various landscape configurations and forest harvesting intensities on the persistence of two species (an invertebrate and a mammal) with vastly different life-history attributes, home ranges, and dispersal characteristics within the same landscape structure. Again, the management scale was approximately 50–100 ha. The persistence of the invertebrate was highly correlated with available habitat, while the mammal, although sensitive to available habitat, was also sensitive to changes in landscape configuration. Fahrig (1998) also predicted that habitat pattern can affect population persistence only at particular scales relative to the movement range of the organism.

These observations lead us to suggest a general framework for when we need to consider heterogeneity in conservation planning—when the scale of

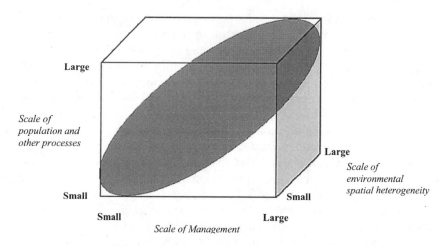

Figure 19.6. The circumstances where configurational heterogeneity is most important in applied conservation occurs when the scales of management, heterogeneity, and ecosystem/population processes overlap. This is found in the darker shaded region of the diagram.

management coincides most closely with the scales of spatial heterogeneity and the scale of population and ecosystem processes (Figure 19.6).

This finding is consistent with hierarchy and scaling theory (O'Neill et al. 1986; Levin 1992; Naveh and Lieberman 1994), which states that levels above the focal level of study in a (spatial, temporal, or organizational) hierarchy constrain and control the lower levels. The level below the focal level provides the details needed to explain the observed behavior of the system. The dynamics of the level above the focal level are so slow, or the spatial lag so great, that variables at that level appear constant from the perspective of the focal level. The dynamics or spatial variation at the lower level are so high-frequency that the average value is experienced at the focal level. Further, as elaborated in Turner et al. (2001), a shift in the relative importance of variables that influence a process, or even a change in the direction of the relationship, often occurs when scales are changed.

Finally, we note two recent developments in population viability analysis modeling that add important realism and will challenge our ability to understand complex interactions in a spatially heterogeneous world. First, there is increasing interest in models that expand on the patch-matrix dichotomy to include consideration of poor but suitable habitat that may, or may not, be a sink (Pulliam 1988). Theoretically and empirically we sometimes need to pay attention to the spatial relationships between habitats of differing quality, not just within good-quality habitat. Second, population modelers have begun to incorporate habitat dynamics into population models (Amarasekare and Possingham 2001). This means that the compositional and configurational habitat heterogeneity can change through time, and the temporal patterns of configuration will become important. Under these circumstances, not only is the spatial scale of the disturbances crucial, but also how the habitat changes following those disturbances (Pickett and White 1985). Although simulation models easily incorporate patch and metapopulation dynamics overlayed on a landscape with several habitat types (Akçakaya et al. 2004), the real challenge will be developing a conceptual understanding of the whole system. This may require a complex system style approach where we seek generalizations, emergent properties, and convenient simplifications that enable us to disentangle the consequences of several spatially and temporally varying factors.

## Conclusions

For conservation planning problems, there is an interplay of spatial scales that is rarely appreciated. The spatial scale of the planning unit, the underlying environmental heterogeneity, and spatially explicit ecosystem processes (including population connectivity) all interact. Although we are aware of these interactions empirically and theoretically, we are only just beginning to treat them explicitly in the more applied subfields of conservation biology, like reserve system design and population viability analysis. Understanding the importance of scale and heterogeneity in conservation planning is in its infancy. Solutions to the problems of dealing with spatial heterogeneity have been *ad hoc*; for example, by minimizing the boundary length of a reserve system. In particular, our capacity to deal with ecosystem and population processes is poorly developed. Adding compositional and configurational heterogeneity to population models is well advanced, but we have a limited conceptual understanding of the consequences of adding such complexity. This chapter has raised some of these issues and suggested some ways forward.

*Acknowledgments.* We are grateful for ideas and contributions shamelessly stolen from Sandy Andelman, Bob Pressey, Ian Ball, Mick McCarthy, Emily Nicholson, and Rom Stewart.

## *References*

Akçakaya, H.R., Radeloff, V.C., Mladenoff, D.J., and He, H.S. 2004. Integrating landscape and metapopulation modeling approaches: viability of the sharp-tailed grouse in a dynamic landscape. Conservation Biol. 18(2): 526–537.
Amarasekare, P., and Possingham, H.P. 2001. Patch dynamics and metapopulation theory: the case of successional species. J. Theor. Biol. 209: 333–344.
Andelman, S.J., and Fagan, W.F. 2000. Umbrellas and flagships: efficient conservation surrogates or expensive mistakes? Proc. Natl. Acad. Sci. U.S.A. 97: 5954–5959.

Araújo, M.B., and Williams, P.H. 2000. Selecting areas for species persistence using occurrence data. Biol. Conservation 96: 331–345.

Bailey, R.G., Avers, P.E., King, T., and McNab, W.H. 1994. Ecoregions and subregions of the United States (map, 1: 750,000, color), prepared by USDA Forest Service. Washington DC: U.S. Geological Survey.

Bedward, M., Pressey, R.L., and Keith, D. 1992. A new approach for selecting fully representative reserve networks: addressing efficiency, reserve design and land suitability with an iterative analysis. Biol. Conservation 62: 115–125.

Beier, P. 1993. Determining minimum habitat areas and habitat corridors for cougars. Conservation Biol. 7: 94–108.

Cabeza, M., and Moilanen, A. 2001. Design of reserve networks and the persistence of biodiversity. Trends Ecol. Evolution 16: 242–248.

Cowling, R., Pressey, R.L., Lombard, A., Desmet, P., and Ellis, A. 1999. From representation to persistence: requirements for a sustainable system of conservation areas in the species-rich Mediterranean-climate desert of southern Africa. Diversity Distributions 5: 51–71.

Davis, F.W., Quattrochi, D.A., Ridd, M.K., Lam, N.S.-N., Walsh, S.J., Michaelson, J.C., Franklin, J., Stow, D.A., Johannsen, C.J., and Johnston, C.A. 1991. Environmental analysis using integrated GIS and remotely sensed data: some research needs and priorities. Photgrammetric Eng. Remote Sensing 57: 689–697.

Elith, J., and Burgman, M.A. 2002. Predictions and their validation: rare plants in the Central Highlands, Vistoria, Australia. In Predicting species occurrences: issues of scale and accuracy, eds. J.M. Scott, P.J. Heglund, M.L. Morrison, J.B. Haufler, M.G. Raphael, W.A. Wall, and F.B. Samson, pp. 303–313. Covelo, CA: Island Press.

Estes, J.E., and Mooneyhan, D.W. 1994. Of maps anf myths. Photogrammetric Eng. Remote Sensing 60: 517–524.

Fagan, W., Cantrell, R., and Cosner, C. 1999. How habitat edges change species interactions. Am. Naturalist 153: 165–182.

Fahrig, L. 1998. When does fragmentation of breeding habitat affect population survival? Ecol. Modelling 105: 273–292.

Fahrig, L. 2003. Effects of habitat fragmentation on biodiversity. Annu. Rev. Ecol. Systematics 34: 487–515.

Ferrier, S. 2002. Mapping spatial pattern in biodiversity for regional conservation planning: Where to from here? Systematic Biol. 51: 331–363.

Ferrier, S., and Watson, G. 1996. An evaluation of the effectiveness of environmental surrogates and modelling techniques in predicting the distribution of biological diversity. Canberra, Australia: NSW National Parks and Wildlife Service for the Department of Environment, Sport and Territories.

Fox, J.C., Regan, T.J., Bekessy, S.A., Wintle, B.A., Brown, M.J., Meggs, J.M., Bonham, K., Mesibov, R., McCarthy, M.A., Munks, S.A., et al. 2004. Linking landscape ecology and management to population viability analysis: Report 2: population viability analyses for eleven forest dependent species. Unpublished report to Forestry Tasmania.

Frank, K., and Wissel, C. 2002. A formula for the mean lifetime of metapopulations in heterogeneous landscapes. Am. Naturalist 159: 530–552.

Franklin, J. 2001. Geographic information science and ecological assessment. An integrated ecological assessment protocols guidebook, eds. P. Bourgeron, M. Jensen, and G. Lessard, pp. 151–161. New York: Springer-Verlag.

Franklin, J., and Woodcock, C.E. 1997. Multiscale vegetation data for the mountains of Southern California: spatial and categorical resolution. Scale in remote sensing and GIS, eds. D.A. Quattrochi and M.F. Goodchild, pp. 141–168. Boca Raton, FL: CRC/Lewis Publishers Inc.

Goodchild, M.F. 1994. Integrating GIS and remote sensing for vegetation analysis and modeling: methodological issues. J. Vegetation Sci. 5: 615–626.

Gustafson, E.J. 1998. Quantifying landscape spatial pattern: What is the state of the art? Ecosystems 1: 143–156.

Haines-Young, R., and Chopping, M. 1996. Quantifying landscape structure: a review of landscape indices and their application to forested landscapes. Progr. Phys. Geogr. 20: 418–445.

JANIS. 1997. Nationally agreed criteria for the establishment of a comprehensive, adequate and representative reserve system for forests in Australia. Joint ANZECC/MCFFA National Forest Policy Statement Implementation Sub-committee. National forest conservation reserves: Commonwealth proposed criteria. Canberra, Australia: Commonwealth of Australia.

Kuchler, A.W., and Zonneveld, I.S. eds. 1998. Vegetation mapping. Boston: Kluwer Academic Publishers.

Levin, S.A. 1992. The problem of pattern and scale in ecology. Ecology 73: 1943–1967.

Lindenmayer, D.B., and Possingham, H.P. 1996. Effectiveness of alternative heuristic algorithms for identifying indicative minimum requirements for conservation reserves. Conservation Biol. 10: 235–251.

Margules, C., and Pressey, R.L. 2000. Systematic conservation planning. Nature 405: 243–253.

McAlpine, C.A., Lindenmayer, D.B., Eyre, T.J., and Phinn, S.R. 2002. Landscape surrogates of forest fragmentation: synthesis of Australian Montreal Process case studies. Pacific Conservation Biol. 8: 108–120.

McGarigal, K. 2002. Landscape pattern metrics. In Encyclopedia of environmetrics, eds. A.H. El-Shaarawi and W.W. Piergorsch, pp. 1135–1142. Sussex, UK: Wiley.

McNeely, J. 1994. Protected areas for the 21st Century: working to provide benefits for society. Biodiversity Conservation 3: 3–20.

Naveh, Z., and Lieberman, A. 1994. Landscape ecology: theory and applications, 2nd edition. New York: Springer-Verlag.

Noss, R., and Csuti, B. 1994. Habitat fragmentation. In Principles of Conservation Biology, eds. G. Meffe and C. Carrol, pp. 237–264. Sunderland, MA: Sinnauer Associates Inc.

O'Neill, R.V., DeAngelis, D.L., Waide, J.B., and Allen, T.H.F. 1986. A hierarchical concept of ecosystems. Princeton, NJ: Princeton University Press.

O'Neill, R.V., Krummel, J.R., Gardner, R.H., Sugihara, G., Jackson, B.J., DeAngelis, D.L., Milne, B.T., Turner, M.G., Zygmut, B., Christensen, S. et al. 1988. Indices of landscape pattern. Landscape Ecol. 1: 153–162.

Pickett, S.T.A., and White, P.S. 1985. The ecology of disturbances and patch dynamics. New York: Academic Press.

Possingham, H.P., Ball, I.R., and Andelman, S.J. 2000. Mathematical methods for identifying representative reserve networks. In Quantitative methods for conservation biology, eds. S. Ferson and M. Burgman. New York: Springer-Verlag.

Pressey, R.L., Possingham, H.P., and Day, J.R. 1997. Effectiveness of alternative heuristic algorithms for identifying indicative minimum requirements for conservation reserves. Biol. Conservation 80: 207–219.

Pressey, R., and Logan, V. 1998. Size of selection units for future reserves and its influence on actual vs targeted representation of features: a case study in western New South Wales. Biol. Conservation 85: 305–319.

Pressey, R.L., Cowling, R.M., and Rouget, M. 2003. Formulating conservation targets for biodiversity pattern and process in the Cape Floristic Region, South Africa. Biol. Conservation 112: 99–127.

Pulliam, H.R. 1988. Sources, sinks and population regulation. American Naturalist 132:(5): 652–661.

Regan, T.J., Regan, H.M., Bonham, K., Taylor, R.J., and Burgman, M.A. 2001. Modelling the impact of timber harvesting on a rare carnivorous land snail (*Tasmaphena lamproides*) in northwest Tasmania, Australia. Ecol. Modelling 139: 253–264

Schadt, S., Knauer, F., Kaczensky, P., Revilla, E., Wiegand, T., and Trelp, L. 2002. Rule-based assessment of suitable habitat and patch connectivity for the Eurasian lynx. Ecol. Applications 12: 1469–1483.

Scott, J.M., Davis, F., Csuti, B., Noss, R., Butterfield, B., Groves, C., Anderson, H., Caicco, S., D'Erchia, F., Edwards, Jr., T. C., et al. 1993. Gap analysis: a geographical approach to protection of biological diversity. Wildlife Monogr. 123: 1–41.

Siitonen, P., Tanskanen, A., and Lehtinen, A. 2002. Method for selection of old-growth reserves. Conservation Biol. 16: 1398–1408.

Smith, R. 2000. An investigation into the relationship between anthropogenic forest disturbance patterns and population viability. M.S. Thesis, School of Botany, University of Melbourne, Australia.

Smith, R., Regan, T.J., Burgman, M.A., Brown, M., Wells, P., and Taylor, R. 2000. Evaluation of landscape planning tools for forest management and conservation in a fragmented environment: a case study of *Tasmaphena lamproides* in north west Tasmania. Unpublished report to Forestry Tasmania.

Turner, M.G., Gardner, R.H., and O'Neill, R.V. 2001. Landscape ecology in theory and practice. New York: Springer-Verlag.

# Section V

## Synthesis

# Editors' Introduction to Section V: Synthesis

What is the state of the science in understanding landscape heterogeneity and ecosystem function? What lessons have been learned, and where are the exciting and important new lines of inquiry? This last section of the book includes the reflections of four scientists who participated in the Tenth Cary Conference and were assigned the task of seeking commonalities and challenges that integrated the diverse presentations. These chapters balance synthesis and new directions in essays that nicely reflect the state of the science. The section concludes with a brief summary chapter prepared by the editors.

In Chapter 20, Dave Strayer discusses five essential components of heterogeneity and then addresses the challenging topic of determining when spatial heterogeneity must be considered and when it might be safely ignored. Recognizing that ecosystem functions are always heterogeneous, the question about when spatial heterogeneity really must be included in our studies remains a practical challenge in both modeling and empirical studies. His suggestions on when and how to incorporate heterogeneity will provide substantive food for thought as ecologists design studies to enhance understanding of how ecosystems function.

In a thoughtful commentary on the state of the science, Jerry Franklin (Chapter 21) discusses some important "take-home" messages he extracted from the conference and from his personal experience. He sees a scarcity of general principles for understanding the effects of landscape heterogeneity on ecosystem function, and he recognizes the question-dependent nature of the problem. He encourages us to use a broader definition of ecosystem functioning than has traditionally been used (e.g., biodiversity support function, not just energy and nutrient flow) to embrace multiple scales of inquiry "... from the rhizosphere to the globe ..." and to see the world in shades of gray (or green) as opposed to the black and white view of patches and corridors. In his view, the lack of relevant empirical data sets is limiting progress on this subject at least as much as the lack of theoretical constructs.

Gus Shaver (Chapter 22) nicely synthesizes several consistent themes that emerged at the conference and provides a unique historical perspective on the issues that were considered. From his reflections on the conference, Gus puts forth four questions designed both to remind ecologists of some key intellectual roots and to stimulate thinking on future directions of research linking spatial heterogeneity and ecosystem function. He emphasizes the development of new spatial networks that collect and synthesize data at larger scales, the search for broad spatial patterns to simplify the bewildering complexity of natural systems, and the development of theory that merges ideas of sustainability with the nonequilibrium, spatially heterogeneous view of the world that dominates modern ecology. As Shaver puts it, "A reconciliation of the need for sustainable environmental goods and services within a global ecosystem that is heterogeneous in both space and time is one of the greatest intellectual challenges in ecology today."

Judy Meyer (Chapter 23) explores four themes that emerged from the conference: distinguishing between the influences of landscape composition and configuration on ecosystem function; employing a network perspective to enhance understanding of landscape heterogeneity; incorporating the diversity of human activities and history into our conceptualizations; and using this understanding to enhance interactions of humans with nature. Her synthesis nicely links the more abstract consideration of spatial heterogeneity and ecosystem function with the practical applications of this understanding.

Finally, Lovett et al. (Chapter 24) discuss one of the principal goals of the book by reviewing the varied conceptual frameworks provided in the book's chapters. They metaphorically compare current knowledge to a half-built house, with some well constructed rooms and some empty spaces. They then suggest a series of questions akin to a dichotomous tree that might assist the ecologist in determining whether and how to deal with spatial heterogeneity; that is, providing help in navigating through the half-built house.

# 20
# Challenges in Understanding the Functions of Ecological Heterogeneity

DAVID L. STRAYER

## Abstract

Ecological systems usually are heterogeneous, and this heterogeneity has important functional consequences. Nevertheless, it is not always necessary for ecologists to explicitly include this heterogeneity in their studies and models of ecological systems. Heterogeneity may be safely ignored if its grain size is much smaller than the spatial extent over which measurements are integrated or much larger than the spatial extent of the study area. Heterogeneity may be functionally unimportant if the vectors connecting patches are small or slow relative to the time span of the study or if the system is governed by processes with linear dynamics. Further, the heterogeneity expressed by some ecological systems may be amenable to analysis using simplified models. Finally, it may not be efficient to include heterogeneity in study designs or models, even if including heterogeneity would improve the study performance. Despite these considerations, ecologists will need to address heterogeneity explicitly in many cases to achieve a satisfactory understanding of ecosystem functioning, particularly for regional to global scales.

Several other general issues concerning the functional consequences of heterogeneity arose at the Tenth Cary Conference. Human-caused heterogeneity probably has different characteristics and functional consequences than heterogeneity arising from other sources and therefore needs special attention. Models of heterogeneity developed in other disciplines that deal with heterogeneous, reactive systems (e.g., economics) may have application in ecology. At least some heterogeneous ecological systems appear to evolve in predictable ways because the functional consequences of heterogeneity feed back onto the structure of the system; these feedbacks need further study.

## Introduction

*All models are wrong, but some are useful.*
            —G.E.P. Box

The subject of ecological heterogeneity encompasses a diverse collection of scientific, management, and policy issues, many of which are important to ecology and difficult to address. The diversity of issues and rapid pace of conceptual and empirical progress on ecological heterogeneity make it difficult to summarize the current state of the field, and I will not try to provide such a summary based on the Tenth Cary Conference. Instead, I will offer brief impressions of some interesting issues that arose at the conference, lay out research challenges, and, where possible, suggest directions in which answers might lie.

## A Model of Heterogeneity

It may be useful to introduce a simple conceptual model of a functionally heterogeneous system to provide a context for a discussion of the issues that arose at the conference. Consider a system (shown as two-dimensional in Figure 20.1 but more often three-dimensional in ecological contexts) consisting of a series of patches (Figure 20.1A) with different functional attributes (such as denitrification rate, prey abundance, leaf area index,

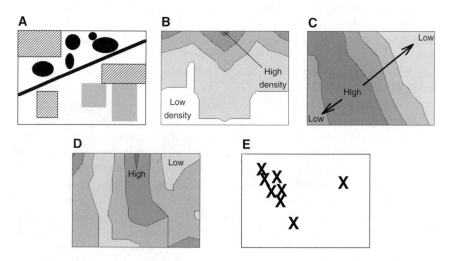

FIGURE 20.1. General model of a heterogeneous system, emphasizing five aspects of heterogeneity occurring in the same hypothetical geographical area. (A) Patch structure, (B) vector mass-density, (C) potential field, (D) resistance, (E) location of externally driven disturbances.

permeability, etc.). The system might be conceived of as continuous rather than discrete, although discrete models are more often used by ecologists and are easier to describe by simple drawings.

The patches are connected by vectors that move across this heterogeneous landscape. The vectors carry reactive objects (material, energy, information) across the landscape, where these objects interact differentially with the different patches. Ecological systems contain many kinds of vectors; familiar examples include wind, the flow of water, diffusion, and the movement of animals. Reiners (this volume; see also Reiners and Dreise 2001) described and categorized the kinds of vectors that are important in ecological systems. The flux rate (direction and magnitude) of a vector is jointly determined by the mass density of the vector (e.g., the amount of water, the density of animals moving nutrients; Figure 20.1B), differences in the potential field that drives vector movement (e.g., the movement of water downhill or down hydraulic gradients; the movement of air down pressure gradients; the movement of animals from areas of low food abundance to high; Figure 20.1C), and resistance to vector movement through the various patches (Figure 20.10). Often, more than one substance (e.g., water, nitrogen, and organic matter) or vector (water and animals) needs to be considered simultaneously to satisfactorily understand the process or function of interest (Fisher and Welter this volume).

Finally, the system may be affected by forces arising from outside the system (e.g., lightning strikes, inputs of water from streams and precipitation) whose influence typically is spatially heterogeneous (Figure 20.1E).

This model thus identifies five essential components of heterogeneity: (1) the patch structure, (2) the spatial pattern of vector mass-density, (3) the potential field, (4) the spatial pattern of resistance to the vector, and (5) the spatial distribution of external influences on the system. Typically, ecological systems contain heterogeneity over a very wide range of spatial scales, so that maps of heterogeneity at any given scale mask heterogeneity that occurs at finer scales. It may be a daunting task to describe adequately all of these components of heterogeneity and then construct a model that mimics the behavior of the system at one time. But of course, we often are interested in the behavior of the system over a period of time, not just at a single time. Therefore, we must add to our already complicated conceptual model the possibility that the function of the system feeds back to change the patch structure, vector mass, potential field, and resistance over time. Likewise, external influences on the system (such as disturbances) may be affected by the patch structure. Explicit consideration of heterogeneity presents three formidable difficulties: (1) conceptualizing such a complicated system, (2) gathering the spatially referenced data to describe adequately the system, and (3) building and evaluating models of the function of dynamic, heterogeneous systems.

Of course, there are alternative ways to conceptualize heterogeneous systems (e.g., Reiners and Dreise 2001). It is not necessary to accept the particular conceptualization of Figure 20.1, though, to appreciate the difficulty of conceptualizing, describing, understanding, and modeling the behavior of temporally dynamic, heterogeneous, reactive systems like ecosystems.

## When Does Heterogeneity Matter?

The central question of the conference was "When and how does spatial heterogeneity matter for ecosystem processes and functions?" This question can be interpreted in two ways. The first interpretation might be phrased as, "When and how does heterogeneity affect processes and functions in real ecosystems?" Briefly, whether considered in the abstract (Strayer et al. 2003) or through empirical studies (below, and elsewhere in this volume), heterogeneity nearly always affects processes and functions in ecosystems, and in diverse ways. All five aspects of heterogeneity identified in Figure 20.1 may affect ecosystem function, although only two have received much attention. There are many compelling examples showing that the patch structure of the ecological system may have important consequences for its function. Turestky et al. (this volume) showed that different parts of the boreal ecosystem accumulate carbon at very different rates, and even transient patchiness in the apparently homogeneous open ocean may substantially increase nitrogen uptake by phytoplankton and reduce phytoplankton-zooplankton encounters (Mahadevan this volume). Many other examples presented at the conference (e.g., Fisher and Welter this volume; Tague this volume; Tongway and Ludwig this volume) and elsewhere show that heterogeneity in patch structure often affects ecosystem function. Likewise, the effects of external disturbances may be distributed heterogeneously in ecosystems, either because the disturbance is heterogeneous in occurrence or because it is propagated unevenly through the system. Fires in forests in the western United States are patchy in occurrence and have different ecosystem effects because ignition sources are patchy (e.g., Gosz et al. 1995), because the different parts of the ecosystem are differentially susceptible to the initiation and propagation of wildfires, and because the nature and severity of fire's effects vary across ecosystems (Romme this volume). Presumably, spatial variation in the mass-density of vectors, potential fields, or resistance may affect ecosystem function as well, although these seem not to have been studied much. Regardless of the details, the importance of heterogeneity to ecosystem function is indisputable.

The second interpretation of "When and how does spatial heterogeneity matter for ecosystem processes and functions?" is "When should we explicitly consider heterogeneity when we study, model, or manage processes and functions of ecosystems?" If we accept that heterogeneity nearly always affects the functioning of ecological systems, it might seem obvious that we should nearly always explicitly incorporate that heterogeneity in our studies and models. However, as we have seen, it may be exceedingly difficult to incorporate fully the multiple heterogeneities that occur in ecological systems into our research. Therefore, we must carefully consider when it is

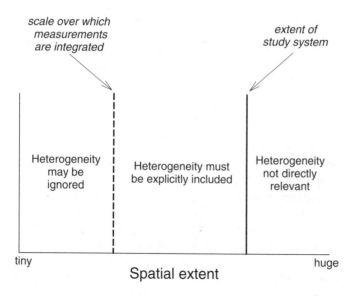

FIGURE 20.2. The relative spatial scales of ecological heterogeneity, measurements, and the extent of the study area determine whether spatially explicit studies or models are required. See text for discussion.

really helpful or necessary to explicitly include heterogeneity in our studies of ecological systems. Given the ubiquity and importance of functional heterogeneity in ecological systems, it probably is simplest to begin by listing the conditions under which it is *not* helpful or necessary to explicitly consider heterogeneity.

First, heterogeneity may safely be ignored if its grain size (or dominant length scale, Mahadevan this volume) is much smaller than the spatial extent over which measurements are integrated or much larger than the spatial extent of the study area (Figure 20.2). This recommendation follows the conclusion of hierarchy theorists (e.g., O'Neill et al. 1986) that processes operating at levels higher than the scale of observations change slowly and may be treated as constants, whereas processes operating at a level lower than the scale of observation change so rapidly that they may be treated as averages. All ecological measurements integrate over some spatial (and temporal) scale. The net functional effects of all heterogeneity finer than this scale of integration are implicitly included in any measurement we make and need not be further considered in our analysis of the system. Thus, a typical pH measurement measures the average pH in several cubic centimeters of water, a free-water estimate of stream metabolism integrates the net functional effects of a large area (perhaps $10^2$–$10^4$ m$^2$) of patchy streambed, and gas flux measurements from a soil chamber integrate the function of the heterogeneous system enclosed in the chamber. As long as our focus is on the stream ecosystem rather than the patches it contains, free-water productivity measurements allow us to ignore the fine-scale interactions that occur among the various parts of the streambed and together determine the productivity of the stream ecosystem. Of course, a scientist may choose for various reasons to study these fine-scale interactions (by using finer-scale techniques that integrate over smaller spatial scales), but it is not necessary to engage in reflexive reductionism by including finer and finer heterogeneity merely because we know it exists.

The scale of measurements may deliberately be chosen to minimize problems in dealing with heterogeneity. Consider the problem of estimating nutrient loss from a patchy forest. We could measure nutrient losses from a series of lysimeters placed in the different habitats of the forest and then try to integrate these measurements by studying the interactions among patches that govern nutrient cycling. Alternatively, we could measure nutrient losses at a weir on a stream that drains a large section of forest. This latter measurement already implicitly includes the results of interactions among patches and probably would provide simpler, less expensive, and more accurate estimates of nutrient losses from the forest than the lysimeter study.

Any influence of heterogeneity much larger than the study area will be expressed through external inputs to the study system and need not be considered explicitly. Again, the extent of the study area may deliberately be chosen to minimize problems with heterogeneity. Indeed, many classic studies of ecosystem function were based on study areas deliberately defined to exclude large-scale heterogeneity (e.g., lakes, relatively homogeneous watersheds).

It is no longer always possible for ecologists to choose relatively small, homogeneous study areas, though. Regional- and global-scale management issues have increasingly forced ecologists to work on large, heterogeneous study areas (e.g., Possingham et al. this volume), thereby moving the solid line in Figure 20.2 to the right. At the same time, the rapid rise of landscape ecology (Turner et al. 2001) has provided the intellectual impetus to understand large, heterogeneous landscapes. Indeed, the move by ecologists to embrace regional and global problems has probably been one of the important motivations for bringing the subject of the functional consequences of heterogeneity to the fore.

Second, we may safely disregard heterogeneity in our studies if that heterogeneity truly has small effects on ecosystem function. There are at least three classes of circumstances in which heterogeneity is most likely to have small functional effects. If the vectors connecting patches are small or slow (relative to the time span of the study), then the mosaic or quasidistributed approach described by Turner and Chapin (this volume) and Tague (this volume) may be adequate, especially over short timescales. Note that vectors will be small if the contrast across patches is small (or equivalently, if gradients in a continuous system are short or shallow). If the system is governed by nearly linear dynamics, then models based on the mean values of variables (rather than the spatial distribution of variables) will adequately predict the function of the system (Strayer et al. 2003). This result follows because the mean of a linear function evaluated at a series of values of independent variables gives the same value as that function evaluated at the mean values of the independent variables. Nevertheless, truly linear ecological systems probably are rare, in part because interactions among controlling ·variables produce nonlinearities. Finally, we can disregard the heterogeneity we measure across patches if it has no functional significance. That is, heterogeneity in sulfate in a strongly light-limited wetland will probably have little effect on primary production even if we can readily measure variations in sulfate concentrations. Kolasa and Rollo (1990) made a similar distinction between functional and what we might call measurable but functionally neutral heterogeneity.

Third, there are special cases in which a greatly reduced model of heterogeneity may be adequate to capture the behavior of a functionally heterogeneous system. For example, if a single patch or element of the landscape strongly dominates system function, then it may be permissible to study only the properties of this master element and ignore the heterogeneity elsewhere in the system. If we are studying vertical water movement through a layered aquifer and a layer of clay has hydraulic conductivity several orders of magnitude lower than that of the other materials in the aquifer, we can concentrate our attention on the properties of that clay layer and disregard heterogeneity above that layer. Systems with regular heterogeneity (which are discussed below in more detail) may also be amenable to simplified approaches.

Fourth, even if the explicit consideration of heterogeneity improves our abilities to predict or understand the function of an ecological system, it may not be *efficient* to explicitly include that heterogeneity in our studies. It may not be parsimonious to add a lot of detailed information describing the heterogeneity of a system if that information improves only slightly our understanding or predictive power. In cases where models are fitted to data (i.e., the number of data points is much larger than the number of parameters), information theoretic criteria can be used formally to choose the most parsimonious of several competing models (Burnham and Anderson 2002). Smith (this volume) described the application of this approach to epidemiological models. Such an approach can help ecologists in some circumstances to decide whether it is efficient to include heterogeneity.

In cases where models cannot be statistically fitted, increasing model complexity to account for heterogeneity may introduce serious problems with error propagation and model selection. It has long been recognized that errors associated with parameterizing a complex model may outweigh those associated using aggregated parameter estimates (O'Neill 1973; Rastetter et al. 1992). Further, as the number of variables rises, the number of possible (or even likely) functional connections among variables rises sharply. The investigator must then choose among a large number of competing model structures by intuition or by somehow testing the various parts of the model. Thus, it may be preferable to accept a simple model, even if it is biased and incomplete, than to build a complicated model whose structure and accuracy must either be accepted on faith or subjected to extensive testing [see further debate by DeAngelis (2003) and Hakanson (2003) about whether complex models are prone to error amplification].

Finally, it may not be efficient from a cost/benefit perspective to explicitly include heterogeneity, even if its inclusion undeniably improves understanding or predictive power (Figure 20.3). In science, we often think our goal is to maximize predictive power, but other goals probably are closer to our actual needs. For instance, our goal may be to achieve some fixed level of predictive power (say the coefficient of variation of a prediction <20%) at minimum cost (lines $P_1$ and $P_2$ in Figure 20.3). Alternatively, we may want to maximize predictive power for a given fixed cost (lines $C_3$ and $C_4$ in Figure 20.3). In both of these cases, it may be desirable to disregard heterogeneity in the frequent situations in which simpler approaches initially cost less per unit understanding than explicitly heterogeneous approaches (Figure 20.3), especially if our predictive needs or available budgets are modest (lines $P_1$ and $C_3$). Many ecologists believe that spatially explicit approaches will ultimately allow us to achieve greater understanding by giving us the mechanistic understanding needed to extrapolate across sites and scales (Turner and Chapin this volume), so if our predictive needs are great or if we have a large budget, spatially explicit approaches may be preferable.

Despite these considerations, which allow ecologists to ignore heterogeneity in many studies of ecological function, it seems clear that it will be necessary to address heterogeneity explicitly in many cases if we are to achieve a satisfactory understanding of ecosystem functioning. This is particularly true for regional to global studies, in which the grain size of functionally important heterogeneity is larger than the scale of measurement but smaller than the size of the study area. The increasing importance of understanding the functioning of these large ecosystems means that ecologists will have to learn to incorporate heterogeneity into their studies and models of ecosystems, however knotty the problem.

# How Do We Best Include Heterogeneity in Studies of Ecosystem Function?

Reaching the conclusion that heterogeneity often will need to be included explicitly in studies and models of ecosystem function immediately raises the question of how best to do so. I expect that a large effort will be devoted to answering this question in the near future. Already at the conference there were discussions of technical issues such as the use of discrete *versus* continuous models (Turner and Chapin this volume), the use of network

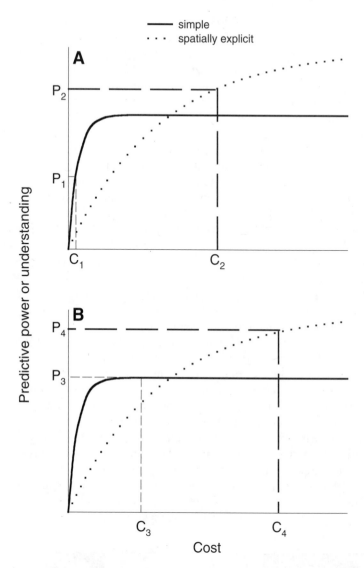

FIGURE 20.3. Hypothetical cost-benefit curves for simple and spatially explicit approaches. (A) Minimizing cost to achieve some predetermined level of predictive power in cases where predictive needs are modest ($P_1$) or stringent ($P_2$). (B) Maximizing predictive power for a given fixed cost where the budget is small ($C_3$) or large ($C_4$). If available funds are low ($C_3$) or requirements for understanding are modest ($P_1$), then simple approaches will be preferable. If more funds are available ($C_4$) or if more complete understanding is needed ($P_2$), the spatially explicit approach will be preferable.

models (Swanson and Jones 2003), the adequacy of mosaic *versus* interactive models (Fisher and Welter this volume; Tague this volume; Turner and Chapin this volume), and the best mathematical and statistical approaches to describe and analyze heterogeneity (Fortin et al. 2003; Mahadevan this volume; Possingham et al. this volume; Reiners this volume; Reiners and Dreise 2001; Smith this volume; Tague this volume). Here, I will raise just a few general issues about approaches to understanding the functional consequences of ecological heterogeneity.

Careful selection of study systems can speed up progress in understanding the functional consequences of ecological heterogeneity. The heterogeneity contained in many ecological systems is more regular (and therefore simpler to study) than that shown in Figure 20.1. For example, ecological heterogeneity often is *directional*, in which conditions change monotonically across the study area (Figure 20.4); *periodic*, in which the units of heterogeneity are predictably repeated; or *fractal* (Brown and White this volume). All of these kinds of regular heterogeneity are common in nature. Soil catenas and elevational gradients are familiar examples of directional heterogeneity; sediment waves and the kind of patterned vegetation described by Tongway and Ludwig (this volume), Aguiar and Sala (1999), and Armesto et al. (2003) represent periodic heterogeneity; and forest patches, Minnesota lakes, and shrub patches in New Mexico have fractal-like properties (Brown and White this volume). Systems can contain more than one kind of regular heterogeneity: streams combine the periodic heterogeneity of the repeated riffle-pool sequence with the directional heterogeneity of headwaters-to-mouth succession. The ability to detect and describe regularity in heterogeneity depends on the study extent and grain; if the spatial components are large, for example, the regularity will not be detected unless the study area is very large.

It should be much easier to model and design studies of systems with regular heterogeneity than those with irregular heterogeneity. As Tongway and Ludwig (this volume) showed, studies of relatively simple, regularly heterogeneous systems can give rise to general hypotheses about heterogeneity that can be extended to or tested in other systems.

 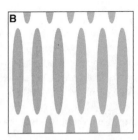 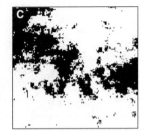

FIGURE 20.4. Examples of regular heterogeneity. (A) Directional heterogeneity, (B) periodic heterogeneity, (C) fractal heterogeneity in a simulated landscape (from Hargrove et al. 2002).

Also, it will be useful to choose study systems whose actual integrated function is measurable, so that we can test our models of the functional consequences of heterogeneity. If we do not have an independent measure of the function of the entire heterogeneous system, it will be difficult to assess how well our models work or to compare the performance of competing models. As a result, systems such as watersheds, whose actual integrated function is readily measurable, will continue to be valuable.

There are many kinds of heterogeneous reactive systems other than ecological systems. Scientists working on these nonecological systems have developed models and methods for understanding their systems that may be helpful to ecologists. For example, discussions of the functional implications of economic and cultural heterogeneity in human societies (e.g., Löfgren and Robinson 1999; Sen 2004) are reminiscent of the discussions at the Tenth Cary Conference. The formal models used to analyze this economic heterogeneity (e.g., Löfgren and Robinson 1999; Vargas et al. 1999; Devarajan et al. 2004) may be inspirational to or usable by ecologists. Likewise, chemical engineers (e.g., Smith 1981; Oran and Boris 1987) have developed models to describe the function of solid catalysts, which formally resemble some kinds of ecological boundaries, and their models of multiphase flow may have ecological counterparts. There must be many other examples of disciplines that have to deal quantitatively with the function of heterogeneous systems. In view of the widespread occurrence of heterogeneous, reactive systems outside of ecology, we might ask if there is even such a thing as a separate theory of ecological heterogeneity, as distinguished from a general theory of heterogeneity. If so, what characterizes such a distinctively ecological theory? That is, to what extent must ecologists develop their own body of knowledge about the functional consequences of heterogeneity, as opposed to using or adapting theories from other disciplines or working jointly with scientists from other disciplines to develop and test truly general theories of heterogeneity? I would guess that ecologists and scientists in other disciplines could benefit from closer communication about the functional consequences of heterogeneity.

# Does Anthropogenic Heterogeneity Have Distinctive or Strong Functional Consequences?

Human activities are among the many sources of heterogeneity in ecological systems. With the increasing focus on humans as parts of ecosystems, we might ask if anthropogenic heterogeneity has the same functional consequences as heterogeneity arising from other sources or is distinctive in some way. I suggest that anthropogenic heterogeneity may both have different actual consequences for ecosystem functioning and be harder to ignore than heterogeneity arising from other sources, for three reasons.

On one hand, humans often create and maintain sharp boundaries and high-contrast landscapes through heavy subsidies of material and energy. For instance, Band et al. (this volume) noted that lawn watering, leaky pipes, and an intensive drainage network give cities and suburbs high hydrologic contrast, where wet and dry areas may be closely juxtaposed. Such high contrast and steep gradients should lead to strong interactions among patches, one of the key conditions under which heterogeneity has strong functional consequences. On the other hand, humans create nearly impermeable barriers (e.g., highways and dams that block animal movement) or patches that are entirely inhospitable to certain ecological processes (e.g., pavement that supports no primary production or denitrification), which would reduce patch interactions below natural levels. As a result, landscape interactivity may vary over a wider range in human-dominated landscapes than in landscapes without humans.

Management issues involving humans often occur at regional or subregional scales ($>100$ km$^2$), so that study areas are necessarily large. This leads to a large range over scales for which heterogeneity must be considered explicitly (Figure 20.2).

Finally, although anthropogenic heterogeneity occurs across a range of spatial scales, it is my impression that humans create a lot of heterogeneity at a scale of 0.1–1000 ha (i.e., housing lots, farm fields, parking lots), and often obliterate heterogeneity at smaller scales (Cumming 2003; Fraterrigo et al. 2005). This scale is larger than the scale of integration of many ecological measurements but smaller than that of many kinds of study areas—just the scale most likely to force us to consider explicitly heterogeneity in our studies. These considerations suggest that ecologists who are interested in the ecological roles of humans will need to consider explicitly heterogeneity more often than other ecologists.

## How Do the Functional Consequences of Heterogeneity Feed Back into the Temporal Dynamics of Heterogeneous Systems?

Because the functional consequences of heterogeneity can feed back onto the structure of the ecological system, the structure or function of heterogeneous systems can evolve over time in a predictable way. Such feedbacks could affect any of the five aspects of heterogeneity (Figure 20.1), and in complex ways.

At least some heterogeneous ecological systems do appear to evolve in predictable ways as a result of these feedbacks. For example, Meinders and van Breemen (this volume) described several examples of ecological systems in which strong positive feedbacks result in self-organizing heterogeneity. Thus, interactions between litter quality, soil nutrients, and the nutrient-driven growth and survival of trees may reinforce or alter spatial patterning of tree species in forests of the northeastern United States over time (e.g., Bigelow and Canham 2002). Likewise, patches may move across a landscape in a predictable way. There are many examples of regular patch movement driven by physical forces (e.g., dunes, sediment waves), but ecological interactions may also drive such regular patch movement, as in the case of forested patches moving across the patterned landscape of Fray Jorge (Armesto et al. 2003). Naiman et al. (this volume) showed that interactions between a stream channel and the surrounding riparian forest produce debris jams, which initiate a predictable development of channel form and vegetation. It would be interesting to know how general such cases are and whether there are simple rules for identifying systems whose spatial structure changes predictably over time.

## Conclusions

It is apparent even from this brief survey that the subject of ecological heterogeneity encompasses a diverse collection of scientific, management, and policy problems in ecosystem science, some of which are difficult. These problems are likely to become increasingly important in the future, as ecologists strive to address regional- to global-scale problems and incorporate humans into their studies of ecosystem functioning. Ecologists must learn both to develop effective solutions to these difficult problems and to know when to avoid the problem of explicitly including heterogeneity in their studies and models. Presentations at the conference showed that there is a broad front of progress on understanding the importance of ecological heterogeneity to ecosystem functioning, as well as many promising avenues to follow into the future.

*Acknowledgments.* I thank the conference organizers for the opportunity to express these thoughts; the conference participants for their many ideas, which I hope I haven't misrepresented too badly; Gautam Sethi for introducing me to economic models; and Holly Ewing, Seth Bigelow, and my other colleagues at the Institute of Ecosystem Studies for helping me to think about heterogeneity. Holly's critical review of an early version of the manuscript forced me to sharpen my thinking, and Tim Kratz, Gary Lovett, Monica Turner, and an anonymous reviewer offered helpful comments on the manuscript.

## *References*

Aguiar, M., and Sala, O.E. 1999. Patch structure, dynamics, and implications for the functioning of arid ecosystems. Trends Ecol. Evolution 14: 273–277.

Armesto, J.J., Barbosa, O., Christie, D., Gutierrez, A.G., Jones, C.G., Marquet, P.A., and Weathers, K.C. 2003. Fog capture and structural attributes of cloud-dependent forest patches in the Coastal range of semiarid Chile. Poster presentation at the Tenth Cary Conference, April 29-May 1, 2003, Millbrook, NY.

Bigelow, S.G., and Canham, C.D. 2002. Community organization of tree species along soil gradients in a north-eastern USA forest. J. Ecol. 90: 188–200.

Burnham, K.P., and Anderson, D.R. 2002. Model selection and multimodel inference: a practical information-theoretic approach, Second edition. New York: Springer-Verlag.

Cumming, G.S. 2003. Measures of "eroading" ecology. Frontiers Ecol. Environ. 1: 233.

DeAngelis, D.L., and Mooij, W.M. 2003. In praise of mechanistically rich models. In Understanding ecosystems: the role of quantitative models in observation, synthesis, and prediction, eds. C.D. Canham, J.J. Cole, and W.K. Lauenroth, pp. 63–82. Princeton, NJ: Princeton University Press.

Devarajan, S., Lewis, J.D., and Robinson, S. 2004. Getting the model right: the general equilibrium approach to adjustment policy. Cambridge, UK: Cambridge University Press.

Fortin, M.-J., Boots, B., Csillag, F., and Remmel, T.K. 2003. On the role of spatial stochastic models in understanding landscape indices in ecology. Oikos 102: 203–212.

Fraterrigo, J.M., Turner, M.G., Pearson, S.M., and Dixon, P. 2005. Effects of past land use on spatial heterogeneity of soil nutrients in southern Appalachian forests. Ecological Monographs 75: 215–230.

Gosz, J.R., Moore, D.I., Shore, G.A., Grover, H.D., Rison, W., and Rison, C. 1995. Lightning estimates of precipitation location and quantity on the Sevilleta LTER, New Mexico. Ecol. Applications 5: 1141–1150.

Hakanson, L. 2003. Propagation and analysis of uncertainty in ecosystem models. In Understanding ecosystems: the role of quantitative models in observation, synthesis, and prediction, eds. C.D. Canham, J.J. Cole, and W.K. Lauenroth, pp. 139–167. Princeton, NJ: Princeton University Press.

Hargrove, W.W., Hoffman, F.M., and Schwartz, P.M. 2002. A fractal landscape realizer for generating synthetic maps. Conservation Ecology 6 (1): article number 2. Available at http://www.consecol.org/vol6/iss1/art2/index.html.

Kolasa, J., and Rollo, C.D. 1991. Introduction: the heterogeneity of heterogeneity: a glossary. In Ecological heterogeneity, eds. J. Kolasa and S.T.A. Pickett, pp. 1–23. New York: Springer-Verlag.

Löfgren, H., and Robinson, S. 1999. Spatial networks in multi-region computable general equilibrium models. TMD Discussion Paper 35: 1–28, Trade and Macroeconomics Division, International Food Policy Research Institute. Available at http://www.ifpri.org/divs/tmd/dp/papers/tmdp35.pdf.

O'Neill, R.V. 1973. Error analysis of ecological models. In Radionuclides in ecosystems. CONF-710501, ed. D.J. Nelson, pp. 898-908. Springfield, VA: National Technical Information Service.

O'Neill, R.V., DeAngelis, D.L., Waide, J.B., and Allen, T.F.H. 1986. A hierarchical concept of ecosystems. Princeton, NJ: Princeton University Press.

Oran, E.S., and Boris, J.P. 1987. Numerical simulation of reactive flow. Amsterdam: Elsevier.

Rastetter, E.B., King, A.W., Cosby, B.J., Hornberger, G.M., O'Neill, R.V., and Hobbie, J.E. 1992. Aggregating fine-scale ecological knowledge to model coarser-scale attributes of ecosystems. Ecol. Applications 2: 55–70.

Reiners, W.A., and Dreise, K.L. 2001. The propagation of ecological influences through heterogeneous environmental space. BioScience 51: 939–950.

Sen, A. 2004. How does culture matter? In Culture and public action, ed. V. Rao and M. Walton. Stanford University Press.

Smith, J.M. 1981. Chemical engineering kinetics, Third edition. New York: McGraw-Hill.

Strayer, D.L., Ewing, H., and Bigelow, S. 2003. What kinds of spatial and temporal detail are required in models of heterogeneous systems? Oikos 102: 954–662.

Swanson, F.J., and Jones, J.A. 2003. Landscape heterogeneity—a network perspective. Poster presentation at the Tenth Cary Conference, April 29-May 1, 2003, Millbrook, NY.

Turner, M.G., Gardner, R.H., and O'Neill, R.V. 2001. Landscape ecology in theory and practice. New York: Springer-Verlag.

Vargas, E.E., Schreiner, D.F., Tembo, G., and Marcouiller, D.W. 1999. Computable general equilibrium modeling for regional analysis. The Web Book of Regional Science. Available at http://www.rri.wvu.edu/webbook/schreiner/contents.htm.

# 21
# Spatial Pattern and Ecosystem Function: Reflections on Current Knowledge and Future Directions

JERRY F. FRANKLIN

## Abstract

Relationships between spatial patterns and ecosystem function are briefly reviewed with regard to the current state of the science and its application and some important challenges. Ecosystem functions that are affected by heterogeneity include maintenance of species diversity (habitat) as well as material and energy cycles. Structural diversity and spatial heterogeneity play an important role in all of these functions and require increased attention. Spatial pattern or heterogeneity is important to ecosystem function at all spatial scales from centimeters to kilometers, not just the larger scales. The relevance of spatial patterns to ecosystem function, including the statistical patterns and significance of the relationships, depend on the function or parameters chosen and the spatial and temporal scales of interest. As a consequence, few, if any, general principles exist for interpreting the effects of landscape heterogeneity on ecosystem function. For example, heterogeneity does tend to increase the number of niches available and, hence, the diversity of environmental conditions that are present. Whether the effects of this increased diversity are positive or negative depends on the processes or organisms of interest. One important conceptual challenge in studying landscape heterogeneity is to move beyond the classic patch-corridor-matrix model to approaches that incorporate networks and gradients.

## Introduction

Spatial patterns—in resources, in populations of organisms, and in structure—play crucial roles in ecosystem functioning. We have known this for a long time, although we often have chosen to ignore the effects of heterogeneity. Indeed, field biologists traditionally have been trained to avoid complex or heterogeneous circumstances in selecting study sites for reasons of simplification and reduced variance. At other times, heterogeneity is present but is not at a relevant spatial scale (Strayer this volume). Despite

427

the tendency to ignore it, we have known intuitively that heterogeneity is important over a very broad range of spatial scales—not just at the larger, landscape level—and that interpreting the effects of heterogeneity depends largely on the ecosystem processes of interest. This is why heterogeneity is emerging today as a topic of major interest!

My objective here is to provide some general observations on the relationships between spatial heterogeneity and ecosystem function and to identify challenges for scientists and managers in extending the frontiers of this topic. The scientific literature, including papers in this volume, provide a wealth of diverse viewpoints, concepts, and examples. My observations will draw on these sources as well as my own experiences in research on ecosystems and landscapes and the application of this knowledge to management of natural resources. I will also consider the effects of spatial heterogeneity on ecosystem functions at scales smaller than landscapes—such as environmental and structural heterogeneity within forest stands.

In lecturing students and general audiences, I generally find that most important general ecological concepts are so obvious in retrospect as to appear trivial—akin to such profound discoveries as "water runs downhill" (except when it doesn't, of course). Consequently, I always feel a bit like the village idiot when attempting generalizations about ecological topics, such as relationships between spatial heterogeneity and ecosystem function. Note that these represent the viewpoint of a forest ecologist who has learned such profound things during his career as:

- All forest patches are not created equal (as habitat or anything else), and much of the inequality is related to structural, including spatial, complexity;
- Dead trees and down boles have important ecological functions;
- Forests and streams are highly integrated; and
- Landscape (larger spatial scale) perspectives are imperative!

So, at risk of appearing simplistic, my commentary follows.

## General Observations

### *Habitat Is an Ecosystem Function and Structure as Its Measure*

I want to begin by commending ecosystem scientists that consider "maintenance of species diversity" an ecosystem function on a par with more traditional functions, such as material cycles (Lovett et al. 2005). This is an important extension of the list of recognized ecosystem functions, one not yet adopted by many ecosystem scientists—traditionally generally limiting ecosystem functions to the nature of and controls on energy and material cycles. Yet, much of the "ecosystem management" underway today is focused specifically on provision of habitat for organisms. A science of ecosystems that does not incorporate the habitat role is incomplete and risks missing the important relationships between biotic composition and ecosystem processes (Jones and Lawton 1995).

Accepting habitat provision as an ecosystem function increases the level of attention that needs to be given to ecosystem structure *vis-à-vis* process functions and composition. This is because structural features are generally and appropriately used to describe and prescribe habitat for species (Lindenmayer and Franklin 2003). Furthermore, structure here refers not only to such obvious individual structural elements, such as trees, standing dead trees (snags), and logs, but also the three-dimensional spatial arrangements of these structures. The high level of diversity in an old-growth Douglas-fir forest is not just due to a rich array of structures but also to the varied environmental niches provided by the spatial arrangement of these structures (Franklin et. al 2002; Franklin and Van Pelt 2004).

Hence, structure is an important attribute of ecosystems on par with the attributes of composition and process or function. Unfortunately, I think that many ecosystem and organismal scientists still have not fully recognized the collateral importance of structure. This appears to be the case even though much of the observed spatial heterogeneity involves spatial patterns in structures, whether within a patch (stand) or at the level of a landscape. I will have more to say about the importance of spatial patterns in structure later in the paper.

## Spatial Pattern Is Important at All Spatial Scales

The importance of spatial heterogeneity to ecosystem function is not confined to larger spatial scales. Ecosystem function is influenced by spatial heterogeneity across the full range of spatial scales of interest to ecologists—from the rhizosphere to the globe (Lindenmayer and Franklin 2003; Lovett et al. 2005).

An important related issue is whether the heterogeneity present at smaller spatial scales can be subsumed when working on ecosystem processes at a larger spatial scale. This issue occasionally has explicitly been addressed, sometimes to good effect (Strayer this volume), but has been most often ignored. I think that there are often circumstances where smaller scale spatial heterogeneity will have to be either implicitly or explicitly recognized when dealing with spatial pattern-functional relationships at larger scales.

Assessing the level of ecosystem function in heterogeneous landscapes—such as their ability to provide species habitat—provides a general example of where it is usually essential to consider spatial heterogeneity at smaller spatial scales. In effect, determining the level of functionality within landscape patches often requires substantial knowledge of the spatial patterns within the patches. Forest landscapes provide an excellent example of where it is necessary to have information on the compositional and structural character of each patch in order to be able to compare functionality.

An assessment of old-growth forest habitat in the national forests of the Sierra Nevada in California illustrates this principle (Franklin and Fites-Kaufmann 1993). Structural features, such as the density of large, old trees, standing dead trees (snags), and down boles on the forest floor, were the basis for assessing the degree to which the forests would sustain habitat for old-growth related species. In the Pacific Northwest, assessing old-growth function was relatively simple because forests generally were readily categorized as either "old growth" or "non–old growth"; this is because the dominant disturbances have been stand-replacement fires and clearcutting (Forest Ecosystem Management Assessment Team 1993). In the Sierra Nevada, most existing forest stands still retained structural elements of the old-growth forests—such as large old trees —so that a simple dichotomy of "old growth" and "non–old growth" was not possible. These structures were present because the dominant natural (periodic low- to moderate-intensity wildfire) and human (selective logging) disturbance regimes retained legacies of old-growth habitat, albeit of varying quality and quantity.

Consequently, in the Sierra Nevada old-growth analysis, it was necessary to assess the structural complexity within each mapped forest patch (polygon) to quantify its contribution to old-growth habitat function in the Sierra Nevada national forests. This was accomplished by quantifying various internal attributes of the polygons—including numbers of large trees, canopy cover, and the spatial pattern in forest structure—on a scale from 0 (no contribution to old-growth habitat function) to 5 (very high level of contribution) (Figure 21.1) (Franklin and Fites-Kaufmann 1993). Ultimately, the polygons that had high ratings in old-growth characteristics became core elements in a proposed network of late-successional forest reserves.

## Relevance of Spatial Pattern Depends on the Topic

The relevance of spatial patterns to ecosystem function, including the type and importance (e.g., statistical significance) of the relationships, depends on the topic of interest. The fact that the relevance of spatial patterns to ecosystem functions are conditional—determined by the specifics of a study or application—should not be a surprise to anyone but often seems to be overlooked.

Hence, at the outset of any discussion or presentation it is crucial to stipulate the function(s) or parameter(s) that are of interest as well as the relevant spatial and temporal scales. This is the only way to avoid making inappropriate comparisons or commentary about analytic techniques or interpretations. The necessity of being explicit about parameters and temporal and spatial scales should be obvious to all participants in discussions about the science and, especially, in the application of ecological science to environmental issues.

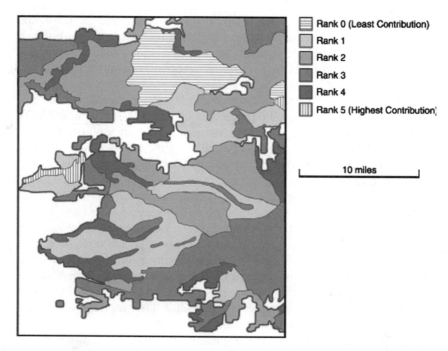

FIGURE 21.1. Landscape polygons on a portion of the Eldorado National Forest, California, rated for their contributions to old-growth function in the Sierra Nevada based on forest structural conditions occurring within the polygons (Franklin and Fites-Kaufmann 1996).

## Spatial Heterogeneity Has Multiple Causes

Most spatial heterogeneity is generated by:

- variability in physical template or environmental conditions;
- biota, including internal community processes; and
- disturbances.

The relative contribution of these three factors to spatial heterogeneity at the landscape level varies substantially, and much of the variability in their relative dominance is associated with major biomes. For example, the physical template is prominent in many extreme environments, such as desert and tundra, landforms forming the primary template. The spatial pattern or patch mosaic at the landscape scale (e.g., hundreds or thousands of meters) often changes little in the short- to mid-term, although biotically driven spatial heterogeneity may be prominent within patches.

Biotic influences are profound in many North American grassland landscapes and may be the result of indigenous organisms, such as bison (*Bison bison*), or exotic organisms, such as an introduced annual grass (cheatgrass, *Bromus tectorum*). Biota may either increase or decrease heterogeneity in the landscape, depending on the nature of the organism and the landscape in which it is operating. For example, bison grazing typically contributes to higher levels of spatial heterogeneity in grasslands than would otherwise be present without grazing (Turner et al. 2001). Cheatgrass introduction typically has resulted in landscape simplification; the original grassland mosaics—composed of patches of varied composition and structure—are replaced by large monospecific patches of cheatgrass (Daubenmire 1970; Mack 1981). Furthermore, landscape homogenization in the form of large patches of cheatgrass also alters the wildfire regime, allowing for more frequent and much larger wildfires, which further sustains and extends the cheatgrass community.

Disturbances are dominant processes creating spatial heterogeneity in many forested landscapes (Mitchell 2003). Intense, stand-replacing disturbances, such as wildfire or large-scale windstorms, create new patch mosaics. Some of the patches that are created may be very large and often have high levels of internal patch heterogeneity. Stand-replacement fires in montane lodgepole pine stands in the western US are an example (Romme this volume). Pacific coastal stands of Douglas-fir, western hemlock, and other conifers provide another example where intense stand-replacement events are characteristic (Franklin and Halpern 2000).

Chronic disturbances are very important in creating and maintaining spatial heterogeneity. Periodic windthrow of individuals or small groups of trees fall into this category. There are many excellent examples in the literature, such as the southern beech forests of Tierra del Fuego (Veblen 1991; Rebertus et al., 1997) and coniferous forests in the Pacific Northwest (Lertzman and Krebs 1991; Franklin et al. 2002). Frequent low- to moderate-intensity wildfire is another example of a chronic disturbance and, again, examples are numerous, such as the pine and mixed conifer forests in western North America (e.g., Agee 1993; Franklin and Fites-Kaufmann 1996).

Surprisingly, chronic disturbances can play as important a role in forests characterized by stand-replacement disturbance regimes as they play in forest ecosystems characterized by chronic, low- to moderate-intensity disturbance regimes (Mitchell et al. 2003). In both cases they act to create and maintain stands (patches) with spatially heterogeneous structures. There is nothing profoundly new about this observation, which can be viewed as an elaboration of the shifting mosaic model of Bormann and Likens (1979), but foresters typically ignore the similarities that exist between old forests of widely different origin (Franklin and Van Pelt 2004). I will return to this topic later.

## General Principles Are Scarce

Few general principles exist for interpreting the effects of landscape heterogeneity on ecosystem function, either positive or negative. Again, I note that stipulating the ecosystem function or species and the spatial and temporal scales of interest is an essential first step in any interpretation. I would assert that although some generalizations can be made about the effects of heterogeneity on variability in a landscape, the consequences are going to be idiosyncratic depending on processes or organisms of interest, a common phenomenon in ecological theory (Schrader-Frechette and McCoy 1993).

For example, spatial heterogeneity generally will result in:

- more niches; and, consequently,
- higher levels of species diversity; and a
- greater diversity in rates and types of ecosystem processes.

Hence, spatial heterogeneity tends to increase richness in both species and processes and, in some respects, increase redundancy and resilience.

The consequences of this heterogeneity for a process, parameter, or organism of interest will vary, however. Heterogeneity at inappropriate spatial scales can be highly dysfunctional for certain processes or species. For example, the dispersed clearcut patch system adopted for federal forest lands in the Pacific Northwest after World War II created patches that were 15 to 25 ha in size (Franklin and Forman 1987). This patch size did not meet the needs of many native vertebrate species that require interior forest habitat conditions, such as the Northern Spotted Owl. Timber cutting at this scale also created large amounts of edge-effected habitat. Edge effects on the environment of the residual forest patches were extreme because of the high level of structural contrast between tall old-growth forests and clearcuts. Consequently, the residual forest patch sizes were insufficient to provide even for microclimatic conditions characteristic of interior forest (Chen et al. 1995). Ultimately, creating a mosaic of small forest patches produced landscapes that had more heterogeneity but were dysfunctional for many organisms and ecosystem processes.

Spatial homogeneity will maximize particular functions and flows, on the other hand, and this is a primary reason why many domesticated landscapes—such as corporate timberlands—tend to be homogeneous. A forest landscape of young, dense, uniformly stocked stands may maximize net primary productivity and rates of carbon accumulation; that is, wood production! However, this homogenous landscape also may have some significant downsides, such as a high susceptibility to insects, diseases, and other disturbances, while providing little habitat for native species.

My point is simply that generalizations about the relative merits of heterogeneity or homogeneity are not possible; the ecosystem function(s) or species that are of interest must be stipulated and the conclusions will vary dramatically. Definition of function/species and of spatial and temporal dimensions of interest seems to me to be imperative when one is determining the functionality of a landscape.

# Looking Forward

## Conceptual and Technical Bases for Advances

Do the necessary theories, empirical databases, analytic techniques, and technologies exist that are needed to advance our understanding of the relationships between spatial heterogeneity and ecosystem function? In my view, such limitations are not likely, given the energetic dialogue on such topics as conceptual and theoretical constructs and analytic approaches (e.g., Lovett et al. 2005).

Many existing techniques can be applied to analytic and measurement issues, including modeling approaches. New and relevant technologies with outstanding potential are emerging. One example is LIDAR, a remote sensing technology using laser radar that can be used, among other things, to assess the three-dimensional structure of aboveground vegetation (Lefsky et al. 2002). Another example is the development of dense, intelligent sensing arrays that can be deployed to provide truly comprehensive data on spatial and temporal patterns in environmental parameters.

Nevertheless, there are currently some important gaps in our technical capacities to quantify spatial heterogeneity and its effect on ecosystem function. My prime candidate for a serious gap is a capacity for nonintrusive sensing of spatial patterns in belowground systems, a capacity that would allow us to study heterogeneity in structure, function, and composition. Belowground portions of ecosystems generally have high levels of organismal diversity and high rates of turnover, often consuming the majority of the photosynthate produced aboveground. Based on what we already know about spatial complexity in biotic composition and processes, it seems clear that belowground complexity will make that of the aboveground appear comparatively simple. Consider simply the energetic and nutrient gradients associated with root systems. Yet, we are very limited in our ability to conduct nonintrusive, spatially explicit studies of patterns in processes and the distribution of organisms. The hyporheic zone is an important subset of our understudied substrates, although students of the hyporheic seem to have made considerable progress on the topic of spatial heterogeneity (e.g., Stanford et al. 1994; Edwards 1998; Fisher and Welter this volume). In any case, the belowground is a realm where technological advances are needed to advance substantially the science.

Relevant empirical databases may, in fact, be more limiting to understanding the relationship between spatial heterogeneity and ecosystem function than the theoretical base or analytic techniques. At a Cary Conference 17 years ago, I asserted that:

> ...because most important questions in ecology ultimately deal with predicting ecosystem responses, testing the correctness of ecological concepts and predictions by observing the future is essential. There are many sophisticated predictive models and general constructs, but few have actually been tested against data. In the final analysis the most convincing validation comes only from such tests against reality. (Franklin 1989, p. 3)

I continue to believe that the absence of definitive empirical databases remains a real limitation to advances in ecosystem science and its application after nearly 50 years of ecosystem research. Empirical data are the ultimate test of any ecological theory, and such data can be very difficult to come by, particularly at the spatial and temporal scales necessary to provide credible tests of general theory. Much current ecological theory seems to me to be applicable to special cases and not general theory at all, and I am not alone in that view (Shrader-Frechette and McCoy 1993).

## Expanding Beyond the Patch-Matrix View

I see extending our view of landscapes beyond the patch-matrix model as an important challenge in advancing our understanding of the relation between spatial pattern and function at larger spatial scales. Based on a review of current textbooks, the patch-matrix model appears to be the dominant conceptual model used in landscape ecology and conservation biology. This model has many limitations in ecosystem science and management, however (Lindenmayer and Franklin 2002). Other constructs, such as those based on networks and gradients, are often more relevant to particular topics or issues.

Network perspectives can be very important in dealing with highly connected landscape elements, such as aquatic (especially riverine) ecosystems on the natural side, and transportation systems, such as road networks, on the manipulated side. Although networked elements (patches) of this type can be dealt with using a patch-matrix model, the network model is more direct and emphasizes the important elements of connectivity and flow (Forman 1995). Network perspectives also make it easier to identify important ecosystem issues associated with interactions between key landscape networks, such as between stream and river systems on the one hand and road and transportation networks on the other.

The patch-matrix view of the landscape encourages scientists and citizens to take a dichotomous view of the landscape (i.e., to divide it into patches that are black and white, habitat and nonhabitat, or functional and nonfunctional). I think that this dichotomous view of the world as habitat and nonhabitat is largely inherited from conservation biology and, ultimately, from island biogeographic theory. Scientists focused on ecosystem processes have been much less prone to make this kind of mistake, recognizing that key ecosystem processes are operative, albeit at different levels, in all landscape patches.

With regard to both function and habitat for biodiversity, we need to increase our emphasis on landscapes as gradients or at least recognize patchworks as being composed of patches with varying levels of functionality. I describe this as viewing landscapes as shades of gray rather than as a patchwork of black and white (Figure 21.2). Or, as in assessing old-growth forest conditions in the Sierra Nevada, viewing the landscape as varying degrees of old-growth function ("shades of green") (Figure 21.1).

**Do we view landscapes as black and white (habitat and non-habitat)...**

**... or as shades of gray (a range of habitat or other values)?**

FIGURE 21.2. The patch-matrix model often encourages a dichotomous view of the landscape (e.g., habitat and nonhabitat or "black and white"). As we deal with the effect of landscape heterogeneity on the array of ecosystem functions, including habitat, it will be increasingly important to recognize gradients of function or conditions within the polygons (i.e., view the landscape as "shades of gray").

If we intend to understand landscapes as shades of gray or gradients, an important corollary is the relevance of information on the content of the patches. As I tell my students, **"patch content matters"** in interpreting landscape function, once you depart from a simple patch dichotomy.

## *Importance of Structure*

Ecosystem structure is a topic that I think deserves more attention from ecosystem scientists as we move forward. I am surprised that there is not more explicit discussion of ecosystem structure in discussions of landscape function, although it does occur more often as a central element in discussions of riverine ecosystems (Naiman et al. this volume).

Structure is an aspect of ecosystems that is on par with composition and processes (or function, as I usually refer to it) in its fundamental importance. Structure is what we most often observe and manipulate in forest ecosystems and provides much of the habitat within which species exist and functions occur. Structural complexity—including spatial heterogeneity—results in more numerous niches and,consequently, greater diversity in species and in types and rates of processes.

Much innovative research is underway on the structure of natural forests and on key processes in structural development. For example, we understand how stand development processes in temperate forests move stands toward higher levels of structural complexity, including spatial heterogeneity, during succession (Franklin et al. 2002; Mitchell et al. 2003). In contrast with young forests, late-successional (old-growth) forests not only have a much higher diversity in individual structures (Spies and Franklin 1991) but also high levels of structural spatial heterogeneity. For example, in late-successional forests, canopies are essentially continuous, from ground to top of tree crowns, whereas young, closed forest stands typically have single layered canopies (Figure 21.3). Similarly, in the horizontal dimension, late-successional forests typically are spatially heterogeneous as a result of such processes as canopy gap formation; spatial homogeneity is the rule in dense young forest stands (Figure 21.3). In fact, temperate late-successional forest stands typically exist as fine-scale structural mosaics.

Understanding and incorporating such structural concepts in management of forest and other natural and seminatural landscapes is crucial to achieving our goals, including maintenance and, where necessary, restoration of ecosystem functions (Lindenmayer and Franklin 2002).

## Conclusions

Why is understanding the relationship of spatial heterogeneity to ecosystem function important other than as an esoteric intellectual pursuit?

(a)

(b)

(c)

FIGURE 21.3. Structural cross-sections of forest stands illustrating (a) the spatial homogeneity of a typical early successional stand and (b and c) the spatial heterogeneity characteristic of late-successional stands. (a) A 100-year-old Douglas-fir forest at the Wind River Experimental Forest in the southern Washington Cascade Range; (b) 650-year-old forest of Douglas-fir, western red cedar, and western hemlock in the Cedar Flats Research Natural Area in the southern Washington Cascade Range; and (c) 400+-year-old forest of ponderosa pine in the Bluejay Springs Research Natural Area in the eastern Oregon Cascade Range. (Diagrams by and courtesy of Robert Van Pelt.)

Mankind is busily engaged in modifying the world, its biota, and its environment, and has been for some time. We have been domesticating and homogenizing our landscapes as well as simplifying the internal structure of the polygons that make up those landscapes. Much of the natural heterogeneity that existed has been viewed as either an obstacle or irrelevant to our perceived objectives.

Forests provide a great example, as the goal in recent times has typically been the efficient production of wood. Heterogeneity is an impediment to that singular goal, at least in the short term and particularly when "efficient" is measured by return on investment, and competition is at the global level (Franklin 2003). Consequently, the pattern has been to convert structurally and compositionally complex native forest stands to dense, "fully stocked," even-aged stands of a single species. With the elimination of the heterogeneity has been a concomitant loss of ecosystem function and habitat to sustain biodiversity.

We need to understand the relationship of spatial heterogeneity to ecosystem function so that we can understand and predict the consequences of alterations in spatial patterns. We need to be able to explain to the public, decision makers, and resource managers the effects of homogenizing the internal patterns within landscape patches and of eliminating landscape heterogeneity, and identify the spatial patterns that make essential contributions to ecosystem function.

Almost certainly, we will be called to go beyond such assessments and contribute to the business of designing new ecosystems and landscapes! We are already deeply engaged in attempting to restore heterogeneity and function to simplified, dysfunctional land- and riverscapes throughout North America. Sophisticated and predictive knowledge about the relationships of spatial heterogeneity to ecosystem function, including provision of habitat for biodiversity, will be essential to meeting these challenges.

## References

Agee, J.K. 1993. Fire ecology of Pacific Northwest forests. New York: John Wiley & Sons. 493 pp.

Bormann, F.H., and Likens, G.E. 1979. Pattern and process in a forested ecosystem. New York: Springer-Verlag. 253 pp.

Chen, J., Franklin, J.F., and Spies, T.A. 1995. Growing season microclimatic gradients extending into old-growth Douglas-fir forests from clearcut edges. Ecol. Applications 5: 74–86.

Daubenmire, R.F. 1970. Steppe vegetation of Washington. Pullman, WA: Washington Agricultural Experiment Station Technical Bulletin 62.

Edwards, R.T. 1998. The hyporheic zone. In River ecology and management, eds. R.J. Naiman and R.E. Bilby, pp. 399–429. New York: Springer-Verlag.

Forest Ecosystem Management Assessment Team. 1993. Forest ecosystem management: an ecological, economic, and social assessment. Various pagination. Portland, OR: USDA Forest Service.

Franklin, J.F. 1989. Importance and justification of long-term studies in ecology. In Long-term studies in ecology, ed. G.E. Likens, pp. 3–19. New York: Springer-Verlag.

Franklin, J.F. 2003. Challenges to temperate forest stewardship—focusing on the future. In Towards forest sustainability, eds. D.B. Lindenmayer and J.F. Franklin, pp. 1–13. Washington, DC: Island Press.

Franklin, J.F., and Fites-Kaufmann, J. 1996. Assessment of late-successional forests of the Sierra Nevada. In Sierra Nevada Ecosystem Project: final report to Congress, Vol. II, Assessments and scientific basis for management options, pp. 627–656. Devis, CA: Centers for Water and Wildland Resources, University of California, Davis.

Franklin, J.F., and Forman, R.T. 1987. Creating landscape patterns by forest cutting: ecological consequences and principles. Landscape Ecol. 1: 5–18.

Franklin, J.F., and Halpern, C.B. 2000. Pacific Northwest forests. In North American terrestrial vegetation, 2nd edition. eds. M.G. Barbour and W.D. Billings, pp. 123–159. New York: Cambridge University Press.

Franklin, J.F., and Van Pelt, R. 2004. Spatial aspects of structural complexity in old-growth forests. J. Forestry 102(3): 22–27

Franklin, J.F., Spies, T.A., Van Pelt, R., Carey, A.B., Thornburgh, D.A., Berg, D.R., Lindenmayer, D.B., Harmon, M.E., Keeton, W.S., Shaw, D.C., et al. 2002. Disturbances and structural development of natural forest ecosystems with silvicultural implications, using Douglas-fir forests as an example. Forest Ecol. Manage. 155: 399–423.

Jones, Clive, G., and Lawton, J.H. 1995. Linking species and ecosystems. 387 p. New York: Chapman & Hall.

Lefsky, M.A., Cohen, W.B., Parker, G.G., and Harding, D.J. 2002. LIDAR remote sensing for ecosystem studies. BioScience 52: 19–30.

Lertzman, K.P., and Krebs, C.J. 1991. Gap-phase structure of a subalpine old-growth forest. Can. J. Forest Res. 21: 1730–1741.

Lindenmayer, D.B., and Franklin, J.F. 2003. Conserving forest biodiversity: a comprehensive multiscaled approach, 351 pp. Washington, DC: Island Press.

Lovett, G.M., Jones, C.G., Turner, M.G., and Weathers, K.C. editors, 2005. Ecosystem function in heterogeneous landscapes. New York: Springer-Verlag.

Mack, R.N. 1981. Invasion of Bromus tectorum L. into western North America: an ecological chronicle. Agro-Ecosystems 7: 145–165.

Mitchell, R.J., Franklin, J.F., Palik, B.J., Kirkman, L.K., Smith, L.L., Engstrom, R.T., and Hunter, M.L. 2003. Natural disturbance-based silviculture for restoration and maintenance of biological diversity. Available at http://www.ncseonline.org/ewebeditpro/items/O62F3299.pdf.

Reburtus, A.J., Kitzberger, T., Veblen, T.T., and Roovers, L.M. 1997. Blowdown history and landscape patterns in the Andes of Tierra del Fuego, Argentina. Ecology 78: 678–692.

Ruggiero, L.F., Aubry, K.B., Carey, A.B., and Huff, M.H. technical editors 1991. Wildlife and vegetation of unmanaged Douglas-fir forests. USDA Forest Service General Technical Report PNW-GTR-285. Portland, OR: Pacific Northwest Research Station.

Spies, T.A., and Franklin, J.F. 1991. The structure of natural young, mature and old-growth forests in Washington and Oregon. In Wildlife and vegetation of unmanaged Douglas-fir forests. USDA Forest Service General Technical Report PNW-GTR-285, technical, eds. L.F. Ruggiero, K.B. Aubry, A.B. Carey, and M.H. Huff, pp. 91–121. Portland, OR: Pacific Northwest Research Station.

Schrader-Frechette, K.S., and McCoy, E.D. 1993. Method in ecology. 328 pp. Cambridge, UK: Cambridge University Press.

Stanford, J.A., Ward, J.V., and Ellis, B. K. 1994. Ecology of the alluvial aquifers of the Flathead River, Montana. In Groundwater ecology, eds. J. Gilbert, D.L. Danielopol, and J.A. Stanford, pp. 367–390. San Diego: Academic Press.

Veblen, T.T., Hadley, K.S., and Reid, M.S. 1991. Disturbance and stand development of a Colorado subalpine forest. J. Biogeogr. 18: 707–716.

# 22
# Spatial Heterogeneity: Past, Present, and Future

Gaius R. Shaver

## Abstract

Explanation and interpretation of spatial heterogeneity in nature have been central concerns in ecology since long before the word *ecosystem* was first defined. As ecological knowledge has developed during the past century, it has become clear that the problems of heterogeneity are diverse and can be studied as a number of component issues; together, the chapters of this book represent a state-of-the-art summary of current understanding. Simplifying assumptions of spatial homogeneity and temporal stability are still used widely in ecological research and will continue to be used for the foreseeable future, but long-term, global understanding requires a multiscale approach in which homogeneity and heterogeneity are parts of one continuum and change never stops. Future priorities for research include dynamic approaches to heterogeneity in complex spatial networks, the search for broad patterns in heterogeneity across spatial scales, and reconciling the goal of environmental sustainability with the fact that we live in a patchy, constantly changing world.

## Introduction

A central goal of ecology is to explain the causes and interpret the implications of spatial variability and spatial patterns in nature. Spatial heterogeneity is inevitable and unavoidable at all levels of ecological organization, because no two places on Earth can have exactly the same chemical, physical, and biological environment and no two organisms can occupy exactly the same place at the same time. Although the search for generality and the need to simplify often lead to assumptions of homogeneity in ecological processes or patterns in particular instances or at a particular spatial scale, any global understanding must acknowledge the importance of spatial heterogeneity.

A reassessment of the current state of the art in understanding of spatial heterogeneity, as summarized in the chapters of this book, is both timely and highly appropriate. As the scale and complexity of human domination of the earth's ecosystems increase, simple models that assume homogeneity over large areas or constant environments over long periods are less and less useful to understanding or management. New techniques and new conceptual models are also being developed at a rapid rate and show great promise for major advances. The aim of this brief chapter is to provide a personal response to the state of the art, as presented at the Tenth Cary Conference on "Ecosystem Function in Heterogeneous Landscapes" (Lovett et al. this volume). The response is organized around four questions that might be asked by a student, a nonspecialist, or an outsider approaching the issues of spatial heterogeneity with the aim of catching up on current knowledge and future directions.

## Is a Concern for Spatial Heterogeneity New to Ecology and Ecosystem Science?

Spatial variation and spatial patterns in nature were a dominant concern of early ecologists, but they lacked both the long history of empirical studies and the rich array of conceptual and mathematical models described in this book. This was particularly true in early studies of ecological succession, in which ecologists like Henry Chandler Cowles (1899) tried to explain spatial patterning and temporal change in vegetation as the result of a dynamic interaction among plants, soils, and the physical environment. The classic debate between Gleason and Clements (e.g., Gleason 1926; Clements 1936) was largely about spatial patterns in vegetation and the factors that cause them. Early studies of niche differentiation in both plants and animals also commonly focused on spatial differences in species distributions, often in relation to resource availability; Weaver's (1919) descriptions of rooting patterns in grasses and Grinnell's (1917) studies of bird distributions are examples.

At the ecosystem level, it is true that spatial heterogeneity has often been ignored in practice, at least within the particular ecosystem under study. This occurred despite the fact that Tansley's (1935) original definition of the word *ecosystem* came as a contribution to the same global discussion of vegetation succession and spatial variation that dominated plant ecology in the early twentieth century. The classic "black box" approach to biogeochemistry of ecosystems, in which the difference between inputs and outputs is analyzed as a means of inferring how the whole system works, does not necessarily require recognition of any internal spatial structure (Likens and Bormann 1972). On the other hand, the simple fact that outputs differ from inputs in a "black box" ecosystem model means that ecosystem processes *generate* spatial heterogeneity at the landscape scale. The overwhelming importance of the spatial context in which ecosystems lie led Swanson and Sparks (1990) to define "The Invisible Place" to reflect the fact that what we know about ecosystem function depends strongly on spatial interactions with neighboring systems.

# Is Spatial Heterogeneity One Problem?

Spatial heterogeneity is a core concern of ecology and ecosystem science and has been since ecology was first recognized as a distinct discipline. As shown in many of the chapters in this book, however, as knowledge has increased and new methods developed, it has become clear that heterogeneity is a multidimensional problem and that it is often more efficient to partition it into several components. For example, the recognition of patch dynamics and periodic disturbances as a distinct set of system characteristics that differ among communities and ecosystems and are often self-generated (e.g., Watt 1947; Pickett and White 1985; Meinders and van Breemen this volume) has led to major advances in understanding the broader problems of vegetation change and stability that Cowles, Clements, Gleason, and others wrestled with a century ago. Spatial heterogeneity now has a range of explicit, quantitative definitions and metrics, and heterogeneity itself is seen as part of the set of variables that describe and control ecological processes and patterns (Turner and Chapin this volume). A host of new methods are available, ranging from those used in quantitative description and measurement at multiple scales (such as remote sensing and GIS methods) to new techniques of modeling and prediction.

Most of the chapters in this book describe different ways of partitioning the questions of heterogeneity into components that can be investigated more efficiently and more explicitly (e.g., White and Brown; Fahrig and Nuttle; Fisher and Welter; Pastor; Reiners; Turner and Chapin; all this volume). The identification of these as distinct issues has not come all at once but rather in a stepwise or saltatory fashion as a result of research insights and new methods developed during the past century—we could not have predicted this particular partitioning a century ago. My own classification of the issues includes the following:

- Heterogeneity as a problem in spatial averaging or aggregation: A classic example is the integration of photosynthesis through a forest canopy, where simple linear averaging may lead to large errors in prediction of overall productivity (e.g., Rastetter et al. 1992).
- Heterogeneity as a problem in spatial complementarity of functions or processes: Essential components of a single system may be located in different places with different controls over their functions, as in the use by birds of one kind of patch for feeding, another for nesting, or as in the spatial separation of leaf functions from root functions in vegetation (Fahrig and Nuttle; Naiman et al.; Tongway and Ludwig; all this volume). In cities, these complementary ecosystem functions may be spatially separated by design (Band et al. this volume).
- Heterogeneity and the simultaneous regulation of system components at different spatial or temporal scales: For example, distribution and abundance of many plant species is regulated simultaneously by both regional

climate and narrow, local soil conditions or fine-scale disturbances. Productivity of ocean systems is simultaneously regulated at fine and coarse scales (Mahadevan this volume). In forest systems, the functioning of long-lived tree species also may depend on much shorter-term responses of herbivores to yearly weather variation (Franklin this volume).

- Heterogeneity, disturbance, and patch dynamics as maintenance processes of overall system properties: Local, fine-scale succession constantly changes both the scale and the pattern of heterogeneity in nearly all ecosystems and landscapes, as well as their average properties. The importance of disturbance, from local to very large scales, in maintaining the overall characteristics of ecosystems, landscapes, and whole biomes is by now well established (Pickett and White 1985; Pastor this volume; Romme this volume).
- Heterogeneity in which configuration as well as composition matters: The particular arrangement of patches in space may be at least as important as the relative abundance of different kinds of patches; examples range from animal use of multiple habitats to the processing of nitrogen in streams (Fahrig and Nuttle; Fisher and Welter; Kratz et al.; Naiman et al.; Tongway and Ludwig; all this volume).
- Heterogeneity and the propagation of energy, matter, and information in space and time: The management of complex, heterogeneous landscapes requires integration of all of these components of heterogeneity. Applications range from the design of nature reserves to the analysis of how disturbances spread across a network of interacting ecosystems and/or ecosystem components (Reiners this volume). Heterogeneity itself may be self-organized over time, at scales ranging from root-soil interactions (Meinders and van Breeman this volume) to large-scale landscape patterns (Pastor this volume).

Each of the above represents a legitimate approach to understanding the problems of ecosystem function in heterogeneous landscapes. There is, however, a potential future problem in communication among researchers about exactly how "the problem of heterogeneity" should be defined. It is clearly important to develop a unified conceptual framework that permits efficiency of communication and development of applications of our understanding.

## Has the Assumption of Homogeneity Outlived Its Usefulness?

The assumption of internal homogeneity is one of the most commonly used simplifying assumptions in ecology and will no doubt continue to be so. There can be no denying that it is often useful and appropriate to ignore spatial heterogeneity in development of generalizations about individual populations, communities, ecosystems, or landscapes. In the initial phases of research on a new site, it often is simply more efficient to assume spatial homogeneity rather than getting lost in "details." On the other hand, as our understanding increases and the expectations of accuracy and precision of our predictions also increase, heterogeneity must be dealt with explicitly. Heterogeneity is not randomness but has characteristics of scale, pattern, connectedness, and other features that can be quantified and compared among the objects of study just as more homogeneous variables can (Turner and Chapin this volume). As new, efficient methods and conceptual models are developed, it is not only appropriate but also easier to incorporate questions of heterogeneity into ecological research.

Many variables that are homogeneous at one spatial scale are heterogeneous at other scales, making the homogeneity-heterogeneity contrast more of a continuum than a sharp divide (e.g., Mahadevan this volume). Most classifications of populations, communities, ecosystems, and landscapes focus on scales at which the objects of classification are relatively homogeneous; examples include the taxonomic hierarchy (species/genus/family/order), soil and vegetation classification schemes, and biome or ecozone classifications. The fact that these are often hierarchical reflects the multiple scales at which these objects are more or less homogeneous. Such classification is needed because much of the time, the goal of ecological research is to develop generalities that can be used to predict processes and patterns beyond the particular location being studied. Only fairly recently, however, have ecologists begun to work out what determines the peaks and valleys in this homogeneity-heterogeneity continuum and the implications for the regulation of ecological systems at multiple scales (Pastor this volume).

Finally, the shift away from approaches that assume homogeneity is also related to fundamental shifts in the theory of regulation of populations, communities, and ecosystems that have occurred over recent decades (e.g., Watt 1947; Pickett and White 1985). As ecologists have begun to appreciate the importance of coarse- and fine-scale disturbances, patch dynamics, and other nonequilibrium processes, it has become clear that assumptions of spatial and temporal homogeneity in environmental and other drivers are simply incompatible with research on such issues. Research on climate change and on long-term vegetation change has shown clearly that equilibrium conditions are more often the exception than the rule and that understanding of uniform, stable, homogeneous environments and ecosystems is useful but at best incomplete. As a result, ecologists have been forced to develop theory and methods that explicitly compare and describe systems that are spatially and temporally heterogeneous; this book provides many fine examples.

## What Lies in the Future?

Future advances in understanding of spatial heterogeneity will follow conceptual insights, theoretical developments, and methodological advances, as they always have. The perceived importance of spatial heterogeneity will continue to be a function of the question under study and the accuracy and precision needed for an appropriate answer. Three areas of research appear particularly ripe for progress at the present time, including:

- Dynamic heterogeneity and complex spatial networks: We are just beginning to develop many of the conceptual models and the computational tools that are needed to understand and predict long-term changes in complex networks of interacting populations, communities, and biogeochemical fluxes (Pastor; Reiners; Turner and Chapin; all this volume). In addition to limited theoretical development, one of the major limiting factors is still a lack of data, particularly data on biogeochemical processes and other data that are labor-intensive to collect and require special expertise. New developments in remote sensing are helping to alleviate this problem, though, and the current explosion in use of tools like geographic information systems is making these efforts much more efficient than in the past. Further development of theory, tools, and approaches is key to a long list of real-world applications ranging from the design and management of nature reserves, to the quantification of the importance and meaning of "biodiversity," to the management of the spread of human disturbances over landscapes.

- Broad patterns among heterogeneous ecosystems and landscapes: The need to generalize our understanding so that it can be applied globally requires us to continue to search for homogeneity of patterns in nature (White and Brown this volume). This includes improved definition and quantification of exactly what spatial heterogeneity is, so that heterogeneity itself can be compared across ecosystems and landscapes. It also includes the development of what might be called ecosystem and landscape *allometry* (or "macroecology" of ecosystems and landscapes, cf. Brown 1995). This kind of understanding is essential to development of our ability to extrapolate predictions over large regions and the globe.

- Heterogeneity and ecological sustainability: As human domination of the earth and its resources continues, there is increasing concern for the sustainability of ecological goods and services including clean air, clean water, and adequate food (Lubchenko et al. 1981; Clark and Dickson 2003). At present, there is little quantitative theory developed that might be useful in guaranteeing or predicting the sustainability of these good and services, particularly theory that incorporates the emerging nonequilibrium, spatially heterogeneous paradigm of ecological regulation over large areas and long timescales (Kates et al. 2001; Turner et al. 2003). A reconciliation of the need for sustainable environmental goods and services within a global ecosystem that is heterogeneous in both space and time is one of the greatest intellectual challenges in ecology today.

## References

Brown, J.H. 1995. Macroecology. Chicago: Chicago University Press. 269 pp.
Clark, W.C., and Dickson, N.M. 2003. Sustainability science: the emerging research program. Proc. Natl. Acad. Sci. U.S.A. 100: 8059–8061.

Clements, F.E. 1936. Nature and structure of the climax. J. Ecol. 24: 252–284.

Cowles, H.C. 1899. The ecological relations of the vegetation on the sand dunes of Lake Michigan. Botanical Gazette 27: 95–391.

Gleason, H.A. 1926. The individualistic concept of the plant association. Bull. Torrey Botanical Club 53: 7–26.

Grinnell, J. 1917. The Niche-Relationships of the California Thrasher. The Auk 34: 427–433.

Kates, R.W., Clark, W.C., Corell, R., Hall, J.M., Jaeger, C.C., Lowe, I., McCarthy, J.J., Schellnhuber, H.J., Bolin, B., Dickson, N.M., et al. 2001. Sustainability science. Science 292: 641–642.

Likens, G., and Bormann, F.H. 1972. Nutrient Cycling in Ecosystems. In Ecosystem structure and function, ed. J. Weins, pp. 25–67. Corvallis: Oregon State University Press.

Lubchenco, J., Olson, A.M., Brubaker, L.B., Carpenter, S.R., Holland, M., Hubbell, S.P., Levin, S.A., MacMahon, J.A., Matson, P.A., Melillo, J.M., et al. 1991. The Sustainable biosphere initiative: an ecological research agenda. Report from the Ecological Society of America. Ecology 72: 371–412.

Pickett, S.T.A., and White, P.S. eds. The Ecology of Natural disturbance and Patch Dynamics. London: Academic Press. 472 pp.

Rastetter, E.B., King, A.W., Cosby, B.J., Hornberger, G.M., O'Neill, R.V., and Hobbie, J.H. 1992. Aggregating Fine-scale Ecological Knowledge to Model Coarser-Scale Attributes of Ecosystems. Ecol. Applications 2: 55–70.

Swanson, F.J., and Sparks, R.E. 1990. Long-Term Ecological Research and the Invisible Place. BioScience 40: 502–508.

Tansley, A.G. 1935. The Use and Abuse of Vegetational Terms and Concepts. Ecology 16: 284–307.

Turner, B.L., Kasperson, R.E., Matson, P.A., McCarthy, J.J., Corell, R.W., Christensen, L., Eckley, N., Kasperson, J.X., Luers, A., Martello, M., Polsky, C., Pulsipher, A., and Schiller, A. 2003. A framework for vulnerability analysis in sustainability science. Proc. Natl. Acad. Sci. U.S.A. 100: 8074–8079.

Watt, A.S. 1947. Pattern and process in the plant community. J. Ecology 35: 1–22.

Weaver, J.E. 1919. The ecological relations of roots. Carnegie Institution of Washington Publication No. 286.

# 23
# Heterogeneity and Ecosystem Function: Enhancing Ecological Understanding and Applications

Judy L. Meyer

## Abstract

Four themes emerging from papers presented at the Tenth Cary Conference are discussed in this synthesis. First, conditions are considered in which both landscape composition and configuration influence ecosystem function. Where there is some vector of flow (e.g., water, wind, animal) between landscape components that differ in type or rate of ecosystem processes, configuration will influence ecosystem function. The second theme is that a network perspective offers opportunities to advance both theoretical and applied analyses of landscape heterogeneity. Analyses of the interaction of human and natural networks and of networks and patches are fruitful areas for future research. A third theme is the need to reflect more fully the diversity of human activities and their history and institutions when assessing landscape heterogeneity and ecosystem function. Finally, an enhanced understanding of landscape heterogeneity and ecosystem function will improve four aspects of human interactions with nature: the framework of environmental regulations, management of land and water resources, environmental design, and ecosystem restoration. Realizing these improvements requires that we find a vocabulary to express to the public the influence of landscape heterogeneity on ecosystem function and hence the ecosystem services that society values.

## Introduction

This paper provides a synthesis of the Tenth Cary conference papers and discussions from the perspective of an aquatic ecologist. I explore four themes that emerged as conference participants discussed the consequences of landscape heterogeneity on ecosystem function: (1) Both composition and configuration impact ecosystem function under certain conditions. (2) Network analysis offers opportunities to explore the impact of configuration on ecosystem function. (3) Expanding ecological horizons to include heterogeneity of humans and their institutions into ecological analyses of landscape heterogeneity offers exciting research opportunities. (4) Application of an enhanced understanding of the consequences of spatial heterogeneity on ecosystem function could improve environmental regulations, management, design, and restoration.

## Composition and Configuration

Conference participants considered the question of when and how spatial heterogeneity impacts ecosystem function. Turner and Chapin (this volume) identified a series of conditions in which heterogeneity would have an impact. These included situations where the impact was a consequence of variation in landscape composition, such as where spatial heterogeneity in process rates resulted from differences in community composition (e.g., vegetation patches differing in time since last fire). They also identified situations where both composition and configuration mattered, such as when horizontal transfers occur between patches that exhibit different processing rates.

Population ecologists have a long history of considering the impacts of spatial heterogeneity on population persistence (Fahrig and Nuttle this volume). Landscape composition is often of importance in controlling birth and death rate of populations, and hence population persistence. Landscape configuration is of importance where animals show a strong response to boundaries resulting in interpatch movement, which, combined with high probability of local extinction or use of multiple habitat types, alters birth and death rates and hence population persistence.

In her presentation at the Cary Conference, Lenore Fahrig used a flow diagram to illustrate these ideas. I have modified that figure (Figure 23.1) to consider how composition and configuration impact ecosystem function. As illustrated in the figure, ecosystem function is the net result of a number of ecosystem processes. Rates of these processes are influenced directly by landscape composition. For example, nitrogen delivery by a river to an estuary is an example of an ecosystem function resulting from processes such as leaching from soils, terrestrial and aquatic plant uptake rates, and microbial processes in soils and streams. There is an extensive literature documenting how landscape composition (e.g., proportion of land in crops *vs.* forest *vs.* urban) alters the amount of N delivered to rivers and hence its concentration (e.g., citations in Gergel et al. 2002; Turner and Rabelais 2003). The concentration of N in rivers regulates the rate of ecosystem processes such as N uptake and transformation by aquatic plants and microbes (e.g., Dodds et al. 2002). In situations where process rates vary little between landscape patches or where there is little exchange between patches, ecosystem function reflects landscape composition, and configuration is relatively unimportant (arrows on the left in Figure 23.1). But, as illustrated by arrows on

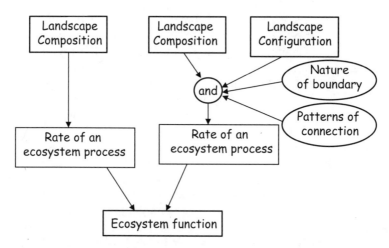

Figure 23.1. Diagram adapted from Fahrig (presentation at the Tenth Cary Conference) to illustrate how landscape composition and configuration impact ecosystem function. Influence is indicated by arrows, and ellipses enclose aspects of configuration that influence process rates. See text for an illustrative example.

the right half of Figure 23.1, both composition and configuration will play a role in determining the rate of an ecosystem process depending on the nature of the boundary between landscape components with different process rates and on the patterns of connection between components. For example, the extent and composition of riparian ecosystems can modulate the delivery of N from fields or forests to the river (Gergel et al. 2002); and position in the river network will influence the surface: volume ratio of water and benthos as well as the hydrologic retention time, both of which can impact N uptake rate (Peterson et al. 2001). Hence the right half of Figure 23.1 illustrates a situation where both composition and configuration influence ecosystem function.

Fisher and Welter (this volume) argue that if the vector integrating the ecosystem (e.g., water) influences ecosystem function (e.g., nitrogen transformation rates at the interfaces between patches), the function of the whole will be greater than the sum of its parts. For example, calculating nitrogen retention in such an ecosystem is not a simple addition of the retention rates associated with each patch; rather, it is a nonlinear function whose form depends on the processes occurring at the interface between patches (Fisher and Welter this volume). This form of landscape heterogeneity offers research challenges that include analyzing the consequences of temporal variation in integrators (e.g., drying and wetting of drainage networks) and exploring the impacts of changing network structure on nutrient dynamics resulting in altered stoichiometry (Fisher and Welter this volume). Understanding the consequence of changing network structure is particularly relevant because anthropogenically altered networks (e.g., stream networks with headwaters replaced by pipes) are common features of the modern landscape (Meyer and Wallace 2001).

As Turner and Chapin (this volume) reminded us at the beginning of the conference, ecosystem analysis is about analyzing flows of energy and matter. As a discipline interested in flows, ecosystem studies incorporate existing flows between landscape components; and when these patches differ in processing rates, configuration must be considered when analyzing ecosystem function. Several papers in the conference provided examples of ecosystems in which flows between landscape components were important drivers, and hence where landscape configuration influences ecosystem function. In these examples, the most common vector of flow was water. In semiarid landscapes, pathways taken by water in the landscape alter nitrogen concentrations, resulting in changes in algal assemblages (Fisher and Welter this volume). In Australian semiarid ecosystems, plant biomass and productivity is higher when rainfall is distributed heterogeneously than if it is distributed uniformly; the interaction between infiltration rate and water movement across the landscape alters the type, biomass, and productivity of vegetation (Tongway and Ludwig this volume). Lake characteristics reflect their position in the landscape because of groundwater flowpaths (Kratz et al. this volume). Other ecosystems where flow between patches that differ in processing rates and hence where configuration was identified as a significant driver include oceans (Mahadevan this volume), riparian ecosystems (Naiman et al. this volume), boreal forests (Turetsky et al. this volume), and urban ecosystems (Band et al. this volume).

Although most of the examples of flow between patches involved water, there are many other vectors (Reiners and Driese 2001). These include wind and animal activity (Meinders and van Breeman this volume; Pastor this volume). For example, in Isle Royale, foraging behavior of moose (i.e., patterns of movement and when feeding starts or stops) alters sustainability of the forest ecosystem (Pastor this volume). Pathways taken by fires profoundly influence ecosystem processes (Romme this volume). Paths taken by other disturbances such as disease (Smith this volume) or insect outbreaks (Franklin this volume) alter tree mortality rates and hence the structure and function of the forest ecosystem.

The significance of configuration, or flows between patches, can also be overwhelmed by disturbances. Romme (this volume) shows an example of this with fire, where intense crown fires triggered by climatic events swept across all landscape components regardless of patch structure or configuration. A similar phenomenon was noted in desert streams, where floods overwhelm previously existing patch structure and connections (Fisher and Welter this volume). In both of these examples, a massive disturbance changes the entire ecosystem in a manner that is not influenced by patch composition or configuration.

In conclusion, both composition and configuration can impact ecosystem function. Spatial configuration is important when there is a vector or integrator flowing between landscape components that have different processing rates. The papers in this volume provide numerous examples of landscapes in which this occurs.


## A Network Perspective

A network is a set of interconnected nodes; both the patterns of connection and the magnitude of the flux between nodes characterize the network. Viewing landscapes as networks allows one to consider not only spatial arrangement of landscape components but also magnitude of flows between them. The natural world provides us with a diversity of examples of networks, such as the branched, hierarchical network of rivers. River networks are ecosystems with distinct longitudinal, lateral, and vertical connections, and much of current research in lotic ecology is directed at understanding and quantifying those linkages. Although a network perspective offers opportunities to advance both theoretical and applied aspects of landscape ecology, networks are but one part of heterogeneous landscapes, which include patches, networks and gradients (Swanson and Jones 2003).

Ecologists have traditionally viewed the landscape from one of these perspectives, but considering how these components interact offers an opportunity to enhance understanding of ecosystem function in heterogeneous landscapes. For example, network structure interacts with patch structure to alter ecosystem function in forests of the Pacific Northwest, where networks of forest roads fragment the landscape, while stream networks generate distinct habitats in riparian zones (Swanson et al. 1997). The structure of a river network influences the pattern of flood disturbance and recovery in stream and riparian habitats in steep forested landscapes (Swanson et al. 1998). Habitat patches in stream ecosystems (pools, riffles, debris dams) are the product of differential material transport and storage along stream networks. The position of a road network in a landscape and the arrangement of the road network with respect to the stream network alters the nature of debris flows and their effects on the landscape (Jones et al. 2000). It also influences recovery from disturbance because unimpacted tributaries provide biotic refuges (Jones et al. 2000). In urban ecosystems, network structure has been changed by construction of road networks, storm drains, pipes that remove drinking water from streams and return wastewater; humans also alter water and nutrient loading at the patch scale (e.g., fertilizing and watering lawns) (Band et al. this volume). These actions have profoundly altered the flowpaths of water and hence urban ecosystem function (Band et al. this volume). These examples illustrate that analyses of the interaction of networks and patches and of human and natural networks are fruitful areas for research.

Studies of spatial heterogeneity in both landscapes and riverscapes have commonly taken either a patchwork or gradient perspective (e.g., Pringle et al. 1988; Vannote et al. 1980; Turner and Chapin this volume); network analysis may offer new insights (Fisher 1997). For example, Poole (2002) presents theoretical models illustrating how changes in network branching pattern could influence patterns of solute concentration and species distribution of aquatic insects along the length of a river. Further work along these lines is warranted as are efforts in methods development. Methods and metrics for landscape analysis and spatial modeling developed from a patchwork perspective may not be applicable to dendritic networks (Poole 2002). For example, Fagan (2002) demonstrates that neither linear nor two-dimensional frameworks are appropriate for capturing the dynamics of metapopulations in dendritic networks. Populations in dendritic networks (e.g., fishes in desert streams) differ from those in linear landscapes in their connectivity, response to fragmentation, and risk of extinction (Fagan 2002). A conceptual model based on interactions between road and stream networks produced a different picture of the spatial distribution of ecological responses to disturbance than predicted by the "zone of influence" approach used to assess effects on terrestrial ecosystems (Jones et al. 2000). Further development of analytical methods to explore interactions among networks, gradients, and patches is likely to further our understanding of the impact of spatial heterogeneity on ecosystem function.

## The Diversity of Human Influence

Humans have created new networks (e.g., roads) and altered existing networks. In fact, human actions have affected all components of the flow diagram in Figure 23.1: landscape composition, landscape configuration, the nature of boundaries, and the pattern of connections. Humans have increased spatial heterogeneity by fragmenting both landscapes and riverscapes (Dynesius and Nilsson 1994; Pringle et al. 2000). Humans have also created more homogeneous ecosystems in agriculture and silviculture and by allowing excess sedimentation in aquatic ecosystems. They have profoundly altered landscape configuration and the nature of boundaries between patches, such as simplifying riparian zones by planting single species (e.g., crops or willows) along stream banks. The pattern of connections has been altered by tile drains, stormwater pipes, and stream burial (e.g., Meyer and Wallace 2001). Humans concentrate resources and thereby alter both composition and configuration of the landscape (Band et al. this volume).

Ecologists have long recognized the changes in landscape composition and configuration caused by human action. Ecologists commonly refer to anthropogenic effects; this simplification ignores the diversity of human activities and their social, cultural, and economic context. Clearer recognition of the heterogeneity of human actions could benefit ecological science, just as recognition of landscape heterogeneity has increased our current understanding of ecosystem function. Researchers working with social scientists in urban ecosystems have incorporated human diversity into their study design (see Band et al. this volume). For example, human impacts are likely to vary based on socioeconomic status, cultural attitudes, age distribution of the human population, age of the development, and many other factors. Variations in past human activity (i.e., history) help explain current patterns or reveal hidden heterogeneity. Persistent land-use legacies have been shown to influence ecosystem structure and function in both terrestrial and aquatic ecosystems (Foster et al. 2003). Future landscape trajectories are influenced by the response of humans to environmental conditions, and there is great diversity in the nature of those human responses. Truly incorporating the richness and complexity of the human dimension into ecosystem research offers exciting research opportunities and is likely to provide more effective approaches to improving environmental conditions.

## Practical Benefits

Several presentations at this conference addressed the significant practical benefits resulting from an improved understanding of the linkage between spatial heterogeneity and ecosystem function. "Understanding heterogeneity directly aids management and rehabilitation" in semiarid landscapes of Australia (Tongway and Ludwig this volume). The ability of managers to maintain and improve ecosystem services provided by aquatic ecosystems is enhanced by incorporating considerations of spatial heterogeneity (Steinman and Denning this volume). In the following paragraphs, I discuss how an improved understanding of heterogeneity will improve four aspects of the human interaction with nature: the framework of environmental regulations, management of land and water resources, environmental design, and ecosystem restoration.

If spatial heterogeneity influences ecosystem function, then uniform regulations and standards across diverse environmental zones make little sense. Romme (this volume) discusses this for fire management in the West, where a uniform national fire management policy does not account for the spatial heterogeneity in fire susceptibility or historical pattern of fire; hence, this uniform policy does not result in sustainable forests. Statewide water quality standards are another example where ignoring heterogeneity could lead to standards that are either too lenient or too harsh for regionally varying conditions.

Connections between landscape components are not always obvious to regulators or to the legal profession. An example of this is the recent U.S. Supreme Court decision that eliminated Clean Water Act jurisdiction for intrastate isolated wetlands whose only connection to the landscape is migratory birds. In response to this decision, U.S. Environmental Protection Agency and U.S. Army Corps of Engineers proposed removing Clean Water Act protection from wetlands without visible surface water connections and from intermittent stream channels. The scientific literature has documented the significance of these small ecosystems to the larger river network (e.g., Meyer and Wallace 2001), but the significance of configuration and connection has been inadequately incorporated into the regulatory arena.

Land managers make decisions that would benefit from an understanding of the way in which landscape composition and configuration impact ecosystem function (Franklin this volume). Decisions on size and location of cuts, or where to build logging roads, would benefit from a better understanding of the significance of these alterations in landscape structure on ecosystem processes. Fausch et al. (2002) illustrate how fish conservation would benefit from a riverscape perspective that takes into account location and linkages between habitat patches. Steinman and Denning (this volume) provide examples of decisions that are currently being made in south Florida on where to place water storage reservoirs, where on the landscape wetlands should be constructed to maximize nutrient retention, and where confined animal feeding operations should be targeted for improvements. Clearly, these decisions benefit from an understanding of the significance of landscape composition and configuration on these processes. In conference discussion sessions, K.B. Jones (Environmental Protection Agency, personal communication) noted that agencies like the Natural Resources Conservation Service are making decisions on how to invest dollars to establish conservation reserves; where should these be placed on the landscape to derive maximum benefit from them? Possingham et al. (this volume) provide a clear illustration of how reserve design benefits from considerations of spatial heterogeneity.

Although most of this conference was on spatial heterogeneity, a recognition of its linkage with patterns of temporal heterogeneity was also present. Stream ecologists have dealt with these issues with respect to flow regulation, and the approach they have taken offers suggestions for how issues of temporal and spatial heterogeneity might be incorporated in a management context. Flood control dams have reduced the temporal variability of river discharge (lower flood peaks and higher baseflows) as well as altering the timing of high flows (e.g., Richter et al. 2003). To reduce the environmental impacts of dams, altered dam operations are being considered. Desired river flows were initially determined by considering the flow needs of individual species of interest; not too surprisingly, species differ in their needs and hence this approach can result in contradictory recommendations (Poff et al. 1997). An alternative approach refers to the natural flow regime (Poff et al. 1997) and identifies ways in which the flows have been altered; key components of the natural flow regime are identified and flows are recommended that mimic those key components (Poff et al. 2003). A similar approach may be appropriate as ecologists seek to incorporate lessons learned in spatial heterogeneity into landscape management. If key components of the natural patterns of spatial heterogeneity can be identified and linked with ecosystem services of concern, then management schemes that seek to mimic those can be implemented. An example of this was offered by

Romme (this volume). If historical patterns of fire frequency are understood for different components of the landscape, then these can be used to establish meaningful differential responses to fire outbreaks.

Some of the most exciting applications of the advances in this discipline will be in the area of environmental design. Human societies are occupying ever increasing areas of the landscape; insights from this research have the opportunity to impact the way those human habitations are designed so that they have less impact on the ecosystem services that society values. Possible applications are numerous, and here I suggest only a few. Local governments make decisions on land management when they pass zoning laws. How should zoning laws or land-use planning be designed to have the least impact on ecosystem function? Can we provide some suggestions for subdivision design or even golf course design that will foster sustainable ecosystems? As human demands for water increase, many new water supply reservoirs are being built. If reservoir construction is going to be one of society's answers to increasing water availability, where should reservoirs be built on the landscape and in the stream network? Is it better to build one large or many small reservoirs? Considerations of landscape composition and configuration are essential for creating these designs and making these decisions.

Considerable sums are being spent to restore or rehabilitate damaged land- and riverscapes. Insights from spatial heterogeneity could benefit these efforts, not only in establishing desired patterns for a rehabilitated landscape but also in setting priorities for what components of the landscape would provide the greatest benefit for ecosystem services. In the absence of these kinds of guidelines, money can be wasted in projects that are less effective or in some cases even harmful to the ecosystem they seek to restore.

Achieving the practical benefits described will require not only development of the underlying theory and science, but also communication of this understanding to practitioners and the public. Theoretical insights from landscape ecology need to be expressed in terms and placed into a framework that can help guide architects, civil engineers, city planners, managers, and the public, for these are the individuals who are determining the design of urban and suburban landscapes. As suggested by one of the conference participants, we need a vocabulary to convey the concept of heterogeneity and its benefits to decisions makers and the public. Application of ideas from this conference and subsequent research requires both outreach and collaboration with those individuals and institutions shaping the modern landscape. This provides both the greatest challenge and the most promising opportunity for the future of this discipline.

*Acknowledgments.* I thank the symposium organizers for challenging me to think about these issues and providing a stimulating environment in which to do so. This paper benefited from comments on an earlier draft by Monica Turner, Julia Jones, Fred Swanson, Gary Lovett, and two anonymous reviewers.

## References

Dodds, W.K., Lopez, A.J., Bowden, W.B., Gregory, S., Grimm, N.B., Hamilton, S.K., Hershey, A.E., Marti, E., McDowell, W.H., Meyer, J.L., et al. 2002. N uptake as a function of concetration in streams. J. North Am. Benthol. Soc. 21: 206–220.

Dynesius, M., and Nilsson, C. 1994. Fragmentation and flow regulation of river systems in the northern third of the world. Science 266: 753–762.

Fagan, W.F. 2002. Connectivity, fragmentation, and extinction risk in dendritic metapopulations. Ecology 83: 3243–3249.

Fausch, K.D., Torgersen, C., Baxter, C., and Li, H. 2002. Landscapes to riverscapes: bridging the gap between research and conservation of stream fishes. BioScience 52: 483–498.

Fisher, S.G. 1997. Creativity, idea generation and the functional morphology of streams. J. North Am. Benthol. Soci. 16: 305–318.

Foster, D., Swanson, F., Aber, J., Burke, I., Brokaw, N., Tilman, D., and Knapp, A. 2003. The importance of land-use legacies to ecology and conservation. BioScience 53: 77–88.

Gergel, S.E., Turner, M.G., Miller, J.R., Melack, J.M., and Stanley, E.H. 2002. Landscape indicators of human impacts to riverine systems. Aquatic Sci. 64: 118–128.

Jones, J.A., Swanson, F.J., Wemple, B.C., and Snyder, K.U. 2000. Effects of roads on hydrology, geomorphology and disturbance patches in stream networks. Conservation Biol. 14: 76–85.

Meyer, J.L., and Wallace, J.B. 2001. Lost linkages and lotic ecology: rediscovering small streams. In Ecology: Achievement and Challenge, eds. M.C. Press, N. Huntly, and S. Levin, pp. 295–317. Blackwell Science.

Peterson, B.J., Wolheim, W., Mulholland, P.J., Webster, J.R., Meyer, J.L., Tank, J.L., Grimm, N.B., Bowden, W.B., Vallet, H.M., Hershey, A.E., McDowell, W.B., Dodds, W.K., Hamilton, S.K., Gregory, S., and Morrall. D.J. 2001. Stream processes alter the amount and form of nitrogen exported from small watersheds. Science 292: 86–90.

Poff, N.L., Allan, J.D., Bain, M.B., Karr, J., Presegaard, K.L., Richter, B.D., Sparks, R.E., and Stromberg. J.C. 1997. The natural flow regime: a paradigm for river conservation and restoration. BioScience 47: 769–784.

Poff, N.L., Allan, J.D., Palmer, M.A., Hart, D.D., Richter, B.D., Arthington, A.H., Rogers, K.H., Meyer, J.L., and Stanford. J.A. 2003. River flows and water waters: emerging science for environmental decision-making. Frontiers Ecol. Environ. 1: 298–306.

Poole, G.C. 2002. Fluvial landscape ecology: addressing uniqueness within the river discontinuum. Freshwater Biol. 47: 641–660.

Pringle, C.M., Freeman, M.C., and Freeman. B.J. 2000. Regional effects of hydrologic alterations on riverine macrobiota in the New World: Tropical-temperate comparisons. BioScience 50: 807–823.

Pringle, C.M., Naiman, R.J., Bretschko, G., Karr, J.R., Oswood, M.W., Webster, J.R., Welcomme, R.L., and Winterbourne, M.J. 1988. Patch dynamics in lotic systems: the stream as a mosaic. J. North Am. Benthol. Soc. 7: 503–524.

Reiners, W.A., and Driese, K.L. 2001. The propagation of ecological influences through heterogeneous environmental space. BioScience 51: 939–950.

Richter, B.D., Mathews, R., Harrison, D.L., and Wigington, R. 2003. Ecologically sustainable water management: managing river flows for ecological integrity. Ecol. Applications 13: 206–224.

Swanson, F.J., Johnson, S.L., Gregory, S.V., and Acker, S.A. 1998. Flood disturbance in a forested mountain landscape. BioScience 48: 681–689.

Swanson, F.J., and Jones, J.A. 2003. Landscape heterogeneity: a network perspective. Available at http://www.fsl.orst.edu/lter/pubs/webdocs/posters/cary2003_files/.

Swanson, F.J., Jones, J.A., and Grant, G.E. 1997. The physical environment as a basis for managing ecosystems. In Creating a Forestry for the 21st Century, eds. K.A. Kohm, and J.F. Franklin, pp. 229–238. Washington, DC: Island Press.

Turner, R.E., and Rabelais, N.N. 2003. Linking landscape and water quality in the Mississippi River Basin for 200 years. BioScience 53: 563–572.

Vannote, R.L., Minshall, G.W., Cummins, K.W., Sedell J.R., and Cushing, C.E. 1980. The river continuum concept. Can. J. Fisheries Aquatic Sci. 37: 130–137.

# 24
# Conceptual Frameworks: Plan for a Half-Built House

GARY M. LOVETT, CLIVE G. JONES, MONICA G. TURNER, and
KATHLEEN C. WEATHERS

## Abstract

The consideration of spatial heterogeneity in ecosystem science is a challenging problem both empirically and conceptually. Although conceptual frameworks have been developed for some aspects of the problem, there is as yet no overarching framework that links them together. In this paper, we review many of the conceptual frameworks used in the chapters of this book. We discuss how the ecosystem concept can be extended to the "landscape system." Like the ecosystem, the landscape system must have defined boundaries so that inputs and outputs can be distinguished from internal circulation. Given the delineation of the landscape system and its component ecosystems, a series of questions is posed that allow the investigator to determine what aspects of heterogeneity are likely to be important and what kind of model (homogeneous, mosaic, or interactive) most appropriately captures the behavior of the system.

## Conceptual Frameworks

One of the principal goals of this book is to advance the development of conceptual frameworks for consideration of spatial heterogeneity in ecosystem science. In science, conceptual frameworks provide an intellectual structure on which to hang empirical observations and hypotheses, and within which to design empirical studies (Pickett et al. 1994). Like the joists and rafters of a wood-frame house, the conceptual framework provides the bounds and constraints for the structure within. The construction of a house usually starts with an architectural plan, but science rarely proceeds that way because the form of the completed structure is not known when the building begins. It appears that the house for spatial heterogeneity and ecosystem processes has some well constructed rooms, almost ready to live in; a few rooms with bare framing where the wind still whistles through; and some empty spaces where no structure is yet apparent. Our purpose in this

463

chapter is to provide a plan that will at least show how the rooms fit together and to begin constructing a roof that will encompass them all.

With many types of entities (e.g., mass, energy, information, organisms) moving simultaneously within and between ecosystems, and many different ecosystems juxtaposed in a landscape, incorporating spatial heterogeneity into an understanding of ecosystem function can get exceedingly complex. One way to simplify is to search for pattern in the spatial heterogeneity that can inform us about important processes. This approach is discussed by White and Brown (Chapter 3), Pastor (Chapter 4), Tongway and Ludwig (Chapter 10), and Meinders and van Breemen (Chapter 11), among others. This approach is particularly appealing to the mathematically inclined, because pattern lends itself to mathematical description. But not all spatial heterogeneity generates recognizable patterns, so this approach, while valuable, has limitations. Another simplifying approach is the use of probabilistic models, where heterogeneity is expressed as a statistical distribution (see Tague, Chapter 7; Band et al., Chapter 13). This is a useful shortcut for some applications but does not allow explicit spatial interactions between ecosystems within a landscape, so it cannot shed any light on the potential importance of configurational heterogeneity. Our instincts as ecologists often tell us that much of the heterogeneity we observe is noise that is not important to the overall functioning of the ecosystem. Parsimony tells us that we should use simple models until they are proven inadequate, and economy tells us we cannot afford to measure all the heterogeneity in every property of an ecosystem (Smith, Chapter 8; Strayer, Chapter 20). So, the central question is, when do we need to deal with all this heterogeneity and when can we safely ignore it?

Strayer (Chapter 20) answers this question directly by proposing four situations in which it might be acceptable, or even wise, to ignore spatial heterogeneity: when the heterogeneity is unimportant functionally, when it is at too small a scale to be appropriate for the analysis, when it is not parsimonious scientifically (i.e., a simpler model yields adequate accuracy), or when it is not cost-effective (i.e., even if a simpler model doesn't work as well, it is all you can afford). Other chapters of this book and other recent publications provide conceptual models that can help ecologists understand how heterogeneity might be important in their study systems and how to deal with it if it is. For instance, the simple scheme proposed by Shugart (1998) to divide spatial models into homogeneous, mosaic, and interactive approaches has been very useful (see Turner and Chapin, Chapter 2; Lovett et al., Chapter 1). Turner and Chapin (Chapter 2) describe an important distinction between what they term "point" processes, for which horizontal fluxes among ecosystems on a landscape are unimportant, and "lateral" processes, for which they are important. Reiners (Chapter 5) discusses a related conceptual framework that has been extensively developed—the factors that regulate transport within heterogeneous environmental space. He analyzes the factors that control the rate and extent of propagation of mass, energy, and information in the environment. Steinman and Denning (Chapter 18) discuss a framework in which the service or function desired of the ecosystem determines what aspects of heterogeneity are likely to be important. Network theory, as discussed briefly by White and Brown (Chapter 3) and Meyer (Chapter 23) provides another conceptual framework when flows, rather than states or pools, are the main focus of study. For some processes, transport is regulated by the dynamics of the boundary between patches, and a nascent conceptual framework has recently been proposed for ecological boundaries (Cadenasso et al. 2003; Strayer et al. 2003). These various conceptual frameworks are the rooms in our half-built house. A long-term goal might be to unify these frameworks in some sort of overarching theory ( i.e., build the roof), but our shorter term objective is to understand what room we need to be in for any given problem and to learn to navigate among them.

## The Landscape System

When building a house, one always starts with the foundation; here, the foundation is the ecosystem concept. Ecosystem analysis involves careful definition of the boundaries of the system under study and measurement of the inputs and outputs of mass and energy across those boundaries and the circulation within the system. This standard analysis does not explicitly address spatial heterogeneity, but it is obvious that the inputs come from somewhere and the outputs go to somewhere. The donor and recipient ecosystems can be viewed as embedded in a larger, landscape system (Figure 24.1), which is

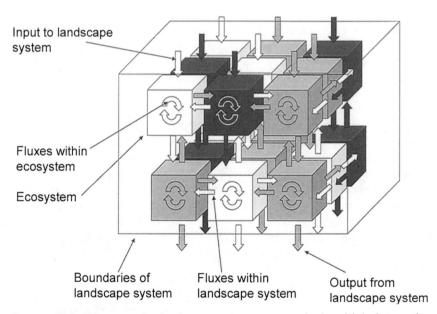

FIGURE 24.1. Diagram of a landscape system composed of multiple interacting ecosystems. Each ecosystem has boundaries defined by the small boxes, and the landscape system is delineated by the outer box.

the collection of interconnected ecosystems under study. Transfer of mass, energy, and information between ecosystems may be important to the functioning of both the individual ecosystems and the landscape system.

Like an ecosystem, a landscape system must have defined boundaries (Figure 24.1). Loreau et al (2003) proposed the "meta-ecosystem" concept for the study of connected ecosystems exchanging materials or energy, parallel to meta-populations that exchange organisms. They define the meta-ecosystem as a closed system in which sources in some component ecosystems must balance sinks in others. This is unrealistic because all ecological systems are open to inputs and outputs, nonetheless the meta-ecosystem is a useful concept, akin to the landscape system we diagram in Figure 24.1. In our view, the landscape system is open to inputs and outputs, and therefore internal sources and sinks need not be in balance. The landscape system is subject to the same constraints of mass and energy conservation as any ecosystem and can be analyzed with similar conservation equations.

## Analyzing the Landscape System: When Does Heterogeneity Matter?

Before we can analyze the landscape system, we need to specify the ecosystem process(es) of interest. Are we interested in primary productivity, denitrification, or the movement of salmon to spawning areas? Because we define an ecosystem process as the transfer of some entity between pools in the system (see Chapter 1), this involves specifying what entity is being transferred—(e.g., carbon, nitrogen and salmon, in the above examples). Next, we need to carefully delineate the ecosystems in the study area, because we cannot study flows between ecosystems unless we know precisely where those ecosystems are. The delineation of the ecosystems is at the discretion of the investigator, but they are usually relatively homogeneous areas or patches within the larger landscape system (see definition in Chapter 1). We also need to specify the boundaries of the landscape system, which is a volume of space that encompasses the ecosystems of interest (Figure 24.1). Given these specifications of the system, it is then possible to ask a few simple questions that can guide our consideration of heterogeneity in the system. (We note that the simple questions do not necessarily have simple answers, and the answers presume substantial knowledge of the process and the system.) These questions allow us to navigate a decision tree that can help us understand how to conceptualize, model, and scale heterogeneous landscapes (Figure 24.2).

First, one should ask,

1. Are there significant fluxes of the entity across ecosystem boundaries?

If the answer is yes, one should further ask,

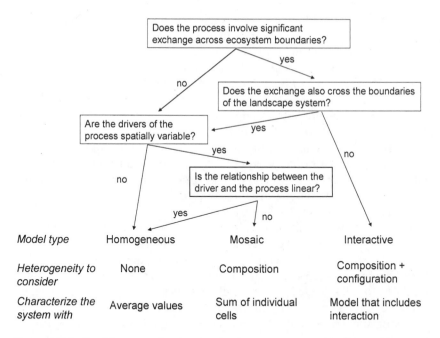

FIGURE 24.2. Decision tree allowing user to determine what type of model is necessary to represent heterogeneous landscape systems. See text for explanation.

2. Do the fluxes that cross the ecosystem boundaries also cross the boundary of the landscape system? (That is, is the flux an input or an output to the landscape system?)

If the answer to question 2 is no, then the system has significant internal exchanges and is best analyzed with an "interactive" model, which we discuss below. An example of an exchange that also crosses the boundary of the landscape system is the vertical exchange of $CO_2$ between a forest canopy and the atmosphere, assuming the atmosphere is not included in the landscape system. Turner and Chapin (Chapter 2) discuss these exchanges in terms of point processes and lateral processes, but in a more general sense it does not matter if the cross-boundary exchanges are lateral (or horizontal, say between a field and a forest) or vertical (say, between the epilimnion and the hypolimnion of a lake), what matters is whether they cross the boundaries of the defined landscape system.

If the answer to question 1 is no or the answer to question 2 is yes, then one should ask,

3. Are the principal drivers of the process spatially variable?

If the answer is no, then a homogeneous characterization of the system should suffice. If the answer is yes, then it is necessary to ask,

4. Is the relationship between the divers and the process linear?

If the answer is yes, then again a homogeneous model may still suffice, in that mean values of parameters should be sufficient to characterize the process within the system. If the answer to question 4 is no, then one should use a mosaic model, where the behavior of the process in individual ecosystems is modeled separately, and the results are summed to yield the whole-system behavior.

To elaborate the previous example, suppose we want to model the carbon budget of a forested watershed composed of forest patches growing on different soil types. We might presume that the principal flux of carbon in the system is the exchange between the canopy and the atmosphere, and that intrapatch transfers of carbon (say, transport by animals carrying seeds from one place to another) are insignificant. The exchange of $CO_2$ with the atmosphere is a cross-boundary exchange, but it also crosses the boundary of the landscape system (defining the upper boundary as the top of the canopy), so the answer to question 2 is yes. Further, suppose we know that the main control on photosynthesis in this system is the soil moisture status, that the moisture varies between soil types, and that the response of photosynthesis to soil moisture is nonlinear. These facts lead us through the decision tree to recommend a mosaic model (Figure 24.2).

The distinction between homogeneous, mosaic, and interactive models has several important consequences. In the homogeneous model, one does not need to consider heterogeneity at all, and the system is characterized by average values of its pools and fluxes [e.g., Equation (13.1) in Band et al., Chapter 13]. To determine the response of a process in this system to a change in one of its drivers, one need only use an average value of the driver to determine an average value of the process for the landscape system. This approach essentially redefines the landscape system as an ecosystem in its traditional, homogeneous sense.

On the other hand, if there is a significant spatial variation in the drivers, then ecosystem processes will vary spatially as well. Turner and Chapin (Chapter 2) point out several reasons why it may be important to understand and quantify that heterogeneity. In this case, one needs to consider only compositional heterogeneity—the number, types and sizes of patches. In this type of system, modeling the response of a process to a change in drivers is best done by determining the value of the driver for each patch within the system, modeling the response, and summing across all patches [e.g., Equation (13.2) in Band et al., Chapter 13]. It is necessary to use summation, rather than an average value for the system, because nonlinearities in the response may make averages inaccurate (Strayer et al. 2003). Mahadevan (Chapter 9) gives an excellent example of this phenomenon, in which gas exchange from the ocean surface is a nonlinear function of wind speed, so using an average value of wind speed over the ocean to calculate an average gas exchange rate yields a biased answer.

If there are significant fluxes between patches within the landscape ecosystem, an interactive model is usually the best approach. Both compositional and configurational heterogeneity should be considered. In this case, the behavior of the process in the landscape system cannot be predicted from an average value or from a summation of the individual patches, but instead requires a more complex model that incorporates the interpatch exchanges.

The design of such a model of course depends on the question being asked. However, one can glean some advice from papers presented in this book and elsewhere. Several chapters (Smith, Chapter 8; Strayer, Chapter 20) recommend parsimony—including only the amount of complexity necessary to get an adequate answer to the question of interest. Strayer et al. (2003) point out that the amount and type of information needed to model the system depends on the complexity of the interactions, but that relatively simple models often work adequately in ecology because the scale of variation in ecological systems is often much smaller than the scale of analysis (see also Possingham et al., Chapter 19), and empirical parameterization of larger-scale models can average across this small-scale variation.

In some cases, however, substantial complexity is necessary to capture the important functions of the system. More complex models often need to consider multiple "currencies"—different types of mass moving in different directions, perhaps controlled by signals (information flow) from different ecosystems or outside the system (Shachak and Jones 1995; Band et al., Chapter 13). For instance, consider a stream in which water and dissolved elements are moving downstream, while salmon and the elements they are composed of are moving upstream. Moreover, the path and timing of the salmon movement may be controlled by chemical cues that impart no significant mass flux. This situation, with multiple currencies moving via multiple vectors, partially controlled by spatial transport of information, would certainly require quite a complex model. The oft-heard phrase at the Cary Conference when considering this type of situation was "thinking about this makes my head hurt."

Reiners (Chapter 5) provides a general conceptual framework for understanding transport in heterogeneous systems that may be useful in modeling movement between patches. Other conceptual frameworks that may be useful for particular types of spatial interactions are those concerning patch dynamics (Pickett and White 1985), boundaries (Cadenasso et al. 2003), and river system gradients (Vannote et al. 1980). These "rooms" in our house are relatively well constructed, and it remains for the investigator to determine their relevance to the particular questions being asked.

In summary, understanding and modeling heterogeneity in ecosystem function can be a very complex problem, providing a boon to aspirin manufacturers. As yet, there is no overall conceptual framework—that is, there is no roof for our house—and perhaps we should not expect one given the multifaceted nature of the problem. Nonetheless, there are useful tools and conceptual constructs that can help us deal with pieces of the problem, and some of those tools, such as those concerned with boundaries and transport processes, are fairly well developed. In this paper, we presented a plan for navigating in this "half-built house." The plan is grounded in the ecosystem concept and requires careful definition of the ecosystems and the landscape system that are involved. Given those definitions, the answers to a few questions allow us to decide the most appropriate way to conceptualize and model the system, at least as far as whether to use homogeneous, mosaic, or interactive models. For those problems requiring interactive models, the potential complexity is mind-boggling, and more development of frameworks that allow us to clarify and simplify problems is sorely needed. Our hope is that ecosystem scientists of all disciplines will contribute to the further development of these frameworks that will allow full consideration of spatial heterogeneity in ecosystem science.

*Acknowledgments.* We thank the contributors to this book and the attendees at the 2003 Cary Conference for discussion that inspired this paper. This work was supported by the A.W. Mellon Foundation. This is a contribution to the program of the Institute of Ecosystem Studies.

## References

Cadenasso, M.L., Pickett, S. T. A., Weathers, K.C., and Jones, C.G. 2003. A framework for a theory of ecological boundaries. BioScience 53: 750–758.

Loreau, M., Mouquet, N., and Holt, R.D. 2003. Meta-ecosystems: a theoretical framework for a spatial ecosystem ecology. Ecol. Lett. 6: 673–679.

Pickett, S.T.A., and White, P.S. 1985. The ecology of natural disturbance and patch dynamics. Orlando, FL: Academic Press.

Pickett, S.T.A., Kolasa, J., and Jones, C.G. 1994. Ecological understanding: the nature of theory and the theory of nature. San Diego: Academic Press.

Shachak, M., and Jones, C.G. 1995. Ecological flow chains and ecological systems: concepts for linking species and ecosystem perspectives. In Linking species and ecosystems, eds. C.G. Jones and J.H. Lawton, pp. 280–294. New York: Chapman and Hall.

Shugart, H.H. 1998. Terrestrial ecosystems in changing environments. Cambridge, UK: Cambridge University Press.

Strayer, D.L., Ewing, H.A., and Bigelow, S. 2003. What kind of spatial and temporal details are required in models of heterogeneous systems? Oikos 102: 654–662.

Vannote, R.L., Minshall, G.W., Cummins, K.W., Sedell, J.R., and Cushing, C.E. 1980. River continuum concept. Can. J. Fisheries Aquatic Sci. 37: 130–137.

# Index